Florencio Zaragoza Dörwald

Side Reactions in Organic Synthesis

Further Reading from Wiley-VCH

Sierra, M. A., de la Torre, M. C.

Dead Ends and Detours

2004, ISBN 3-527-30644-7

de Meijere, A., Diederich, F. (Eds.)

Metal-Catalyzed Cross-Coupling Reactions

2nd Ed., 2 Vols.

2004, ISBN 3-527-30518-1

Mahrwald, R. (Ed.)

Modern Aldol Reactions

2 Vols.

2004, ISBN 3-527-30714-1

Nicolaou, K. C., Snyder, S. A.

Classics in Total Synthesis II

2004, ISBN 3-527-30685-4 (Hardcover)
2004, ISBN 3-527-30684-6 (Softcover)

Florencio Zaragoza Dörwald

Side Reactions in Organic Synthesis

A Guide to Successful Synthesis Design

WILEY-VCH Verlag GmbH & Co. KGaA

Author

Dr. Florencio Zaragoza Dörwald
Medicinal Chemistry
Novo Nordisk A/S
Novo Nordisk Park
2760 Måløv
Denmark

Library of Congress Card No.:
applied for

British Library Cataloguing-in-Publication Data
A catalogue record for this book is available from the British Library.

Bibliographic information published by Die Deutsche Bibliothek
Die Deutsche Bibliothek lists this publication in the Deutsche Nationalbibliografie; detailed bibliographic data is available in the Internet at <http://dnb.ddb.de>.

Printed in the Federal Republic of Germany.

Printed on acid-free paper.

Typesetting Kühn & Weyh, Satz und Medien, Freiburg
Printing Strauss GmbH, Mörlenbach
Bookbinding Litges & Dopf Buchbinderei GmbH, Heppenheim

ISBN 3-527-31021-5

Contents

Side Reactions in Organic Synthesis. Florencio Zaragoza Dörwald

ISBN: 3-527-31021-5

Preface

Most non-chemists would probably be horrified if they were to learn how many attempted syntheses fail, and how inefficient research chemists are. The ratio of successful to unsuccessful chemical experiments in a normal research laboratory is far below unity, and synthetic research chemists, in the same way as most scientists, spend most of their time working out what went wrong, and why.

Despite the many pitfalls lurking in organic synthesis, most organic chemistry textbooks and research articles do give the impression that organic reactions just proceed smoothly and that the total synthesis of complex natural products, for instance, is maybe a labor-intensive but otherwise undemanding task. In fact, most syntheses of structurally complex natural products are the result of several years of hard work by a team of chemists, with almost every step requiring careful optimization. The final synthesis usually looks quite different from that originally planned, because of unexpected difficulties encountered in the initially chosen synthetic sequence. Only the seasoned practitioner who has experienced for himself the many failures and frustrations which the development (sometimes even the repetition) of a synthesis usually implies will be able to appraise such work.

This book attempts to highlight the competing processes and limitations of some of the most common and important reactions used in organic synthesis. Awareness of these limitations and problem areas is important for the design of syntheses, and might also aid elucidation of the structure of unexpected products. Two chapters of this book cover the structure–reactivity relationship of organic compounds, and should also aid the design of better syntheses.

Chemists tend not to publish negative results, because these are, as opposed to positive results, never definite (and far too copious). Nevertheless, I have ventured to describe some reactions as difficult or impossible. A talented chemist might, however, succeed in performing such reactions anyway, for what I congratulate him in advance. The aim of this book is not to stop the reader from doing bold experiments, but to help him recognize his experiment as bold, to draw his attention to potential problems, and to inspire, challenge, and motivate.

Side Reactions in Organic Synthesis. Florencio Zaragoza Dörwald

ISBN: 3-527-31021-5

I wish to express my thanks to Ullrich Sensfuss, Bernd Peschke, and Kilian W. Conde-Frieboes for the many helpful discussions and for proofreading parts of the manuscript, and to Jesper Lau (my boss) for his support.

Smørum, Denmark
May 2004

Florencio Zaragoza Dörwald

Glossary and Abbreviations

Ac	acetyl, MeCO
acac	pentane-2,4-dione
AIBN	azobis(isobutyronitrile)
All	allyl
Alloc	allyloxycarbonyl
Amberlyst 15	strongly acidic, macroporous ion exchange resin
aq	aqueous
Ar	undefined aryl group
9-BBN	9-borabicyclo[3.3.1]nonane
BHT	2,6-di-*tert*-butyl-4-methylphenol
bimim	*N*-butyl-*N*′-methylimidazolium
BINAP	2,2′-bis(diphenylphosphino)-1,1′-binaphthyl
Bn	benzyl
Boc	*tert*-butyloxycarbonyl
Bom	benzyloxymethyl
Bs	4-bromobenzenesulfonyl
BSA	*N,O*-bis(trimethylsilyl)acetimidate
Bt	1-benzotriazolyl
Bu	butyl
Bz	benzoyl
CAN	ceric ammonium nitrate, $(NH_4)_2Ce(NO_3)_6$
cat	catalyst or catalytic amount
Cbz	Z, benzyloxycarbonyl, $PhCH_2OCO$
CDI	carbonyldiimidazole
celite	silica-based filter agent
COD	1,5-cyclooctadiene
coll	collidine, 2,4,6-trimethylpyridine
conc	concentrated
Cp	cyclopentadienyl
CSA	10-camphorsulfonic acid

Side Reactions in Organic Synthesis. Florencio Zaragoza Dörwald

ISBN: 3-527-31021-5

Cy	cyclohexyl
D	bond dissociation enthalpy
DABCO	1,4-diazabicyclo[2.2.2]octane
DAST	(diethylamino)sulfur trifluoride
dba	1,5-diphenyl-1,4-pentadien-3-one
DBN	1,5-diazabicyclo[4.3.0]non-5-ene
DBU	1,8-diazabicyclo[5.4.0]undec-5-ene
DCC	*N,N'*-dicyclohexylcarbodiimide
DCE	1,2-dichloroethane
DCP	1,2-dichloropropane
DDQ	2,3-dichloro-5,6-dicyano-1,4-benzoquinone
de	diastereomeric excess
DEAD	diethyl azodicarboxylate, $EtO_2C{-}N{=}N{-}CO_2Et$
Dec	decyl
DIAD	diisopropyl azodicarboxylate, $i PrO_2C{-}N{=}N{-}CO_2 i Pr$
DIBAH	diisobutylaluminum hydride
DIC	diisopropylcarbodiimide
diglyme	bis(2-methoxyethyl) ether
dipamp	1,2-bis[phenyl(2-methoxyphenyl)phosphino]ethane
DIPEA	diisopropylethylamine
DMA	*N,N*-dimethylacetamide
DMAD	dimethyl acetylenedicarboxylate, $MeO_2C{-}C{\equiv}C{-}CO_2Me$
DMAP	4-(dimethylamino)pyridine
DME	1,2-dimethoxyethane, glyme
DMF	*N,N*-dimethylformamide
DMI	1,3-dimethylimidazolidin-2-one
DMPU	1,3-dimethyltetrahydropyrimidin-2-one
DMSO	dimethyl sulfoxide
DMT	4,4'-dimethoxytrityl
DNA	deoxyribonucleic acid
Dnp	2,4-dinitrophenyl
DPPA	diphenylphosphoryl azide, $(PhO)_2P(O)N_3$
dppb	1,2-bis(diphenylphosphino)butane
dppe	1,2-bis(diphenylphosphino)ethane
dppf	1,1'-bis(diphenylphosphino)ferrocene
dppp	1,3-bis(diphenylphosphino)propane
dr	diastereomeric ratio
E	undefined electrophile
EDC	*N*-ethyl-*N'*-[3-(dimethylamino)propyl]carbodiimide hydrochloride
EDT	1,2-ethanedithiol
ee	enantiomeric excess
EEDQ	2-ethoxy-1-ethoxycarbonyl-1,2-dihydroquinoline
eq	equivalent
er	enantiomeric ratio
Et	ethyl

Fmoc	9-fluorenylmethyloxycarbonyl
FVP	flash vacuum pyrolysis
Hal	undefined halogen
Hep	heptyl
Hex	hexyl
HMPA	hexamethylphosphoric triamide, $(Me_2N)_3PO$
hν	light
HOAt	3-hydroxy-3*H*-[1,2,3]triazolo[4,5-*b*]pyridine, 4-aza-3-hydroxybenzotriazole
HOBt	1-hydroxybenzotriazole
HOSu	*N*-hydroxysuccinimide
HPLC	high pressure liquid chromatography
HSAB	hard and soft acids and bases
*i*Pr	isopropyl
IR	infrared
L	undefined ligand
LDA	lithium diisopropylamide
M	molar, mol/l; undefined metal
MCPBA	3-chloroperbenzoic acid
Me	methyl
MEK	2-butanone
MES	2-(4-morpholino)ethanesulfonic acid
MMT	monomethoxytrityl
MOM	methoxymethyl
Mos	4-methoxybenzenesulfonyl
mp	melting point
Ms	methanesulfonyl
MS	molecular sieves
nbd	norbornadiene
NBS	*N*-bromosuccinimide
NCS	*N*-chlorosuccinimide
NIS	*N*-iodosuccinimide
NMM	*N*-methylmorpholine
NMO	*N*-methylmorpholine-*N*-oxide
NMP	*N*-methyl-2-pyrrolidinone
NMR	nuclear magnetic resonance
Nos	nosyl, 4-nitrobenzenesulfonyl
Nu	undefined nucleophile
Oct	octyl
oxone™	$2\,KHSO_5 \cdot KHSO_4 \cdot K_2SO_4$, potassium peroxymonosulfate
PEG	poly(ethylene glycol)
Pent	pentyl
PG	protective group
Ph	phenyl
Pht	phthaloyl

Piv	pivaloyl, 2,2-dimethylpropanoyl
PMDTA	*N,N,N′,N″,N″*-pentamethyldiethylenetriamine
PNB	4-nitrobenzoyl
Pol	undefined polymeric support
PPTS	pyridinium tosylate
Pr	propyl
PTC	phase transfer catalysis
PTFE	poly(tetrafluoroethylene)
R	undefined alkyl group
Red-Al™	sodium bis(2-methoxyethoxy)aluminum hydride
satd	saturated
sec	secondary
L-Selectride™	lithium tri(2-butyl)borohydride
SET	single electron transfer
SN1	monomolecular nucleophilic substitution
SN2	bimolecular nucleophilic substitution
SNR1	monomolecular radical nucleophilic substitution
st. mat.	starting material
Su	*N*-succinimidyl
TBAF	tetrabutylammonium fluoride
TBDPS	*tert*-butyldiphenylsilyl
TBS	*tert*-butyldimethylsilyl
*t*Bu	*tert*-butyl
Tentagel™	PEG-grafted cross-linked polystyrene
tert	tertiary
Teoc	2-(trimethylsilyl)ethoxycarbonyl
Tf	trifluoromethanesulfonyl
TFA	trifluoroacetic acid
TfOH	triflic acid, trifluoromethanesulfonic acid
thd	2,2,6,6-tetramethyl-3,5-heptanedione
THF	tetrahydrofuran
THP	2-tetrahydropyranyl
TIPS	triisopropylsilyl
TMAD	*N,N,N′,N′*-tetramethyl azodicarboxamide
TMEDA	*N,N,N′,N′*-tetramethylethylenediamine
TMG	*N,N,N′,N′*-tetramethylguanidine
TMP	2,2,6,6-tetramethylpiperidin-1-yl
TMPP	tris(2,4,6-trimethoxyphenyl)phosphine
TMS	trimethylsilyl, Me_3Si
Tol	4-tolyl, 4-methylphenyl
Tr	trityl
Triton™ X-100	polyoxyethylene isooctylcyclohexyl ether
Ts	tosyl, *p*-toluenesulfonyl
Tyr	tyrosine
UV	ultraviolet

Wang resin	cross-linked polystyrene with 4-benzyloxybenzyl alcohol linker
X	undefined leaving group for nucleophilic displacement
X, Y	undefined heteroatoms with unshared electron pair
Z	Cbz, benzyloxycarbonyl; undefined electron-withdrawing group

1 Organic Synthesis: General Remarks

1.1 Introduction

Organic reactions almost never yield exclusively the desired product. Students learn this when they perform their first synthesis in the laboratory, for example the synthesis of anisole from phenol. Although the starting materials, the intermediates, and the product are all colorless, the reaction mixture will turn uncannily dark. This darkening shows that in reality much more is going on in addition to the expected process, and that obviously quite complex chemistry must be occurring, giving rise to extended conjugated polyenes from simple starting materials. Fortunately these dyes are usually formed in minute amounts only and the student will hopefully also learn not to be scared by color effects, and that even from pitch-black reaction mixtures colorless crystals may be isolated in high yield.

Because most reactions yield by-products and because isolation and purification of the desired product are usually the most difficult parts of a preparation, the work-up of each reaction and the separation of the product from by-products and reagents must be carefully considered while planning a synthesis. If product isolation seems to be an issue, the work-up of closely related examples from the literature (ideally two or three from different authors) should be studied. Many small, hydrophilic organic compounds which should be easy to prepare are still unknown, not because nobody has attempted to make them, but because isolation and purification of such compounds can be very difficult. Therefore the solubility of the target compound in water and in organic solvents, and its boiling or melting point, should be looked up or estimated, because these will aid choice of the right work-up procedure.

The chemical stability of the target compound must also be taken into account while planning its isolation. Before starting a synthesis one should also have a clear idea about which analytical tools will be most appropriate for following the progress of the reaction and ascertaining the identity and purity of the final product. Last, but not least, the toxicity and mutagenicity of all reagents, catalysts, solvents, products, and potential by-products should be looked up or estimated, and appropriate precautionary measures should be taken.

Side Reactions in Organic Synthesis. Florencio Zaragoza Dörwald

ISBN: 3-527-31021-5

1.2
Synthesis Design

The synthesis of a structurally complex compound requires careful retrosynthetic analysis to identify the shortest synthetic strategies which are most likely to give rapid access to the target compound, ideally in high yield and purity. It is critical to keep the synthesis as short as possible, because, as discussed throughout this book, each reaction can cause unexpected problems, especially when working with structurally complex intermediates. Also for synthesis of "simple-looking" structures several different approaches should be considered, because even structurally simple compounds often turn out not to be so easy to make as initially thought.

1.2.1
Convergent vs Linear Syntheses

If a target compound can be assembled from a given number of smaller fragments, the highest overall yields will usually be obtained if a convergent rather than linear strategy is chosen (Scheme 1.1). In a convergent assembly strategy the total number of reactions and purifications for all atoms or fragments of the target are kept to a

convergent strategy:

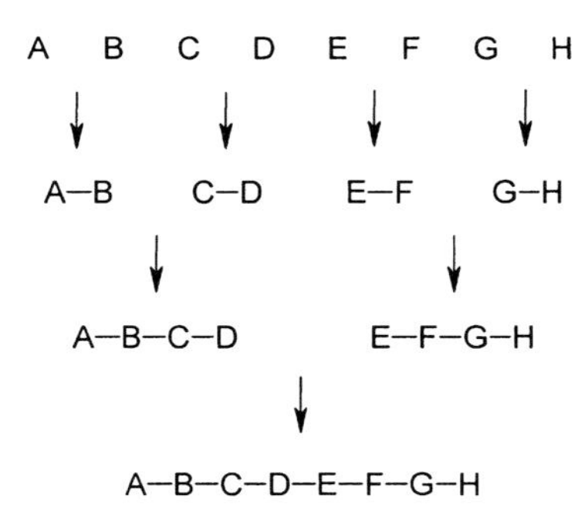

7 reactions, total yield with respect to monomer A: 51%
(for 80% yield per coupling step)

linear strategy:

A → A–B → A–B–C → A–B–C–D → A–B–C–D–E →
A–B–C–D–E–F → A–B–C–D–E–F–G → A–B–C–D–E–F–G–H

7 reactions, total yield with respect to monomer A: 21%
(for 80% yield per coupling step)

Scheme 1.1. Convergent and linear assembly strategies.

minimum. If a linear strategy is chosen the first fragment (A in Scheme 1.1) will be subjected to a large number of reactions and purifications, and the total yield with regard to this first fragment will be rather low. Syntheses should be organized in such a way that expensive and/or structurally complex fragments are subjected to the fewest possible number of transformations.

1.2.2
Retrosynthetic Analysis

1.2.2.1 Introduction

When planning a synthesis, the most suitable starting materials should be chosen. These should be structurally and/or stereochemically as closely related to the target as possible, to keep the synthesis brief. The first steps of a good synthesis may even be low-yielding (if the products are easy to purify), because at these early stages little work and reagents have been invested and the intermediates are still cheap. Poor yields at later stages of a multistep synthesis, however, strongly reduce its usefulness, because most steps of the synthesis will have to be run on a large scale, using large amounts of solvents and reagents, to obtain a small amount only of the final product, which will, accordingly, be rather expensive.

In a retrosynthesis the easiest bonds to make are often cleaved first (i.e. these bonds will be made at the end of the synthesis), yielding several fragments which can be joined together at late stages of the synthesis, using straightforward and high-yielding chemistry. Such reactions would usually be condensations, for example acetal, amide, or ester formation, or the formation of carbon–heteroatom bonds, but might also be high-yielding C–C bond-forming reactions if the required reaction conditions are compatible with all the structural elements of the final product.

If the target contains synthetically readily accessible substructures (e.g. cyclic elements accessible by well established cycloaddition or cyclization reactions), these might be chosen as starting point of a disconnection [1]. If such substructures are not present, their generation by introduction of removable functional groups (e.g. by converting single bonds into double bonds or by formal oxidation of methylene groups to carbonyl groups, Scheme 1.5) should be attempted. If this approach fails to reveal readily accessible substructures, the functional groups present in the target structure which might assist the stepwise construction of the carbon framework must be identified, and the bonds on the shortest bond paths between these groups should be considered as potential sites of disconnection (Scheme 1.3). Retro-aldol or Mannich reactions, optionally combined with the "Umpolung" of functional groups, have been the most common and successful tools for disconnection of intricate carbon frameworks, but any other, high-yielding C–C bond-forming reaction can also be considered. As illustrated by the examples discussed below, a good retrosynthesis requires much synthetic experience, a broad knowledge of chemical reactivity, and the ability to rapidly recognize synthetically accessible substructures.

1.2.2.2 Shikimic Acid

In Scheme 1.2 one possible retrosynthetic analysis of the unnatural enantiomer of shikimic acid, a major biosynthetic precursor of aromatic α-amino acids, is sketched. Because *cis* dihydroxylations can be performed with high diastereoselectivity and yield, this step might be placed at the end of a synthesis, what leads to a cyclohexadienoic acid derivative as an intermediate. Chemoselective dihydroxylation of this compound should be possible, because the double bond to be oxidized is less strongly deactivated than the double bond directly bound to the (electron-withdrawing) carboxyl group.

Despite being forbidden by the Baldwin rules (5-*endo*-trig ring opening; see Section 9.2), cyclohexadienoic acid derivatives such as that required for this synthesis can be prepared by base-induced ring scission of 7-oxanorbornene derivatives, presumably because of the high strain-energy of norbornenes. The required 7-oxanorbornene, in turn, should be readily accessible from furan and an acrylate via the

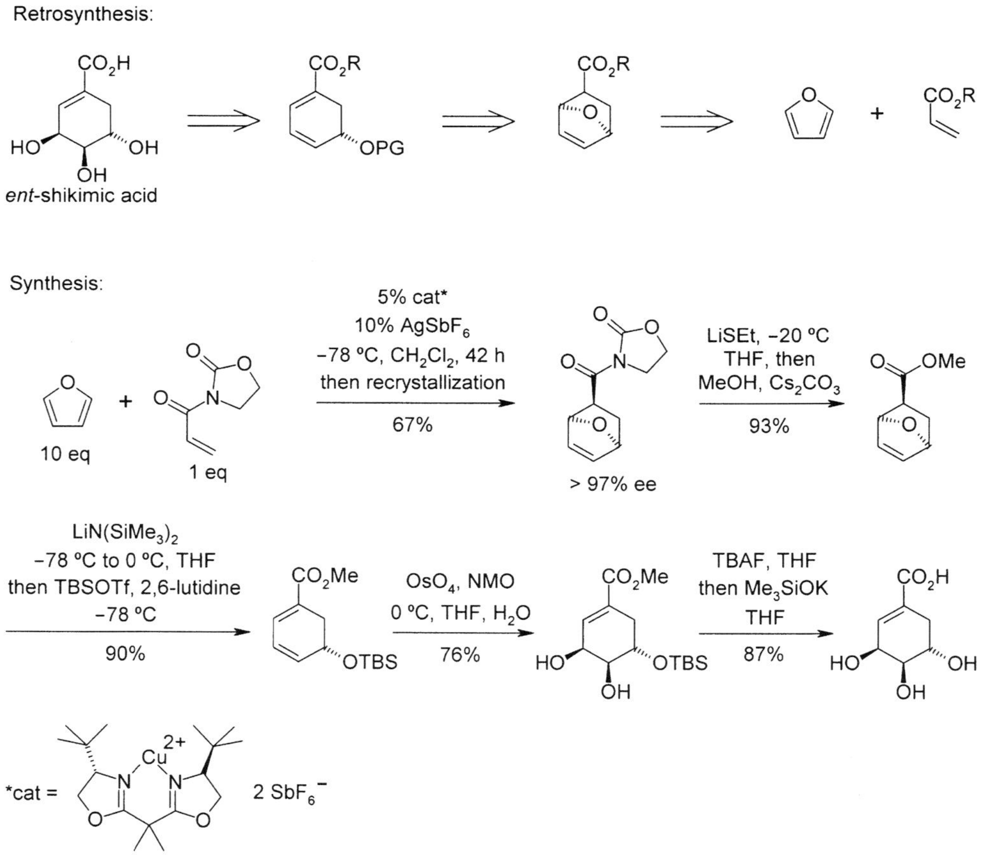

Scheme 1.2. Retrosynthetic analysis and synthesis of *ent*-shikimic acid [2].

Diels–Alder reaction. With the aid of an enantiomerically pure Lewis acid this Diels–Alder reaction yields a highly enantiomerically enriched 7-oxanorbornene, so that the remaining steps of this elegant synthesis only need to proceed diastereoselectively and without racemization.

1.2.2.3 **Lycopodine**

A further target which contains a readily accessible and easily recognizable substructure is the alkaloid lycopodine. Being a β-amino ketone, a possible retrosynthesis could be based on an intramolecular Mannich reaction, as outlined in Scheme 1.3. In this case two of the targets four rings would be generated in one step by a Mannich condensation; this significantly reduces the total number of steps required. A robust, intramolecular *N*-alkylation was chosen as last step. Realization of this synthetic plan led to a synthesis of racemic lycopodine in only eight steps with a total yield of 13 % [3]. Fortunately the Mannich reaction yielded an intermediate with the correct relative configuration.

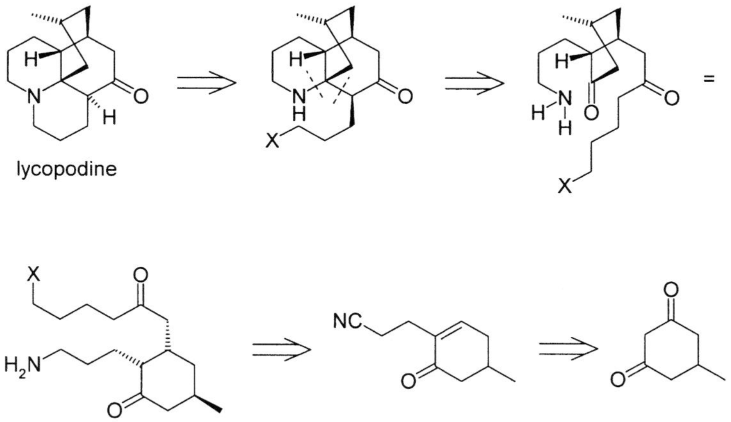

Scheme 1.3. Retrosynthesis of lycopodine based on an intramolecular Mannich reaction [3].

1.2.2.4 **The Oxy-Cope Rearrangement**

Less obvious than the retrosyntheses discussed above are those based on intramolecular rearrangements, because these often involve a major change of connectivity between atoms. For instance, exploitation of oxy-Cope rearrangements as synthetic tools requires some practice and the ability to recognize the substructures accessible via this reaction from readily available starting materials. Oxy-Cope rearrangements yield 4-penten-1-yl ketones by formal allylation of a vinyl ketone at the β position or γ-vinylation of an allyl ketone (Scheme 1.4). This rearrangement can be used to prepare decalins [4] or perhydroindenes [5, 6] from bicyclo[2.2.2]octenones or norbornenones, respectively, which can be prepared by using the Diels–Alder reaction. Moreover, oxy-Cope rearrangements may be used for ring expansions or contractions.

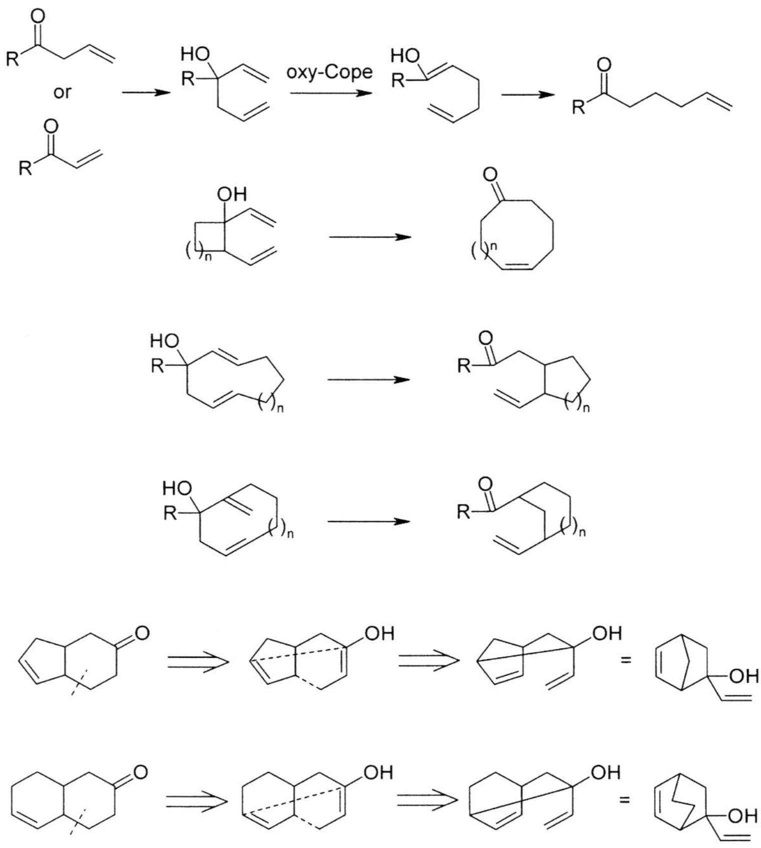

Scheme 1.4. The oxy-Cope rearrangement.

Numerous natural products have been prepared using the oxy-Cope rearrangement as the key step [5], in particular, and with high virtuosity, by the group of L.A. Paquette [4, 6, 7]. Three examples of retrosynthetic analyses of natural products or analogs thereof based on the oxy-Cope rearrangement are shown in Scheme 1.5. Because all the products are devoid of a keto group, the required 4-penten-1-yl ketone substructure (i.e. the oxy-Cope retron [1]) must be introduced during the retrosynthesis in such a way that accessible starting materials result.

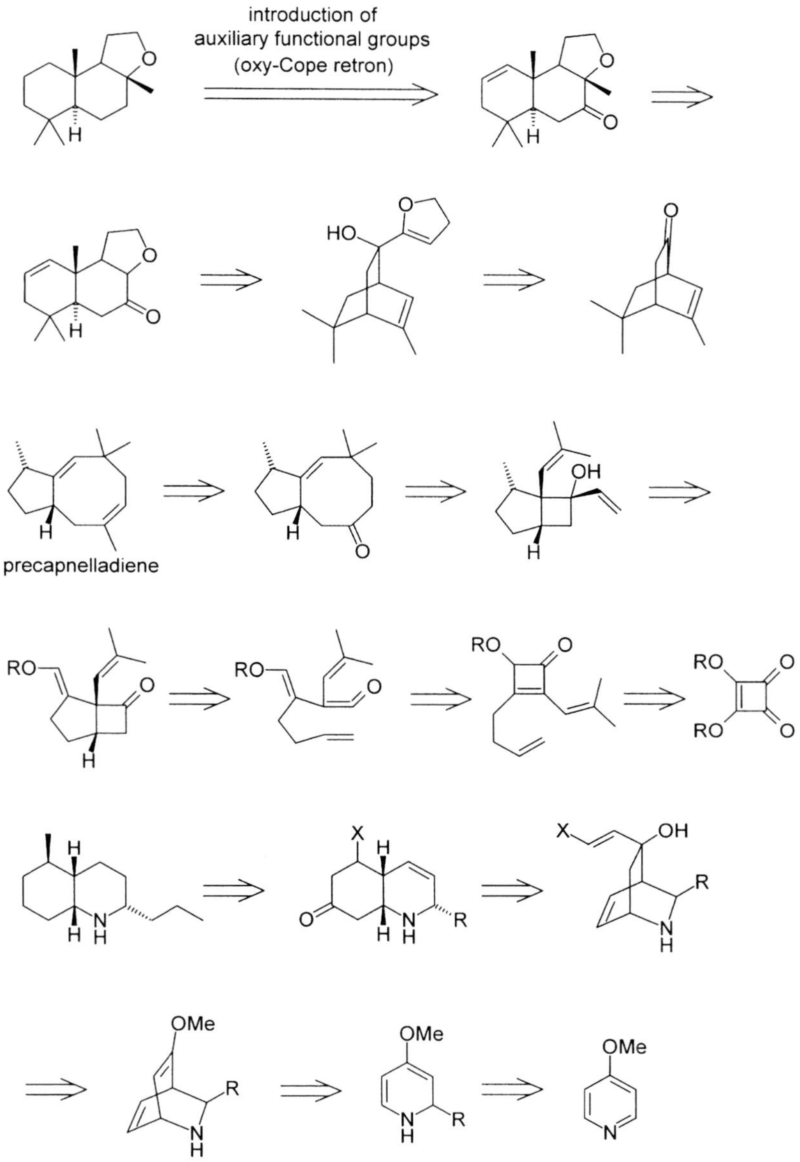

introduction of auxiliary functional groups (oxy-Cope retron)

precapnelladiene

Scheme 1.5. Retrosynthesis of an ambergris-type ether, of precapnelladiene, and of an alkaloid based on the oxy-Cope rearrangement [8–10].

1.2.2.5 Conclusion

As will be shown throughout this book, the outcome of organic reactions is highly dependent on all structural features of a given starting material, and unexpected products may readily be formed. Therefore, while planning a multistep synthesis, it is important to keep the total number of steps as low as possible.

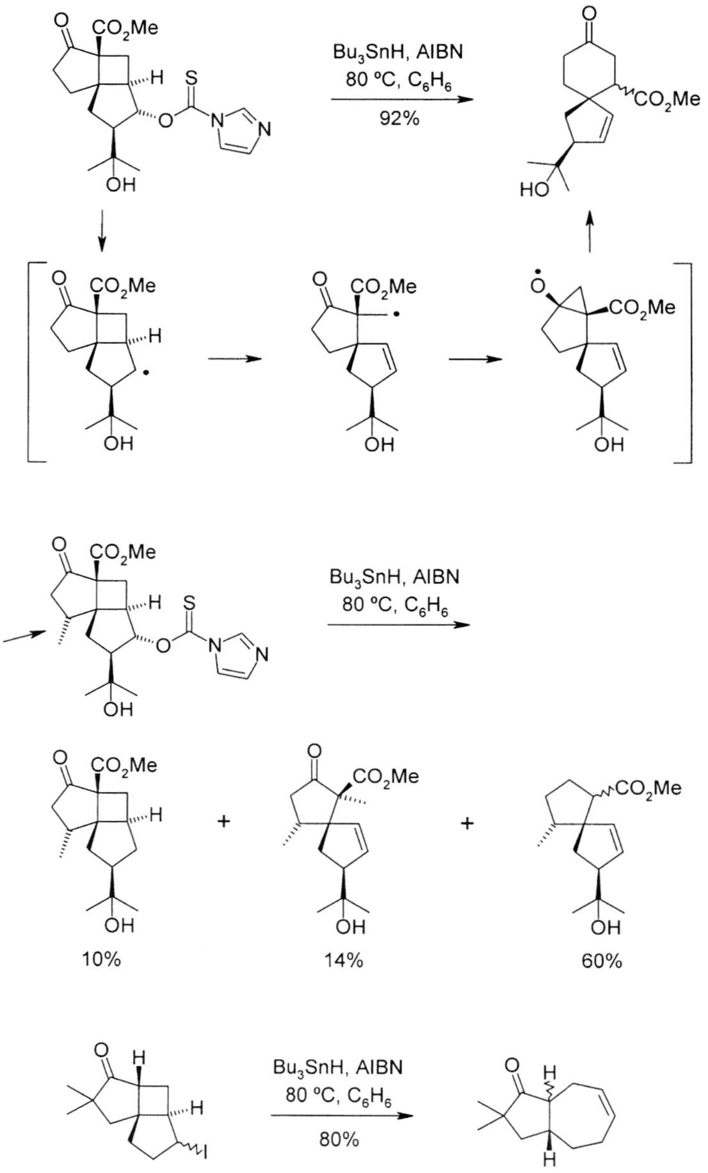

Bu_3SnH, AIBN
80 °C, C_6H_6
92%

Bu_3SnH, AIBN
80 °C, C_6H_6

10% + 14% + 60%

Bu_3SnH, AIBN
80 °C, C_6H_6
80%

Scheme 1.6. Rearrangement of polycyclic cyclobutylmethyl radicals [11, 12].

Even the most experienced chemist will not be able to foresee all potential pitfalls of a synthesis, specially so if multifunctional, structurally complex intermediates must be prepared. The close proximity or conformational fixation of functional groups in a large molecule can alter their reactivity to such an extent that even simple chemical transformations can no longer be performed [11]. Small structural variations of polyfunctional substrates might, therefore, bring about an unforeseeable change in reactivity.

Examples of closely related starting materials which upon treatment with the same reagents yield completely different products are sketched in Scheme 1.6. The additional methyl group present in the second starting material slows addition to the carbonyl group of the radical formed by ring scission of the cyclobutane ring, and thus prevents ring expansion to the cyclohexanone. Removal of the methoxycarbonyl group leads to cleavage of a different bond of the cyclobutane ring and thereby again to a different type of product [12].

The understanding and prediction of such effects and the development of milder and more selective synthetic transformations, applicable to the synthesis of highly complex structures or to the selective chemical modification of proteins, DNA, or even living cells will continue to be the challenge for current and future generations of chemists.

1.3
Hard and Soft Acids and Bases

One of the most useful tools for predicting the outcome of chemical reactions is the principle of hard and soft acids and bases (HSAB), formulated by Pearson in 1963 [13–15]. This principle states that hard acids will react preferentially with hard bases, and soft acids with soft bases, "hard" and "soft" referring to sparsely or highly polarizable reactants. A selection of hard and soft Lewis acids and bases is given in Table 1.1.

Several chemical observations can be readily explained with the aid of the HSAB principle. For instance, the fact that the early transition metals in high oxidation states, for example titanium(IV), do not usually form complexes with alkenes, carbon monoxide, or phosphines, but form stable oxides instead can be attributed to their hardness. The late transition metals, on the other hand, being highly polarizable, because of their almost completely filled *d* orbitals, readily form complexes with soft bases such as alkenes, carbanions, and phosphines, and these complexes are often unreactive towards water or oxygen. For the same reason, in alkali or early transition metal enolates the metal is usually bound to oxygen, whereas enolates of late transition metals usually contain M–C bonds [17, 18]. While alkali metal alkyls or Grignard reagents react with enones presumably by initial coordination of the metal to oxygen followed by transfer of the alkyl group to the carbonyl carbon atom [16, 19], organocuprates or organopalladium compounds preferentially coordinate and transfer their organic residue to soft C–C double bonds.

Table 1.1. Hard and soft Lewis acids and bases [13, 15, 16] (Z = electron-withdrawing group, M = metal). The acidic or basic centers in molecules are in italics.

Hard acids (non-metals)	Borderline acids (non-metals)	Soft acids (non-metals)
H^+, $B(OR)_3$, BF_3, BCl_3, RCO^+, CO_2, NC^+, R_3Si^+, Si^{4+}, RPO_2^+, $ROPO_2^+$, As^{3+}, RSO_2^+, $ROSO_2^+$, SO_3, Se^{3+}, Cl^{7+}, I^{7+}, I^{5+}	BR_3, R^+ (softer $CH_3^+ > RCH_2^+ > R_2CH^+ > R_3C^+ >$ vinyl$^+ \approx C_6H_5^+ \approx RC{\equiv}C^+$ harder), $RCHO$, R_2CO, $R_2C{=}NR$, NO^+, SO_2	BH_3, Ar–Z, C=C–Z, quinones, carbenes, HO^+, RO^+, RS^+, RSe^+, RTe^+, Br_2, Br^+, I_2, I^+
Hard acids (metals)	**Borderline acids (metals)**	**Soft acids (metals)**
Li^+, Na^+, K^+, $BeMe_2$, Be^{2+}, $RMgX$, Mg^{2+}, Ca^{2+}, Sr^{2+}, $AlCl_3$, $AlMe_3$, AlH_3, $Al(OR)_3$, Al^{3+}, $GaMe_3$, Ga^{3+}, $InMe_3$, In^{3+}, SnR_3^+, $SnMe_2^{2+}$, Sn^{2+}, Sc^{3+}, La^{3+}, $Ti(OR)_4$, Ti^{4+}, Zr^{4+}, VO_2^+, Cr^{3+}, Fe^{3+}, Co^{3+}, Ir^{3+}, Th^{4+}, UO_2^{2+}, Pu^{4+}, Yb^{3+}	GaH_3, $Sn(OR)_4$, $SnCl_4$, Pb^{2+}, Sb^{3+}, Bi^{3+}, $Sc(OTf)_3$, $ScCl_3$, Fe^{2+}, Co^{2+}, Ni^{2+}, Cu^{2+}, RZn^+, Zn^{2+}, $Yb(OTf)_3$, $YbCl_3$	Cs^+, $TlMe_3$, Tl^+, Tl^{3+}, $Pd(PAr_3)_2$, $Pd(PAr_3)_2^{2+}$, Pd^{2+}, Pt^{2+}, Cu^+, Ag^+, Au^+, CdR^+, Cd^{2+}, HgR^+, Hg^+, Hg^{2+}, M^0
Hard bases	**Borderline bases**	**Soft bases**
NH_3, RNH_2, R_2N^-, N_2H_4, H_2O, OH^-, ROH, RO^-, R_2O, RCO_2^-, CO_3^{2-}, NO_3^-, PO_4^{3-}, SO_4^{2-}, ClO_4^-, F^-, Cl^-	AlH_4^-, N_2, N_3^-, $PhNH_2$, R_3N, C_5H_5N, $R_2C{=}NR$, NO_2^-, SO_3^{2-}, Br^-	H^-, BH_4^-, R^- (softer $RC{\equiv}C^- >$ vinyl$^- > R_3C^-$ harder), C_6H_6, $R_2C{=}CR_2$, $RC{\equiv}CR$, CN^-, RNC, CO, PR_3, $P(OR)_3$, AsR_3, RS^-, SCN^-, RSH, R_2S, $S_2O_3^{2-}$, RSe^-, I^-

HSAB is particularly useful for assessing the reactivity of ambident nucleophiles or electrophiles, and numerous examples of chemoselective reactions given throughout this book can be explained with the HSAB principle. Hard electrophiles, for example alkyl triflates, alkyl sulfates, trialkyloxonium salts, electron-poor carbenes, or the intermediate alkoxyphosphonium salts formed from alcohols during the Mitsunobu reaction, tend to alkylate ambident nucleophiles at the hardest atom. Amides, enolates, or phenolates, for example, will often be alkylated at oxygen by hard electrophiles whereas softer electrophiles, such as alkyl iodides or electron-poor alkenes, will preferentially attack amides at nitrogen and enolates at carbon.

2-Pyridone is *O*-alkylated more readily than normal amides, because the resulting products are aromatic. With soft electrophiles, however, clean *N*-alkylations can be performed (Scheme 1.7). The Mitsunobu reaction, on the other hand, leads either to mixtures of *N*- and *O*-alkylated products or to *O*-alkylation exclusively, probably because of the hard, carbocation-like character of the intermediate alkoxyphosphonium cations. Electrophilic rhodium carbene complexes also preferentially alkylate the oxygen atom of 2-pyridone or other lactams [20] (Scheme 1.7).

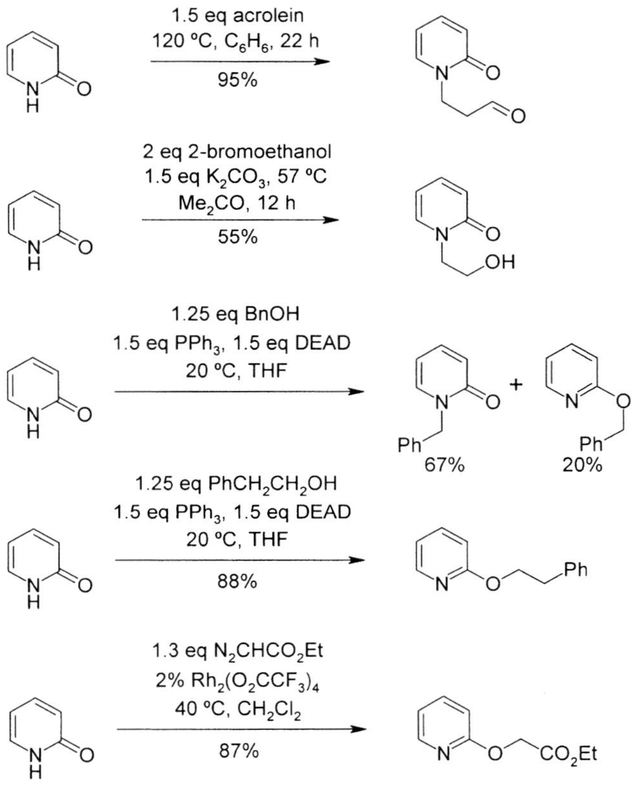

1.5 eq acrolein
120 °C, C_6H_6, 22 h
95%

2 eq 2-bromoethanol
1.5 eq K_2CO_3, 57 °C
Me_2CO, 12 h
55%

1.25 eq BnOH
1.5 eq PPh_3, 1.5 eq DEAD
20 °C, THF
67% + 20%

1.25 eq $PhCH_2CH_2OH$
1.5 eq PPh_3, 1.5 eq DEAD
20 °C, THF
88%

1.3 eq N_2CHCO_2Et
2% $Rh_2(O_2CCF_3)_4$
40 °C, CH_2Cl_2
87%

Scheme 1.7. Regioselective alkylation of 2-pyridone [20–22].

Lactams and some non-cyclic, secondary amides (RCONHR) can be alkylated with high regioselectivity either at nitrogen (Section 6.6) or at oxygen. *N*-Alkylations are generally conducted under basic reaction conditions whereas *O*-alkylations are often performed with trialkyloxonium salts, dialkyl sulfates, or alkyl halides/silver salts without addition of bases. Protonated imino ethers are formed; these are usually not isolated but are converted into the free imino ethers with aqueous base during the work-up. Scheme 1.8 shows examples of the selective alkylation of lactams and of the formation of 2-pyrrolidinones or 2-iminotetrahydrofurans by cyclization of 4-bromobutyramides.

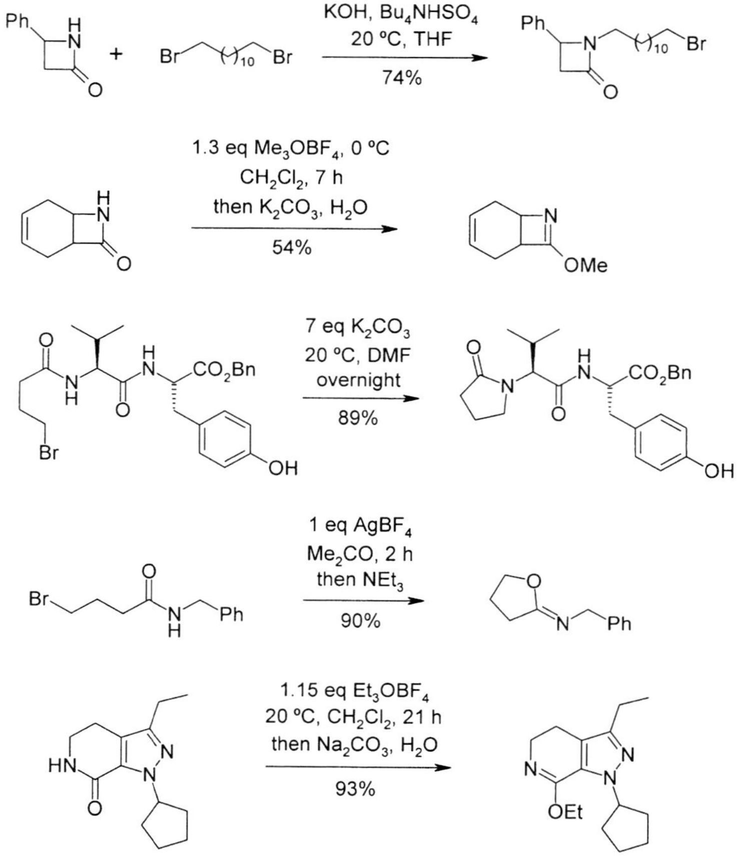

Scheme 1.8. Regioselective alkylation of amides [23–27].

The triflate sketched in Scheme 1.9 mainly alkylates the amide at oxygen, instead of alkylating the softer, lithiated phosphonate. Selective *C*-alkylation can be achieved in this instance by choosing a less reactive mesylate as electrophile and by enhancing the acidity of the phosphonate.

The regioselectivity of the alkylation of enolates can also be controlled by the hardness of the alkylating agent [29]. As illustrated by the examples in Scheme 1.10, allyl, propargyl, or alkyl bromides or iodides mainly yield *C*-alkylated products, whereas the harder sulfonates preferentially alkylate at oxygen.

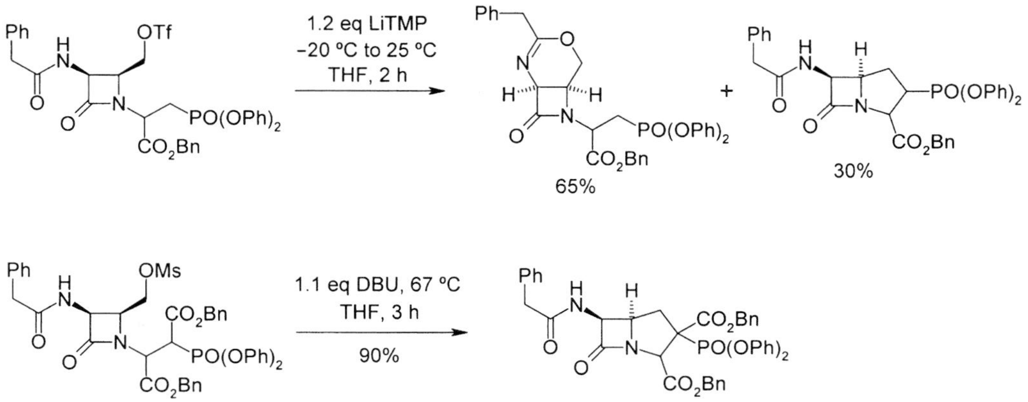

1.2 eq LiTMP, −20 °C to 25 °C, THF, 2 h: 65% + 30%

1.1 eq DBU, 67 °C, THF, 3 h: 90%

Scheme 1.9. Intramolecular alkylation of amides and phosphonates [28].

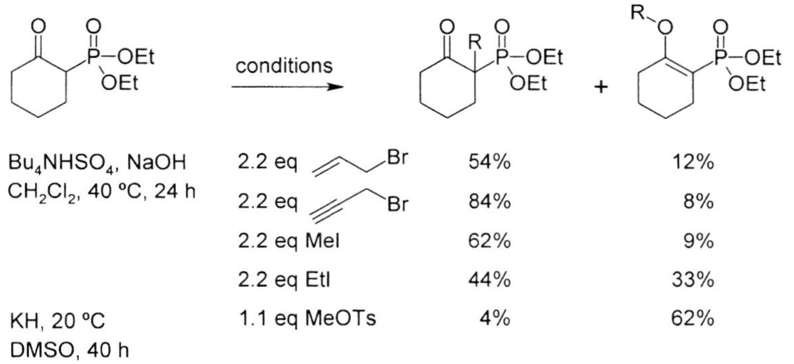

conditions		C-alkylation	O-alkylation
Bu_4NHSO_4, NaOH, CH_2Cl_2, 40 °C, 24 h	2.2 eq allyl bromide	54%	12%
	2.2 eq propargyl bromide	84%	8%
	2.2 eq MeI	62%	9%
	2.2 eq EtI	44%	33%
KH, 20 °C, DMSO, 40 h	1.1 eq MeOTs	4%	62%

Scheme 1.10. Regioselective alkylation of enolates [30, 31].

1.4 The Curtin–Hammett Principle

In the 1940s the idea was prevalent among chemists that the conformation of a reactant could be determined from the structure of a reaction product, i.e. the major conformer would yield the major product. This assumption was shown to be incorrect by Curtin and Hammett in the 1950s [32].

For a reaction in which a starting material A is an equilibrium mixture of two conformers (or diastereomers, tautomers, rotamers, etc.) A^1 and A^2 (Eq. 1.1), two extreme situations can be considered – one in which equilibration of A^1 and A^2 is slow if compared with their reaction with B ($k^1, k^2 \ll k^C, k^D$), and one in which equilibration of A^1 and A^2 is much faster than their reaction with B ($k^1, k^2 \gg k^C, k^D$).

$$\mathrm{C} \xleftarrow[k^C]{\mathrm{B}} \mathrm{A}^1 \underset{k^1}{\overset{k^2}{\rightleftharpoons}} \mathrm{A}^2 \xrightarrow[k^D]{\mathrm{B}} \mathrm{D} \qquad \text{(Eq. 1.1)}$$

If equilibration of A^1 and A^2 is slow, the product ratio [C]/[D] will be equal to the ratio of conformers of the starting material A ($[A^1]/[A^2]$) and independent of the ratio k^C/k^D. If equilibration is rapid, however, the amount of C and D formed will depend both on the ratio of starting materials ($[A^1]/[A^2]$) and on the ratio of the two reaction rate constants k^C and k^D: $[D]/[C] = [A^2]/[A^1] \times k^D/k^C$ [32].

The main implication of these derivations is that if equilibration is rapid, the product ratio cannot always be intuitively predicted if the reaction rates k^C and k^D are unknown. Because energy-rich conformers, present in low concentrations only, are often more reactive than more stable conformers, it is not unusual for the main product of a reaction to result from a minor conformer which cannot even be observed.

Two examples of such situations are sketched in Scheme 1.11. Quaternization of tropane occurs mainly from the less hindered "pyrrolidine side" (equatorial attack at the piperidine ring), even though the main conformer of tropane has an equatorial methyl group. Similarly, 1-methyl-2-phenylpyrrolidine yields mainly an *anti* alkylated product via alkylation of the minor *cis* conformer when treated with phenacyl bromide [33]. In both instances the less stable conformer is more reactive to such an extent that the major product of the reaction results from this minor conformer. A further notable example of a reaction in which the main product results from a minor but more reactive intermediate is the enantioselective hydrogenation of α-acetamidocinnamates with a chiral rhodium-based catalyst [34].

This does, however, not need to be so. Oxidation of 1-methyl-4-*tert*-butylpiperidine, for example, yields mainly the amine *N*-oxide derived from the most stable conformer (Scheme 1.12). In this example the more energy-rich (less stable) conformer reacts more slowly than the major conformer.

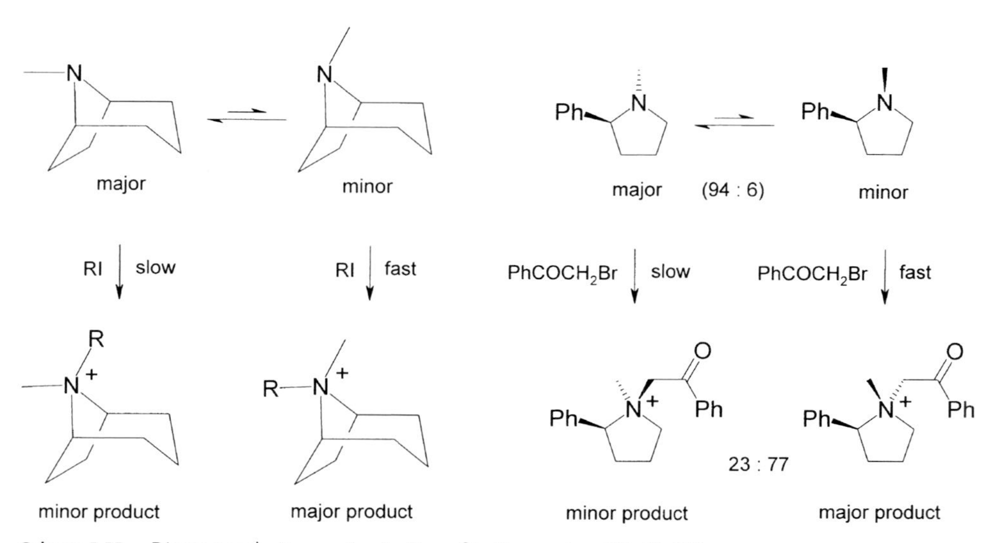

Scheme 1.11. Diastereoselective quaternization of tertiary amines [32, 33, 35].

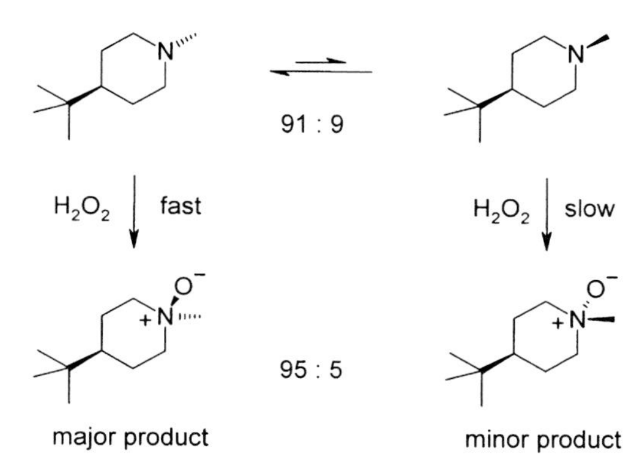

Scheme 1.12. Diastereoselective oxidation of 4-*tert*-butyl-1-methylpiperidine [32, 36, 37].

To conclude, the Curtin–Hammett principle states that the relative amounts of products formed from two interconverting conformers depend on the reactivity of these two conformers if the interconversion of these conformers is rapid, and cannot always be intuitively predicted.

References

1 Corey, E. J.; Cheng, X.-M. *The Logic of Chemical Synthesis*; Wiley: New York, 1989.

2 Evans, D. A.; Barnes, D. M. Cationic bis(oxazoline)Cu(II) Lewis acid catalysts. Enantioselective furan Diels–Alder reaction in the synthesis of *ent*-shikimic acid. *Tetrahedron Lett.* **1997**, *38*, 57–58.

3 Heathcock, C. H.; Kleinman, E. F.; Binkley, E. S. Total synthesis of lycopodium alkaloids: (±)-lycopodine, (±)-lycodine, and (±)-lycodoline. *J. Am. Chem. Soc.* **1982**, *104*, 1054–1068.

4 Oplinger, J. A.; Paquette, L. A. Synthesis of the forskolin skeleton via anionic oxy-Cope rearrangement. *Tetrahedron Lett.* **1987**, *28*, 5441–5444.

5 Bérubé, G.; Fallis, A. G. An intramolecular cycloaddition-sigmatropic rearrangement approach to (±) gascardic acid. *Tetrahedron Lett.* **1989**, *30*, 4045–4048.

6 Paquette, L. A.; Romine, J. L.; Lin, H.-S. Diastereoselective π-facially controlled nucleophilic additions of chiral vinylorganometallics to chiral β,γ-unsaturated ketones. 2. A practical method for stereocontrolled elaboration of the decahydro-*as*-indacene subunit of ikarugamycin. *Tetrahedron Lett.* **1987**, *28*, 31–34.

7 Paquette, L. A.; Zhao, M. Enantiospecific total synthesis of natural (+)-taxusin. 1. Retrosynthesis, advancement to diastereomeric *trans*-$\Delta^{9,10}$-tricyclic olefinic intermediates, and the stereocontrol attainable because of intrinsic rotational barriers therein. *J. Am. Chem. Soc.* **1998**, *120*, 5203–5212.

8 Maleczka, R. E.; Paquette, L. A. Adaptation of oxyanionic sigmatropy to the convergent enantioselective synthesis of ambergris-type odorants. *J. Org. Chem.* **1991**, *56*, 6538–6546.

9 MacDougall, J. M.; Santora, V. J.; Verma, S. K.; Turnbull, P.; Hernandez, C. R.; Moore, H. W. Cyclobutenone-based syntheses of polyquinanes and bicyclo[6.3.0]undecanes by tandem anionic oxy-Cope reactions. Total synthesis of (±)-precapnelladiene. *J. Org. Chem.* **1998**, *63*, 6905–6913.

10 Polniaszek, R. P.; Dillard, L. W. Stereospecific total synthesis of decahydroquinoline alkaloids (±)-195A and (±)-2-*epi*-195A. *J. Org. Chem.* **1992**, *57*, 4103–4110.

11 Sierra, M. A.; de la Torre, M. C. Dead ends and detours en route to total syntheses of the 1990s. *Angew. Chem. Int. Ed.* **2000**, *39*, 1538–1559.

12 Crimmins, M. T.; Dudek, C. M.; Cheung, W.-H. A fragmentation–rearrangement sequence of cyclobutylcarbinyl radicals. *Tetrahedron Lett.* **1992**, *33*, 181–184.

13 Pearson, R. G. *Chemical Hardness*; Wiley–VCH: Weinheim, 1997.

14 Ho, T.-L. The hard soft acids bases (HSAB) principle and organic chemistry. *Chem. Rev.* **1975**, *75*, 1–20.

15 Woodward, S. HSAB matching and mismatching in selective catalysis and synthesis. *Tetrahedron* **2002**, *58*, 1017–1050.

16 March, J. *Advanced Organic Chemistry*, John Wiley and Sons: New York, 1992.

17 Albanese, J. A.; Staley, D. L.; Rheingold, A. L.; Burmeister, J. L. Synthesis and molecular structure of a dinuclear phosphorus ylide complex: μ-dichlorobis[chloro(benzoylmethylenetri-*n*-butylphosphorane)palladium(II). *J. Organomet. Chem.* **1989**, *375*, 265–272.

18 Culkin, D. A.; Hartwig, J. F. C–C Bond forming reductive elimination of ketones, esters, and amides from isolated arylpalladium(II) enolates. *J. Am. Chem. Soc.* **2001**, *123*, 5816–5817.

19 Hoffmann, R. W.; Hölzer, B. Concerted and stepwise Grignard additions, probed with a chiral Grignard reagent. *Chem. Commun.* **2001**, 491–492.

20 Busch-Petersen, J.; Corey, E. J. A Rh(II) catalytic approach to the synthesis of ethers of a minor component in a tautomeric set. *Org. Lett.* **2000**, *2*, 1641–1643.

21 Somekawa, K.; Okuhira, H.; Sendayama, M.; Suishu, T.; Shimo, T. Intramolecular [2 + 2] photocycloadditions of 1-(ω-alkenyl)-2-pyridones possessing an ester group on the olefinic carbon chain. *J. Org. Chem.* **1992**, *57*, 5708–5712.

22 Comins, D. L.; Jianhua, G. *N*- vs *O*-Alkylation in the Mitsunobu reaction of 2-pyridone. *Tetrahedron Lett.* **1994**, *35*, 2819–2822.

23 Crombie, L.; Jones, R. C. F.; Haigh, D. Transamidation reactions of β-lactams: a synthesis of (±)-dihydroperiphylline. *Tetrahedron Lett.* **1986**, *27*, 5151–5154.

24 Paquette, L. A.; Kakihana, T.; Hansen, J. F.; Philips, J. C. π-Equivalent heterocyclic congeners of cyclooctatetraene. The synthesis and valence isomerization of 2-alkoxyazocines. *J. Am. Chem. Soc.* **1971**, *93*, 152–161.

25 Reid, R. C.; Kelso, M. J.; Scanlon, M. J.; Fairlie, D. P. Conformationally constrained macrocycles that mimic tripeptide β-strands in water and aprotic solvents. *J. Am. Chem. Soc.* **2002**, *124*, 5673–5683.

26 Alanine, A. I. D.; Fishwick, C. W. G.; Szantay, C. Facile preparation of 2-imino tetrahydrofurans, pyrans and oxepans. *Tetrahedron Lett.* **1989**, *30*, 6571–6572.

27 Urban, F. J.; Anderson, B. G.; Orrill, S. L.; Daniels, P. J. Process research and large-scale synthesis of a novel 5,6-dihydro-(9*H*)-pyrazolo[3,4-*c*]-1,2,4-triazolo[4,3-*a*]pyridine PDE-IV inhibitor. *Org. Process Res. Dev.* **2001**, *5*, 575–580.

28 Hakimelahi, G. H.; Moosavi-Movahedi, A. A.; Tsay, S.-C.; Tsai, F.-Y.; Wright, J. D.; Dudev, T.; Hakimelahi, S.; Lim, C. Design, synthesis, and SAR of novel carbapenem antibiotics with high stability to xanthomonas maltophilia oxyiminocephalosporinase type II. *J. Med. Chem.* **2000**, *43*, 3632–3640.

29 Damoun, S.; Van de Woude, G.; Choho, K.; Geerlings, P. Influence of alkylating reagent softness on the regioselectivity in enolate ion alkylation: a theoretical local hard and soft acids and bases study. *J. Phys. Chem. A* **1999**, *103*, 7861–7866.

30 Ruder, S. M.; Kulkarni, V. R. Phase transfer catalyzed alkylation of 2-(diethoxyphosphinyl)-cyclohexanone. *Synthesis* **1993**, 945–947.

31 Ruder, S. M.; Ding, M. [2 + 2] Cycloaddition of cyclic vinyl phosphonates with ketenes. *J. Chem. Soc. Perkin Trans. 1* **2000**, 1771–1776.

32 Seeman, J. I. Effect of conformational change on reactivity in organic chemistry. Evaluations, applications, and extensions of Curtin–Hammett/Winstein–Holness kinetics. *Chem. Rev.* **1983**, *83*, 83–134.

33 Solladié-Cavallo, A.; Solladié, G. Etude de la stéréosélectivité de la quaternisation de pyrrolidines-1,2 disubstituées. *Tetrahedron Lett.* **1972**, 4237–4240.

34 Halpern, J. Mechanism and stereoselectivity of asymmetric hydrogenation. *Science* **1982**, *217*, 401–407.

35 Seeman, J. I.; Secor, H. V.; Hartung, H.; Galzerano, R. Steric effects in conformationally mobile systems. The methylation of 1-methyl-2-arylpyrrolidines related to nicotine. *J. Am. Chem. Soc.* **1980**, *102*, 7741–7747.

36 Shvo, Y.; Kaufman, E. D. Configurational and conformational analysis of cyclic amine oxides. *Tetrahedron* **1972**, *28*, 573–580.

37 Crowley, P. J.; Robinson, M. J. T.; Ward, M. G. Conformational effects in compounds with 6-membered rings XII. The conformational equilibrium in *N*-methylpiperidine. *Tetrahedron* **1977**, *33*, 915–925.

2
Stereoelectronic Effects and Reactivity

2.1
Hyperconjugation with σ Bonds

Stereoelectronic effects can be defined as effects on structure and reactivity determined by the efficiency of orbital overlap as a function of molecular conformation. Interactions involving sp^3 hybrid orbitals are usually referred to as "hyperconjugation", whereas interactions of p orbitals of sp^2 hybridized atoms are called "conjugation". Hyperconjugation will stabilize or destabilize certain conformations, strengthen or weaken bonds, and can increase or reduce the energy of lone electron pairs, and thereby modulate the nucleophilicity and basicity of a given compound. Those stereoelectronic effects with highest impact on the reactivity of compounds generally result from interaction of vicinal orbitals.

In Scheme 2.1 the orbital interactions between two sp^3 hybridized, tetravalent atoms X and Y are sketched in the staggered conformation. This conformation enables efficient transfer of electrons from the (bonding) σ_{X-A} orbital to the empty (antibonding) σ^*_{X-A} orbital; this leads to longer and weaker X–A bonds and a shorter, stronger X–Y bond. The net effect is lowering of the ground state energy (i.e. stabilization) of the molecule. This form of hyperconjugation can also be illustrated by the two canonical forms sketched in Scheme 2.1.

In principle a $\sigma_{X-A} \rightarrow \sigma^*_{Y-A}$ charge-transfer interaction would also be possible when the two vicinal X–A and Y–A bonds adopt a synperiplanar conformation. However, in this latter conformation the overlap integral $\sigma_{X-A} \rightarrow \sigma^*_{Y-A}$ and thus the

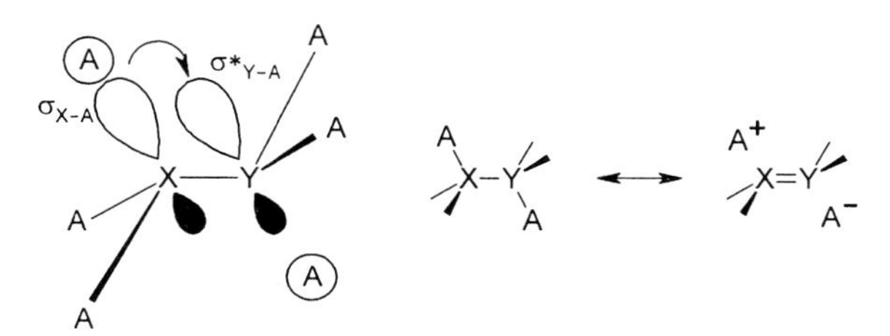

Scheme 2.1. Orbital interaction and canonical forms for hyperconjugation between σ bonds.

Side Reactions in Organic Synthesis. Florencio Zaragoza Dörwald

ISBN: 3-527-31021-5

stabilization achieved by hyperconjugation is smaller than with the antiperiplanar orientation of the two interacting bonds.

The relative abilities of σ^*_{C-X} bonds to accept electrons from a vicinal C–H bond in ethanes have been calculated (Table 2.1), and were found not to correlate well with the electronegativity of X. Thus, within each group of the periodic table the energies of the $\sigma_{C-H} \rightarrow \sigma^*_{C-X}$ interaction decrease with decreasing atomic weight of X, although the electronegativity increases. The reason for this is that the energy of the σ^*_{C-X} orbitals decreases when going to the heavier elements within one group; this leads to a smaller energy gap between the bonding and antibonding orbitals, and thereby to greater stabilization [1].

Table 2.1. Energies E_{hyp} (kcal/mol) of hyperconjugative ($\sigma_{C(2)-H} \rightarrow \sigma^*_{C(1)-X}$) interaction in CH_3–CH_2–X [1].

X	I	N_2^+	Br	Cl	SH	F	OH	SeH		
E_{hyp}	7.81	7.61	6.29	6.20	5.36/4.70[a]	5.09	4.74/4.22[a]	4.68		
X	PH_2	AsH_2	NH_2	CMe_3	GeH_3	SiH_3	CF_3	CH_3	H	
E_{hyp}	4.61/4.01[a]	4.55	4.46/3.82[a]	3.93	3.80	3.63	3.59	3.38	3.17	

a The two values refer to two different orientations of the lone pairs.

The energies given in Table 2.1 are valid for substituted ethanes only, and a different ranking might result with other compound classes. These values are highly sensitive to small structural variations and should, therefore, be used as a rough guideline only. The organic chemist can use these values to estimate how strongly a C,H group is acidified by a group X in compounds with the substructure H–C–C–X.

Hyperconjugation between sp^3 hybridized atoms can have important implications for the ground-state conformation of organic compounds. It has, for example, been suggested that the energy difference between the staggered and the eclipsed conformations of ethane is due to both hyperconjugation and repulsion [2–5]. The fact that 1,2-difluoroethane [6, 7] or *N*-(2-fluoroethyl)amides [8] preferentially adopt a *gauche* conformation is also thought to result from hyperconjugation between the σ_{C-H} orbital and the σ^*_{C-F} orbital (Scheme 2.2). The *anti* conformation is, despite the mutually repulsive C–F dipoles pointing into opposite directions, less favorable for 1,2-difluoroethane. Because the C–F bond is a poorer electron donor than the C–H bond, the *gauche* conformation, which enables two $\sigma_{C-H} \rightarrow \sigma^*_{C-F}$ interactions, is approximately 0.7 kcal/mol more stable than the *anti* conformation.

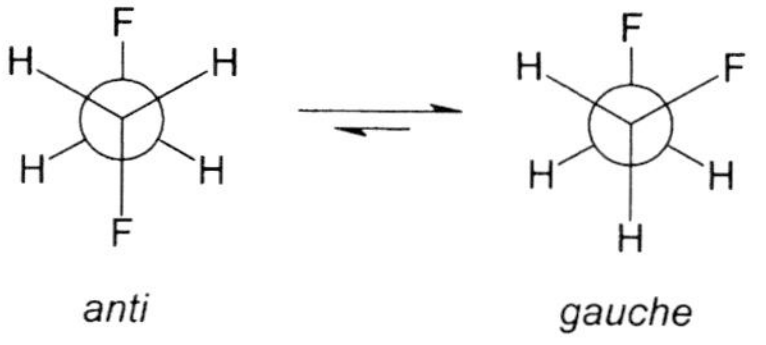

anti *gauche*

Scheme 2.2. Conformations of 1,2-difluoroethane.

Calculations have shown that the rotational barrier of the C–O bond in methanol (1.1 kcal/mol) is significantly lower than the corresponding rotational barrier of methyl hypofluorite (MeOF, 3.7 kcal/mol) or methyl hypochlorite (MeOCl, 3.5 kcal/mol), in which a strong $\sigma_{C-H} \rightarrow \sigma^*_{O-Hal}$ hyperconjugation is possible [9]. Similarly, in 1,2-dihaloethenes such as 1,2-difluoroethene, 1-chloro-2-fluoroethene, or 1,2-dichloroethene the *cis* isomers are more stable than the corresponding *trans* isomers [10, 11].

2.2 Hyperconjugation with Lone Electron Pairs

2.2.1 Effects on Conformation

Lone electron pairs can donate electron density to the antibonding orbitals of antiperiplanar σ bonds more efficiently than most σ_{C-X} bonds. Consequently, these interactions lead to the most conspicuous stereoelectronic effects. Examples of the interaction of lone pairs with antibonding orbitals (negative hyperconjugation) include the stabilization of the *anti* conformation of ethylamine [1] or α-fluoroamines [12], or the stabilization of the *gauche* conformation of hydrazine [13] (Scheme 2.3). In ethylamine the σ^*_{C-C} orbital is a slightly better acceptor than the σ^*_{C-H} orbital, whereas in hydrazine hyperconjugation between the lone pairs and vicinal antibonding orbitals is only possible in the *gauche* conformation.

gauche anti anti gauche

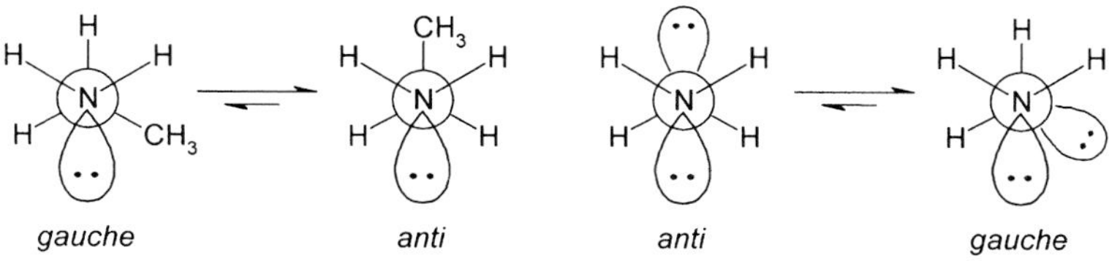

Scheme 2.3. Conformations of $EtNH_2$ (left [1]) and H_2NNH_2 (right [13]).

Because the precise energies of charge-transfer interactions are sensitive to small structural modifications, purely intuitive predictions often turn out to be wrong. In tetrafluorohydrazine, for instance, hyperconjugation of the type $n_N \rightarrow \sigma^*_{N-F}$ should be even stronger than in hydrazine, and a preferred *gauche* conformation would be expected. The *anti* conformer of tetrafluorohydrazine is, though, slightly more stable than the *gauche* conformer, because of efficient hyperconjugation between the nonbonding electrons on fluorine and σ^*_{N-F} [13] (Scheme 2.4).

Other compounds for which the most stable conformation is probably because of negative hyperconjugation include difluorodiazene [10], hydrogen peroxide, dioxygen difluoride [14], and bis(trifluoromethyl) peroxide [15] (Scheme 2.5).

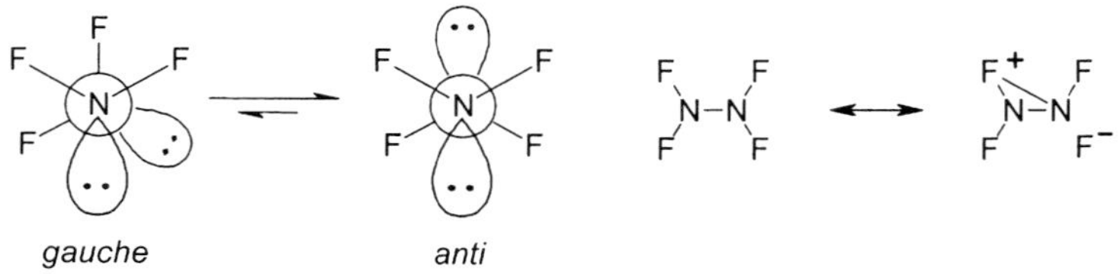

Scheme 2.4. Conformations and canonical forms of F_2NNF_2 [13].

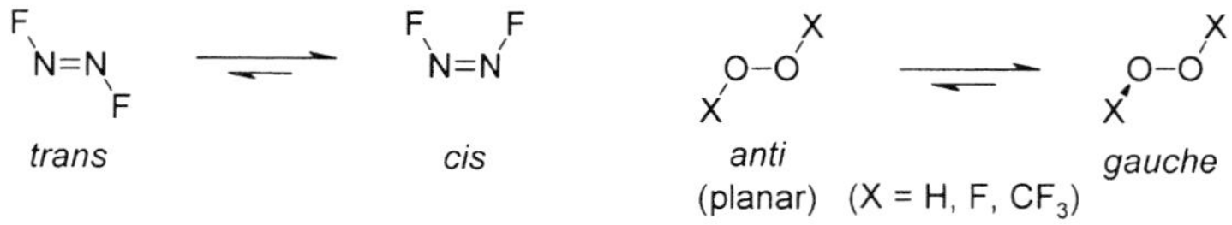

Scheme 2.5. Conformations of diazenes and peroxides.

2.2.2
The Anomeric Effect

The anomeric effect [16], i.e. the tendency of some groups at the C-2 position of tetrahydropyrans to adopt preferentially an axial position, can also be rationalized as a consequence of negative hyperconjugation (Scheme 2.6). Efficient overlap of the antibonding C–X orbital and one lone pair at oxygen is only possible in pyrans with the substituent X positioned axially; if σ^*_{C-X} is a good acceptor hyperconjugation may stabilize this conformation, which otherwise (i.e. in the corresponding cyclohexyl derivative) would normally be unfavorable. A further observation which points toward hyperconjugation as reason for the anomeric effect is that axially positioned substituents X have longer C–X bonds than similar compounds with X in the equatorial position [16]. The groups listed in Table 2.1 which have good ability to accept electrons from σ_{C-H} bonds also will tend to have a strong anomeric effect when bound to C-2 of a pyran.

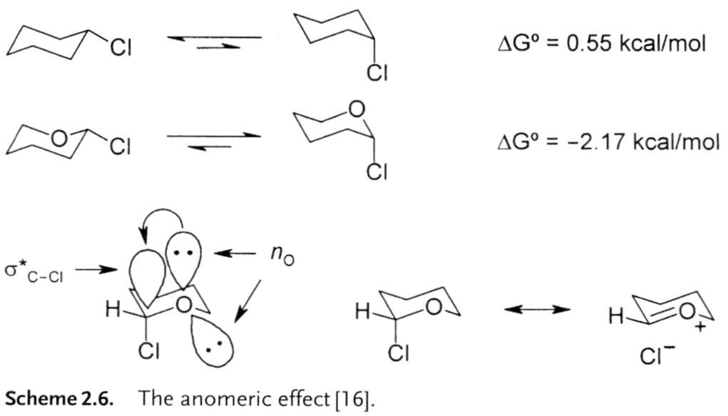

Scheme 2.6. The anomeric effect [16].

2.2.3
Effects on Spectra and Structure

The weakening of σ bonds by negative hyperconjugation with lone electron pairs also reveals itself in IR and NMR spectra. Thus, C–H, N–H, or O–H bonds oriented *trans* or antiperiplanar to an unshared, vicinal electron pair are weakened and have therefore a significantly reduced IR vibrational frequency [17]. The C–H vibrational frequency in aldehydes is, for example, lower than that in alkenes (Scheme 2.7). Polycyclic amines with at least two hydrogen atoms antiperiplanar to the lone pair on nitrogen have characteristic absorption bands at 2800–2700 cm^{-1} which have been used to infer the relative configuration of such amines [18].

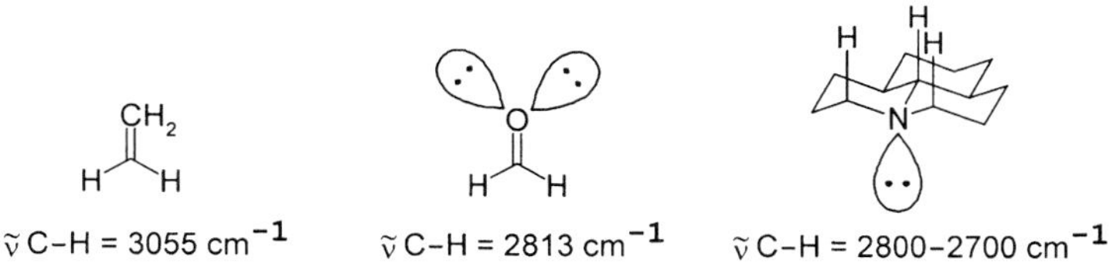

CH_2 H H $\quad$ O H H $\quad$ H H H N

$\tilde{\nu}$ C–H = 3055 cm^{-1} $\quad$ $\tilde{\nu}$ C–H = 2813 cm^{-1} $\quad$ $\tilde{\nu}$ C–H = 2800–2700 cm^{-1}

Scheme 2.7. Antiperiplanar lone pairs weaken C–H bonds and reduce their IR wavenumber [17, 18].

Negative hyperconjugation can also be used to explain some of the structural and spectroscopic features of simple carbonyl group-containing compounds such as aldehydes, ketones, or carboxylic acid derivatives,. As illustrated by the canonical forms shown in Scheme 2.8, the strength of the C=O bond in carbonyl compounds should increase with increasing electron-withdrawing capability of substituents directly attached to the carbonyl group. The hybridization of the carbonyl carbon atom should at the same time become more *sp*-like, and the angle R–C=O should become larger than 120°. This has, for instance, been observed in X-ray structural analyses of acyl halides [19, 20], trihalomethyl ketones [21], lactones, and lactams [22, 23] (Scheme 2.8).

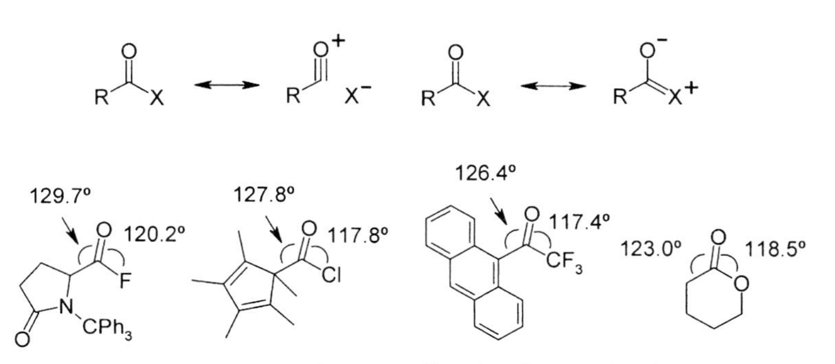

Scheme 2.8. Hyperconjugative distortion of bond angles in carbonyl compounds [19–22].

Because the strength of the C=O bond correlates with its stretching frequency, the latter increases when the substituent X becomes more electronegative (Table 2.2). Complete abstraction of the group X (e.g. for X = BF_4^-) leads to the formation of acylium cations with a short and strong C≡O bond, as revealed by the high vibrational IR-frequency of these compounds [24] (Table 2.2).

Table 2.2. C=O Wavenumbers $\tilde{\nu}$(C=O) of acetyl derivatives MeCOX.

X	CH_3	H	CN	CCl_3	CF_3
$\tilde{\nu}$(CO) (cm^{-1})	1720	1729	1730	1765	1781
Ref.	25	26	27	28	29
X	I	Br	Cl	F	BF_4^-
$\tilde{\nu}$(CO) (cm^{-1})	1808	1812	1800	1848	2299
Ref.	30	31	24	24	24

If the group X has lone electron pairs, conjugation of these with the carbonyl group will not lead to a distortion of the O=C–X and R–C=O angles, but will weaken the C=O bond (Scheme 2.8). This conjugation might be the reason for the small changes in C=O stretching frequency in the series of acyl halides (Table 2.2). The decrease of electronegativity when going from fluorine to iodine (which should lead to a strong decrease of the C=O stretching frequency) is partly compensated by less efficient conjugation between the carbonyl group and the unshared electron pairs on the larger, heavier halogen atoms.

In NMR spectra, hyperconjugative electron transfer into σ^* orbitals can manifest itself as a diminished chemical shift of protons located antiperiplanar to unshared electron pairs [32, 33]. Thus, the chemical shifts of the methine proton of tris(dialkylamino)methanes vary strongly as a function of their orientation toward the lone electron pairs at nitrogen [34] (Scheme 2.9). In compound **B** the lone pairs are synperiplanar to the C–H bond, and efficient negative hyperconjugation is not possible. The inductive (electron-withdrawing) effect of nitrogen leads to deshielding of this proton compared with the methine proton in compound **A**, in some conformers of which hyperconjugation is possible. Strong shielding of this proton is observed in compound **C**, with three electron pairs oriented simultaneously antiperiplanar to the methine C–H bond [34].

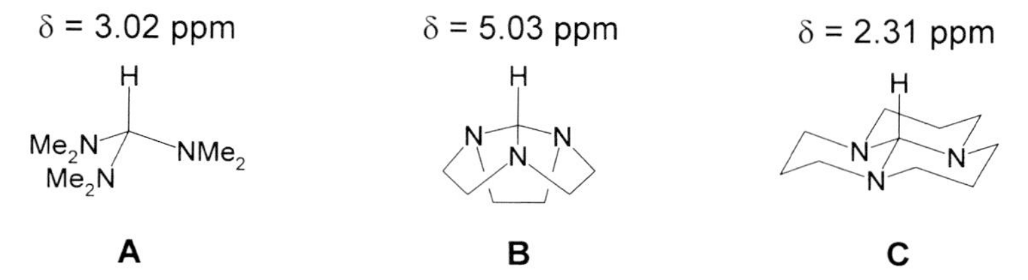

Scheme 2.9. Magnetic shielding of protons by antiperiplanar lone electron pairs.

The weakening of C–H bonds by hyperconjugation can also lead to lower one-bond NMR coupling constants. Calculations [35] have shown that in tetrahydropyran 2-H_{ax} has a lower coupling constant to C-2 (129.5 Hz) than 2-H_{eq} (140.7 Hz). These coupling constants correlate as expected with the calculated bond lengths [35]. These C–H bond lengths have, however, relatively little dependence on the orientation of vicinal lone pairs (Scheme 2.10), because the $\sigma^*_{\mathrm{C-H}}$ orbital is a poor electron acceptor. The similar coupling constants of the axial and equatorial C(2)–H bonds in tetrahydrothiopyran can be explained by assuming that the electron transfer $\sigma_{\mathrm{C(6)-S}} \rightarrow \sigma^*_{\mathrm{C(2)-H}}$ is more efficient than $n_{\mathrm{S}} \rightarrow \sigma^*_{\mathrm{C(2)-H}}$ [35]. This would imply that no anomeric effect should be observed in tetrahydrothiopyrans, but this is not so [16]. Certain heterocycles can, however, have an anomeric effect which is not due to negative hyperconjugation but to other factors such as steric and electrostatic effects [36, 37].

The coupling constants and calculated bond lengths for axial and equatorial C–H bonds in cyclohexane (Scheme 2.10) have been interpreted as a result of the superior electron-donating capacity of C–H bonds compared with C–C bonds (see, e.g., Ref. [38]; calculations (Table 2.1) do not support this idea). Thus, the axial C–H bonds are weaker and longer than equatorial C–H bonds because each of the former undergoes hyperconjugation with two axial C–H bonds.

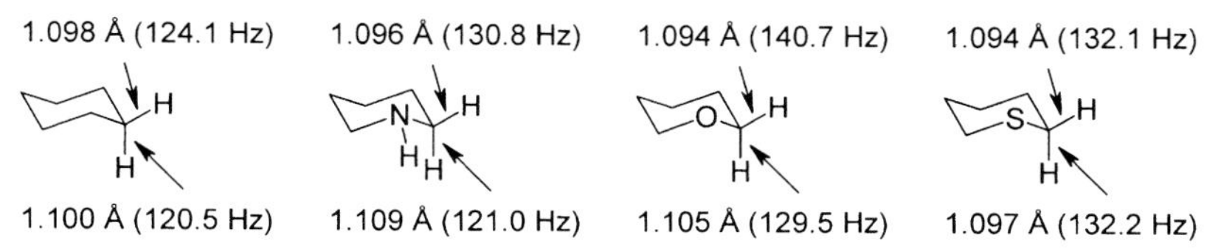

Scheme 2.10. Calculated bond lengths and ^{13}C–^{1}H coupling constants in cyclohexane and six-membered heterocycles [35].

2.3 Hyperconjugation and Reactivity

2.3.1 Basicity and Nucleophilicity

The orbital interactions discussed above not only govern the energy of ground state conformations or configurations but can also modulate the energy of transition states and, therefore, the reactivity of compounds. In conformationally constrained systems it has been observed that orbital overlap can affect the nucleophilicity and basicity of unshared electron pairs. The basicity differences of the amines shown in Scheme 2.11 [39] can, for instance, be interpreted as a result of a more or less efficient overlap between vicinal $\sigma_{\mathrm{C-N}}$ and $\sigma^*_{\mathrm{C-X}}$ orbitals, where X represents an electron-withdrawing group.

The reactivity of cyclic phosphites also depends strongly on the mutual orientation of n_{P} and n_{O} lone pairs. Triethyl phosphite (**1**) (Scheme 2.12), for instance,

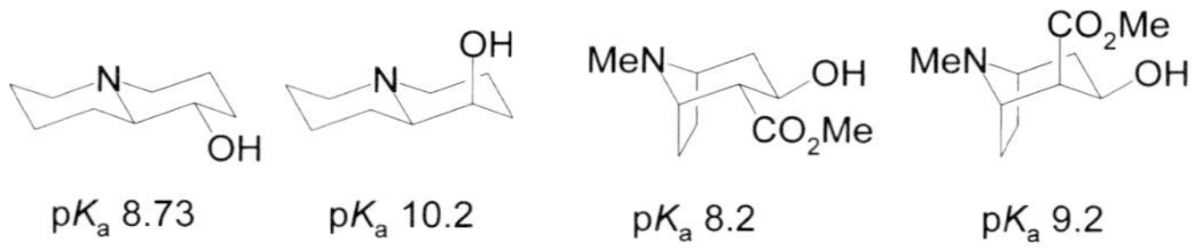

N OH MeN OH CO_2Me MeN CO_2Me OH

pK_a 8.73 pK_a 10.2 pK_a 8.2 pK_a 9.2

Scheme 2.11. Basicity of bicyclic amines substituted with electron-withdrawing groups [39].

undergoes rapid addition to 2-benzylidenepentane-2,4-dione to yield the cyclic, pentavalent phosphorus derivative **2**. The more constrained phosphite **3**, however, reacts much more slowly, despite the easier access to its electron pair [40]. The low reactivity of phosphite **3** is even more puzzling when compared with the relative reactivity of triethylamine (**5**) [41] and quinuclidines such as **6** [42] towards electrophiles. The fixing of the lone pair (inversion of nitrogen is no longer possible) and the alkyl groups in quinuclidines enhances the rate of reactions with electrophiles (but not their basicity) to such an extent that even poor electrophiles such as dichloromethane react swiftly [43]. (The fixing of the lone pair in quinuclidines has no significant effect on their thermodynamic basicity (i.e. their pK_a), which reflects the energy difference between the protonated and the non-protonated forms. Thermodynamic basicity is not directly related to the *rate* of protonation (kinetic basicity), which should in fact be higher for quinuclidines than for acyclic tertiary amines).

The low reactivity of phosphite **3** has been explained as follows [44]. During the reaction of phosphite **3** with an electrophile (E), efficient electron transfer from the lone pairs of oxygen to the incipient antibonding orbital of the P–E bond is not pos-

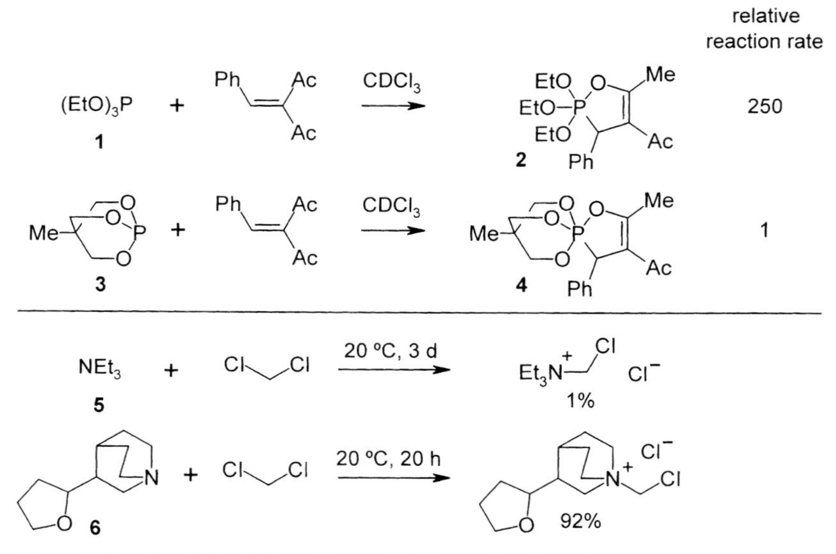

Scheme 2.12. Reactivity of phosphites and tertiary amines towards electrophiles [40–42].

sible, because their mutual orientation is *gauche*. Calculations suggest [44] that an $n_O \rightarrow \sigma^*_{P-E}$ interaction, where σ^*_{P-E} is the antibonding orbital of the incipient P–E bond, which is possible during reaction of triethyl phosphite with an electrophile (left sketch, Scheme 2.13), should stabilize the transition state in a fashion similar to its stabilization of the transition state of the reverse reaction or the final product, and thereby accelerate the reaction. In other words, if a transition state is product-like, and the product is stabilized by hyperconjugation ($n_O \rightarrow \sigma^*_{P-E}$ in this case) the rate of formation of this product should increase [38].

The orientation of the unshared electron pairs of oxygen in compounds containing P–O bonds also modulates the gas-phase basicity [45, 46] and oxidation potentials [47] of these compounds.

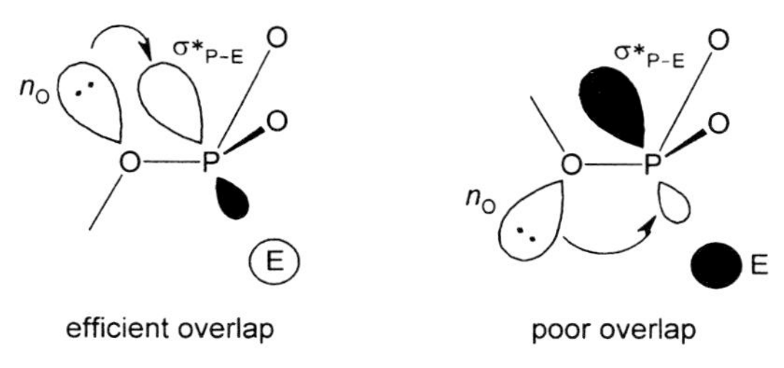

Scheme 2.13. Hyperconjugation in phosphites.

2.3.2
Rates of Oxidation

The rate of oxidation of organic compounds might also be influenced by stereoelectronic effects. Thus, when polycyclic amines are oxidized to iminium ions by treatment with mercury(II) acetate (Scheme 2.14), amines with a lone electron pair antiperiplanar to the reacting hydrogen atom will react most rapidly [48]. Hexahydropyrrolizine (third example of Scheme 2.14), in which the bridgehead hydrogen atom and the lone pair are synperiplanar, could not be oxidized [48]; this might, however, also be partly a result of the strained structure of the product. Similarly, compound **B** in Scheme 2.9 was resistant to oxidants. Tris(dialkylamino)methane **C** (Scheme 2.9), on the other hand, with three lone pairs antiperiplanar to the methine C–H bond, is a strong reducing agent [49], and reacts with acids to yield hydrogen and the corresponding guanidinium salt [50] (fourth reaction, Scheme 2.14).

Some 2-alkoxytetrahydropyrans show a reactivity toward oxidants which parallels the reactivity of polycyclic amines discussed above, and which is in line with the hypothesis that weakening of C–H bonds by hyperconjugation should also increase the rate of C–H bond cleavage. For instance, of the two epimeric pyrans sketched in Scheme 2.15 only that with an axial 2-H is oxidized by ozone [51]. The same selectivity has been observed in the oxidation of methyl α- and β-glucopyranoside with ozone [52], and in homolytic C–H bond cleavage in cyclic ethers [53].

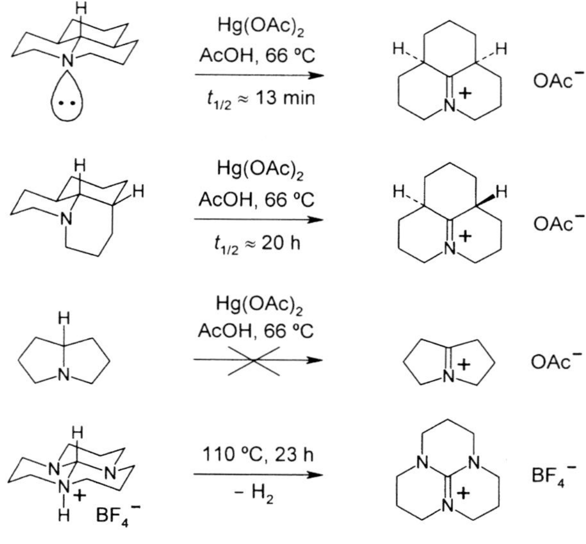

Scheme 2.14. Oxidation of polycyclic amines [48, 50].

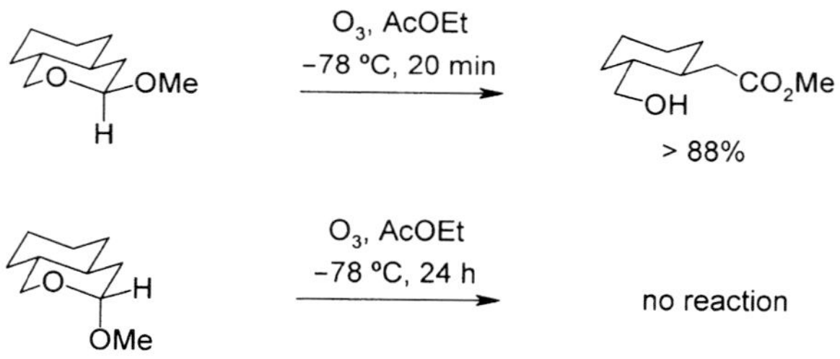

O OMe H
O_3, AcOEt
−78 °C, 20 min
OH CO_2Me
> 88%

O H OMe
O_3, AcOEt
−78 °C, 24 h
no reaction

Scheme 2.15. Oxidation of tetrahydropyrans [51].

2.3.3
Rates of Deprotonation

The α-deprotonation of conformationally constrained thioethers can proceed with high diastereoselectivity (Scheme 2.16) [54]. That equatorial protons are removed much more rapidly than axial protons suggests that stabilization of carbanions $C_\alpha^{(-)}$–S–$C_{\alpha'}$ by sulfur is mainly a result of hyperconjugation between the carbanion lone pair and the antibonding S–$C_{\alpha'}$ orbital [55].

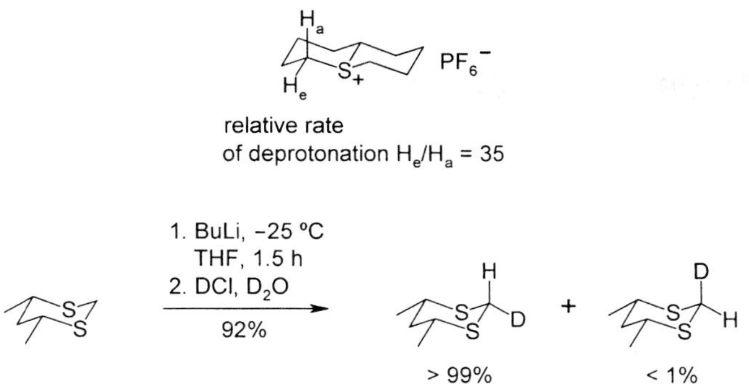

H_a PF_6^- H_e

relative rate
of deprotonation H_e/H_a = 35

1. BuLi, −25 °C THF, 1.5 h
2. DCl, D_2O
92%

> 99% < 1%

Scheme 2.16. Diastereoselective metalation of thioethers and sulfonium salts [55, 56].

2.3.4 Other Reactions

One remarkable example of nucleophilic substitution at a 1,3-dioxane in which the isomer with an axial (and thus weaker) bond to the leaving group reacts more rapidly than the equatorial epimer is shown in Scheme 2.17 [57]. Because the product is formed with retention of configuration, the reaction must proceed by an SN1-like mechanism.

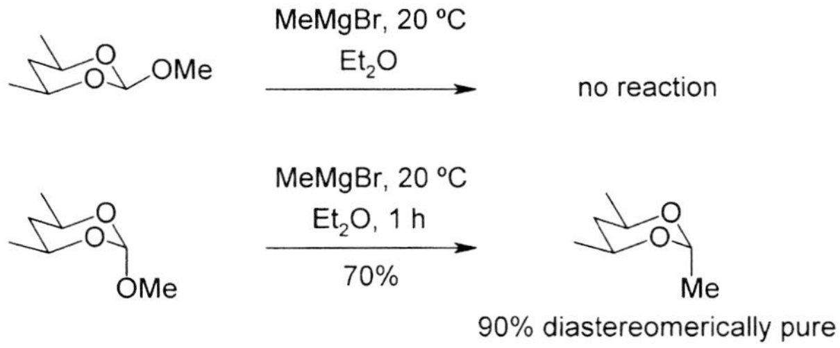

MeMgBr, 20 °C, Et_2O → no reaction

MeMgBr, 20 °C, Et_2O, 1 h, 70% → 90% diastereomerically pure

Scheme 2.17. Reactions of cyclic orthoesters with Grignard reagents [57].

Unfortunately, intuitive predictions of reactivity on the basis of stereoelectronic effects are not always possible, because these effects are subtle and can easily be overridden by steric, inductive, or field effects, or by conformational changes during the reaction [58]. It must also be kept in mind that hyperconjugation in the transition state, and not in the ground state, will be have the largest effect on the reaction rate.

As discussed above, 2-halotetrahydropyrans tend to adopt a conformation in which the halogen is located axially, and which is stabilized by negative hyperconjugation of the type $n_O \rightarrow \sigma^*_{C-Hal}$. X-ray structural analyses have shown that in these compounds the carbon–halogen bond is, as expected, longer than in similar compounds with an equatorial halogen atom [16]. Accordingly, we would expect that 2-halotetrahydropyrans with an axial halogen would react more rapidly with nucleophiles than the corresponding pyrans with an equatorial halogen atom, because C–Hal_{ax} is longer and weaker than C–Hal_{eq}. This, however, is not observed. The rates of alcoholysis of α- and β-pyranosyl halides [59, 60] cannot be used to distinguish between the reactivity of axial and equatorial halides, because of the conformational flexibility of these substrates, which preferentially adopt the conformation with the halide axial, even if the remaining substituents are thereby also forced into the axial position [16]. But even in conformationally constrained substrates the axial leaving groups are not easier to displace [61]. For instance, α-glycosides are not always hydrolyzed more rapidly than the corresponding β-glycosides [58, 62], as is illustrated by the similarly rapid oxidative hydrolysis of the two conformationally constrained anomeric glycosides shown in Scheme 2.18 [63].

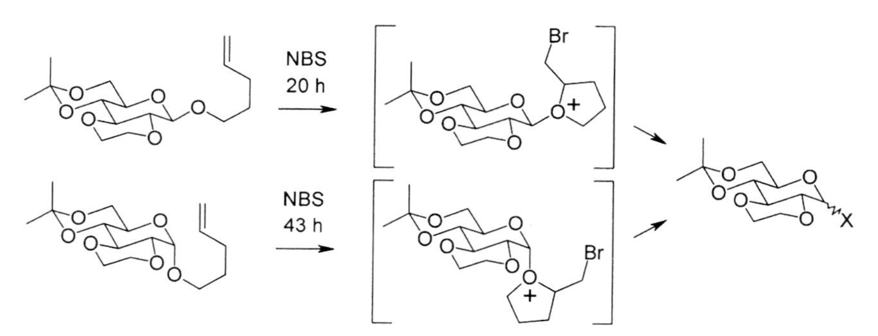

Scheme 2.18. Oxidative cleavage of anomeric pentenyl glycosides [63].

Numerous other examples have, moreover, been reported in which bonds with antiperiplanar lone electron pairs react more slowly than, or at rates similar to, comparable bonds without antiperiplanar lone pairs [52, 63–67].

The diastereoselectivity of the first two reactions shown in Scheme 2.19 [68] can also be interpreted as a result of a stereoelectronic effect. Although the diastereoselectivity of additions to enones is usually governed by steric effects, which lead to an addition of the nucleophile from the sterically less demanding side of the double bond (as in the third reaction in Scheme 2.19; for additional examples, see Refs [69, 70]), the first two reactions shown in Scheme 2.19 are, surprisingly, *syn*-selective. Cyclopentenones [68, 71] and cycloheptenones [72] can also react with the same *syn*-diastereoselectivity.

Hyperconjugation between the incipient antibonding $\sigma^*_{\mathrm{Nuc-C}}$ orbital and the binding $\sigma_{\mathrm{C}(\gamma)-\mathrm{O}}$ and $\sigma_{\mathrm{C}(\gamma)-\mathrm{H}}$ orbitals in the transition states [68] can be invoked as a possible reason for this diastereoselectivity. Because of the high electronegativity of oxygen, C–O bonds have poor electron-donating properties compared with C–H bonds. Accordingly, the transition state leading to the *syn* product might be stabilized by $\sigma_{\mathrm{C}(\gamma)-\mathrm{H}} \rightarrow \sigma^*_{\mathrm{C}(\beta)-\mathrm{Nuc}}$ hyperconjugation, where $\sigma^*_{\mathrm{C}(\beta)-\mathrm{Nuc}}$ is the incipient

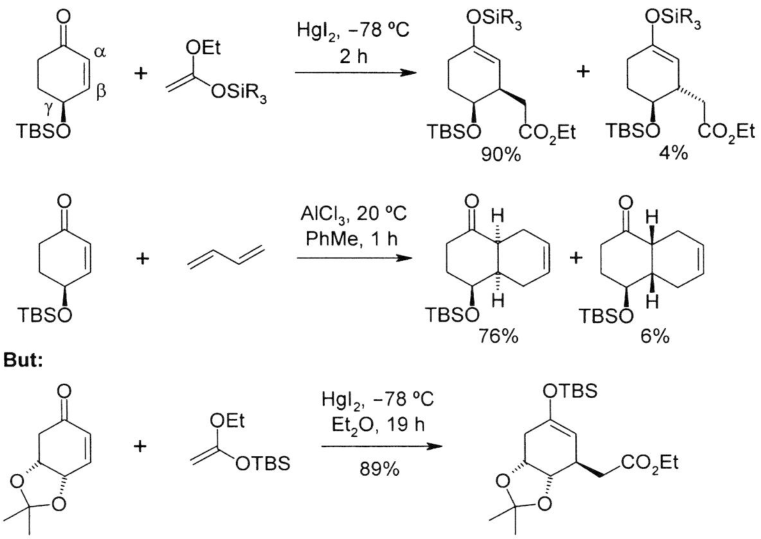

O α β γ TBSO + OEt OSiR3 — HgI_2, −78 °C, 2 h → OSiR3 TBSO CO2Et 90% + OSiR3 TBSO CO2Et 4%

O TBSO + (butadiene) — $AlCl_3$, 20 °C, PhMe, 1 h → O H H TBSO 76% + O H H TBSO 6%

But:

O O O + OEt OTBS — HgI_2, −78 °C, Et_2O, 19 h, 89% → OTBS O O CO2Et

Scheme 2.19. Diastereoselective addition of C-nucleophiles to enones [68, 71, 73].

antibonding orbital between C(β) and the nucleophile. Such hyperconjugation should also stabilize the product, in the way illustrated by the first two canonical forms in Scheme 2.20. For the *anti* diastereomer, however, the third valence bond resonance structure of Scheme 2.20 can also be considered. These canonical forms, which would stabilize the *anti* product, would, however, not resemble the transition

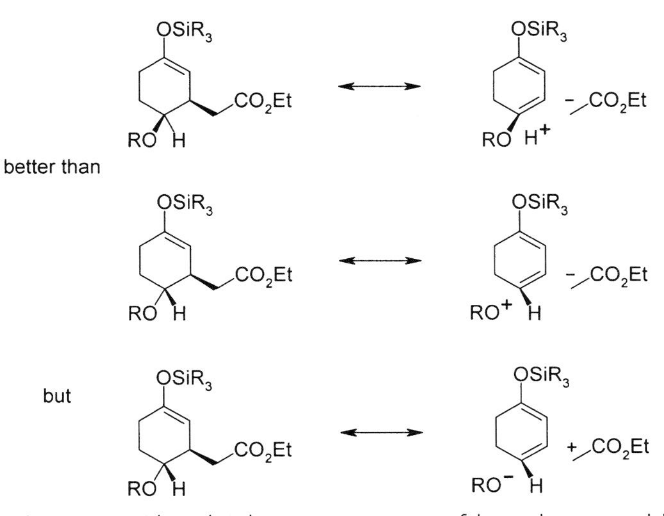

Scheme 2.20. Valence bond resonance structures of the products as models of the transition state leading to these products.

state for addition of an electron-rich nucleophile to an electrophilic double bond, and should, therefore, not be relevant to the diastereoselectivity of this reaction.

The arguments used above to explain the diastereoselectivity of the two first reactions in Scheme 2.19 on the basis of hyperconjugation of incipient antibonding orbitals are analogous to that used to explain the reactivity of phosphites (Scheme 2.12). This form of hyperconjugation was also used by Cieplak to explain π-facial diastereoselection in the addition of nucleophiles to ketones [38, 74]. The third reaction of Scheme 2.19 clearly shows that the diastereoselectivity of these reactions is responsive to small structural changes and are, therefore, hardly predictable.

2.4
Conclusion

Stereoelectronic effects can have a profound effect on the ground-state structure of molecules, and can often help to explain counter-intuitive conformational preferences or spectroscopic features. Their effect on the energy of transition states is, however, less straightforward to predict. As stated by the Curtin–Hammett principle [75] (Section 1.4), reactions will proceed via energetically unfavorable conformers if these are more reactive (as is often the case) than better stabilized conformers. In such instances ground-state stabilization of certain conformers or the weakening of bonds by hyperconjugation will not necessarily be predictive for the outcome of a reaction.

The examples discussed above illustrate that reactivity and stereoselectivity are subject to numerous, often subtle, influences. Continuous improvements in molecular modeling have enabled clarification of many, previously unexplained observations; in the future even solvent effects might, perhaps, be taken into account. Predictive models based on such calculations should, however, always be substantiated by experimental data.

References

1 Alabugin, I. V.; Zeidan, T. A. Stereoelectronic effects and general trends in hyperconjugative acceptor ability of σ bonds. *J. Am. Chem. Soc.* **2002**, *124*, 3175–3185.
2 Pophristic, V.; Goodman, L. Hyperconjugation not steric repulsion leads to the staggered structure of ethane. *Nature* **2001**, *411*, 565–568.
3 Bickelhaupt, F. M.; Baerends, E. J. The case for steric repulsion causing the staggered conformation of ethane. *Angew. Chem. Int. Ed.* **2003**, *42*, 4183–4188.
4 Mo, Y.; Wu, W.; Song, L.; Lin, M.; Zhang, Q.; Gao, J. The magnitude of hyperconjugation in ethane: a perspective from ab initio valence bond theory. *Angew. Chem. Int. Ed.* **2004**, *43*, 1986–1990.
5 Weinhold, F. Rebuttal to the Bickelhaupt–Baerends case for steric repulsion causing the staggered conformation of ethane. *Angew. Chem. Int. Ed.* **2003**, *42*, 4188–4194.
6 Craig, N. C.; Chen, A.; Suh, K. H.; Klee, S.; Mellau, G. C.; Winnewisser, B. P.; Winnewisser, M. Contribution to the study of the *gauche* effect. The complete structure of the *anti* rotamer of 1,2-difluoroethane. *J. Am. Chem. Soc.* **1997**, *119*, 4789–4790.

7 Tavasli, M.; O'Hagan, D.; Pearson, C.; Petty, M. C. The fluorine *gauche* effect. Langmuir isotherms report the relative conformational stability of (±)-*erythro*- and (±)-*threo*-9,10-difluorostearic acids. *Chem. Commun.* **2002**, 1226–1227.

8 Briggs, C. R. S.; O'Hagan, D.; Howard, J. A. K.; Yufit, D. S. The C–F bond as a tool in the conformational control of amides. *J. Fluorine Chem.* **2003**, *119*, 9–13.

9 Wu, Y.-D.; Houk, K. N. Theoretical studies of rotational barriers of heteroatom derivatives of methanol. *J. Phys. Chem.* **1990**, *94*, 4856–4861.

10 Craig, N. C.; Piper, L. G.; Wheeler, V. L. Thermodynamics of *cis–trans* isomerizations. II. The 1-chloro-2-fluoroethylenes, 1,2-difluorocyclopropanes, and related molecules. *J. Phys. Chem.* **1971**, *75*, 1453–1460.

11 Viehe, H. G. Geometrische Isomerenpaare mit bevorzugter *cis*-Struktur. *Chem. Ber.* **1960**, *93*, 1697–1709.

12 Rahman, M. M.; Lemal, D. M.; Dailey, W. P. Negative hyperconjugation. The rotation–inversion barrier in α-fluoroamines. *J. Am. Chem. Soc.* **1988**, *110*, 1964–1966.

13 Wilcox, C. F.; Bauer, S. H. DFT calculations of thermochemical and structural parameters of tetracyanohydrazine and related tetrasubstituted hydrazines. *J. Mol. Struct. (Theochem)* **2003**, *625*, 1–8.

14 Cotton, F. A.; Wilkinson, G. *Anorganische Chemie*; Verlag Chemie: Weinheim, 1982.

15 Gase, W.; Boggs, J. E. The structures of dimethyl peroxide and bis(trifluoromethyl) peroxide. *J. Mol. Struct.* **1984**, *116*, 207–210.

16 Juaristi, E.; Cuevas, G. Recent studies of the anomeric effect. *Tetrahedron* **1992**, *48*, 5019–5087.

17 McKean, D. C. Individual C–H bond strengths in simple organic compounds: effects of conformation and substitution. *Chem. Soc. Rev.* **1978**, *7*, 399–422.

18 Bohlmann, F. Zur Konfigurationsbestimmung von Chinolizidin-Derivaten. *Chem. Ber.* **1958**, *91*, 2157–2167.

19 Jutzi, P.; Schwartzen, K.-H.; Mix, A.; Stammler, H.-G.; Neumann, B. Bis(pentamethylcyclopentadienyl)keton und -thioketon: Kohlenstoff-Verbindungen mit präformierter Diels–Alder-Geometrie. *Chem. Ber.* **1993**, *126*, 415–420.

20 Nastopoulos, V.; Karigiannis, G.; Mamos, P.; Papaioannou, D.; Kavounis, C. (*S*)-*N*-Triphenylmethylpyroglutamyl fluoride. *Acta Cryst.* **1998**, *C54*, 1718–1720.

21 Corey, E. J.; Link, J. O.; Sarshar, S.; Shao, Y. X-Ray diffraction studies of crystalline trihalomethyl ketones ($RCOCX_3$) reveal an unusual structural deformation about the carbonyl group. *Tetrahedron Lett.* **1992**, *33*, 7103–7106.

22 Schweizer, W. B.; Dunitz, J. D. 152. Structural characteristics of the carboxylic ester group. *Helv. Chim. Acta* **1982**, *65*, 1547–1554.

23 Nørskov-Lauritsen, L.; Bürgi, H.-B.; Hofmann, P.; Schmidt, H. R. Bond angles in lactones and lactams. *Helv. Chim. Acta* **1985**, *68*, 76–82.

24 Olah, G. A.; Kuhn, S. J.; Tolgyesi, W. S.; Baker, E. B. Stable carbonium ions II. Oxocarbonium (acylium) tetrafluoroborates, hexafluorophosphates, hexafluoroantimonates and hexafluoroarsenates. Structure and chemical reactivity of acyl fluoride: Lewis acid fluoride complexes. *J. Am. Chem. Soc.* **1962**, *84*, 2733–2740.

25 Springer, G.; Elam, C.; Edwards, A.; Bowe, C.; Boyles, D.; Bartmess, J.; Chandler, M.; West, K.; Williams, J.; Green, J.; Pagni, R. M.; Kabalka, G. W. Chemical and spectroscopic studies related to the Lewis acidity of $LiClO_4$ in Et_2O. *J. Org. Chem.* **1999**, *64*, 2202–2210.

26 Filippini, F.; Susz, B.-P. Etude de composes d'addition d'acides de Lewis XXXIII. Resonance magnetique nucleaire de composes formes par des aldehydes aliphatiques et aromatiques avec des accepteurs electroniques. *Helv. Chim. Acta* **1971**, *54*, 1175–1178.

27 Borch, R. F.; Levitan, S. R.; Van-Catledge, F. A. The reaction of acyl cyanides with Grignard reagents. *J. Org. Chem.* **1972**, *37*, 726–729.

28 Gallina, C.; Giordano, C. Selected methods for the oxidation of 1,1,1-trichloro-2-alkanols. An efficient modification using chromic acid. *Synthesis* **1989**, 466–468.

29 Mello, R.; Fiorentino, M.; Sciacovelli, O.; Curci, R. On the isolation and characterization of methyl(trifluoromethyl)dioxirane. *J. Org. Chem.* **1988**, *53*, 3890–3891.

30 Sustmann, R.; Korth, H. G. Carbonsäure-, Thiocarbonsäurehalogenide bzw. Carbonsäure-halogenid-imide. *Houben-Weyl, Methoden der Organischen Chemie* **1985**, *E5*, 589.

31 Smith, C. W.; Rasmussen, R. S. Spectroscopic evidence concerning the structure of 2-phenyl-4,4-dimethyl-5(4)-oxazolone hydrobromide. *J. Am. Chem. Soc.* **1949**, *71*, 1080–1082.

32 Hamlow, H. P.; Okuda, S.; Nakagawa, N. NMR effects of cyclic tertiary amines. *Tetrahedron Lett.* **1964**, 2553–2559.

33 Hutchins, R. O.; Kopp, L. D.; Eliel, E. L. Repulsion of *syn*-axial electron pairs. The "rabbit-ear effect". *J. Am. Chem. Soc.* **1968**, *90*, 7174–7175.

34 Atkins, T. J. Tricyclic triaminomethanes. *J. Am. Chem. Soc.* **1980**, *102*, 6364–6365.

35 Cuevas, G.; Juaristi, E. Manifestation of stereoelectronic effects on the calculated C–H bond lengths and one bond $^{1}J_{C-H}$ NMR coupling constants in cyclohexane, six-membered heterocycles, and cyclohexanone derivatives. *J. Am. Chem. Soc.* **2002**, *124*, 13088–13096.

36 Kleinpeter, E.; Taddei, F.; Wacker, P. Electronic and steric substituent influences on the conformational equilibria of cyclohexyl esters: the anomeric effect is not anomalous. *Chem. Eur. J.* **2003**, *9*, 1360–1368.

37 Cortés, F.; Tenorio, J.; Collera, O.; Cuevas, G. Electronic delocalization contribution to the anomeric effect evaluated by computational methods. *J. Org. Chem.* **2001**, *66*, 2918–2924.

38 Cieplak, A. S. Stereochemistry of nucleophilic addition to cyclohexanone. The importance of two-electron stabilizing interactions. *J. Am. Chem. Soc.* **1981**, *103*, 4540–4552.

39 Jensen, H. H.; Lyngbye, L.; Jensen, A.; Bols, M. Stereoelectronic substituent effects in polyhydroxylated piperidines and hexahydropyridazines. *Chem. Eur. J.* **2002**, *8*, 1218–1226.

40 Taira, K.; Mock, W. L.; Gorenstein, D. G. Experimental tests of the stereoelectronic effect at phosphorus: nucleophilic reactivity of phosphite esters. *J. Am. Chem. Soc.* **1984**, *106*, 7831–7835.

41 Wright, D. A.; Wulff, C. A. (Chloromethyl)-triethylammonium chloride. A serendipitous preparation. *J. Org. Chem.* **1970**, *35*, 4252

42 Nilsson, B. M.; Sundquist, S.; Johansson, G.; Nordvall, G.; Glas, G.; Nilvebrant, L.; Hacksell, U. 3-Heteroaryl-substituted quinuclidin-3-ol and quinuclidin-2-ene derivatives as muscarinic antagonists. Synthesis and structure–activity relationships. *J. Med. Chem.* **1995**, *38*, 473–487.

43 Almarzoqi, B.; George, A. V.; Isaacs, N. S. The quaternization of tertiary amines with dihalomethanes. *Tetrahedron* **1986**, *42*, 601–607.

44 Taira, K.; Gorenstein, D. G. Stereoelectronic effects on the basicity and nucleophilicity of phosphites and phosphates. Ab initio molecular orbital calculations and the α-effect. *J. Am. Chem. Soc.* **1984**, *106*, 7825–7831.

45 Hodges, R. V.; Houle, F. A.; Beauchamp, J. L.; Montag, R. A.; Verkade, J. G. Effects of molecular structure on basicity. Gas-phase proton affinities of cyclic phosphites. *J. Am. Chem. Soc.* **1980**, *102*, 932–935.

46 Griend, L. J. V.; Verkade, J. G.; Pennings, J. F. M.; Buck, H. M. Structure–basicity relations among phosphate and phosphite esters. CNDO/2 and protonation studies. *J. Am. Chem. Soc.* **1977**, *99*, 2459–2463.

47 Yasui, S.; Tsujimoto, M.; Okamura, M.; Ohno, A. Stereoelectronic effects on the one-electron donor reactivity of trivalent phosphorus compounds. Experimental and theoretical investigations. *Bull. Chem. Soc. Jpn.* **1998**, *71*, 927–932.

48 Bohlmann, F.; Arndt, C. Konfiguration, Synthese und Reaktionen der isomeren Hexahydrojulolidine. *Chem. Ber.* **1958**, *91*, 2167–2175.

49 Erhardt, J. M.; Grover, E. R.; Wuest, J. D. Transfer of hydrogen from orthoamides. Synthesis, structure, and reactions of hexahydro-6b*H*-2a,4a,6a-triazacyclopenta[*cd*]pentalene and perhydro-3a,6a,9a-triazaphenalene. *J. Am. Chem. Soc.* **1980**, *102*, 6365–6369.

50 Erhardt, J. M.; Wuest, J. D. Transfer of hydrogen from orthoamides. Reduction of protons to molecular hydrogen. *J. Am. Chem. Soc.* **1980**, *102*, 6363–6364.

51 Li, S.; Deslongchamps, P. Experimental evidence for a synperiplanar stereoelectronic effect in the ozonolysis of a tricyclic acetal. *Tetrahedron Lett.* **1993**, *34*, 7759–7762.

52 Sinnott, M. L. The principle of least nuclear motion and the theory of stereoelectronic control. *Adv. Phys. Org. Chem.* **1988**, *24*, 113–204.

53 Malatesta, V.; Ingold, K. U. Kinetic applications of electron paramagnetic resonance spectroscopy. 36. Stereoelectronic effects in hydrogen atom abstraction from ethers. *J. Am. Chem. Soc.* **1981**, *103*, 609–614.

54 Solladié-Cavallo, A.; Balaz, M.; Salisova, M.; Welter, R. New 1,3-oxathianes derived from myrtenal: synthesis and reactivity. *J. Org. Chem.* **2003**, *68*, 6619–6626.

55 Barbarella, G.; Dembech, P.; Garbesi, A.; Bernardi, F.; Bottoni, A.; Fava, A. Kinetic acidity of diastereotopic protons in sulfonium ions. A transition state conformational effect. *J. Am. Chem. Soc.* **1978**, *100*, 200–202.

56 Eliel, E. L.; Hartmann, A. A.; Abatjoglou, A. G. Organosulfur chemistry. II. Highly stereoselective reactions of 1,3-dithianes. "Contrathermodynamic" formation of unstable diastereoisomers. *J. Am. Chem. Soc.* **1974**, *96*, 1807–1816.

57 Eliel, E. L.; Nader, F. W. Conformational analysis. XX. The stereochemistry of reaction of Grignard reagents with ortho esters. Synthesis of 1,3-dioxanes with axial substituents at C-2. *J. Am. Chem. Soc.* **1970**, *92*, 584–590.

58 Andrews, C. W.; Fraser-Reid, B.; Bowen, J. P. An ab initio study (6–31 G*) of transition states in glycoside hydrolysis based on axial and equatorial 2-methoxytetrahydropyrans. *J. Am. Chem. Soc.* **1991**, *113*, 8293–8298.

59 Ness, R. K.; Fletcher, H. G.; Hudson, C. S. New tribenzoyl-D-ribopyranosyl halides and their reactions with methanol. *J. Am. Chem. Soc.* **1951**, *73*, 959–963.

60 Fletcher, H. G.; Hudson, C. S. The reaction of tribenzoyl-β-D-arabinopyranosyl bromide and tribenzoyl-α-D-xylopyranosyl bromide with methanol. *J. Am. Chem. Soc.* **1950**, *72*, 4173–4177.

61 Taira, K.; Lai, K.; Gorenstein, D. G. Stereoelectronic effects in the conformation and hydrolysis of epimeric (4aα, 8aβ)-hexahydrobenzo-2-(*p*-nitrophenoxy)-2-oxo-1,3,2λ^5 dioxaphosphorinanes. *Tetrahedron* **1986**, *42*, 229–238.

62 Bennet, A. J.; Sinnott, M. L. Complete kinetic isotope effect description of transition states for acid-catalyzed hydrolyses of methyl α- and β-glucopyranosides. *J. Am. Chem. Soc.* **1986**, *108*, 7287–7294.

63 Ratcliffe, A. J.; Mootoo, D. R.; Andrews, C. W.; Fraser-Reid, B. Concerning the antiperiplanar lone pair hypothesis: oxidative hydrolysis of conformationally constrained 4-pentenyl glycosides. *J. Am. Chem. Soc.* **1989**, *111*, 7661–7662.

64 Perrin, C. L. Is there stereoelectronic control in formation and cleavage of tetrahedral intermediates? *Acc. Chem. Res.* **2002**, *35*, 28–34.

65 Graczyk, P. P.; Mikolajczyk, M. Inapplicability of the antiperiplanar lone pair hypothesis to C–P bond breaking and formation in some S–C–P$^+$ systems. *J. Org. Chem.* **1996**, *61*, 2995–3002.

66 Hosie, L.; Marshall, P. J.; Sinnott, M. L. Failure of the antiperiplanar lone pair hypothesis in glycoside hydrolysis. Synthesis, conformation, and hydrolysis of α-D-xylopyranosyl and α-D-glucopyranosyl pyridinium salts. *J. Chem. Soc. Perkin Trans. 2* **1984**, 1121–1131.

67 Perrin, C. L.; Engler, R. E.; Young, D. B. Bifunctional catalysis and apparent stereoelectronic control in hydrolysis of cyclic imidatonium ions. *J. Am. Chem. Soc.* **2000**, *122*, 4877–4881.

68 Jeroncic, L. O.; Cabal, M.-P.; Danishefsky, S. J.; Shulte, G. M. On the diastereofacial selectivity of Lewis acid catalyzed C–C bond forming reactions of conjugated cyclic enones bearing electron-withdrawing substituents at the γ position. *J. Org. Chem.* **1991**, *56*, 387–395.

69 Iwasawa, N.; Funahashi, M.; Hayakawa, S.; Ikeno, T.; Narasaka, K. Synthesis of medium-sized bicyclic compounds by intramolecular cyclization of cyclic β-keto radicals generated from cyclopropanols using Mn(III) tris(pyridine-2-carboxylate) and its application to total synthesis of 10-isothiocyanatoguaia-6-ene. *Bull. Chem. Soc. Jpn.* **1999**, *72*, 85–97.

70 Johnson, C. R.; Chen, Y.-F. Development of a triply convergent aldol approach to prostanoids. *J. Org. Chem.* **1991**, *56*, 3344–3351.

71 Danishefsky, S. J.; Cabal, M.-P.; Chow, K. Novel stereospecific silyl group transfer reactions: practical routes to the prostaglandins. *J. Am. Chem. Soc.* **1989**, *111*, 3456–3457.

72 Pearson, A. J.; Chang, K. Organoiron-templated stereocontrolled alkylation of enolates: functionalization of cycloheptadienones to give useful synthetic building blocks. *J. Org. Chem.* **1993**, *58*, 1228–1237.

73 Danishefsky, S. J.; Simoneau, B. Total synthesis of ML-236A and compactin by combining the lactonic (silyl) enolate rearrangement and aldehyde–diene cyclocondensation technologies. *J. Am. Chem. Soc.* **1989**, *111*, 2599–2604.

74 Cieplak, A. S.; Tait, B. D.; Johnson, C. R. Reversal of π-facial diastereoselection upon electronegative substitution of the substrate and the reagent. *J. Am. Chem. Soc.* **1989**, *111*, 8447–8462.

75 Seeman, J. I. The Curtin–Hammett principle and the Winstein–Holness equation. *J. Chem. Educ.* **1986**, *63*, 42–48.

3
The Stability of Organic Compounds

3.1
Introduction

The enormous number of known organic compounds (and the existence of life on this planet) is mainly due to the fact that hydrocarbons and many other organic compounds are kinetically stable in air over a broad range of temperatures. If an organic compound is highly strained, electron-rich, or if it contains incompatible functional groups, the activation barrier to undergo rearrangement, fragmentation, or reaction with oxygen can, however, be sufficiently reduced to make its isolation on a preparative scale a difficult if not dangerous task. Because a synthesis will only succeed if the desired product does not undergo further chemical transformations after its formation or during its isolation, the organic chemist should have a clear knowledge of the limits of stability of organic compounds.

In this chapter some classes of unstable compound and their decomposition reactions will be presented, with the aim of sparing the reader unpleasant surprises in the laboratory.

3.2
Strained Bonds

Compounds in which bond angles deviate substantially from the normal values, or in which steric repulsion lengthens bonds significantly, will undergo homolytic or heterolytic C–C bond cleavage more readily and be therefore thermally less stable than comparable, unstrained compounds. Thus, cyclopropanes and ethanes with sterically demanding substituents have little thermal stability (Scheme 3.1). Transition metals readily undergo oxidative addition to C–C σ bonds if these are weakened by strain, and might therefore catalyze the rearrangement or fragmentation of strained compounds (last reaction, Scheme 3.1).

Strained bonds will be cleaved particularly readily if there is a reversible reaction by which these bonds might have been formally formed. Such reactions could be the Diels–Alder reaction, aldol additions, Michael additions, or related processes

Side Reactions in Organic Synthesis. Florencio Zaragoza Dörwald

ISBN: 3-527-31021-5

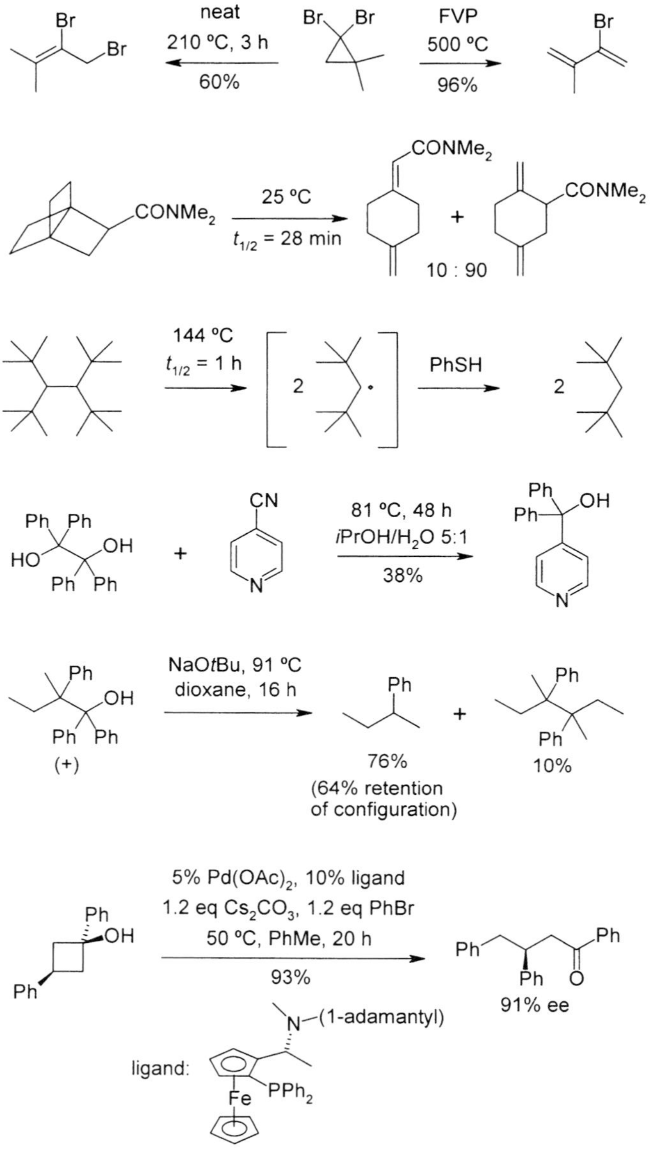

Br
Br
neat
210 °C, 3 h
60%
Br Br
FVP
500 °C
96%
Br

CONMe$_2$
25 °C
$t_{1/2}$ = 28 min
CONMe$_2$
+
CONMe$_2$
10 : 90

144 °C
$t_{1/2}$ = 1 h
2
PhSH
2

Ph Ph
HO OH
Ph Ph
+
CN
N
81 °C, 48 h
iPrOH/H_2O 5:1
38%
Ph
Ph OH
N

Ph
OH
Ph Ph
(+)
NaOtBu, 91 °C
dioxane, 16 h
Ph
76%
(64% retention
of configuration)
+
Ph
Ph
10%

Ph
OH
Ph
5% Pd(OAc)$_2$, 10% ligand
1.2 eq Cs_2CO_3, 1.2 eq PhBr
50 °C, PhMe, 20 h
93%
Ph
Ph
Ph O
Ph
91% ee
ligand:
N–(1-adamantyl)
Fe
PPh$_2$

Scheme 3.1. Thermal decomposition of strained compounds [1–8].

(Scheme 3.2). Reversible reactions are, for this reason, not usually suitable for the formation of strained bonds.

Strained cyclohexenes, such as norbornene derivatives, can undergo retro-Diels–Alder reactions even at relatively low temperatures, and this reaction can be used to prepare 1,3-dienes and alkenes (e.g. synthesis of cyclopentadiene by thermolysis of

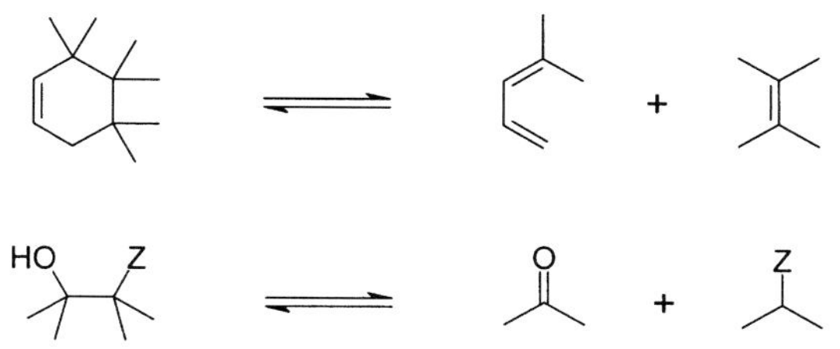

Scheme 3.2. Fragmentation via retro-Diels–Alder cycloaddition and retro-aldol addition.

dicyclopentadiene at approximately 160 °C). In the example shown in Scheme 3.3 (first reaction) an enantiomerically pure bicyclic cyclohexanone is prepared from a readily available natural product. The radical scavenger is added to the starting material to prevent oxidation.

The second example in Scheme 3.3 illustrates the reversibility of aldol additions. The starting bicyclic ketone is a vinylogous aldol which upon treatment with base undergoes retro-aldol addition by cleavage of a strained, hexasubstituted ethane sub-

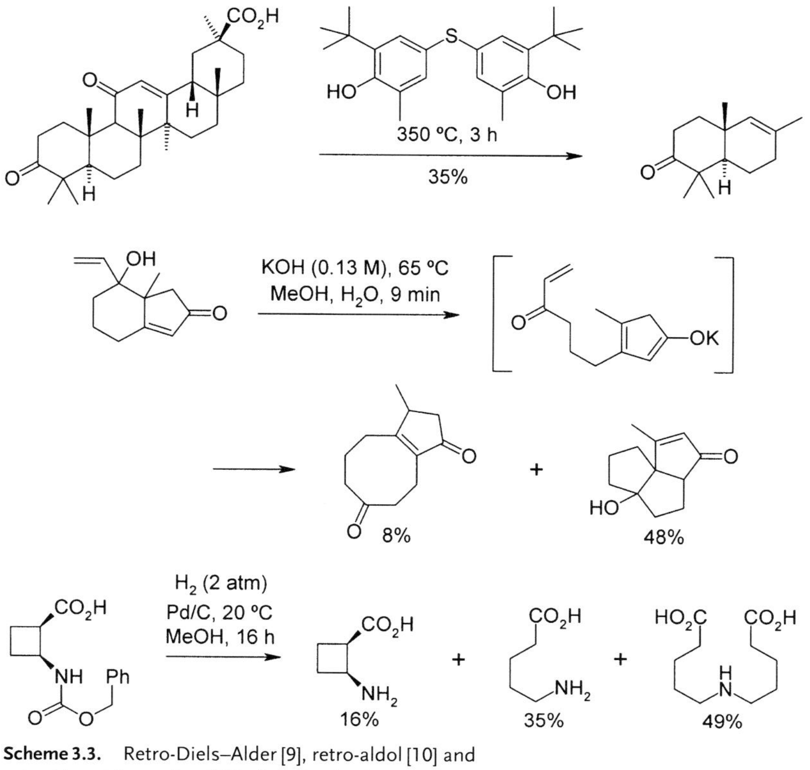

Scheme 3.3. Retro-Diels–Alder [9], retro-aldol [10] and retro-Mannich reaction [11] of strained substrates.

structure. Interestingly, the main product of this reaction also contains a similarly strained bond. Nevertheless, because these transformations are reversible, we must assume that the tricylic ketone is thermodynamically the most stable product.

In the final example in Scheme 3.3 the starting material is the product of a formal intramolecular Mannich reaction. Accordingly, the C–C bond between the amino and the carboxyl group is particularly weak. Upon hydrogenolysis this bond is cleaved, whereby an imine or aldehyde is presumably formed as an intermediate which leads to the formation of a secondary amine as main product by reductive *N*-alkylation of 5-aminopentanoic acid.

Tertiary homoallylic alcohols can formally undergo fragmentation into a ketone and an allyl anion [12, 13]. If the homoallylic alcohol contains several bulky alkyl groups its ground state energy may be sufficiently high to enable C–C bond cleavage to occur under mild reaction conditions. In the example sketched in Scheme 3.4 the tertiary alcoholate undergoes reversible fragmentation into di-*tert*-butyl ketone and an allylic organolithium compound when treated with BuLi at room temperature. Treatment of this alcoholate at low temperatures with an aldehyde activated by a Lewis acid leads to highly diastereoselective transfer of the allyl anion to the aldehyde.

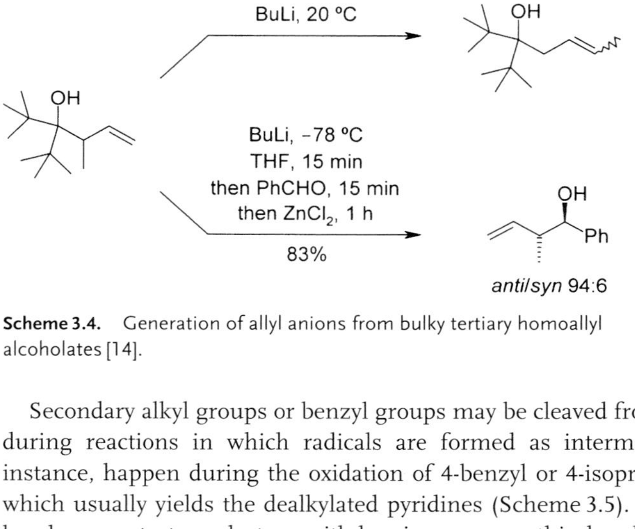

BuLi, 20 °C

OH

OH

BuLi, −78 °C
THF, 15 min
then PhCHO, 15 min
then $ZnCl_2$, 1 h

83%

OH

Ph

anti/*syn* 94:6

Scheme 3.4. Generation of allyl anions from bulky tertiary homoallyl alcoholates [14].

Secondary alkyl groups or benzyl groups may be cleaved from certain compounds during reactions in which radicals are formed as intermediates. This can, for instance, happen during the oxidation of 4-benzyl or 4-isopropyl dihydropyridines, which usually yields the dealkylated pyridines (Scheme 3.5). Similarly, if a strained bond connects two electron-withdrawing groups, this bond might be cleaved by SET-mediated reduction (second reaction, Scheme 3.5) [15].

If alkenes or alkynes are subjected to strain, their π bonds are weakened, and such compounds often behave chemically as diradicals. Their tendency to dimerize or polymerize will be significantly enhanced, and quick reaction with oxygen will occur in air [18, 19]. Reactions of strained alkenes which lead to a decline of strain, for example Michael additions or cycloadditions, can proceed significantly faster than with related, unstrained alkenes (Scheme 3.6).

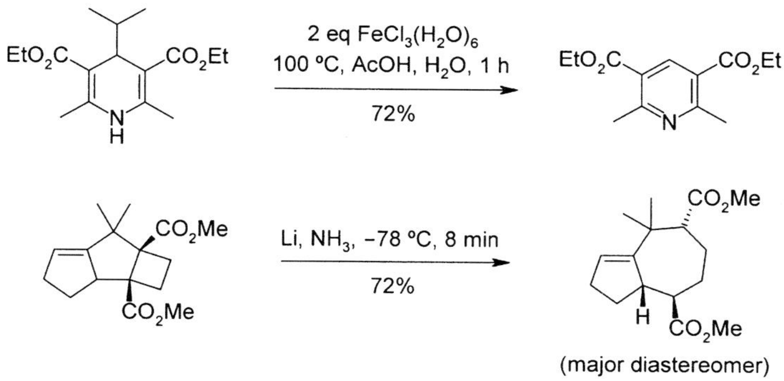

2 eq $FeCl_3(H_2O)_6$
100 °C, AcOH, H_2O, 1 h
72%

Li, NH_3, −78 °C, 8 min
72%

(major diastereomer)

Scheme 3.5. Dealkylative oxidation of dihydropyridines [16] and reductive cleavage of cyclobutanes [17].

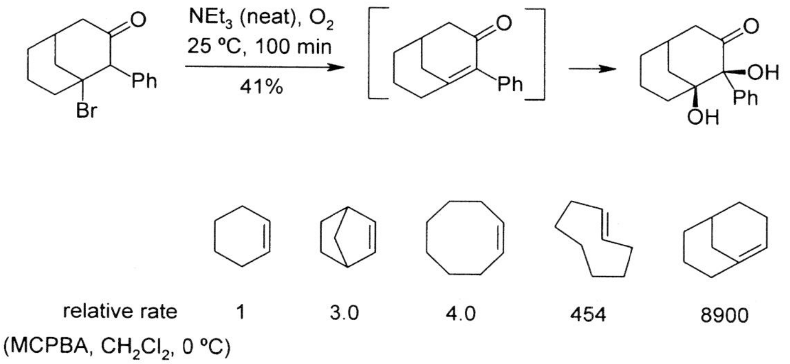

NEt_3 (neat), O_2
25 °C, 100 min
41%

relative rate	1	3.0	4.0	454	8900

(MCPBA, CH_2Cl_2, 0 °C)

Scheme 3.6. Oxidation of a strained alkene by air [20] and relative rates of epoxidation of various cyclic alkenes [21].

Strain can also be used to enhance the reactivity of otherwise unreactive alkenes. Non-cyclic 1,1-dimethyl-1,3-butadienes, for instance, undergo Diels–Alder reactions slowly or not at all. In the first reaction in Scheme 3.7, for example, the diene undergoes rearrangement to a terminal diene before cycloaddition. Because of the harsh reaction conditions the product is oxidized to an enol ether by the starting quinone [22]. If the terminal isopropylidene group is replaced by a cyclopropylidene group, however, the resulting dienes do not isomerize and undergo cycloaddition with dienophiles much more rapidly (Scheme 3.7).

Not all strained compounds are necessarily more reactive than less strained analogs. Reactivity will always depend on the type of reaction under scrutiny, and if the rate determining step of a given reaction is not accelerated by strain, the rate of reaction of strained and unstrained compounds will be similar. One example of such strain-independent reaction rates is the hydrolysis of lactams under basic reaction conditions (Scheme 3.8). Although β-lactams are more strained than six-membered lactams, both are hydrolyzed at approximately the same rate, presumably because the rate determining step is the addition of hydroxide to the amide bond, and not

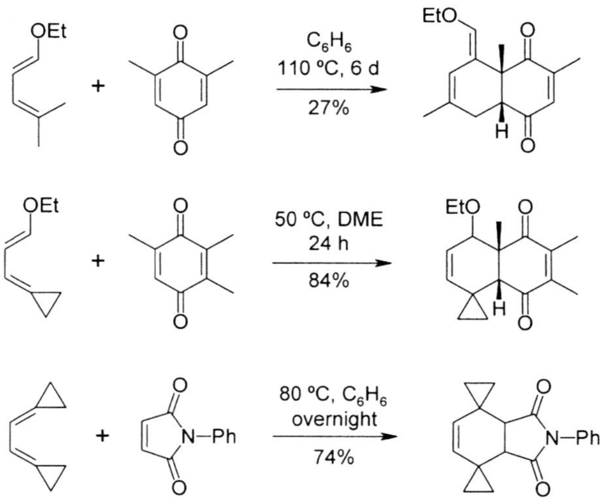

Scheme 3.7. Diels–Alder reactions with cyclopropylidenes [22–24].

ring scission. Eight- and nine-membered lactams are more resistant to nucleophilic hydrolysis than non-cyclic amides; this might be because of an increase in trans-annular repulsive van der Waals interactions during the addition of hydroxide to these amides.

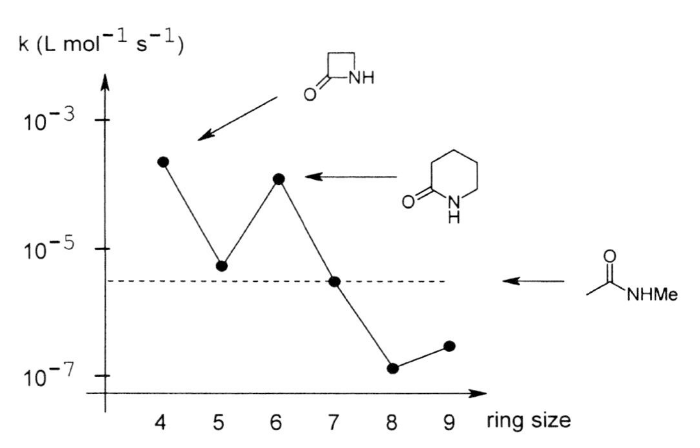

Scheme 3.8. Rates of lactam hydrolysis with aqueous base [25]. For comparison, the rate of hydrolysis of *N*-methylacetamide is shown.

3.3
Incompatible Functional Groups

It goes without saying that a compound will decompose or oligomerize if it contains functional groups which can react with each other. Because intramolecular reactions often proceed at much higher rates than their intermolecular variants, functional group incompatibilities may arise unexpectedly, involving groups which would not react intermolecularly. For instance, many phthalic acid, maleic acid, or 2-(hydroxymethyl)benzoic acid derivatives are notoriously sensitive because of the close proximity of the two functional groups (Scheme 3.9). Similarly, amides of 2-aminoethanols rearrange readily to esters under acidic reaction conditions [26] and undergo hydrolysis under mildly basic conditions, mainly because the hydroxyl group is close to the amide bond (Scheme 3.9). Normal amides usually require treatment with highly concentrated mineral acids at high temperatures to undergo hydrolysis (last example, Scheme 3.9).

Mannich bases (2-aminoethyl ketones) are another class of inherently unstable compounds which often undergo facile thermal elimination of the amine to yield a vinyl ketone [30]. Chemically related to Mannich bases are 2- or 4-(aminomethyl)-

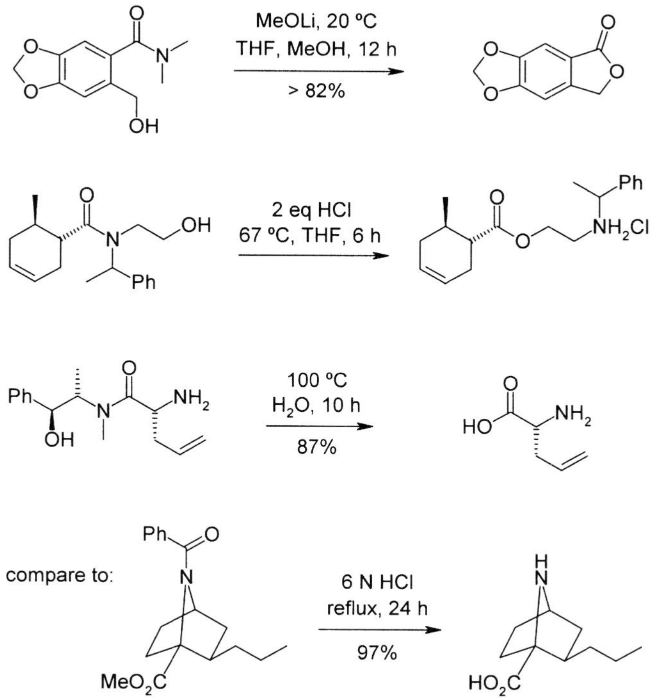

MeOLi, 20 °C
THF, MeOH, 12 h
> 82%

2 eq HCl
67 °C, THF, 6 h

100 °C
H_2O, 10 h
87%

compare to:

6 N HCl
reflux, 24 h
97%

Scheme 3.9. Hydrolysis of amides with and without intramolecular assistance [26–29].

phenols, which also decompose readily. For the same reason compounds of general formula 2/4-(RO/R_2N)$C_6H_4CH_2$(OR/NR_2/Hal) tend to be unstable, and many polymerize on standing at room temperature [31] (Scheme 3.10).

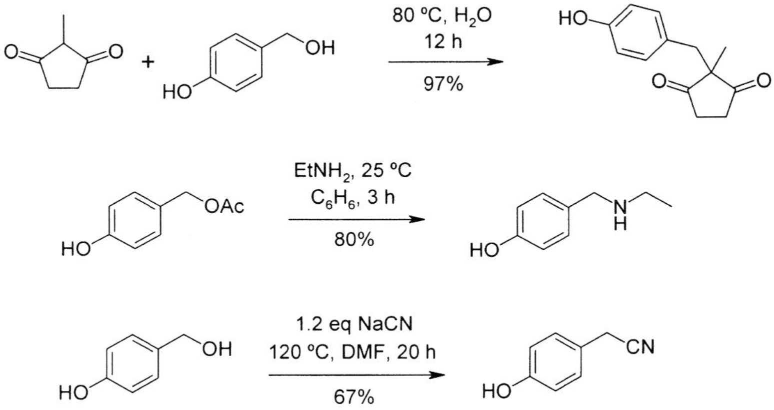

80 °C, H_2O, 12 h, 97%

$EtNH_2$, 25 °C, C_6H_6, 3 h, 80%

1.2 eq NaCN, 120 °C, DMF, 20 h, 67%

Scheme 3.10. Facile nucleophilic substitutions at 4-hydroxybenzyl derivatives [31–33].

3.4 Conjugation and Hyperconjugation of Incompatible Functional Groups

Less obvious than the reactivity discussed above is the chemical instability brought about by hyperconjugation or conjugation of incompatible groups. Methanes substituted with both amino and acyl groups, for example α-amino ketones, tend to form radicals readily (Scheme 5.69), and can undergo facile thermal rearrangements (Scheme 3.11).

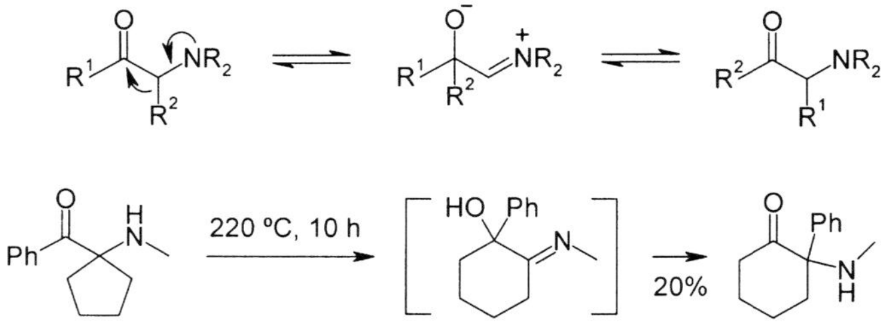

220 °C, 10 h; 20%

Scheme 3.11. Rearrangement of α-amino ketones [34].

Alkenes substituted with an electron-withdrawing (Z) and an electron-donating group (X) will be less reactive than the unsubstituted alkene if the substituents are vicinal (Z–C=C–X; push-pull alkenes), but will generally become highly reactive when these two substituents are bound to the same carbon atom. The latter type of alkene ($ZXC=CR_2$; "1,1-captodative" alkenes) readily dimerize to yield cyclobutanes and can react with 1,3-dienes to yield products of [2 + 2] and/or [2 + 4] cycloaddition [35–37] (Scheme 3.12). This effect is probably because of the stabilization of radicals which results from substitution with both an electron-withdrawing and an electron-donating substituent (captodative effect [38, 39]).

The different reactivity of 1,2- and 1,1-acceptor/donor-substituted alkenes is paralleled by the readiness with which these compounds can be prepared. Thus, β-amino acrylates are often spontaneously formed by mixing amines with β-keto esters, and these derivatives have been used as protective groups for amines because of their low reactivity. α-Amino acrylates can, similarly, be prepared from α-keto esters [40], but this condensation reaction does not proceed spontaneously and requires chemical or azeotropic removal of water [41–43]. α-Amino acrylates are unstable compounds which must be stored at low temperatures [41] or *N*-acylated immediately after their generation [43].

A highly reactive natural product which contains such a geminally donor–acceptor substituted alkene is protoanemonin (Scheme 3.12), a toxic, skin-irritating lactone produced by various plants (ranunculaceae). The natural precursor to this compound is the glucoside ranunculin [44, 45], which yields protoanemonin enzymatically on maceration of plant tissue. Protoanemonin is unstable and quickly polymerizes or dimerizes to the less toxic anemonin.

Ground-state stabilization by hyperconjugation can modify the reactivity of organic compounds significantly, and must be taken into account when estimating the stability of certain types of compound. Di- or polysubstituted methanes can have high ground-state stability if the substituents are simultaneously σ-acceptors and π-donors, as is observed for many heteroatoms. If, on the other hand, the substitu-

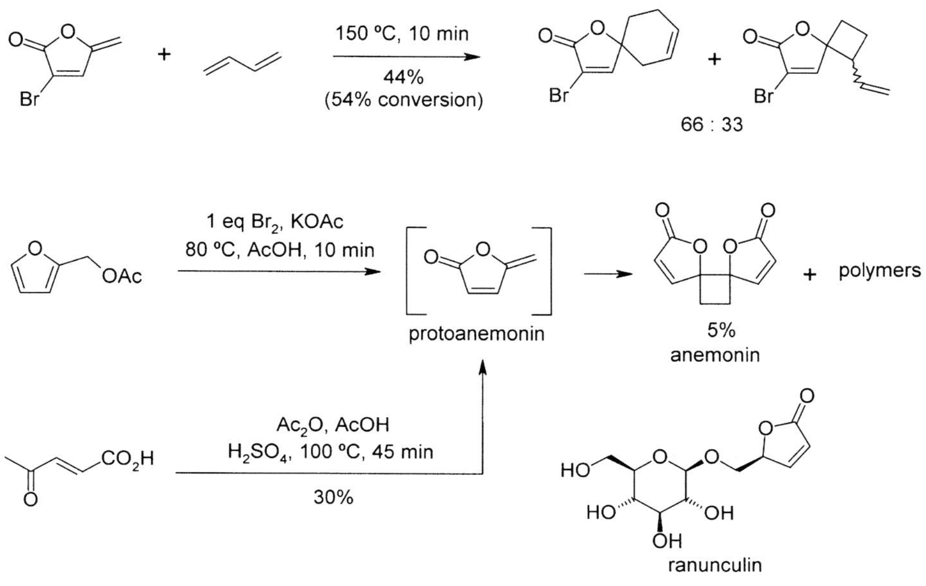

150 °C, 10 min
44%
(54% conversion)
66 : 33

1 eq Br_2, KOAc
80 °C, AcOH, 10 min
OAc
protoanemonin
5%
anemonin
+ polymers

Ac_2O, AcOH
H_2SO_4, 100 °C, 45 min
30%
CO_2H
ranunculin

Scheme 3.12. Generation and reactions of captodative alkenes [35, 44, 46, 47].

ents are both σ- and π-acceptors (as, e.g., in malonodinitrile), strong destabilization will result [48]. The stabilization of heteroatom-substituted methanes results from hyperconjugation between the lone pairs of one heteroatom and the antibonding orbital of the C–X bond to the other heteroatom (σ^*_{C-X}) [49, 50]. This stabilization is the reason for the high exothermicity of the two isodesmic reactions shown in Scheme 3.13 [51], and for the large difference in hydrolysis rates between dimethoxymethane and chloromethyl methyl ether [52]. Dimethoxymethane is stabilized by hyperconjugation whereas in chloromethyl methyl ether this stabilization is less effective [49]. Accordingly, there is a larger difference between the rates of hydrolysis of the disubstituted methanes than between related benzyl derivatives (Scheme 3.13).

For similar reasons the stability of polyalkoxymethanes to acid-catalyzed hydrolysis is unexpectedly dependent on the number of carbon-bound alkoxy groups. Contrary to chemical intuition, orthocarbonates are more difficult to hydrolyze than orthoesters [51] (Scheme 3.13), although a carbocation substituted with three alkoxy groups is stable than one with only two alkoxy groups. Similarly, the tendency of fluoride to add to fluorinated alkenes is much greater than that of other halides [53], despite its low nucleophilicity (see also Section 4.2.2).

These effects, which have also been called symbiosis, double bond-no bond resonance, clustering, or geminal or anomeric effect [54], can also be explained in terms of the HSAB principle: X_3C^+ will be harder than H_3C^+ if X^- is hard. Accordingly, X_3C^+ will have a higher affinity for the hard X^- than H_3C^+ will have [54].

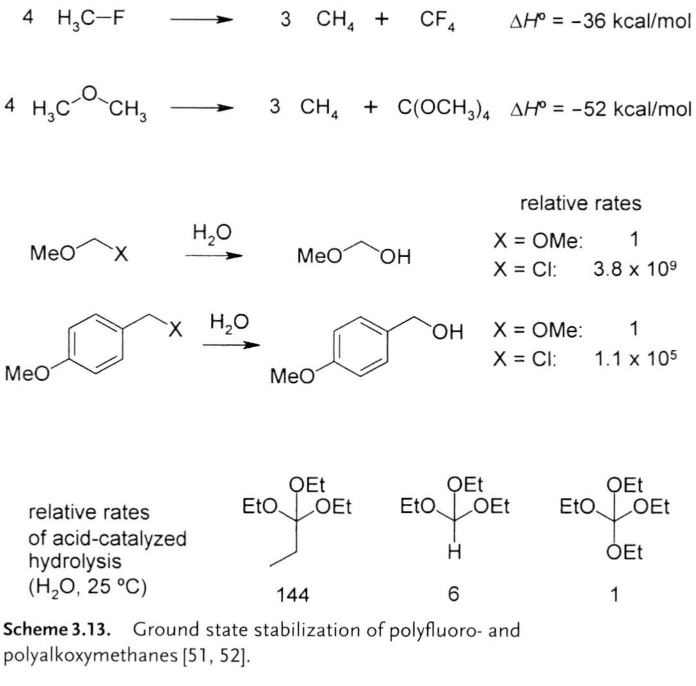

$$4\ H_3C{-}F \longrightarrow 3\ CH_4 + CF_4 \qquad \Delta H^o = -36\ \text{kcal/mol}$$

$$4\ H_3C{-}O{-}CH_3 \longrightarrow 3\ CH_4 + C(OCH_3)_4 \qquad \Delta H^o = -52\ \text{kcal/mol}$$

MeO–CH₂–X $\xrightarrow{H_2O}$ MeO–CH₂–OH

relative rates: X = OMe: 1; X = Cl: 3.8×10^9

4-MeO-C₆H₄–CH₂–X $\xrightarrow{H_2O}$ 4-MeO-C₆H₄–CH₂–OH

X = OMe: 1; X = Cl: 1.1×10^5

relative rates of acid-catalyzed hydrolysis (H_2O, 25 °C):

$EtC(OEt)_3$	$HC(OEt)_3$	$C(OEt)_4$
144	6	1

Scheme 3.13. Ground state stabilization of polyfluoro- and polyalkoxymethanes [51, 52].

3.5 Stability Toward Oxygen

Triplet oxygen is a stable and surprisingly unreactive electrophilic diradical. Its reactions with organic compounds (autoxidation) include abstraction of hydrogen atoms or single electrons and addition to reactive C–C double bonds (Scheme 3.14).

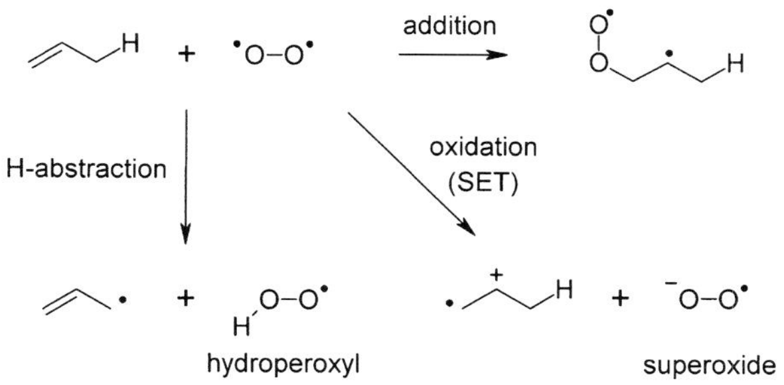

Scheme 3.14. Reactions of organic compounds with triplet oxygen.

3.5.1 Hydrogen Abstraction

The strength of the O–H bond in the hydroperoxyl radical (HO_2) is only 49.6 kcal/mol (298 K) (Table 3.1), and thus much smaller than the strength of most C–H bonds. For this reason the cleavage of C–H bonds by oxygen is highly endothermic (Eq. 3.1) and, accordingly, most organic compounds are stable in the presence of air at moderate temperatures.

$$CH_4 + O_2 \longrightarrow CH_3^{\bullet} + HO_2^{\bullet} \qquad \Delta H°(298\ K) = 55.5\ kcal/mol$$

This remarkably low reactivity of triplet oxygen is in sharp contrast with the reactivity of other oxygen-centered radicals. Hydrogen peroxide (D(O–H) = 87.1 kcal/mol) or aliphatic alcohols such as methanol (D(O–H) = 104 kcal/mol), for instance, have much stronger O–H bonds than the hydroperoxyl radical, and the corresponding oxyl radicals will usually quickly and irreversibly abstract hydrogen atoms from alkanes to yield alkyl radicals (Table 3.1).

Most organic compounds will react with oxygen at high temperatures, but some types of compound react already at room temperature when exposed to the air. Alkanes undergo slow hydrogen abstraction by oxygen and are thereby transformed into hydroperoxides according to the radical chain mechanism sketched in Scheme 3.15. As discussed above, triplet oxygen is not reactive enough to abstract hydrogen atoms from most alkanes, so another radical-forming process will usually be required to initiate the reaction. Initiation might, for instance, be mediated by UV radiation or transition metal salts (which react with peroxides to yield alkoxyl radicals [59]). As soon as carbon-centered radicals are formed they react quickly with oxygen to yield peroxyl radicals, which are much more reactive than oxygen (Table 3.1) and will be able to abstract hydrogen from alkanes and in this way maintain a radical chain reaction. Peroxyl radicals are quite selective, and hydrogen abstraction will occur preferentially at methine groups (R_3CH) and at benzylic or allylic positions.

Table 3.1. Bond dissociation enthalpies *D* (in kcal/mol at 298 K) for selected X–H bonds [$D(X–H) = \Delta H_f(X) + \Delta H_f(H) - \Delta H_f(XH)$] [55–58].

Bond	*D*	Bond	*D*
Me–H	105	O_2–H	49.6
H_2C=$CHCH_2$–H	88.8	HO_2–H	87.8
$PhCH_2$–H	89.8	HO–H	119
Me_3C–H	96.5	MeO–H	104.6
$MeOCH_2$–H	93	AcO–H	112
Ph–H	112.9	PhO–H	90
(1,3-cyclohexadien-5-yl)–H	70	2,6-(tBu)$_2$-4-MeC_6H_2O–H	80.0
PhCO–H	86.9	α-tocopherol (O–H)	78.2
$MeCOCH_2$–H	98.3	Me_2N–H	95
Me_3Si–H	90	PhNH–H	86.4

The direct catalyzed or uncatalyzed oxidation of alkanes with oxygen is an important reaction in the industrial production of carboxylic acids, hydroperoxides (for production of epoxides from alkenes), alcohols, ketones, or aldehydes [60].

Heteroatoms or functional groups can either increase or diminish the rate of autoxidation of alkyl groups. Haloalkanes and alkanes substituted with electron-withdrawing groups are usually more resistant toward homolytic C–H bond cleav-

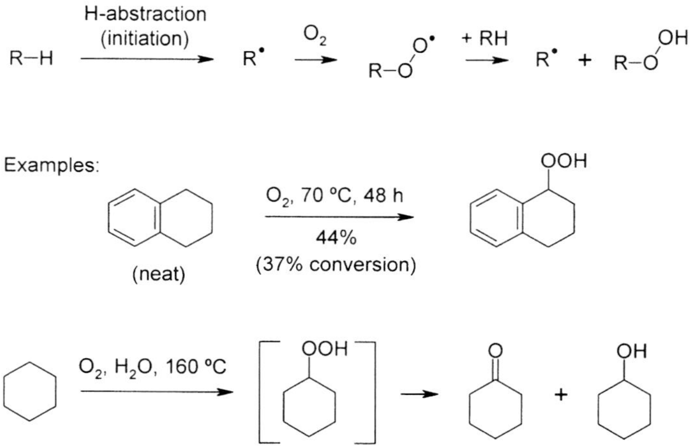

R–H —H-abstraction (initiation)→ R• —O_2→ R–O–O• —+ RH→ R• + R–O–OH

Examples:

(neat) tetralin —O_2, 70 °C, 48 h; 44% (37% conversion)→ 1-hydroperoxytetralin (OOH)

cyclohexane —O_2, H_2O, 160 °C→ [cyclohexyl hydroperoxide (OOH)] → cyclohexanone (O) + cyclohexanol (OH)

Scheme 3.15. Autoxidation of hydrocarbons [61, 62].

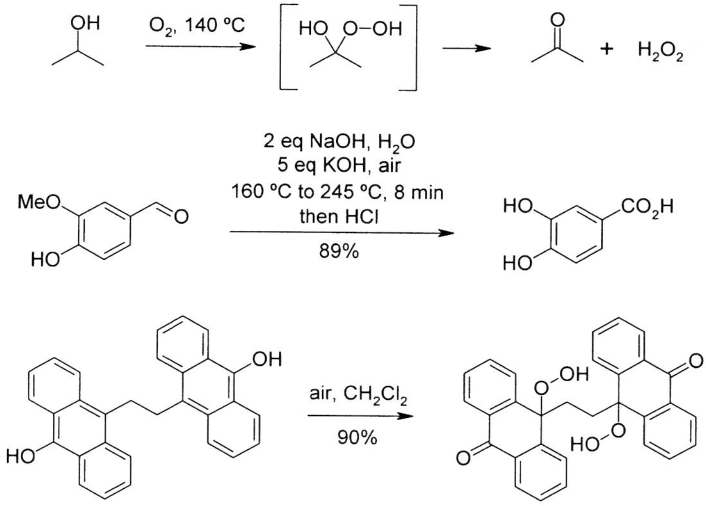

Scheme 3.16. Autoxidation of alcohols, aldehydes, and phenols [60, 63, 64].

age by oxygen or hydroperoxyl radicals whereas ethers or alcohols usually form hydroperoxides more readily than pure hydrocarbons (Table 3.1, Scheme 3.16). Aldehyde-derived acetals, aminals, or hydrates are, accordingly, particularly prone to radical-mediated autoxidation.

Autoxidation can lead to deterioration of food, drugs, cosmetics, or polymers, and inhibition of this reaction is therefore an important technical issue. The most important classes of autoxidation inhibitors are radical scavengers (phenols, sterically demanding amines [65, 66]), oxygen scavengers (e.g. ascorbic acid), UV-light absorbers, and chelators such as EDTA (to stabilize high oxidation states of metals and thereby suppress the metal-catalyzed conversion of peroxides to alkoxyl radicals) [67].

To be effective as autoxidation inhibitors radical scavengers must react quickly with peroxyl or alkyl radicals and lead thereby to the formation of unreactive products. Phenols substituted with electron-donating substituents have relatively low O–H bond dissociation enthalpies (Table 3.1; even lower than arene-bound isopropyl groups [68]), and yield, on hydrogen abstraction, stable phenoxyl radicals which no longer sustain the radical chain reaction. The phenols should not be too electron-rich, however, because this could lead to excessive air-sensitivity of the phenol, i.e. to rapid oxidation of the phenol via SET to oxygen (see next section). Scheme 3.17 shows a selection of radical scavengers which have proved suitable for inhibition of autoxidation processes (and radical-mediated polymerization).

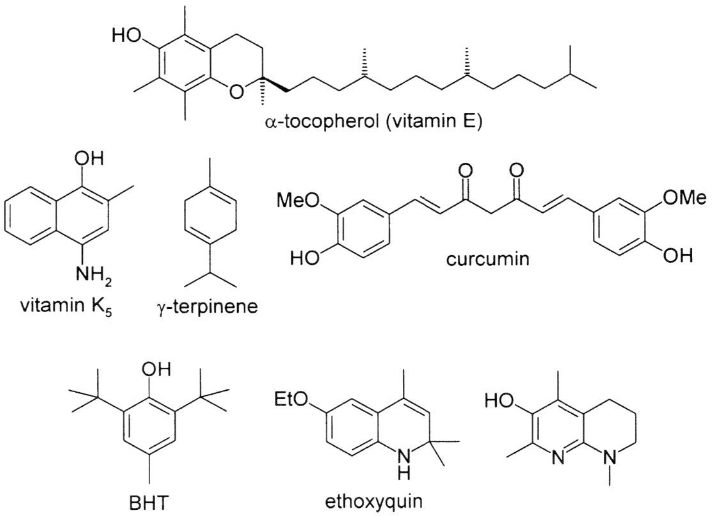

α-tocopherol (vitamin E)

vitamin K_5 γ-terpinene curcumin

BHT ethoxyquin

Scheme 3.17. Representative radical scavengers, suitable for the suppression of autoxidation [69–71].

3.5.2
Oxidation by SET

Treatment of arenes or heteroarenes with oxidants can lead to the formation of radical cations by SET. These radical cations can dimerize, oligomerize, or react with other radicals present in the reaction mixture; deprotonation of the resulting intermediates yields the final products (Scheme 3.18).

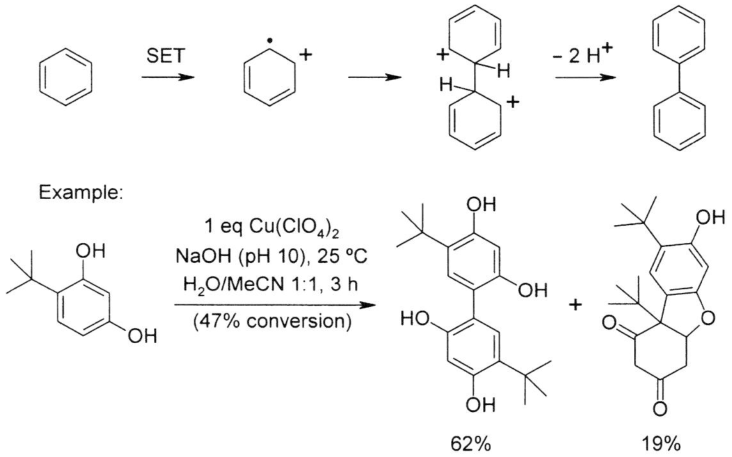

SET $-2\,H^+$

Example:

1 eq $Cu(ClO_4)_2$
NaOH (pH 10), 25 °C
H_2O/MeCN 1:1, 3 h
(47% conversion)

62% 19%

Scheme 3.18. Oxidation of arenes via abstraction of single electrons [72].

Some electron-rich arenes or heteroarenes undergo SET even at room temperature when exposed to the air. Such compounds will usually darken quickly, even if only trace amounts of oligomers are formed by autoxidation, because these oligomers can absorb visible light very efficiently and tend to be oxidized even more readily than the monomer. Thus, older samples of aniline, alkoxyanilines, or aminophenols are usually dark or black, even if analysis by ^{1}H NMR does not reveal any impurities. Particularly air-sensitive are five-membered heteroarenes (pyrroles, furans, thiophenes) with electron-donating substituents. Some of these compounds polymerize on oxidation to yield materials with good electric conductivity (Scheme 3.19).

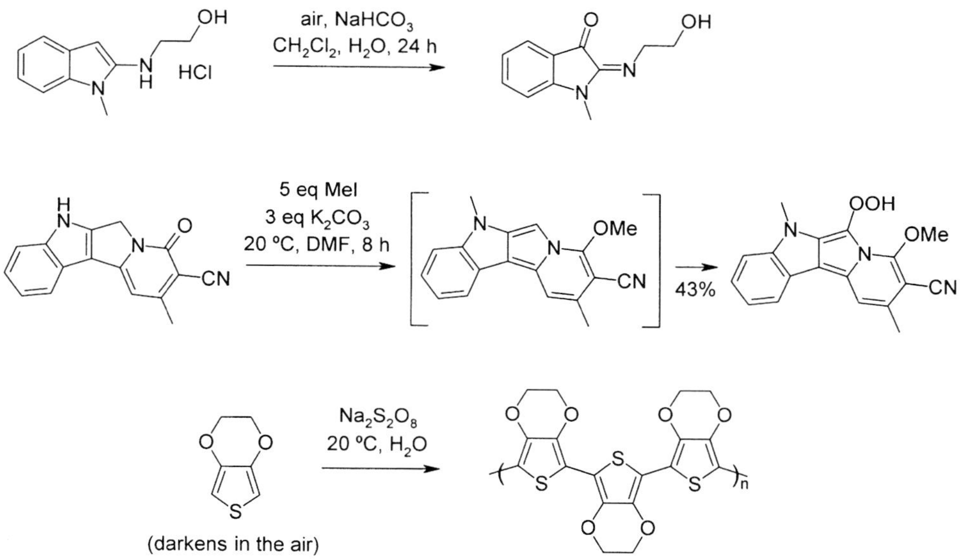

Scheme 3.19. Air-sensitive indoles, pyrroles, and thiophenes [73–75].

Arene-bound alkoxy or amino groups will facilitate oxidation by SET only if their lone electron pairs can interact with the aromatic π-system. For this reason 1,3,5-tris(dialkylamino)benzenes will be significantly more air-sensitive than, for instance, 1,2,3-tris(dialkylamino)benzenes, because steric crowding in the latter will force the lone pairs into the plane of the arene, where efficient conjugation with the aromatic π-system is impossible.

Unfortunately there is no simple correlation between gas-phase ionization potentials and solution-phase oxidation potentials for all classes of compounds, because the energy of solvation is highly dependent on molecular structure. Nevertheless, for closely related compounds there tends to be a linear correlation between ionization potentials in the gas phase and in solution [76, 77]. The air-sensitivity of electron-rich alkenes, arenes, or heteroarenes can therefore be estimated by inspecting either their gas-phase ionization potentials or their oxidation potentials in solution

(Scheme 3.20). Arenes or heteroarenes with an ionization potential smaller than approximately 7 eV or an oxidation potential smaller than 1 V will usually be susceptible to oxidation by air at room temperature. As illustrated by the examples in Scheme 3.20, electron abstraction becomes easier when the number of conjugated C–C double bonds increases. Replacing carbon in an arene or heteroarene by (the more electronegative) nitrogen usually enhances the oxidation potential and thereby the stability towards oxidants. The oxidation potential of the superoxide radical in aprotic solvents (0.98 V) corresponds approximately to the oxidation potential at which arenes become air sensitive.

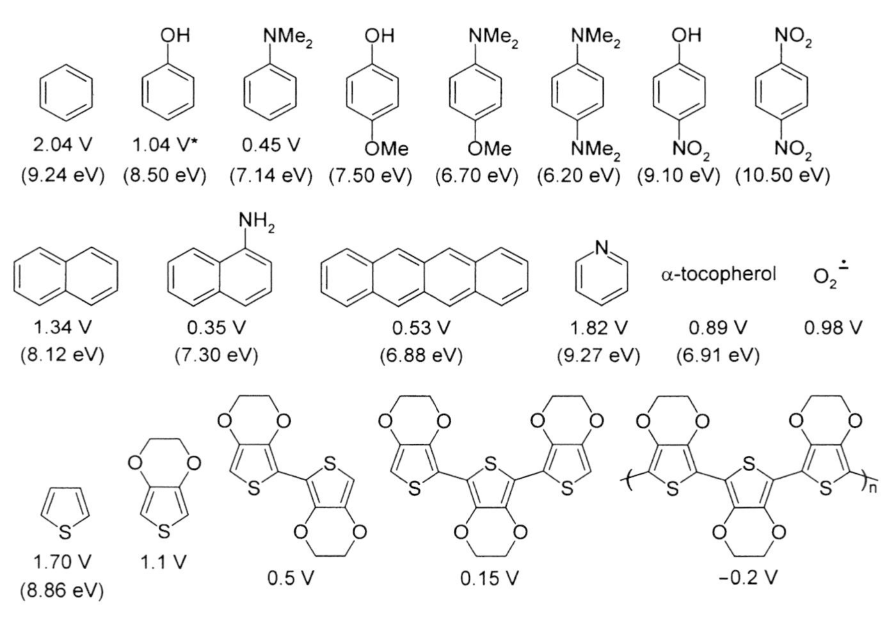

Scheme 3.20. Oxidation potentials (vs Ag/Ag^+) and gas-phase ionization potentials (in parentheses) of some arenes and heteroarenes and triplet oxygen [56, 57, 70, 74, 77–79]. *Ref [56]: 1.83 V.

Not only alkenes and arenes but also other types of electron-rich compound can be oxidized by oxygen. Most organometallic reagents react with air, whereby either alkanes are formed by dimerization of the metal-bound alkyl groups (cuprates often react this way [80]) or peroxides or alcohols are formed [81, 82]. The alcohols result from disproportionation or reduction of the peroxides. Similarly, enolates, metalated nitriles, phenolates, enamines, and related compounds with nucleophilic carbon can react with oxygen by intermediate formation of carbon-centered radicals to yield dimers (Section 5.4.6; [83, 84]), peroxides, or alcohols. The oxidation of many organic compounds by air will, therefore, often proceed faster in the presence of bases (Scheme 3.21).

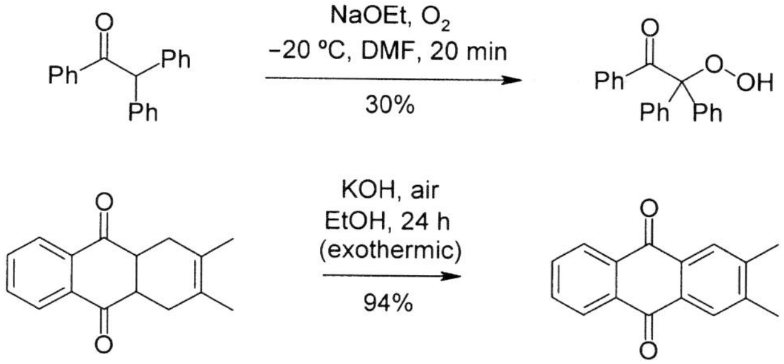

NaOEt, O_2
−20 °C, DMF, 20 min
30%

KOH, air
EtOH, 24 h
(exothermic)
94%

Scheme 3.21. Reaction of electron-rich organic compounds with triplet oxygen [85, 86].

3.5.3
Addition of Oxygen to C–C Double Bonds

Highly strained alkenes often behave chemically as 1,2-diradicals, and can therefore readily react with triplet oxygen (Scheme 3.6). 1,2-Quinodimethanes can also react as diradicals, and can undergo a formal [2 + 4] cycloaddition with triplet oxygen to yield cyclic peroxides (first example, Scheme 3.22). That this reaction also proceeds in the dark strongly indicates that no singlet oxygen is involved [87].

Some cyclopropanes have a high tendency to undergo homolytic ring scission and can react as 1,3-diradicals. Such compounds can be highly air-sensitive and react with triplet oxygen to yield 1,2-dioxolanes (second reaction, Scheme 3.22). 1,3-Dienes conjugated with a carbonyl group react with oxygen, and in the presence of a solvent with benzylic hydrogen atoms epoxides are formed (last example, Scheme 3.22).

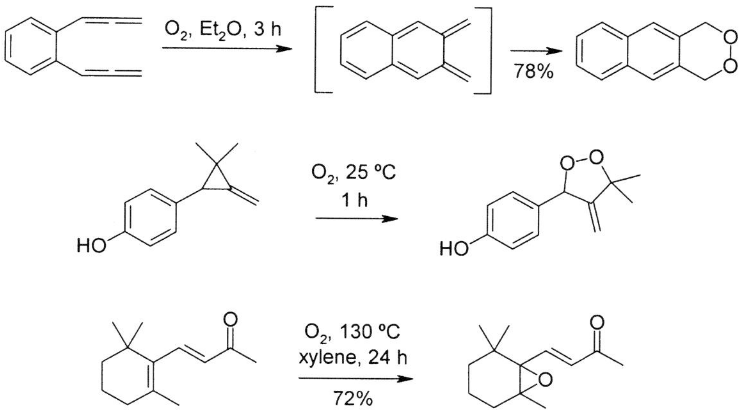

O_2, Et_2O, 3 h
78%

O_2, 25 °C
1 h

O_2, 130 °C
xylene, 24 h
72%

Scheme 3.22. Reaction of triplet oxygen with reactive alkenes and cyclopropanes [87–89].

3.6 Detonations

Most organic compounds are thermodynamically unstable and will burn at the air when heated to a sufficiently high temperature. Some compounds can, however, decompose thermally into thermodynamically more stable molecules without the need for oxygen. Although most of these are nitrogen-containing compounds, alkynes also can undergo such transformations. If such a thermal decomposition is exothermic, under certain conditions the decomposition will not just proceed as a simple, self-sustained reaction (deflagration) but can continuously accelerate and finally lead to a detonation. A detonation is not just a rapid deflagration under high pressure; whereas the propagation of a deflagration is governed by transport phenomena, and will not exceed velocities of about 10 m/s, the propagation of a detonation is mediated by a hypersonic shock wave, with propagation velocities of 5–9 km/s.

Detonations can be initiated by ignition, mechanical impact, friction, or electromagnetic radiation, and physical confinement of the material is often but not always required. Thus, a sample of the potassium salt of 1-tetrazolylacetic acid placed on a surface at room temperature will explode if any part of the sample is heated to >200 °C (e.g. by a hot flint spark or a flame) [90]. If the conditions for detonation are met, the time between shock-initiation and detonation is about 1 μs only [91].

The chemistry behind the detonation of organic compounds is exceedingly complex and poorly understood, because it involves a variety of intermolecular reactions different from those observed during thermolysis in solution or in the gas phase [92–94]. The conditions required to induce detonation of organic explosives vary widely and no clear-cut structure–sensitivity relationship exists. Nevertheless, most known explosives contain characteristic functional groups, and small molecules containing several of these should be handled with great care.

In Scheme 3.23 a selection of currently used explosives is sketched, together with some of their properties. As a measure of impact sensitivity, the height h is given from which a 2.5 kg hammer must hit a 35-mg sample on sand paper to lead to an

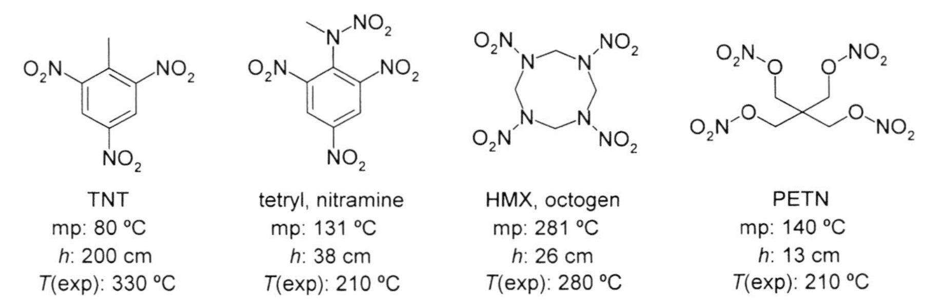

Scheme 3.23. Selection of explosives. h = 50 % impact height [91]; T(exp) = temperature at which the confined substance will explode within 10 s [95].

explosion with 50% probability [91]. The temperature T(exp) at which these compounds will explode within 10 s when confined in a closed container is also given [95].

There is, unfortunately, no obvious correlation between the impact sensitivity of a compound, its thermal behavior, and its structure. Thus, most polynitro arenes can explode on mechanical shock or rapid heating [96], but the shock sensitivity and thermal stability varies widely. TNT, for instance, is rather insensitive toward mechanical shocks or heating, and even pentanitrotoluene and hexanitrobenzene have been prepared, and their melting points (240 °C and 260 °C, respectively) could be determined [97]. The potassium salt of the radical anion of nitrobenzene, on the other hand, is highly shock sensitive, and detonates on simple agitation [92]. Similarly, whereas nitromethane can be handled safely the dry sodium salt of deprotonated nitromethane is potentially explosive [98].

Because impact sensitivity depends on intermolecular reactions in the solid, the sensitivity of explosives depends on their purity. It is often observed that higher purity leads to increased sensitivity, but this is not always so. Small quantities of particular compounds added to an explosive can even increase its sensitivity toward shock- or ignition-induced detonation [99, 100].

Nitro group-containing explosives, such as those shown in Scheme 3.23, have additional properties which render them particularly attractive as commercial explosives. These properties include chemical inertness, high density, low melting points (what enables casting into moulds), and good synthetic accessibility. Compounds devoid of nitro groups can, however, also be potent explosives. Potentially explosive compound classes include alkynes [101], azides, azo compounds, diazo compounds, furazanes (1,2,5-oxadiazoles) [102], ozonides, perchlorates, peroxides, tetrazoles [90], and compounds containing nitrogen–halogen bonds.

Particularly dangerous are small molecules containing these functional groups, because in these the critical functionality makes up a large percentage of the total molecular weight. Thus, whereas most sulfonyl azides are sufficiently stable to be handled without special precautions, methanesulfonyl azide is hazardous and shock-sensitive [103]. Similarly, acetyl nitrate ($MeCOONO_2$, explodes at 60 °C), methyl nitrate ($MeONO_2$, explodes at 65 °C), dimethyl azodicarboxylate [104], diazomethane, methyl azide, vinyl azide, azidoacetonitrile, and N_3–CN are particularly sensitive and dangerous, and should be handled as solutions only. Interestingly, some of these compounds, e.g. small organic azides [105, 106], are highly thermally stable at low pressures in the gas phase, but explode readily at low temperatures when condensed [107]; this indicates the importance of intermolecular reactions in the initiation and progression of a detonation. For this reason reactions with solutions of such reagents should not be performed with cold-finger condensers cooled to temperatures below the boiling point of the explosive reagents, because their condensation might readily lead to an explosion.

References

1 Werstiuk, N. H.; Roy, C. D. Simple and efficient synthesis of bromine-substituted 1,3-dienes and 1,3,5-cycloheptatriene by vacuum pyrolysis of *gem*-dibromocyclopropanes. *Tetrahedron Lett.* **2001**, *42*, 3255–3258.

2 Hellmann, G.; Hellmann, S.; Beckhaus, H.-D.; Rüchardt, C. Thermische Stabilität, Spannungsenthalpie und Struktur symmetrisch tetrasubstituierter Ethane. *Chem. Ber.* **1982**, *115*, 3364–3383.

3 McDevitt, P.; Vittimberga, B. M. The electron transfer reactions of cyano substituted pyridines and quinolines with thermally generated diphenyl ketyl. *J. Heterocyclic Chem.* **1990**, *27*, 1903–1908.

4 Eaton, P. E.; Temme, G. H. The [2.2.2]propellane system. *J. Am. Chem. Soc.* **1973**, *95*, 7508–7510.

5 Flamm-ter Meer, M. A.; Beckhaus, H.-D.; Peters, K.; von Schnering, H.-G.; Fritz, H.; Rüchardt, C. Separation, structure analysis, and thermolysis of a pair of stable rotamers; P-(*R**,*R**)-and M-(*R**,*R**)-D,L-3,4-di-1-adamantyl-2,2,5,5-tetramethylhexane. *Chem. Ber.* **1986**, *119*, 1492–1510.

6 Cram, D. J.; Langemann, A.; Lwowski, W.; Kopecky, K. R. Electrophilic substitution at saturated carbon. IV. Competing radical and anionic cleavage reactions. *J. Am. Chem. Soc.* **1959**, *81*, 5760–5767.

7 Sandler, S. R. Reaction of *gem*-dibromocyclopropanes with morpholine. *J. Org. Chem.* **1968**, *33*, 4537–4539.

8 Matsumura, S.; Maeda, Y.; Nishimura, T.; Uemura, S. Palladium-catalyzed asymmetric arylation, vinylation, and allenylation of *tert*-cyclobutanols via enantioselective C–C bond cleavage. *J. Am. Chem. Soc.* **2003**, *125*, 8862–8869.

9 Falck, J. R.; Chandrasekhar, S.; Manna, S.; Chiu, C.-C. S.; Mioskowski, C.; Wetzel, I. Total synthesis of the spiro-*o*-benzoquinonefuran (–)-stypoldione. *J. Am. Chem. Soc.* **1993**, *115*, 11606–11607.

10 Dauben, W. G.; Hart, D. J. A new entry to the tricyclo[6.3.0.0^{4,8}]undecan ring system. *J. Org. Chem.* **1977**, *42*, 3787–3793.

11 Aitken, D. J.; Gauzy, C.; Pereira, E. Studies on the stability of the cyclobutane β-amino acid skeleton: a cautionary tale. *Tetrahedron Lett.* **2004**, *45*, 2359–2361.

12 Nokami, J.; Ohga, M.; Nakamoto, H.; Matsubara, T.; Hussain, I.; Kataoka, K. The first and highly enantioselective crotylation of aldehydes via an allyl-transfer reaction from a chiral crotyl donor. *J. Am. Chem. Soc.* **2001**, *123*, 9168–9169.

13 Nokami, J.; Nomiyama, K.; Matsuda, S.; Imai, N.; Kataoka, K. Highly enantioselective alk-2-enylation of aldehydes through an allyl-transfer reaction. *Angew. Chem. Int. Ed.* **2003**, *42*, 1273–1276.

14 Jones, P.; Knochel, P. Preparation and reactions of masked allylic organozinc reagents. *J. Org. Chem.* **1999**, *64*, 186–195.

15 Agapie, T.; Diaconescu, P. L.; Mindiola, D. J.; Cummins, C. C. Radical scission of symmetrical 1,4-dicarbonyl compounds: C–C bond cleavage with titanium(IV) enolate formation and related reactions. *Organometallics* **2002**, *21*, 1329–1340.

16 Lu, J.; Bai, Y.; Wang, Z.; Yang, B.; Li, W. Ferric chloride hexahydrate: a convenient reagent for the oxidation of Hantzsch 1,4-dihydropyridines. *Synth. Commun.* **2001**, *31*, 2625–2630.

17 Leonetti, J. A.; Gross, T.; Little, R. D. Cycloaddition–fragmentation as a route to bicyclic ring systems. Use of the intermolecular diyl trapping reaction. *J. Org. Chem.* **1996**, *61*, 1787–1793.

18 Warner, P. M. Strained bridgehead double bonds. *Chem. Rev.* **1989**, *89*, 1067–1093.

19 Borden, W. T. Pyramidalized alkenes. *Chem. Rev.* **1989**, *89*, 1095–1109.

20 House, H. O.; Outcalt, R. J.; Haack, J. L.; VanDerveer, D. Enones with strained double bonds. 9. The 2-phenylbicyclo[3.3.1]non-1-en-3-one system. *J. Org. Chem.* **1983**, *48*, 1654–1661.

21 Shea, K. J.; Kim, J. S. Influence of strain on chemical reactivity. Relative reactivity of torsionally distorted double bonds in MCPBA epoxidations. *J. Am. Chem. Soc.* **1992**, *114*, 3044–3051.

22 Kienzle, F.; Mergelsberg, I.; Stadlwieser, J.; Arnold, W. Diels–Alder Reaktionen mit aktivierten 4-Methyl-1,3-pentadienen. *Helv. Chim. Acta* **1985**, *68*, 1133–1139.

23 Paquette, L. A.; Wells, G. J.; Wickham, G. Silicon in organic synthesis. 26. Expedient synthesis of dicyclopropylideneethane with function-

alized 1-(trimethylsilyl)cyclopropanes. *J. Org. Chem.* **1984**, *49*, 3618–3612.

24 Kienzle, F.; Stadlwieser, J.; Mergelsberg, I. Diels–Alder Reaktionen mit 3-Cyclopropylidenprop-1-enyl-ethyl-ether als 1,3-Dien. *Helv. Chim. Acta* **1989**, *72*, 348–352.

25 Imming, P.; Klar, B.; Dix, D. Hydrolytic stability vs ring size in lactams: implications for the development of lactam antibiotics and other serine protease inhibitors. *J. Med. Chem.* **2000**, *43*, 4328–4331.

26 Sonnet, P. E.; McGovern, T. P.; Cunningham, R. T. Enantiomers of the biologically active components of the insect attractant trimedlure. *J. Org. Chem.* **1984**, *49*, 4639–4643.

27 Buñuel, E.; Gil, A. M.; Díaz-de-Villegas, M. D.; Cativiela, C. Synthesis of constrained prolines by Diels–Alder reaction using a chiral unsaturated oxazolone derived from (*R*)-glyceraldehyde as starting material. *Tetrahedron* **2001**, *57*, 6417–6427.

28 Myers, A. G.; Gleason, J. L.; Yoon, T.; Kung, D. W. Highly practical methodology for the synthesis of D- and L-α-amino acids, *N*-protected α-amino acids, and *N*-methyl-α-amino acids. *J. Am. Chem. Soc.* **1997**, *119*, 656–673.

29 Paleo, M. R.; Lamas, C.; Castedo, L.; Dominguez, D. A new synthesis of phthalides by internal trapping in *ortho*-lithiated carbamates derived from benzylic alcohols. *J. Org. Chem.* **1992**, *57*, 2029–2033.

30 Insuasty, B.; Abonia, R.; Quiroga, J.; Salcedo, A.; Kolshorn, H.; Meier, H. 1-Alkyl and azeto[1,2-*a*][1,5]benzodiazepine derivatives in the reaction of *o*-phenylenediamine with 3-(dimethylamino)propiophenones. *Eur. J. Org. Chem.* **2000**, 1973–1976.

31 Loubinoux, B.; Miazimbakana, J.; Gerardin, P. Reactivity of new precursors of quinone methides. *Tetrahedron Lett.* **1989**, *30*, 1939–1942.

32 Poss, A. J.; Belter, R. K. Quinone methide *p*-hydroxybenzylation of 1,3-diketones. *J. Org. Chem.* **1988**, *53*, 891–893.

33 Schwartz, M. A.; Zoda, M.; Vishnuvajjala, B.; Mami, I. A convenient synthesis of *o*- and *p*-hydroxy substituted phenylacetonitriles and phenethylamines. *J. Org. Chem.* **1976**, *41*, 2502–2503.

34 Stevens, C. L.; Elliott, R. D.; Winch, B. L. Aminoketone rearrangements. II. The rearrangement of phenyl α-aminoketones. *J. Am. Chem. Soc.* **1963**, *85*, 1464–1470.

35 Ochoa de Echagüen, C.; Ortuño, R. M. Thermal 1,2- and 1,4-cycloadditions of butadiene to 3-substituted 5-methylene-2(5*H*)furanones: evidence for a biradical mechanism in some Diels–Alder reactions. *Tetrahedron Lett.* **1995**, *36*, 749–752.

36 Takeda, K.; Shimotani, A.; Yoshii, E.; Yamaguchi, K. A Lewis acid mediated intramolecular [2 + 2] cycloaddition of 3-(9-methylundeca-7,9-dienyl)-9-methylene-2,8-dioxabicyclo[4.3.0]nonane-5,7-dione. *Heterocycles* **1992**, *34*, 2259–2261.

37 Viehe, H. G.; Janousek, Z.; Merényi, R.; Stella, L. The captodative effect. *Acc. Chem. Res.* **1985**, *18*, 148–154.

38 Bordwell, F. G.; Lynch, T.-Y. Radical stabilization energies and synergistic (captodative) effects. *J. Am. Chem. Soc.* **1989**, *111*, 7558–7562.

39 Tanaka, H.; Yoshida, S. Kinetic study of the radical homopolymerization of captodative substituted methyl α-(acyloxy)acrylates. *Macromolecules* **1995**, *28*, 8117–8121.

40 Enders, D.; Meyer, O.; Raabe, G. Diastereo- and enantioselective synthesis of 4-nitrocyclohexanones by [4 + 2] cycloaddition of a chiral 2-aminobutadiene to nitroalkenes. *Synthesis* **1992**, 1242–1244.

41 Arnold, Z. Esters of 2-dialkylamino-2-propenoic acids. *Synthesis* **1990**, 39–40.

42 Koerber-Plé, K.; Massiot, G. Synthesis of an unusual 2,3,4-trisubstituted indole derivative found in the antibiotic nosiheptide. *Synlett* **1994**, 759–760.

43 Goodall, K.; Parsons, A. F. A diastereoselective radical cyclization approach to pyroglutamates. *Tetrahedron Lett.* **1997**, *38*, 491–494.

44 Nachman, R. J.; Olsen, J. D. Ranunculin: a toxic constituent of the poisonous range plant bur buttercup (ceratocephalus testiculatus). *J. Agric. Food Chem.* **1983**, *31*, 1358–1360.

45 Hill, R.; Van Heyningen, R. Ranunculin: the precursor of the vesicant substance of the buttercup. *Biochem. J.* **1951**, *49*, 332–335.

46 D'Alelio, G. F.; Williams, C. J.; Wilson, C. L. Reactions of furan compounds. XVIII. Nuclear acetoxylation. *J. Org. Chem.* **1960**, *25*, 1025–1028.

47 Shaw, E. A synthesis of protoanemonin. The tautomerism of acetylacrylic acid and of penicillic acid. *J. Am. Chem. Soc.* **1946**, *68*, 2510–2513.

48 Wu, Y.-D.; Kirmse, W.; Houk, K. N. Geminal group interactions. *J. Am. Chem. Soc.* **1990**, *112*, 4557–4559.

49 Schleyer, P. v. R.; Jemmis, E. D.; Spitznagel, G. W. Do anomeric effects involving the second-row substituents Cl, SH, and PH_2 exist? Stabilization energies and structural preferences. *J. Am. Chem. Soc.* **1985**, *107*, 6393–6394.

50 Reed, A. E.; Schleyer, P. v. R. The anomeric effect with central atoms other than carbon. I. Strong interactions between nonbonded substituents in polyfluorinated first- and second-row hydrides. *J. Am. Chem. Soc.* **1987**, *109*, 7362–7373.

51 Hine, J. Polar effects on rates and equilibria. VIII. Double bond-no bond resonance. *J. Am. Chem. Soc.* **1963**, *85*, 3239–3244.

52 Richard, J. P.; Amyes, T. L.; Rice, D. J. Effects of electronic geminal interactions on the solvolytic reactivity of methoxymethyl derivatives. *J. Am. Chem. Soc.* **1993**, *115*, 2523–2524.

53 Miller, W. T.; Fried, J. H.; Goldwhite, H. Substitution and addition reactions of the fluoroolefins. IV. Reactions of fluoride ion with fluoroolefins. *J. Am. Chem. Soc.* **1960**, *82*, 3091–3099.

54 Pearson, R. G. *Chemical Hardness*; Wiley–VCH: Weinheim, 1997.

55 Howard, C. J. Kinetic study of the equilibrium $HO_2 + NO \leftrightarrow OH + NO_2$ and the thermochemistry of HO_2. *J. Am. Chem. Soc.* **1980**, *102*, 6937–6941.

56 Valgimigli, L.; Brigati, G.; Pedulli, G. F.; DiLabio, G. A.; Mastragostino, M.; Arbizzani, C.; Pratt, D. A. The effect of ring nitrogen atoms on the homolytic reactivity of phenolic compounds: understanding the radical-scavenging ability of 5-pyrimidinols. *Chem. Eur. J.* **2003**, *9*, 4997–5010.

57 Weast, R. C. *CRC Handbook of chemistry and physics*; 64th edition, CRC Press: Boca Raton, 1983.

58 Blanksby, S. J.; Ellison, G. B. Bond dissociation energies of organic molecules. *Acc. Chem. Res.* **2003**, *36*, 255–263.

59 van Gorkum, R.; Bouwman, E.; Reedijk, J. Fast autoxidation of ethyl linoleate catalyzed by $Mn(acac)_3$ and bipyridine: a possible drying catalyst for alkyl paints. *Inorg. Chem.* **2004**, *43*, 2456–2458.

60 Weissermel, K.; Arpe, H.-J. *Industrial organic chemistry*, VCH: Weinheim, 1993.

61 Chen, J.-R.; Yang, H.-H.; Wu, C.-H. A novel process of autoxidation of cyclohexane using pure oxygen. *Org. Process Res. Dev.* **2004**, *8*, 252–255.

62 Knight, H. B.; Swern, D. Tetralin hydroperoxide. *Org. Synth.* **1963**, *Coll. Vol. IV*, 895–897.

63 Becker, H.-D.; Sanchez, D. Hydrogen transfer by ketyl radicals: the reductive dimerization of quinone methides. *Tetrahedron Lett.* **1975**, 3745–3748.

64 Pearl, I. A. Protocatechuic acid. *Org. Synth.* **1955**, *Coll. Vol. III*, 745–746.

65 Schwetlick, K.; Habicher, W. D. Antioxidant action mechanisms of hindered amine stabilizers. *Polym. Degrad. Stab.* **2002**, *78*, 35–40.

66 Gijsman, P.; Gitton, M. Hindered amine stabilizers as long-term heat stabilizers for polypropylene. *Polym. Degrad. Stab.* **1999**, *66*, 365–371.

67 Richman, J. E.; Chang, Y.-C.; Kambourakis, S.; Draths, K. M.; Almy, E.; Snell, K. D.; Strasburg, G. M.; Frost, J. W. Reaction of 3-dehydroshikimic acid with molecular oxygen and hydrogen peroxide: products, mechanism, and associated antioxidant activity. *J. Am. Chem. Soc.* **1996**, *118*, 11587–11591.

68 Masuda, T.; Inaba, Y.; Takeda, Y. Antioxidant mechanism of carnosic acid: structural identification of two oxidation products. *J. Agric. Food Chem.* **2001**, *49*, 5560–5565.

69 Barclay, L. R. C.; Vinqvist, M. R.; Mukai, K.; Goto, H.; Hashimoto, Y.; Tokunaga, A.; Uno, H. On the antioxidant mechanism of curcumin: classical methods are needed to determine antioxidant mechanism and activity. *Org. Lett.* **2000**, *2*, 2841–2843.

70 Wijtmans, M.; Pratt, D. A.; Valgimigli, L.; DiLabio, G. A.; Pedulli, G. F.; Porter, N. A. 6-Amino-3-pyridinols: towards diffusion-controlled chain-breaking antioxidants. *Angew. Chem. Int. Ed.* **2003**, *42*, 4370–4373.

71 Foti, M. C.; Ingold, K. U. Mechanism of inhibition of lipid peroxidation by γ-terpinene, an unusual and potentially useful hydrocarbon antioxidant. *J. Agric. Food Chem.* **2003**, *51*, 2758–2765.

72 Ling, K.-Q.; Lee, Y.; Macikenas, D.; Protasiewicz, J. D.; Sayre, L. M. Copper(II)-mediated autoxidation of *tert*-butylresorcinols. *J. Org. Chem.* **2003**, *68*, 1358–1366.

73 Hino, T.; Nakagawa, M.; Hashizume, T.; Yamaji, N.; Miwa, Y.; Tsuneoka, K.; Akaboshi, S. 2-Aminoindoles. Preparation from 2-indoli-

nethiones, tautomerism and autoxidation. *Tetrahedron* **1971**, *27*, 775–787.

74 Groenendaal, L. B.; Jonas, F.; Freitag, D.; Pielartzik, H.; Reynolds, J. R. Poly(3,4-ethylenedioxythiophene) and its derivatives: past, present, and future. *Adv. Mater.* **2000**, *12*, 481–494.

75 Bhattacharya, G.; Su, T.-L.; Chia, C.-M.; Chen, K.-T. Synthesis and autoxidation of new tetracyclic 9*H*,10*H*-indolizino[1,2-*b*]indole-1-ones. *J. Org. Chem.* **2001**, *66*, 426–432.

76 Nau, W. M.; Adam, W.; Klapstein, D.; Sahin, C.; Walter, H. Correlation of oxidation and ionization potentials for azoalkanes. *J. Org. Chem.* **1997**, *62*, 5128–5132.

77 Miller, L. L.; Nordblom, G. D.; Mayeda, E. A. A simple, comprehensive correlation of organic oxidation and ionization potentials. *J. Org. Chem.* **1972**, *37*, 916–918.

78 DiLabio, G. A.; Pratt, D. A.; Wright, J. S. Theoretical calculation of ionization potentials for disubstituted benzenes: additivity vs non-additivity of substituent effects. *J. Org. Chem.* **2000**, *65*, 2195–2203.

79 Buzzeo, M. C.; Klymenko, O. V.; Wadhawan, J. D.; Hardacre, C.; Seddon, K. R.; Compton, R. G. Voltammetry of oxygen in the room-temperature ionic liquids 1-ethyl-3-methylimidazolium bis(triflyl)imide and $HexEt_3N^+$ Tf_2N^-: one-electron reduction to form superoxide. Steady-state and transient behavior in the same cyclic voltammogram resulting from widely different diffusion coefficients of oxygen and superoxide. *J. Phys. Chem. A* **2003**, *107*, 8872–8878.

80 Walborsky, H. M.; Banks, R. B.; Banks, M. L. A.; Duralsamy, M. Stability and oxidative coupling of chiral vinyl- and cyclopropylcopper reagents. Formation of a novel dissymmetric diene. *Organometallics* **1982**, *1*, 667–674.

81 March, J. *Advanced Organic Chemistry*, John Wiley and Sons: New York, 1992.

82 Wade, P. A.; D'Ambrosio, S. G.; Murray, J. K. Autoxidation of [2-(1,3-dithianyl)]lithium: a cautionary note. *J. Org. Chem.* **1995**, *60*, 4258–4259.

83 Rodriguez, H.; Marquez, A.; Chuaqui, C. A.; Gomez, B. Oxidation of mesoionic oxazolones by oxygen. *Tetrahedron* **1991**, *47*, 5681–5688.

84 Russell, G. A.; Kaupp, G. Oxidation of carbanions. IV. Oxidation of indoxyl to indigo in basic solution. *J. Am. Chem. Soc.* **1969**, *91*, 3851–3859.

85 Sawaki, Y.; Ogata, Y. Kinetics of the base-catalyzed decomposition of α-hydroperoxy ketones. *J. Am. Chem. Soc.* **1975**, *97*, 6983–6989.

86 Allen, C. F. H.; Bell, A. 2,3-Dimethylanthraquinone. *Org. Synth.* **1955**, *Coll. Vol. III*, 310–311.

87 Bowes, C. M.; Montecalvo, D. F.; Sondheimer, F. *o*-Dipropadienylbenzene and 2,3-dipropadienylnaphthalene. The oxidation of diallenes to cyclic peroxides with triplet oxygen. *Tetrahedron Lett.* **1973**, 3181–3184.

88 Hart, H.; Lavrik, P. B. Synthetically useful epoxidations with molecular oxygen. *J. Org. Chem.* **1974**, *39*, 1793–1794.

89 Creary, X.; Wolf, A.; Miller, K. Facile autoxidation of 2-(4-hydroxyphenyl)-3,3-dimethylmethylenecyclopropane. The radical stabilizing ability of the phenoxide group. *Org. Lett.* **1999**, *1*, 1615–1618.

90 Eizember, R. F.; Krodel, R. M. A cautionary note concerning the isolation of some metal salts of 1-tetrazoleacetic acid. *J. Org. Chem.* **1974**, *39*, 1792–1793.

91 Macek, A. Sensitivity of explosives. *Chem. Rev.* **1962**, *62*, 41–63.

92 Stevenson, C. D.; Garland, P. M.; Batz, M. L. Evidence of carbenes in the explosion chemistry of nitroaromatic anion radicals. *J. Org. Chem.* **1996**, *61*, 5948–5952.

93 Oxley, J. C.; Smith, J. L.; Ye, H.; McKenney, R. L.; Bolduc, P. R. Thermal stability studies on a homologous series of nitroarenes. *J. Phys. Chem.* **1995**, *99*, 9593–9602.

94 Brill, T. B.; James, K. J. Kinetics and mechanisms of thermal decomposition of nitroaromatic explosives. *Chem. Rev.* **1993**, *93*, 2667–2692.

95 Zinn, J.; Rogers, R. N. Thermal initiation of explosives. *J. Phys. Chem.* **1962**, *66*, 2646–2653.

96 Kharasch, N. Explosion hazard in the preparation and use of 2,4-dinitrobenzenesulfenyl chloride. *J. Am. Chem. Soc.* **1950**, *72*, 3322–3323.

97 Nielsen, A. T.; Atkins, R. L.; Norris, W. P.; Coon, C. L.; Sitzmann, M. E. Synthesis of polynitro compounds. Peroxydisulfuric acid oxidation of polynitroarylamines to polynitro aromatics. *J. Org. Chem.* **1980**, *45*, 2341–2347.

98 Barrett, A. G. M.; Dhanak, D.; Graboski, G. G.; Taylor, S. J. (Phenylthio)nitromethane. *Org. Synth.* **1993**, *Coll. Vol. VIII*, 550–553.

99 Singh, G.; Felix, S. P.; Soni, P. Studies on energetic compounds part 28: thermolysis of HMX and its plastic bonded explosives containing estane. *Thermochim. Acta* **2003**, *399*, 153–165.

100 Rogers, R. N. Incompatibility in explosive mixtures. *I & EC Product Research and Development* **1962**, *1*, 169–172.

101 Boese, R.; Matzger, A. J.; Vollhardt, K. P. C. Synthesis, crystal structure, and explosive decomposition of 1,2:5,6:11,12:15,16-tetrabenzo-3,7,9,13,17,19-hexadehydro[20]annulene: formation of onion- and tube-like closed-shell carbon particles. *J. Am. Chem. Soc.* **1997**, *119*, 2052–2053.

102 Ruggeri, S. G.; Bill, D. R.; Bourassa, D. E.; Castaldi, M. J.; Houck, T. L.; Ripin, D. H. B.; Wei, L.; Weston, N. Safety vs efficiency in the development of a high-energy compound. *Org. Process Res. Dev.* **2003**, *7*, 1043–1047.

103 Tuma, L. D. Identification of a safe diazotransfer reagent. *Thermochim. Acta* **1994**, *243*, 161–167.

104 Azodicarboxylates: warning. *Org. Synth.* **1998**, *Coll. Vol. IX*, 837–838.

105 Bock, H.; Dammel, R. Gas-phase pyrolyses of alkyl azides: experimental evidence for chemical activation. *J. Am. Chem. Soc.* **1988**, *110*, 5261–5269.

106 Bock, H.; Dammel, R.; Aygen, S. Gas-phase reactions. 36. Pyrolysis of vinyl azide. *J. Am. Chem. Soc.* **1983**, *105*, 7681–7685.

107 Wiley, R. H.; Moffat, J. Preparation and polymerization of vinyl azide. *J. Org. Chem.* **1957**, *22*, 995–996.

4
Aliphatic Nucleophilic Substitutions: Problematic Electrophiles

4.1
Mechanisms of Nucleophilic Substitution

Nucleophilic substitutions at sp^3-hybridized carbon are among the most useful synthetic transformations, and have been thoroughly investigated [1–3]. The success of these reactions depends mainly on the structures of the nucleophile and electrophile, their concentration, and on the solvent and reaction temperature chosen. The structure of the electrophile in nucleophilic substitutions is of critical importance, because many unwanted side reactions result from unexpected reactivity of the electrophile.

In Scheme 4.1 the mechanisms of typical monomolecular (SN1) and bimolecular (SN2) nucleophilic substitutions at a neutral electrophile with an anionic nucleophile are sketched. SN1 reactions usually occur when the electrophile is sterically

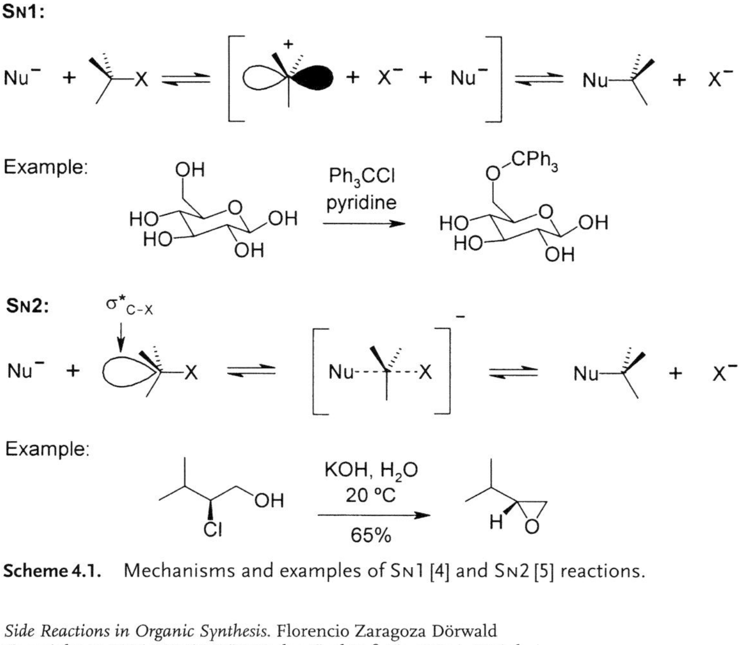

SN1:

Nu⁻ + ⟩—X ⇌ [+ X⁻ + Nu⁻] ⇌ Nu—⟨ + X⁻

Example: Ph_3CCl, pyridine

SN2: σ^*_{C-X}

Nu⁻ + ⟩—X ⇌ [Nu---⟩---X]⁻ ⇌ Nu—⟨ + X⁻

Example: KOH, H_2O, 20 °C, 65%

Scheme 4.1. Mechanisms and examples of SN1 [4] and SN2 [5] reactions.

Side Reactions in Organic Synthesis. Florencio Zaragoza Dörwald

ISBN: 3-527-31021-5

hindered and has a high tendency to form a carbocation. Steric hindrance will increase the activation barrier for the SN2 pathway, and thereby enable other reaction mechanisms to become dominant. Carbocation formation is facilitated by good leaving groups, by secondary, tertiary, α-heteroatom-substituted, or conjugated (benzylic, allylic, or propargylic) alkyl groups, or by solvents of high dielectric constant. Acidic reaction conditions can also promote the formation of carbocations by protonation (i.e. improvement) of the leaving group, and thereby favor the SN1 mechanism. Substitutions in which carbocations are formed as intermediates are usually monomolecular, that is, the rate of reaction will depend on the concentration of one of the starting materials only. Reaction of the nucleophile with the carbocation is usually much faster than the (rate-determining) formation of the carbocation, and the reaction rate will be a function of the concentration of the electrophile only.

SN2 reactions usually occur at primary alkyl groups, and often involve soft electrophiles and nucleophiles. As shown in Scheme 4.1, the SN2 reaction proceeds with (Walden) inversion at the central carbon, and is therefore stereospecific.

When steric or electronic factors interfere with the substitution mechanisms sketched in Scheme 4.1, a substitution may occur anyway. A variety of alternative mechanisms are available and lead to the formation of the products of aliphatic nucleophilic substitution. The most important of the alternative pathways are the elimination–addition mechanism, which is often observed for electrophiles with an electron-withdrawing group in the β position, and the radical nucleophilic substitution (SRN1, Scheme 4.2) [6–10]. The latter, which is particularly important for tertiary substrates, nitroalkanes, nitrobenzyl halides, perfluoroalkyl halides, and some aromatic nucleophilic substitutions, is initiated by single-electron transfer (SET) from the nucleophile to the electrophile. Thereby a radical and a radical anion are formed if the nucleophile was negatively charged and the electrophile uncharged. For some substrates, but not all, SET requires light. The radical anion, which can also be generated electrochemically [11], can now release the anionic leaving group and become a neutral radical; this will react with the nucleophile to form a new radical anion. This last reaction will be particularly fast if the intermediate neutral radical is electrophilic (i.e. when the substituents at the central carbon are electron-withdrawing) and the nucleophile is electron-rich. SET from the newly formed radical anion to the electrophile yields the final product and a new radical anion.

Alternative non-chain reaction pathways have also been discussed [14]. Some radical nucleophilic substitutions, for instance, proceed with inversion at the electrophilic carbon [15]. This observation is not compatible with the formation of a free radical, but can be rationalized by the mechanism sketched in Scheme 4.3 [15], in which the two radicals formed after SET do not diffuse out of the solvent cage but react in an SN2-like manner with Walden inversion.

SET-induced chain reactions are usually much faster than SN2 reactions, and if both pathways are accessible for a given pair of starting materials, products resulting from SRN1 will usually predominate. This is, for example, observed for the first reaction sketched in Scheme 4.4 [11, 16]. Electrochemical or zinc-mediated [17] reduction of the starting pyridinium salt yields a carbanion which, on reaction with Me_2tBuS^+, can either be methylated via SN2 or *tert*-butylated via SRN1. Because the nucleophile

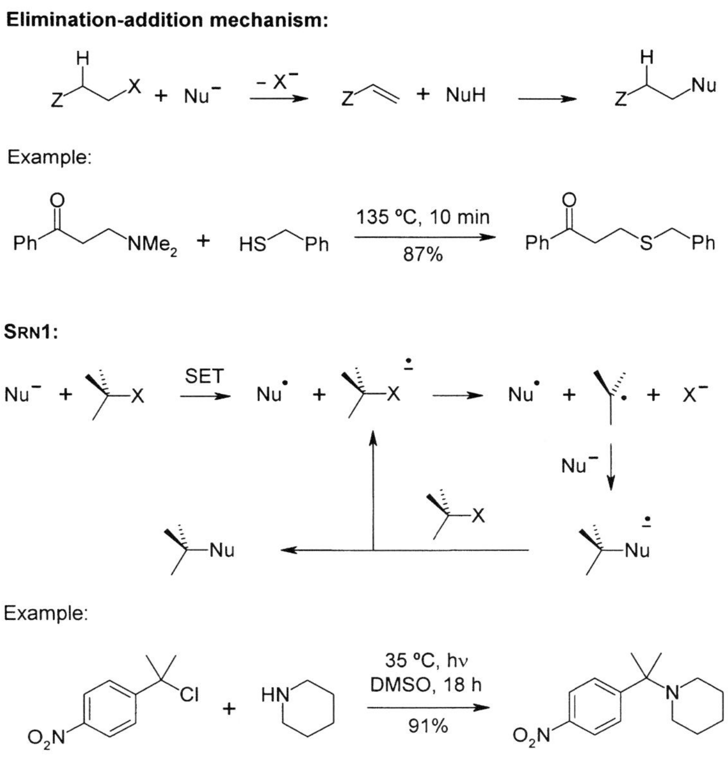

Elimination-addition mechanism:

Example:

135 °C, 10 min
87%

$S_{RN}1$:

SET

Example:

35 °C, hν
DMSO, 18 h
91%

Scheme 4.2. Mechanisms and examples of "elimination–addition" substitution [12] and of radical nucleophilic substitution [13].

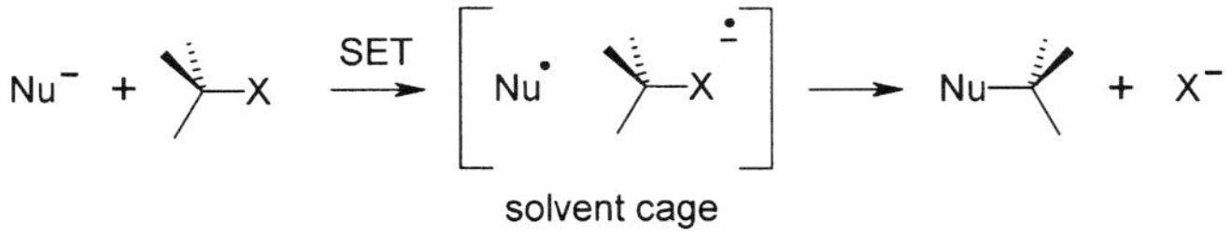

SET

solvent cage

Scheme 4.3. Possible mechanism for stereospecific radical nucleophilic substitutions [15].

is a good electron donor and leads, on oxidation, to a stable radical [18], the radical pathway is preferred in this instance.

Compared with S_N1 or S_N2 reactions, relatively few examples of radical nucleophilic substitutions have been reported [6–8, 16, 22–25]. This suggests that the structural requirements of nucleophile and electrophile for such processes are more stringent than for non-radical nucleophilic substitutions. In the following sections the focus will be on the more generally applicable non-radical S_N1 and S_N2 reactions.

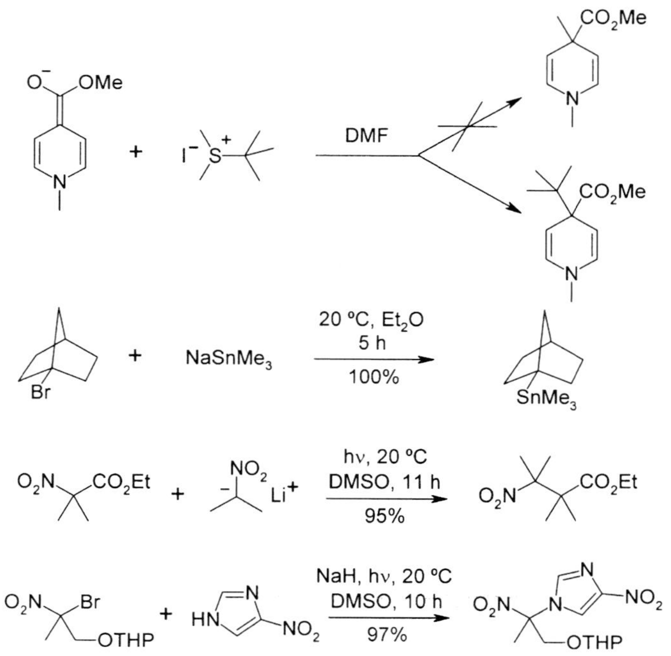

Scheme 4.4. Examples of radical nucleophilic substitutions [11, 19–21].

4.2
Structure of the Leaving Group

4.2.1
Good and Poor Leaving Groups

According to the mechanism of SN1 and SN2 reactions (Scheme 4.1), a good leaving group X should form a highly polarized bond with carbon, prone to heterolytic cleavage, and with a low-energy σ^*_{C-X} orbital. If the displaced leaving group is unreactive and/or forms strong hydrogen bonds with the solvent chosen, its displacement will proceed more readily. Particularly good leaving groups are atoms or molecules of low basicity, such as the anions of strong acids (halides, sulfate, sulfonates, perchlorate), ethers, alcohols, and water (i.e. electrophiles of the general structure $R-OR_2^+$), thioethers ($R-SR_2^+$), aryl halides ($R-HalAr^+$), and nitrogen ($R-N_2^+$). If the basicity of the group increases, its ability to act as a leaving group usually decreases.

The better the leaving group, the more can the mechanism tilt towards a monomolecular (SN1) substitution. With good leaving groups carbocation formation is faster than the SN2 reaction and becomes the rate determining (slowest) step in the new mechanism. The reaction of the nucleophile with the carbocation is the fastest step in SN1 reactions (and much faster than the corresponding SN2 reaction).

Because the SN1 reaction is no longer stereospecific and because carbocations can readily rearrange or be deprotonated (to yield an alkene), it might be advisable not to choose always the best possible leaving group, but one of moderate reactivity, also to keep the electrophilic reagent tractable. Compounds with very good leaving groups, such as alkyl triflates or diazonium salts, will often react with the solvent or undergo elimination instead of yielding the product of substitution. If substitution with a highly reactive or strongly basic nucleophile is intended, it might even be advisable to choose a poor leaving group, for example fluoride [26, 27] (last reaction in Scheme 4.10).

Some of the most common unwanted events during nucleophilic substitutions are β-elimination of HX from the electrophile to yield an alkene, rearrangement of the electrophile, reaction of the electrophile with the solvent, or no reaction at all. Although unwanted β-eliminations usually occur as a result of the structure of the alkyl group in the electrophile or because of the high basicity of the nucleophile, choosing an unsuitable leaving group might also favor the formation of alkenes. Leaving groups with a higher tendency to undergo elimination rather than substitution are those containing a group capable of abstracting a vicinal proton (Scheme 4.5), and which are, therefore, mainly used for thermal eliminations and not for nucleophilic substitutions.

Examples:

X = O,S
Y = R, XR

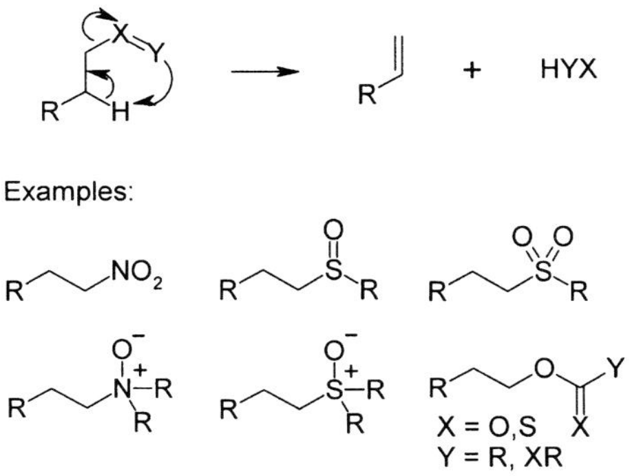

Scheme 4.5. Leaving groups with a high tendency to undergo thermal elimination.

When no β hydrogen is available, however, compounds as those shown in Scheme 4.5 can be useful electrophiles in nucleophilic substitutions. Thus, $Me_3S^+OH^-$ [28], $Me_3S{=}O^+X^-$ [29, 30], $Me_3Se^+OH^-$ [31], $(MeO)_2CO$ [32–34], or $MeNO_2$ [35] can all be used as electrophilic methylating reagents.

Because of the ability of some leaving groups to stabilize an α-carbanion, the pH at which the substitution is performed can be critical. Electrophiles with such leaving groups (e.g. $R{-}NO_2$ [36, 37], $R{-}S({=}O)_2R$ [38, 39], $R{-}S({=}O)R$ [40]) will usually undergo substitution only under neutral or acidic conditions, what limits the choice of suitable nucleophiles. Some nucleophilic displacements of nitro and sulfonyl groups, both under acidic and basic reaction conditions, are shown in Schemes 4.6 and 4.7. Allylic nitro groups can also be readily displaced by catalysis with palladium(0) [36, 41]. The second example in Scheme 4.6 probably proceeds with retention of configuration because of neighboring-group participation (a thiiranium ion is formed as intermediate).

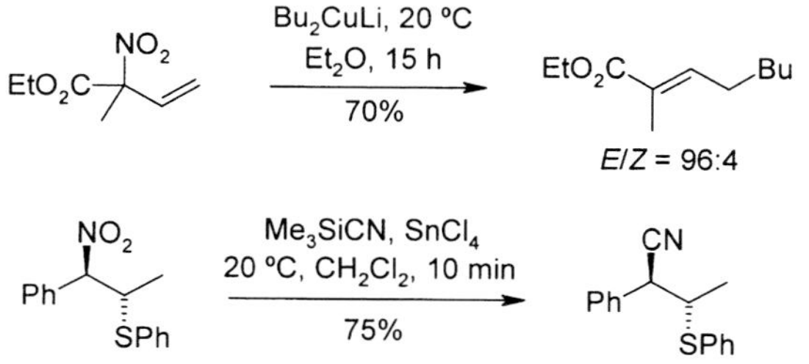

Bu_2CuLi, 20 °C, Et_2O, 15 h, 70%; E/Z = 96:4

Me_3SiCN, $SnCl_4$, 20 °C, CH_2Cl_2, 10 min, 75%

Scheme 4.6. Nucleophilic substitution of nitro groups [37, 42].

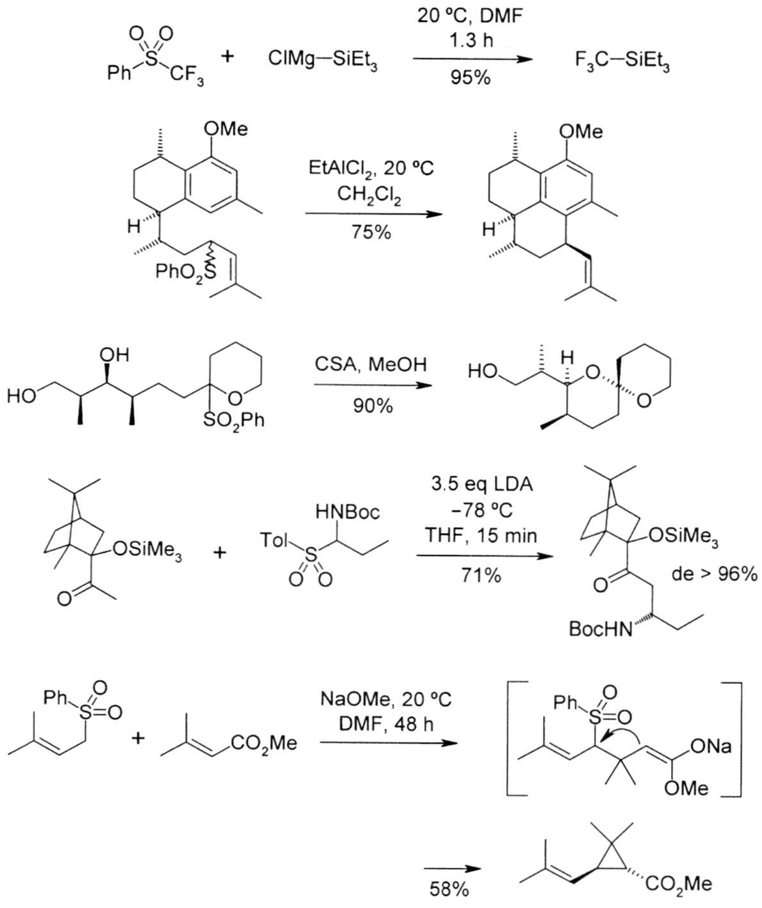

$PhSO_2CF_3$ + $ClMg–SiEt_3$ → (20 °C, DMF, 1.3 h, 95%) $F_3C–SiEt_3$

$EtAlCl_2$, 20 °C, CH_2Cl_2, 75%

CSA, MeOH, 90%

3.5 eq LDA, −78 °C, THF, 15 min, 71%; de > 96%

NaOMe, 20 °C, DMF, 48 h; 58%

Scheme 4.7. Nucleophilic substitution of sulfonyl groups [39, 43–46].

Acidic reaction conditions can also lead to protonation of some leaving groups, thereby increasing their reactivity in nucleophilic substitutions. Such groups include, for instance, alcohols, ethers, amines, amides, or alkyl fluorides.

Nucleophilic substitutions can also fail because the intended leaving group is not good enough, i.e. its bond to carbon is not strongly polarized or the basicity of the leaving group is too high. If this happens either no reaction will occur or, if a strongly basic nucleophile is used, elimination may occur instead. Some potential leaving groups can also form a bond with the nucleophile, and are thereby further deactivated. This occurs, for example, with phosphonium salts, which only rarely undergo substitution or elimination of PR_3 but are, instead, attacked by the nucleophile at phosphorus (Scheme 4.8).

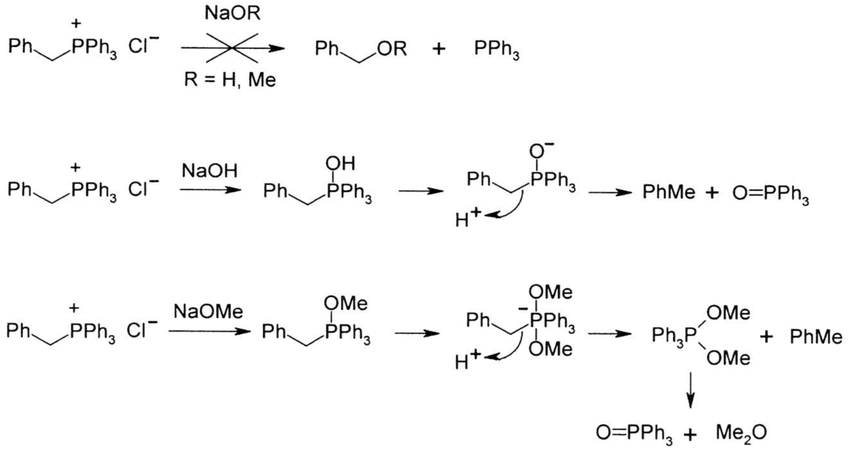

Scheme 4.8. Reaction of benzylphosphonium salts with oxygen nucleophiles [47].

Further poor substrates for nucleophilic substitutions are amines, quaternary ammonium salts, amides, nitriles, and azides; these will usually undergo substitution in particularly reactive substrates only or under harsh reaction conditions. Some rare examples of successful substitutions of these leaving groups are shown in Scheme 4.9. The last reaction is Scheme 4.9 presumably proceeds via an elimination–addition mechanism.

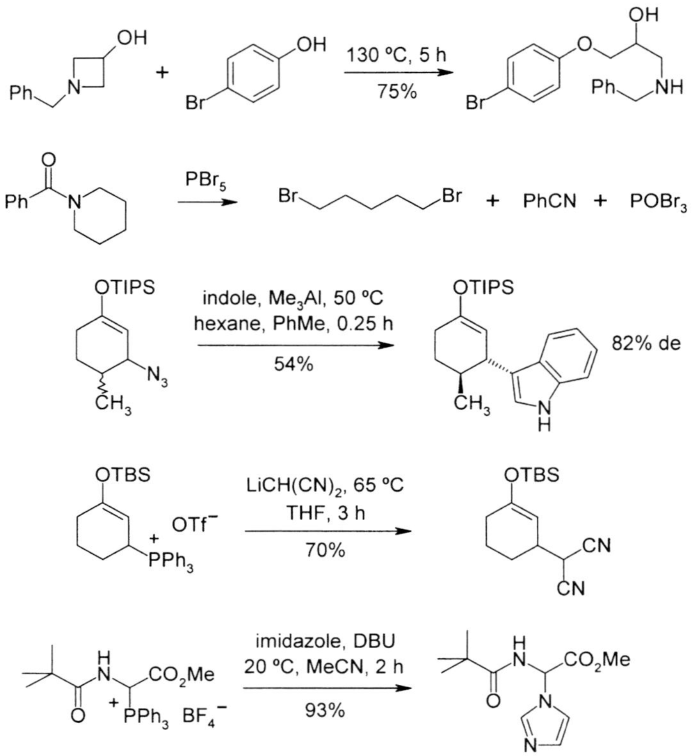

Scheme 4.9. Examples of nucleophilic displacement of unreactive leaving groups [49–53].

4.2.2
Nucleophilic Substitution of Fluoride

Because carbon and fluorine are of similar size, carbon-bound fluorine can undergo hyperconjugation more effectively than any other halogen and can therefore lead more readily to unexpected reactivity [54]. The reactivity of alkyl fluorides towards nucleophiles depends to a large extent on the structure of the alkyl group and on the remaining functional groups. Geminal difluoro and geminal trifluoro compounds are substantially less reactive than monofluoro compounds, because of negative hyperconjugation (Section 3.4).

Simple alkyl fluorides are rather unreactive toward nucleophiles under neutral or basic reaction conditions, as illustrated by the first reaction in Scheme 4.10, and few examples of such reactions have been reported.

These reactions are complicated further by the pronounced tendency of some alkyl fluorides to undergo β-elimination, especially in protic solvents (strong hydrogen-bond formation with fluoride) and in the presence of vicinal electron-withdraw-

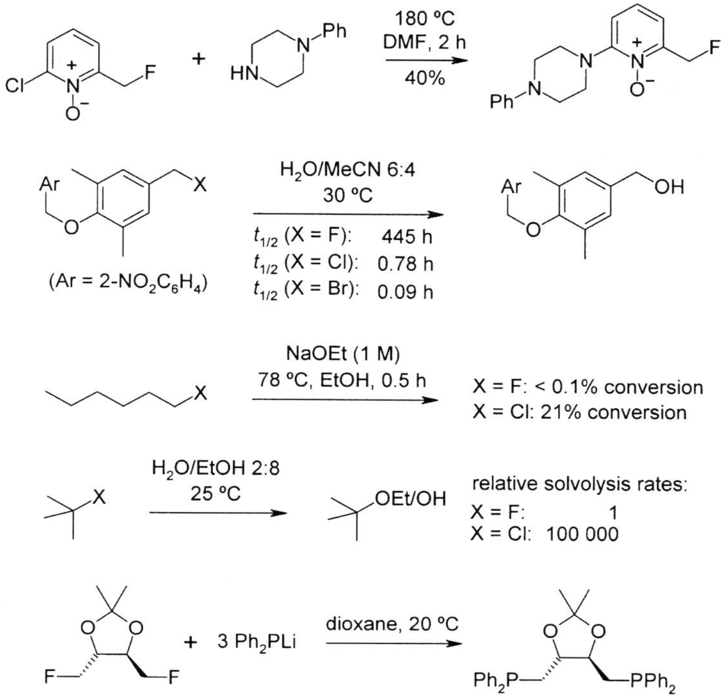

Scheme 4.10. Reactivity of alkyl fluorides toward nucleophiles under neutral or basic reaction conditions [26, 55–58].

ing groups (e.g. FCH_2CH_2Z) [59]. Thus, attempts to prepare β-fluoro ketones or esters from other halides usually lead to the formation of α,β-unsaturated carbonyl compounds. Similarly, 2,2,2-trifluoroethanesulfonates (tresylates) do not always undergo substitution but can eliminate fluoride instead, particularly in protic solvents [60] (Scheme 4.11).

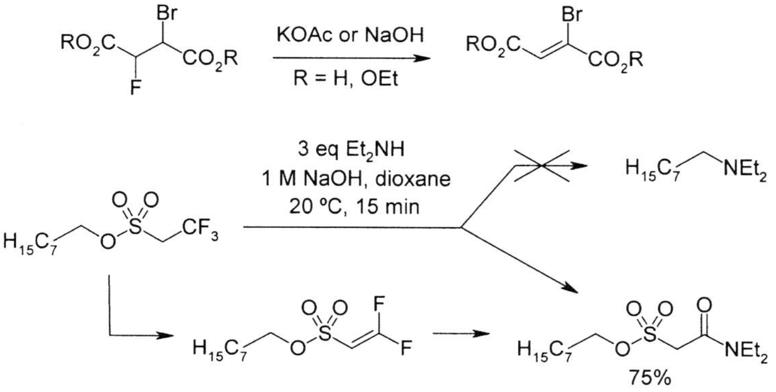

Scheme 4.11. Elimination of hydrogen fluoride from C,H-acidic substrates [61–63].

Some authors occasionally express surprise at the ready β-elimination of fluoride [64] despite the strength of the C–F bond (D_{298} Me–F 115 kcal/mol; D_{298} Me–Cl 83.7 kcal/mol [65]) and despite the lower hyperconjugative energy gain in the interaction of a β-lone pair with σ^*_{C-F} than with the other C–Hal bonds [66]. One should not forget, however, that bond-dissociation enthalpies D refer to *homolytic* bond cleavage (i.e. the formation of radicals, not ions), and do not take into account the energy of solvation of the starting materials and products, which is particularly high for fluoride. Furthermore, fluorine seems to enhance the kinetic acidity of β protons more than that of α protons [67, 68] (and to destabilize β carbocations more than α carbocations [69]). For these reasons β-elimination of fluoride is not an unfavorable process. Some examples of the elimination of hydrogen fluoride from non-C,H-acidic substrates are sketched in Scheme 4.12. Competition experiments in which hydrogen bromide is eliminated more rapidly than hydrogen fluoride have, however, also been reported [70]. When fluoride is used as base for the dehydrohalogenation of 2-haloalkyl fluorides, vinyl fluorides are often obtained [71]; this might, however, be because of prior conversion to a vicinal difluoride by SN2, followed by elimination of HF (second reaction, Scheme 4.12). 2-Fluoroalkyl sulfonates usually also yield vinyl fluorides on treatment with a base [71], as do 1-aryl-1-fluoro-2-haloethanes [72, 73]. As illustrated by the last example in Scheme 4.12, it is difficult to cleave C–F bonds homolytically, because of the strength of this bond.

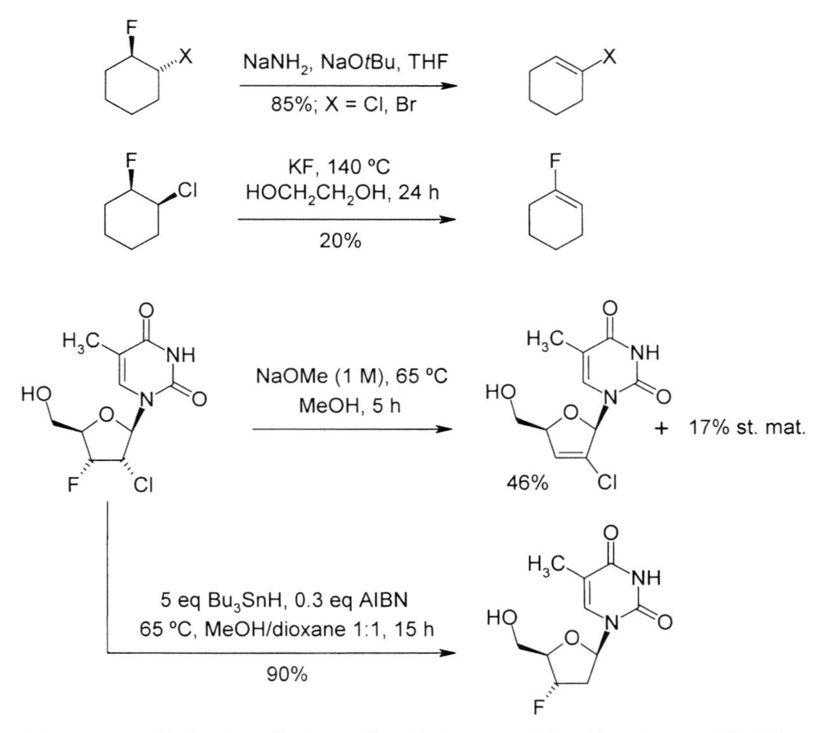

Scheme 4.12. Elimination of hydrogen fluoride from non-C,H-acidic substrates [64, 74].

Under acidic reaction conditions, however, aliphatic fluoride can be readily displaced (Scheme 4.13), especially in alkyl monofluorides. This is probably because of the strong hydrogen bonds formed by fluoride [75]. As illustrated by the second example in Scheme 4.13, for some substrates the fluorides will be even more reactive than the corresponding chlorides or bromides [76]. The first two reactions sketched in Scheme 4.13 can be driven to completion by use of higher temperatures or longer reaction times, to give almost quantitative yields of alkyl iodides.

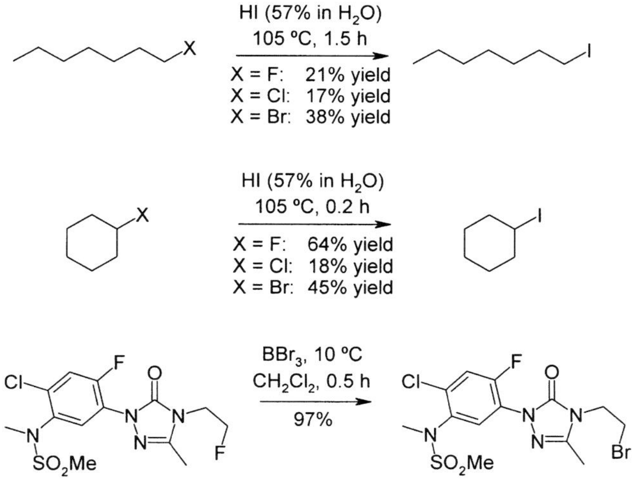

HI (57% in H_2O)
105 °C, 1.5 h
X = F: 21% yield
X = Cl: 17% yield
X = Br: 38% yield

HI (57% in H_2O)
105 °C, 0.2 h
X = F: 64% yield
X = Cl: 18% yield
X = Br: 45% yield

BBr_3, 10 °C
CH_2Cl_2, 0.5 h
97%

Scheme 4.13. Nucleophilic displacement of fluoride under acidic reaction conditions [76, 77].

Lewis acids such as boron trihalides will also form stronger bonds with fluoride than with the other halogens, and can be used to abstract fluoride selectively from halogenated compounds. Some examples of Friedel–Crafts alkylations with fluorohaloalkanes in which only fluoride is displaced are sketched in Scheme 4.14. As shown by the last example, however, hydride migrations can readily occur under such strongly acidic reaction conditions.

PhH, BF_3
20 °C, 4 h
94%

PhH, BF_3
0 °C, 0.25 h
63%

PhH, BF_3
−10 °C, 0.5 h
84%

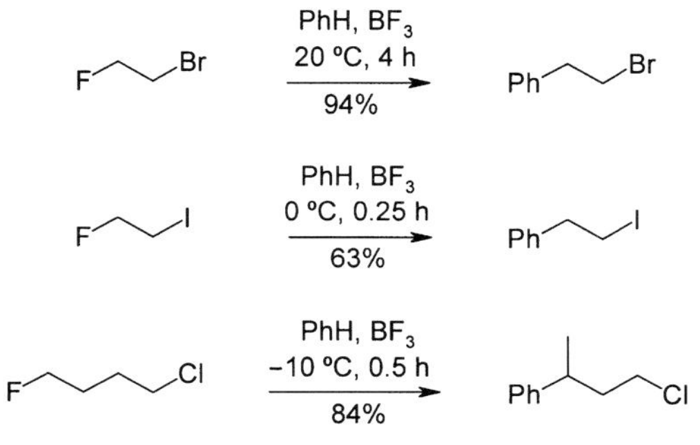

Scheme 4.14. Selective Friedel–Crafts alkylations of benzene with fluorohaloalkanes [78].

4.2.3
Nucleophilic Substitution of Sulfonates

Sulfonates such as mesylates or tosylates are readily prepared from alcohols under mild conditions, and are therefore attractive alternatives to halides as electrophiles. Although sulfonates often undergo clean displacement by nucleophiles, alternative reaction pathways are accessible to these intermediates, which can lead to unexpected results. If the nucleophile used is strongly basic, metalation instead of displacement of the sulfonate can occur. Some potential reactions of such metalated sulfonates include fragmentation into sulfenes and alcoholates, or into sulfinates and carbonyl compounds, or self-alkylation (Scheme 4.15).

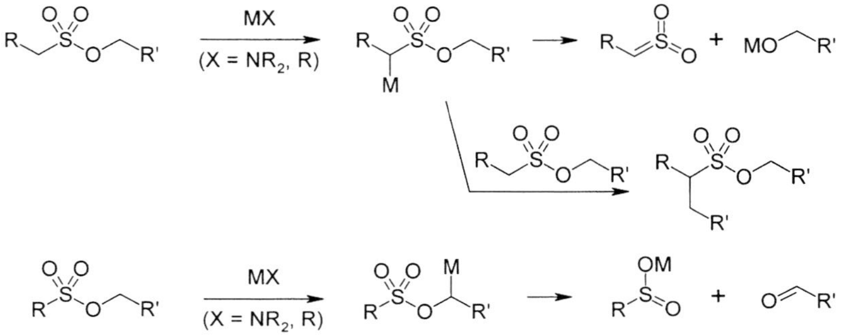

MX
(X = NR_2, R)

MX
(X = NR_2, R)

Scheme 4.15. Reactions of sulfonates with bases.

As mentioned above, 2,2,2-trifluoroethanesulfonates (tresylates) are strongly C,H-acidic and can eliminate fluoride instead of being displaced by a nucleophile (Scheme 4.11), despite the high stability of CF_3 groups. Other alkanesulfonates with a sulfur-bound C,H group can also be deprotonated if the nucleophile is sufficiently basic. Thus, methanesulfonates ($pK_a \approx 25$ in DMSO [79]) can be metalated by treatment with organometallic reagents such as ethynyllithium or BuLi (Scheme 4.16), and deprotonated isopropyl or neopentyl mesylate can be cleanly alkylated at carbon without C–O bond cleavage [80]. Intramolecular aldol-type condensations at mesylates can be mediated by bases as weak as triethylamine [80].

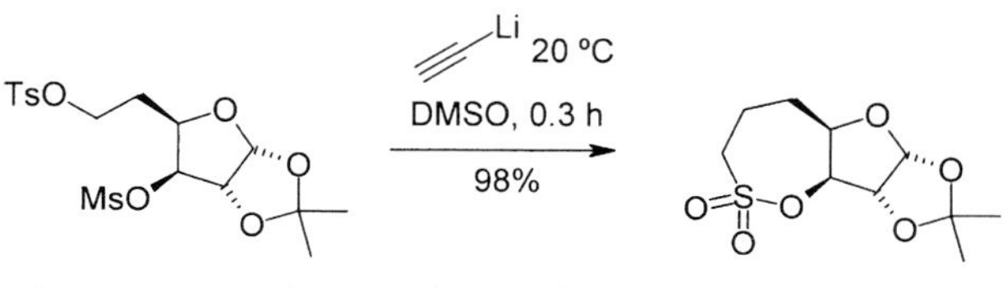

20 °C
DMSO, 0.3 h
98%

Scheme 4.16. Metalation and intramolecular *C*-alkylation of mesylates [81].

Sulfonates with a sulfur-bound C,H group can also eliminate alkoxide when treated with a strong base [82]. Phenylmethanesulfonic esters, for instance, are sufficiently base-labile to be useful protective groups for alcohols (Scheme 4.17).

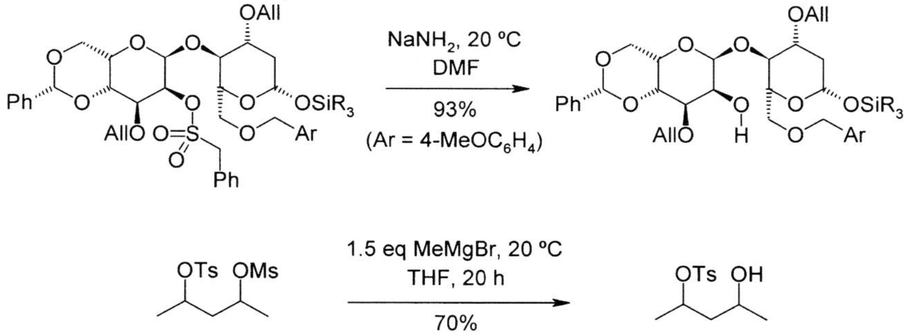

Scheme 4.17. Base-induced cleavage of sulfonates [83, 84].

Some triflates can be deprotonated at the oxygen-bound C,H group when treated with a strong base, and undergo elimination of sulfinate to yield a ketone [85–87]. If the base is an organolithium or Grignard reagent, this will add to the ketone and a tertiary alcohol will finally result (Scheme 4.18).

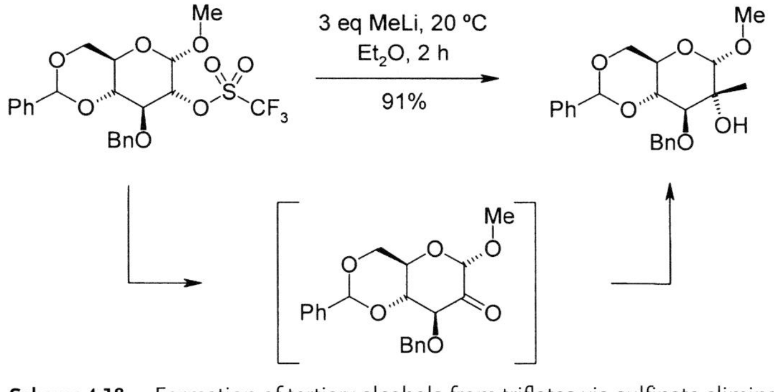

Scheme 4.18. Formation of tertiary alcohols from triflates via sulfinate elimination [88].

Deprotonation of sulfonates by strongly basic nucleophiles can be avoided by using arenesulfonates instead of alkanesulfonates. Arenesulfonates can, however, give rise to another type of side reaction: aromatic nucleophilic substitution. Nitroarenesulfonates are particularly prone to attack by a nucleophile at the arene [89, 90] (Scheme 4.19).

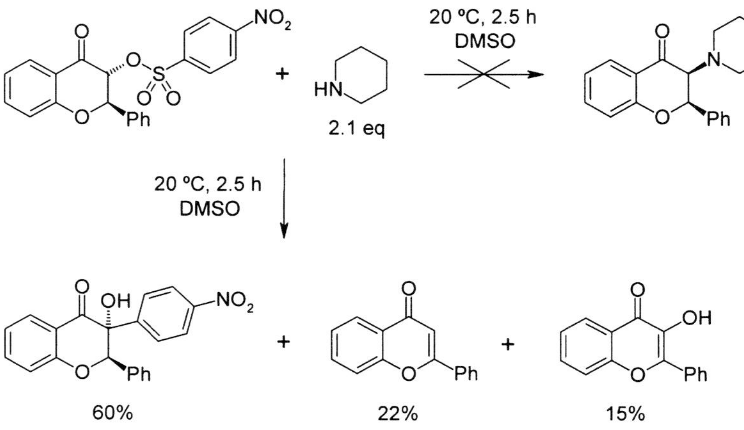

Scheme 4.19. Reaction of a 4-nitrobenzenesulfonate with piperidine [91].

4.3 Structure of the Electrophile

4.3.1 Steric Effects

The reactivity of an electrophile is determined not only by the leaving group, but also to a large extent by steric, inductive, stereoelectronic, and field effects, all of which depend on its precise structure.

As shown in Scheme 4.1, during an SN2 reaction at sp^3-hybridized carbon the nucleophile approaches the electrophile from the side opposite to the leaving group. Hence, if the remaining three substituents at the electrophilic carbon are large, the nucleophile will have to overcome repulsive forces, especially so if the nucleophile also is large. This steric repulsion is believed to be the main reason for the strong dependence of bimolecular substitution rates on the structure of simple alkyl halides (Table 4.1).

Table 4.1. Average relative rates for SN2 reactions of alkyl substrates RX [1, 92, 93].

R	Relative rate	R	Relative rate
Methyl	30	Isobutyl	0.03
Ethyl	1	Neopentyl	0.00001
1-Propyl	0.4	Allyl	40
1-Butyl	0.4	Propargyl	57
2-Propyl	0.025	Benzyl	120

Thus, alkylating agents derived from secondary or β-branched alkyl groups will usually undergo bimolecular substitution only slowly, and SN1 or side reactions

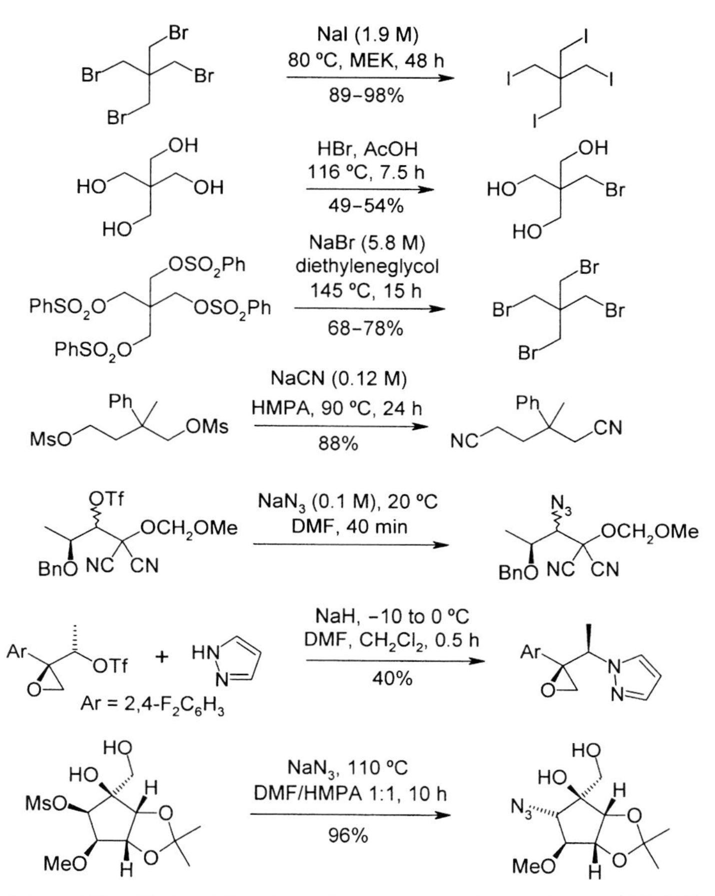

Scheme 4.20. Nucleophilic substitutions at neopentyl-type substrates [96–101].

such as elimination or rearrangement might compete efficiently. For sterically demanding secondary and tertiary substrates, substitution via the SN1 mechanism might become the main reaction pathway, also because of the stabilization of the corresponding carbocations by hyperconjugation and inductive effects [94], and because repulsive interactions between bulky alkyl groups will be smaller in the planar carbocation.

Successful substitutions at neopentyl-type substrates can be performed, but can be accompanied by rearrangements. The best results are obtained with small nucleophiles, for example halides, azide, or cyanide. Some representative examples are shown in Scheme 4.20, to illustrate the reaction conditions required. If electron-rich nucleophiles such as thiolates are used, substitutions at neopentyl derivatives can also occur via SET [95].

Conversion of the hydroxyl group of serine into a leaving group followed by nucleophilic substitution often leads to large amounts of acrylates [102]. In α-alkylated analogs this elimination is no longer possible, but the substrate is neopentyl-like, and substitutions are therefore difficult. Nevertheless, despite these difficulties serine derivatives have proven to be useful intermediates for preparation of α-alkyl α-amino acids. In Scheme 4.21 representative examples of substitutions at such serine-derived substrates are sketched.

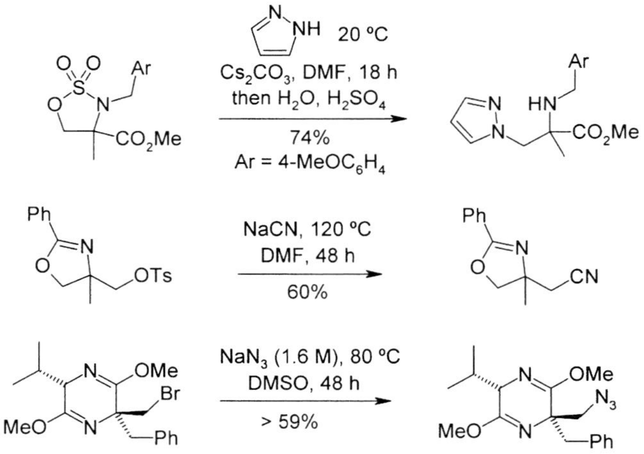

pyrazole, 20 °C, Cs_2CO_3, DMF, 18 h, then H_2O, H_2SO_4; 74%; Ar = 4-$MeOC_6H_4$

NaCN, 120 °C, DMF, 48 h; 60%

NaN_3 (1.6 M), 80 °C, DMSO, 48 h; > 59%

Scheme 4.21. Nucleophilic substitutions at amino acid-derived neopentyl-type substrates [103–105].

During nucleophilic substitutions the hybridization at carbon changes from sp^3 to sp^2; for an electrophile R_3CX this leads to widening of the angle RCR from 109° to 120° in the transition state (SN2) or in the intermediate carbocation (SN1). In cyclic secondary or tertiary electrophiles this required angle widening can lead either to an increase or to a relaxation of strain (Scheme 4.22).

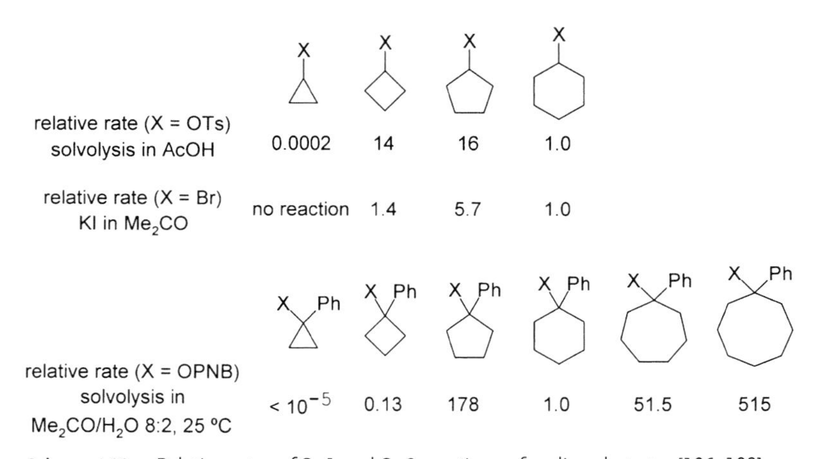

Scheme 4.22. Relative rates of SN1 and SN2 reactions of cyclic substrates [106–108].

If the ring becomes more strained in the transition state nucleophilic substitution should proceed more slowly than with similar, non-cyclic electrophiles. Thus, cyclopropyl derivatives are highly resistant towards nucleophilic substitution because the RCR angle is fixed at 60°, and only rarely can products of an SN2 reaction at cyclopropyl derivatives be obtained [109]. Instead, allyl derivatives are usually the main products (Scheme 4.23).

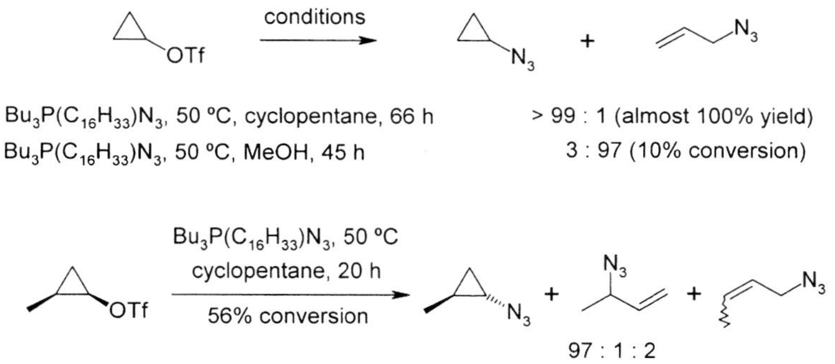

OTf — conditions → N_3 + N_3

$Bu_3P(C_{16}H_{33})N_3$, 50 °C, cyclopentane, 66 h — > 99 : 1 (almost 100% yield)

$Bu_3P(C_{16}H_{33})N_3$, 50 °C, MeOH, 45 h — 3 : 97 (10% conversion)

OTf — $Bu_3P(C_{16}H_{33})N_3$, 50 °C, cyclopentane, 20 h, 56% conversion → N_3 + N_3 + N_3

97 : 1 : 2

Scheme 4.23. Reaction of cyclopropyl triflates with azide [109].

4.3.2 Conjugation

The transition state of an SN2 reaction can be stabilized or destabilized by electronic effects originating from the substituents at the reaction center. As shown in Table 4.1, allylic or benzylic substrates undergo significantly faster SN2 reactions than similar, fully saturated substrates. This acceleration is believed to be due to conjugation of the π-system with the sp^2-hybridized carbon atom in the transition state. Accordingly, constrained benzylic substrates in which such conjugation is not possible are not as reactive as similar, unconstrained systems. Thus, the benzylic C–S bonds in sulfonium salt **A** (Scheme 4.24) are highly unreactive, and this compound reacts with nucleophiles mainly as an ethylating reagent. In compound **A** cleavage of the benzylic C–S bond by a nucleophile has to proceed via a transition state in which the Nuc–C–SR_2^+ trajectory is almost parallel to the arene and no conjugation with the aromatic π-system is possible. The non-cyclic sulfonium salt **B**, on the other hand, can adopt a conformation in which the Nuc–C–SR_2^+ trajectory is orthogonal to the plane of one phenyl group. Salt **B** reacts therefore exclusively as a powerful benzylating reagent.

Conjugation in allylic or benzylic electrophiles can also stabilize the corresponding carbocations and thereby facilitate substitution by the SN1 mechanism, as shown by the relative solvolysis rates given in Table 4.2.

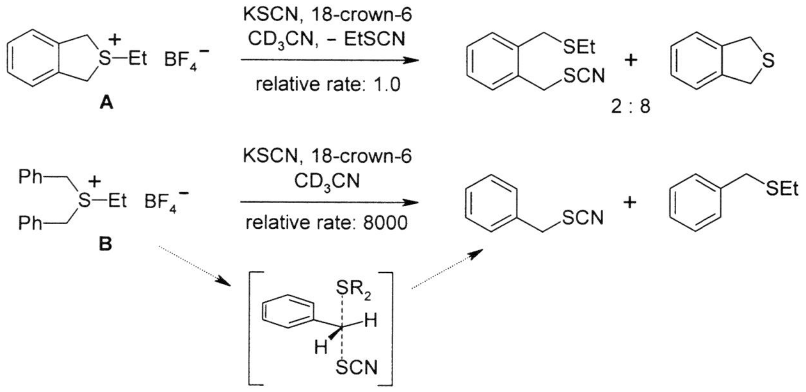

Scheme 4.24. Reactivity of benzyl ethyl sulfonium salts toward thiocyanate [110].

Table 4.2. Relative rates for the SN1 reaction of tosylates R–OTs with ethanol at 25 °C [1].

R	Relative rate	R	Relative rate
Ethyl	0.26	$PhCH_2$	100
2-Propyl	0.69	Ph_2CH	~10^5
Allyl	8.6	Ph_3C	~10^{10}

The effect of substituents at the arene on the reactivity of benzylic electrophiles depends on the type of nucleophile used [111]. Benzylic substrates with electron-withdrawing ring substituents react more rapidly with anionic nucleophiles in bimolecular substitutions than comparable unsubstituted electrophiles, but slower with neutral nucleophiles [111–113]. Because donor-substituted benzylic systems tend to undergo rapid SN1 reactions, the reactivity of benzylic electrophiles toward negatively charged nucleophiles usually reaches a minimum for unsubstituted benzylic electrophiles and increases on substitution with either electron-withdrawing or electron-donating substituents [114–117] (Scheme 4.25).

Donor-substituted benzylic electrophiles usually undergo clean SN2 or SN1 reactions to yield the expected products in high yield. The favorable electronic effect of the aryl group seems to dominate the reaction, and sterically demanding substitu-

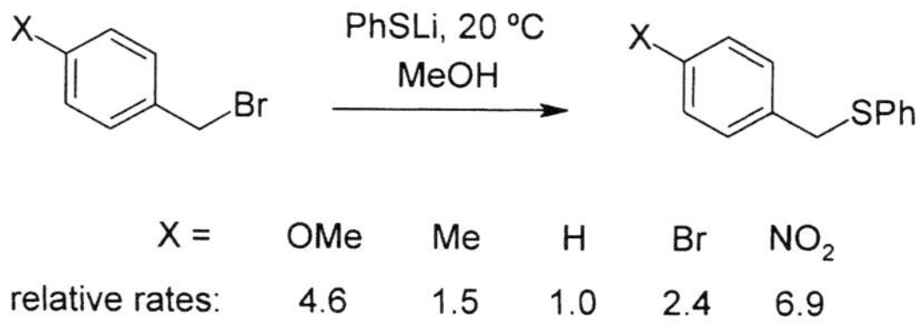

Scheme 4.25. Relative rates of the S-benzylation of thiophenol with benzyl bromides [115].

ents in the *ortho* positions usually have no strong rate-diminishing effect. Even 2,6-diphenyl or 2,6-di-*tert*-butylbenzyl derivatives undergo high-yielding benzylic substitutions (Scheme 4.26). Under acidic conditions, however, reaction of the benzylic cation with the *ortho* substituents can occur (Scheme 4.26).

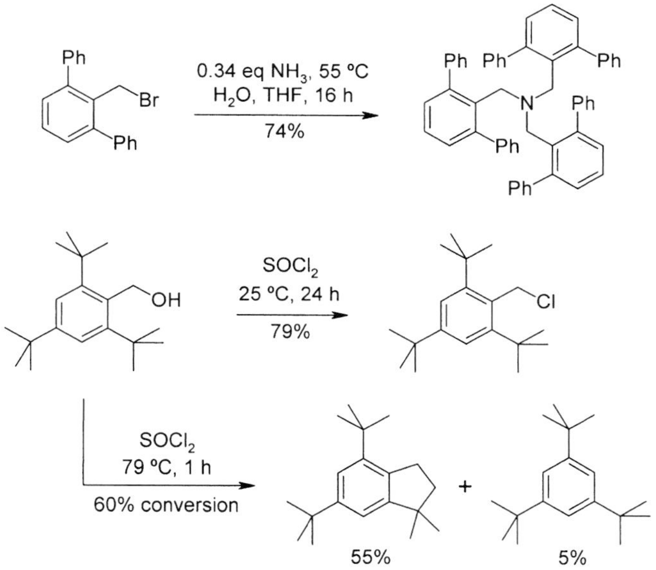

Scheme 4.26. Nucleophilic substitution at sterically demanding benzylic substrates [118, 119].

Benzylic electrophiles bearing electron-withdrawing groups at the arene do not always yield the expected products of nucleophilic substitution on treatment with a nucleophile. One important side reaction is the dimerization of these compounds to yield 1,2-diarylethenes (stilbenes). This dimerization does not require such highly activated systems as the example sketched in Scheme 4.28, but can even occur with, for example, 2- or 4-nitrobenzyl chloride [120, 121]. The latter compounds are converted into the corresponding stilbenes by treatment with KOH in ethanol [120]. Diarylmethyl halides behave similarly and can yield tetraarylethenes on treatment with a base. These reactions presumably proceed via the mechanism sketched in Scheme 4.27, in which the amphiphilic character of the nitro group plays a decisive role (metalated nitroalkanes or 4-nitrobenzyl derivatives can act as nucleophiles **and** as electrophiles).

If the arenes are very electron-deficient direct attack of the nucleophile at the arene might also compete with displacement of the benzylic leaving group, to yield complex structures such as that shown in Scheme 4.28. 2-Nitrobenzyl halides can also react with amines to yield, instead of simple products of nucleophilic substitution, 2*H*-indazoles [122] (Scheme 4.28). 4-Nitrobenzyl halides, however, yield the expected benzyl amines on treatment with amines [123].

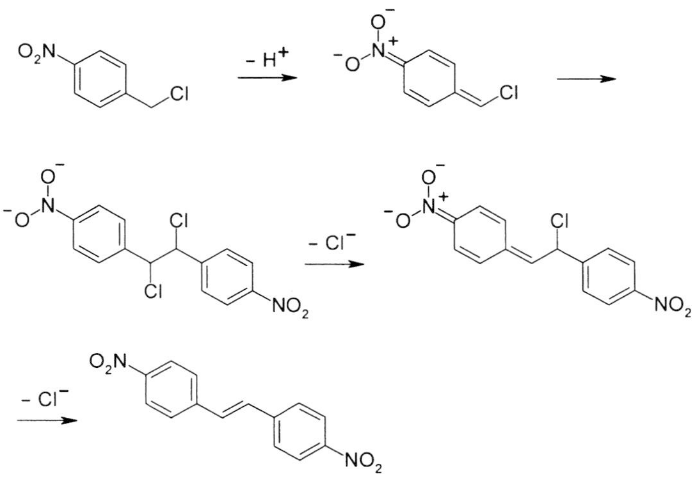

Scheme 4.27. Dimerization of nitrobenzyl halides.

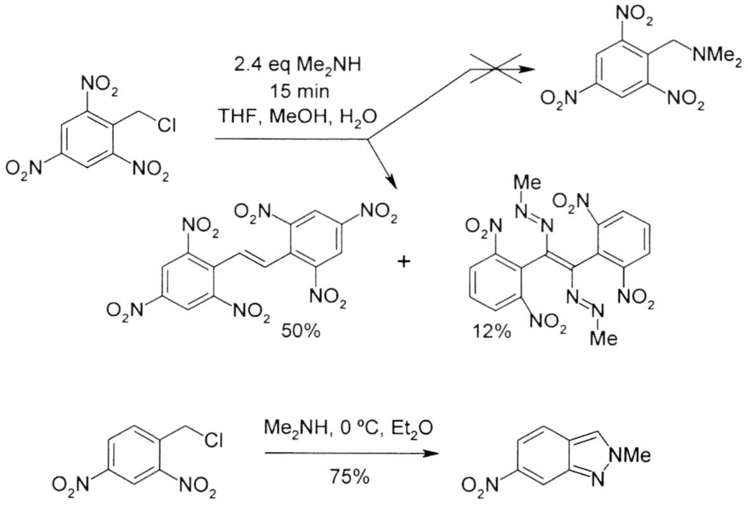

Scheme 4.28. Reactivity of nitrobenzyl halides toward amines [122, 124].

Other benzylic electrophiles which can lead to unexpected products are 1,2- or 1,4-bis(halomethyl)benzenes. On treatment with a nucleophile, oxidation of the nucleophile instead of nucleophilic substitution may occur, followed by the formation of highly reactive quinodimethanes, which can either oligomerize or undergo addition or cycloaddition reactions (Scheme 4.29). The outcome of these reactions can, however, be controlled by choosing the right conditions, as demonstrated by the numerous report of successful SN2 reactions at 1,2- or 1,4-bis(halomethyl)benzenes (see, e.g., Ref. [125]).

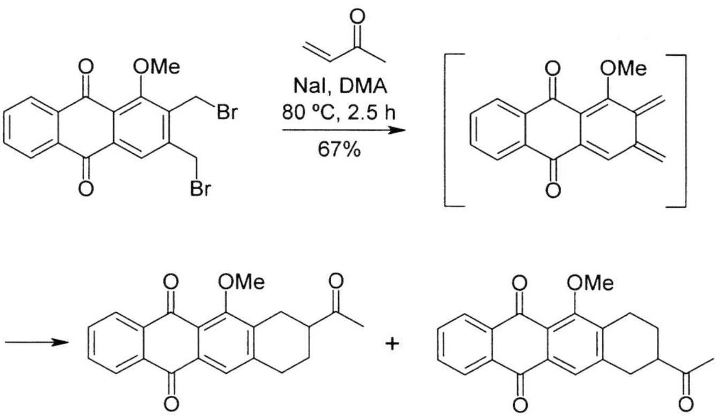

Scheme 4.29. Formation of quinodimethanes from 1,2-bis(bromomethyl)benzenes and iodide [126].

4.3.3 Electrophiles with α-Heteroatoms

4.3.3.1 α-Heteroatoms with Lone Electron Pairs

The rates of SN1 and SN2 reactions are usually strongly enhanced when atoms with an unshared electron pair are directly attached to the reaction center, as in α-halomethyl ethers, thioethers, or amines (Scheme 4.30). Only the halogens do not lead to an enhancement of SN2 reactivity, but to inhibition of bimolecular substitution reactions (Scheme 4.30). In SN1 reactions, however, α-halogens can both increase or reduce the rate of substitution [127].

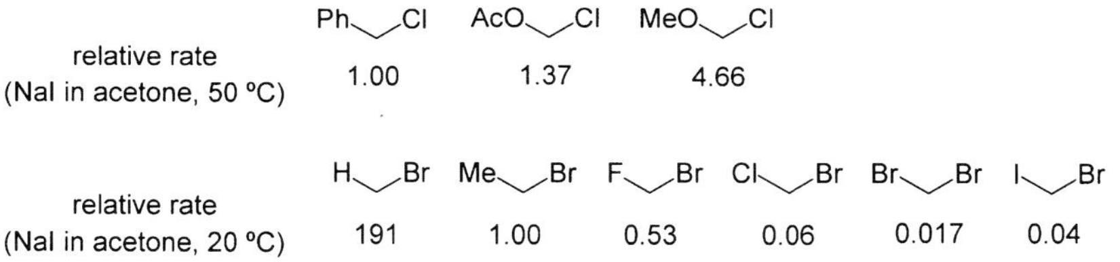

	PhCH$_2$Cl	AcOCH$_2$Cl	MeOCH$_2$Cl
relative rate (NaI in acetone, 50 °C)	1.00	1.37	4.66

	HCH$_2$Br	MeCH$_2$Br	FCH$_2$Br	ClCH$_2$Br	BrCH$_2$Br	ICH$_2$Br
relative rate (NaI in acetone, 20 °C)	191	1.00	0.53	0.06	0.017	0.04

Scheme 4.30. Relative rates of substitution at α-substituted methyl derivatives [112, 128].

An unshared electron pair can usually only assist nucleophilic displacements if this lone pair and the antibonding (σ^*) orbital between carbon and the leaving group are either antiperiplanar or synperiplanar. If such a conformation cannot be achieved, α-heteroatoms can even reduce the rate of SN2 or SN1 reactions. Scheme 4.31 shows some examples of conformationally constrained, tertiary tosylates for which hyperconjugation between the heteroatom lone pairs and the antibonding C–OTs orbital is not possible [129]. Thus, although α-alkoxyalkyl sulfonates usually undergo SN1 reactions much more rapidly than simple *tert*-alkyl sulfonates, compound **2** is much less reactive than its carbon analog **1**. The α-aminoalkyl tosylate **3**, however, reacts faster than **1**, indicating that despite the unfavorable geometry substantial stabilization of the transition state leading to the corresponding carbocation is still possible in this substrate [129].

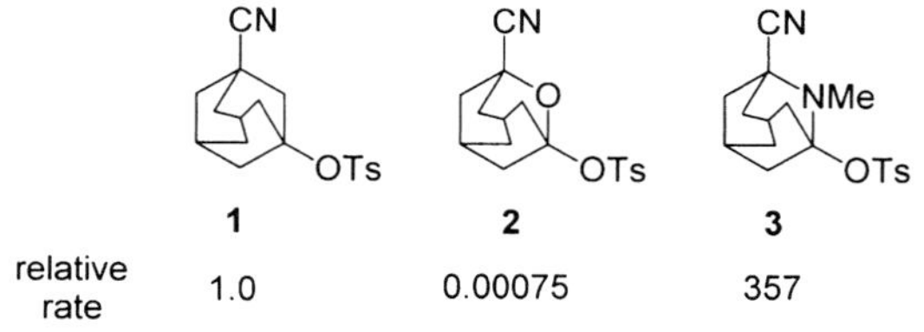

	1	2	3
relative rate	1.0	0.00075	357

Scheme 4.31. Relative rates of solvolysis of tosylates in EtOH/H_2O 8:2 at 25 °C [129].

4.3.3.2 α-Silicon and α-Tin Electrophiles

(Trialkylsilyl)methyl halides and related electrophiles (R_3SiCH_2X) have rather peculiar reactivity which would be difficult to infer by simple comparison with carbon analogs. Although silicon is less electronegative than carbon, and a trimethylsilyl group is therefore a better σ-donor than a *tert*-butyl group, carbocations are strongly destabilized by silyl groups relative to alkyl groups (but stabilized relative to hydrogen) [130–132]. This destabilization is because of the weak hyperconjugation between σ_{C-Si} bonds and the empty *p* orbital on carbon (orbital size and energy mismatch; silicon does not readily form π bonds with carbon or other second-period elements) and the electrostatic destabilization brought about by the partial positive charge on silicon (bound to four more electronegative atoms). For these reasons SN1 reactions of electrophiles such as Me_3SiCR_2X proceed more slowly than with comparable alkyl derivatives (Scheme 4.32).

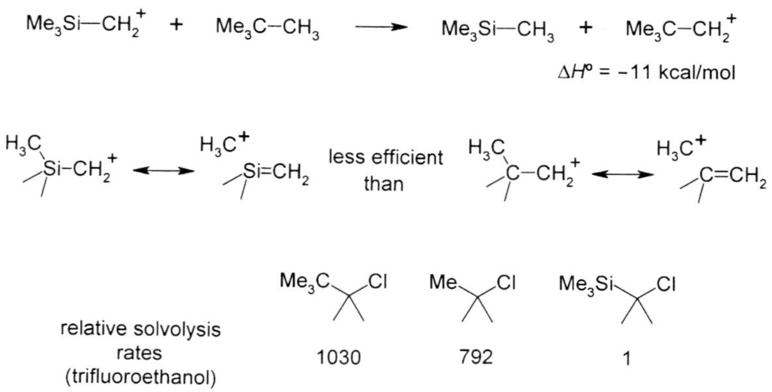

$$Me_3Si{-}CH_2^+ + Me_3C{-}CH_3 \longrightarrow Me_3Si{-}CH_3 + Me_3C{-}CH_2^+ \quad \Delta H^o = -11 \text{ kcal/mol}$$

H_3C–Si–CH_2^+ ⟷ H_3C^+ Si=CH_2 less efficient than H_3C–C–CH_2^+ ⟷ H_3C^+ C=CH_2

	Me_3C–CMe_2–Cl	Me–CMe_2–Cl	Me_3Si–CMe_2–Cl
relative solvolysis rates (trifluoroethanol)	1030	792	1

Scheme 4.32. Destabilization of carbocations by R_3Si groups [130, 131].

Because of the destabilization of α-trialkylsilyl carbocations, electrophiles such as Me_3SiCH_2X undergo rearrangement-free SN2 reactions rather than SN1 reactions, even under conditions where neopentyl derivatives (Me_3CCH_2X) would undergo solvolysis and rearrangement [131]. These SN2 reactions at Me_3SiCH_2X proceed much more rapidly than at comparable neopentyl derivatives (Scheme 4.33), because the long C–Si bonds reduce steric crowding. Electrophiles such as Me_3SiCH_2I can, for instance, be used to *N*-alkylate amines (DMSO, 100 °C, 2–3 h [133]). Similarly, α-trialkylstannyl halides or sulfonates undergo clean SN2 reactions under relatively mild reaction conditions (Scheme 4.33).

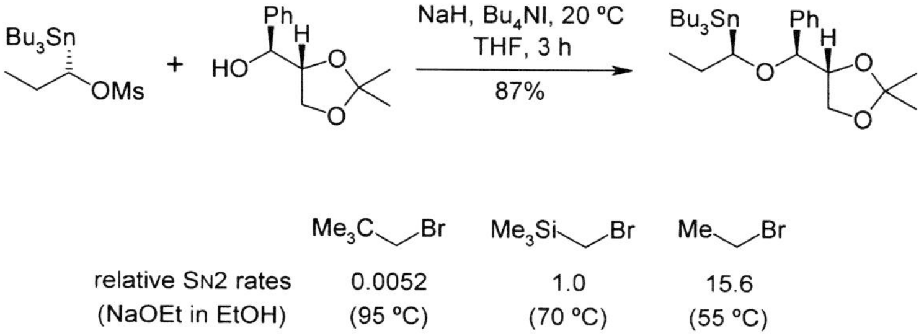

NaH, Bu_4NI, 20 °C
THF, 3 h
87%

	Me_3C–CH_2Br	Me_3Si–CH_2Br	Me–CH_2Br
relative SN2 rates	0.0052	1.0	15.6
(NaOEt in EtOH)	(95 °C)	(70 °C)	(55 °C)

Scheme 4.33. SN2 reactions at α-trialkylsilyl and α-trialkylstannyl electrophiles [134–136].

For electrophiles such as Me_3SiCH_2X strong ground-state destabilization has been observed for X = 4-nitrobenzoate [130]. For X = halide, on the other hand, this ground-state destabilization is significantly smaller, and it may therefore be advisable to choose carboxylates or sulfonates as leaving groups when alkylations with α-silyl-substituted electrophiles are to be performed.

4.3.3.3 α-Boron Electrophiles

The reactivity of electrophiles with an electron-deficient heteroatom such as boron in the position α to the reaction center depends on whether SN1 or SN2 reactions are being considered. By analogy with α-halo ketones or electron-poor benzyl or allyl halides rate enhancement for bimolecular substitutions would be expected, and is indeed found. Thus, α-chloroalkyl boronic esters undergo halide exchange in non-ionizing solvents at significantly higher rates than comparable boron-free compounds [137–139]. These and other bimolecular nucleophilic substitutions at α-haloalkyl boronic esters usually proceed via initial attack of the nucleophile at boron, followed by intramolecular halide displacement [140]. In SN1 reactions, on the other hand, the $B(OR)_2$ group has a slight rate-diminishing effect if compared with a methyl group but a rate-enhancing effect relative to hydrogen (Scheme 4.34), and thus behaves similarly to the SiR_3 group. Amines [141] and carbamates [142] can also be *N*-alkylated with α-haloalkyl boronates, but the products resulting from alkylation of amines are not stable and can rearrange to aminoboronates $R_2NB(OR)_2$ ([141] and references cited therein).

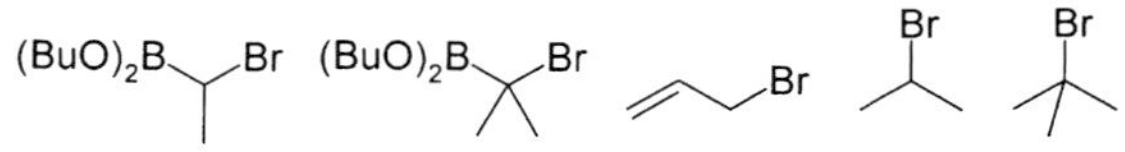

	$(BuO)_2B$–CH(Me)–Br	$(BuO)_2B$–CMe₂–Br	allyl bromide	2-bromopropane	tert-butyl bromide
relative rate SN2 (NaI in Me_2CO)	6000	1600	4000	1.0	
relative rate SN1 (H_2O/EtOH 1:1)		100		1.0	10000

Scheme 4.34. Relative rates of SN1 and SN2 reactions of α-bromo boronic esters [138, 140].

4.3.3.4 α-Nitro Electrophiles

α-Nitroalkyl halides do not usually undergo attack at carbon with simultaneous displacement of the halide, but react as halogenating or oxidizing reagents instead, because of the ability of nitroalkanes to act as leaving groups [143]. Because of the high acidity of nitroalkanes, the precise reaction conditions can have a decisive influence on the course of the reaction. Under basic reaction conditions the nitroalkane will be deprotonated and usually become even less susceptible to nucleophilic attack. As shown in Scheme 4.35, reactions of bromonitromethane with nucleophiles can, nevertheless, yield products of nucleophilic halide displacement, but

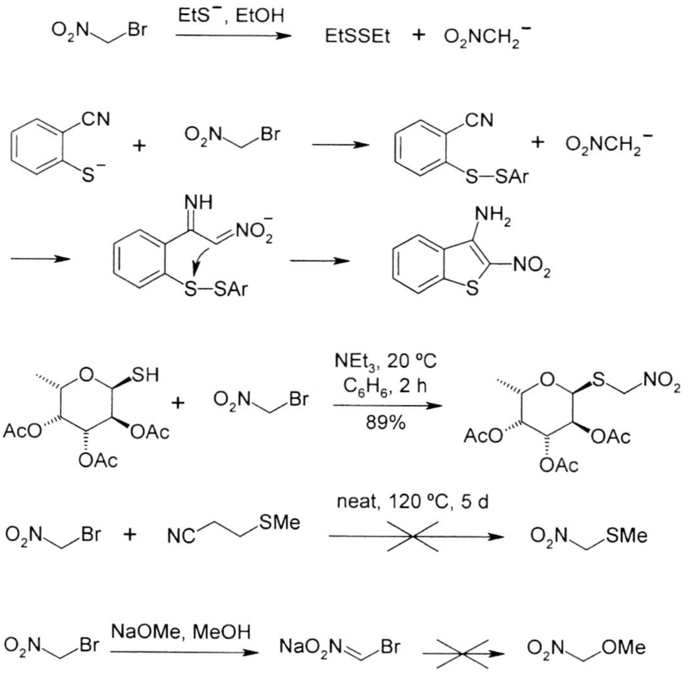

O_2N–CH_2–Br —(EtS⁻, EtOH)→ EtSSEt + $O_2NCH_2^-$

Scheme 4.35. Reactions of 1-halo-1-nitroalkanes with sulfur and oxygen nucleophiles [144–146]. Ar = 2-cyanophenyl.

these reactions might proceed via a more intricate mechanism than a simple bimolecular substitution.

The oxidizing power of bromonitromethane is illustrated further by its reactions with iodide and phosphines [145], which do not yield products of nucleophilic substitution but the oxidized nucleophiles only (Scheme 4.36). Phosphites are also oxidized by bromonitromethane [145], but with higher homologs, for example 1-bromo-1-nitroethane, products of nucleophilic substitution can be isolated [147] (Scheme 4.36).

The fact that products of type Nu–CR_2–NO_2 can be unstable, in particular if the group Nu has an unshared electron pair, and that nitroalkanes are efficient one-electron oxidants with a pronounced tendency to undergo radical reactions, add further uncertainty to the outcome of these reactions. Thus, the reaction of amines with bromonitromethane, for instance, is assumed to yield nitromethylamines initially; these, however, decompose to *N*-nitrosoamines [148] (Scheme 4.36).

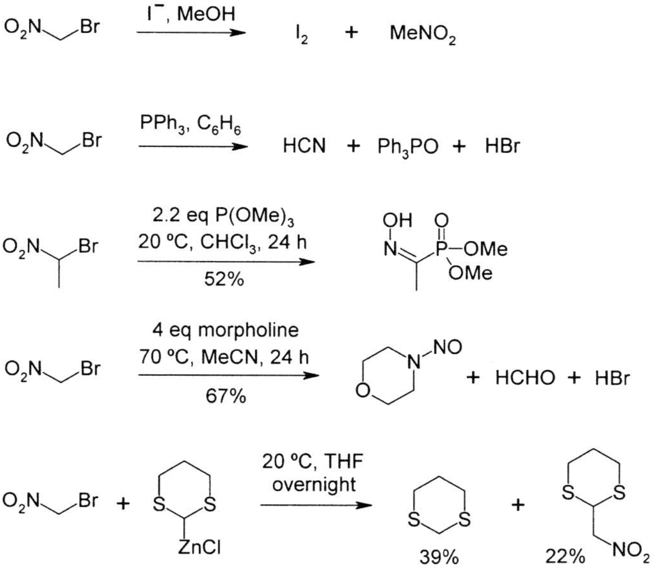

Scheme 4.36. Reactions of 1-halo-1-nitroalkanes with iodide [145] and with phosphorus [145, 147], nitrogen [148], and carbon [149] nucleophiles.

4.3.3.5 Electrophiles with More than one α-Heteroatom

The reactivity of methanes substituted with more than two heteroatoms is not easy to predict. Tetrahalomethanes react with nucleophiles usually at the halogen with displacement of X_3C^-. $CHCl_3$ or $PhCCl_3$, however, undergo clean substitution with strong nucleophiles. Treatment of CCl_4 or $CHCl_3$ with alcoholates, for instance, yields trialkyl orthoformates ($HC(OR)_3$) in both cases [150, 151] (Scheme 4.37). The first step of the reaction of CCl_4 with alcoholates is probably reduction of CCl_4 by the alcoholate to yield $CHCl_3$ and ethyl hypochlorite or diethyl peroxide.

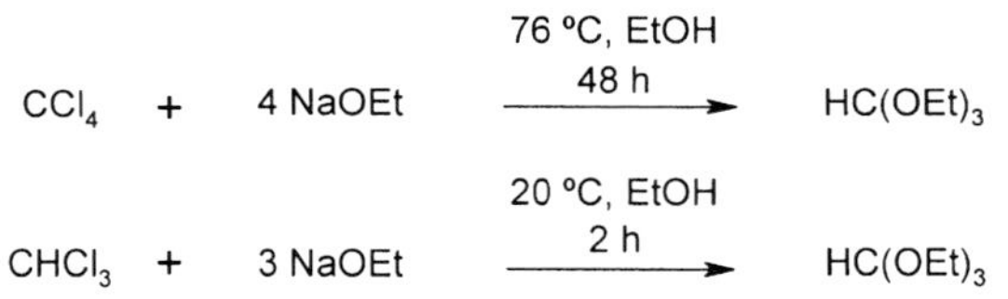

$$CCl_4 + 4\ NaOEt \xrightarrow[48\ h]{76\ °C,\ EtOH} HC(OEt)_3$$

$$CHCl_3 + 3\ NaOEt \xrightarrow[2\ h]{20\ °C,\ EtOH} HC(OEt)_3$$

Scheme 4.37. Preparations of triethyl orthoformate [151, 152].

Electrophiles which enable four substitutions at the same carbon atom include Cl_3CNO_2 and Cl_3CSCl; both compounds yield tetraalkyl orthocarbonates on treatment with alcoholates [150]. In polyhalomethanes selective substitutions of the most reactive halide are possible [153] but not always easy to perform (Scheme 4.38).

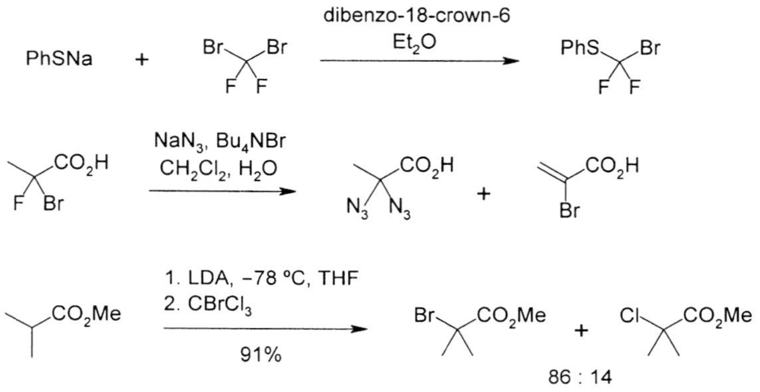

PhSNa + Br_2CF_2 (dibenzo-18-crown-6, Et_2O) → $PhSCF_2Br$

NaN_3, Bu_4NBr, CH_2Cl_2, H_2O

1. LDA, −78 °C, THF
2. $CBrCl_3$

91%

86 : 14

Scheme 4.38. Nucleophilic substitutions at polyhalomethanes [43, 70, 154].

4.3.4
Electrophiles with β-Heteroatoms

Two opposing effects mainly modulate the reactivity of electrophiles with a heteroatom (N, O, S, Hal) in the β position – enhancement of SN1 reactivity by neighboring-group participation (Section 4.3.6) and a reduction of SN1 and SN2 reactivity by the inductive, electron-withdrawing effect through σ bonds. The second of these two effects has only a weak impact on SN2 reactions, and electrophiles with β-heteroatoms usually have a SN2 reactivity similar to that of the related, heteroatom-free electrophiles. In SN1 reactions, however, neighboring-group participation can significantly enhance the reactivity of electrophiles.

1,2-Dihaloethanes are usually slightly less reactive toward nucleophiles than, for example, the corresponding propyl halides [155]. Although several successful nucleophilic substitutions at 1,2-dichloroethane or 1,2-dibromoethane have been performed (Section 10.6), several examples of poor results with these electrophiles have also been reported (Scheme 4.39).

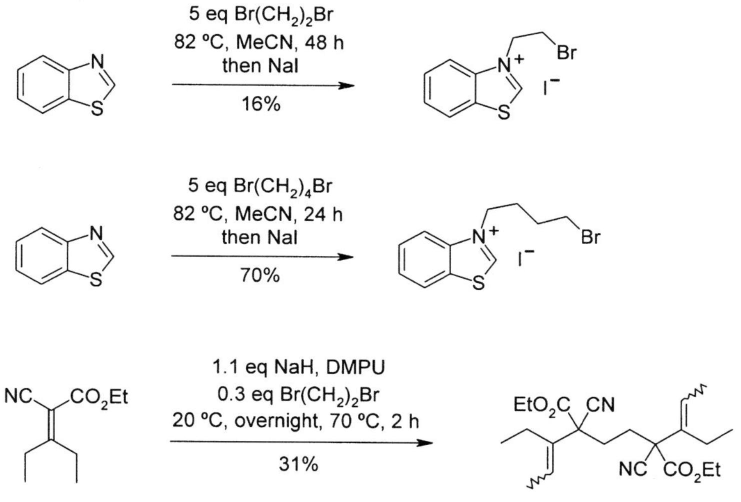

5 eq $Br(CH_2)_2Br$
82 °C, MeCN, 48 h
then NaI
16%

I^- Br

5 eq $Br(CH_2)_4Br$
82 °C, MeCN, 24 h
then NaI
70%

I^- Br

NC CO_2Et

1.1 eq NaH, DMPU
0.3 eq $Br(CH_2)_2Br$
20 °C, overnight, 70 °C, 2 h
31%

EtO_2C CN NC CO_2Et

Scheme 4.39. Alkylations with 1,2-dihaloethanes and 1,4-dihaloalkanes [155, 156].

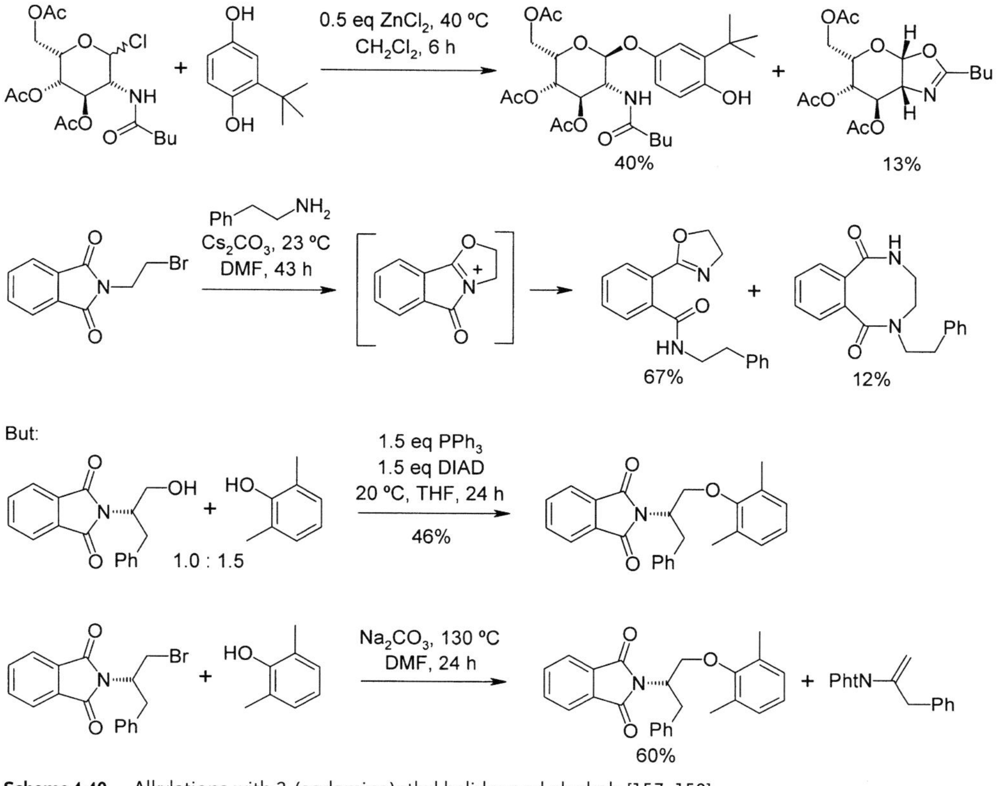

OAc Cl AcO NH AcO O Bu + OH OH

0.5 eq $ZnCl_2$, 40 °C
CH_2Cl_2, 6 h

OAc O AcO NH AcO O Bu OH
40%

+ OAc H O Bu AcO N H AcO
13%

O N Br O

Ph NH$_2$
Cs_2CO_3, 23 °C
DMF, 43 h

O N$^+$ O

O N O HN Ph
67%

+ O H N N Ph O
12%

But:

O N OH O Ph + HO
1.0 : 1.5

1.5 eq PPh_3
1.5 eq DIAD
20 °C, THF, 24 h
46%

O N O O Ph

O N Br O Ph + HO

Na_2CO_3, 130 °C
DMF, 24 h

O N O O Ph
60%

+ PhtN Ph

Scheme 4.40. Alkylations with 2-(acylamino)ethyl halides and alcohols [157–159].

Other problematic electrophiles are 2-(acylamino)ethyl halides and related compounds. Although numerous successful nucleophilic substitutions with such substrates have been described in the literature, occasionally a side reaction becomes dominant. If the leaving group is hard or if the reaction conditions chosen are conducive to the formation of carbocations, intramolecular *O*-alkylation of the electrophile will lead to the formation of oxazolines (Scheme 4.40). This cyclization can sometimes be avoided by choosing a softer leaving group.

4.3.5
Electrophiles with α-Electron-withdrawing Groups

Substitutions at α-nitro (Section 4.3.3.4) and nitrobenzyl electrophiles (Section 4.3.2) have been discussed above.

The rate of substitutions proceeding via free carbocations should decrease when electron-withdrawing groups are linked to the central carbon, and accelerated by carbocation-stabilizing groups. This is, however, not always observed. α-Carbonyl groups, for instance, sometimes have only a small rate-lowering effect, and can even enhance the rate of some SN1 reactions [160, 161] (last row, Scheme 4.41). This

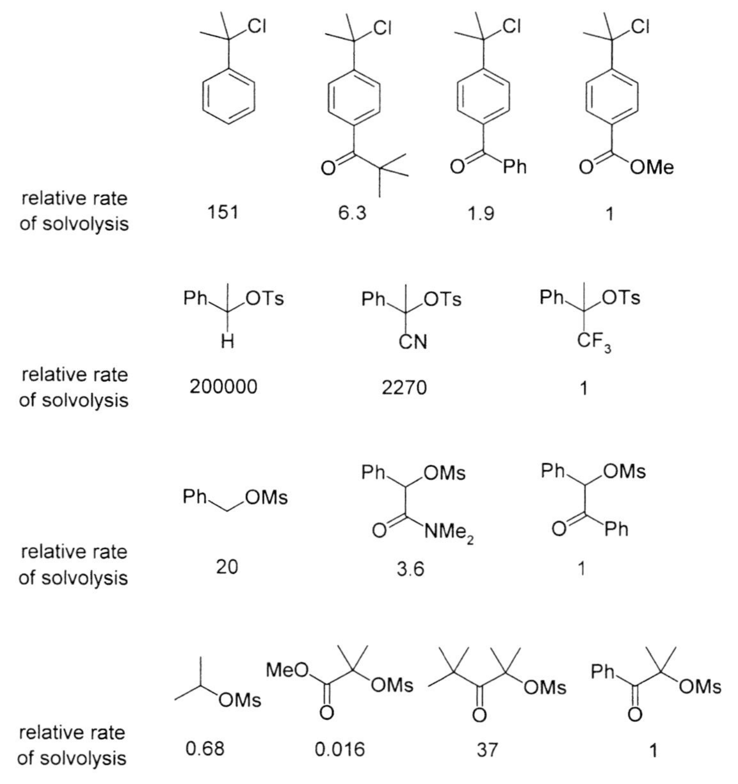

Scheme 4.41. Relative rate of SN1 reactions of α-acceptor-substituted electrophiles [160, 162, 163].

might be because of conjugation of the carbonyl group with the carbocation [161] (R_2C^+–C=O ↔ R_2C=C–O^+) or ground-state destabilization of the electrophile [162] (e.g. by steric crowding or dipole–dipole interaction). Usually, however, α-carbonyl groups reduce the rate of SN1 reactions significantly [161].

For SN2 reactions the effect of electron-withdrawing groups is less easy to predict than for SN1 reactions. α-Halo ketones, α-halo nitriles, or haloacetic acid derivatives undergo bimolecular substitutions at much higher rates than unfunctionalized alkyl halides (Scheme 4.42).

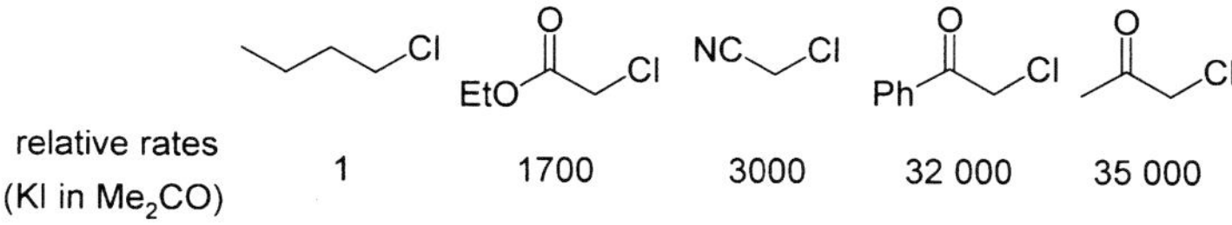

relative rates (KI in Me_2CO): 1, 1700, 3000, 32 000, 35 000

Scheme 4.42. Relative rates of SN2 reactions of electrophiles substituted with electron-withdrawing groups [164, 165].

This effect is believed to result from an enolate-type transition state [166] or from a reaction mechanism in which the nucleophiles initially add to the carbonyl group and then undergo intramolecular 1,2-migration with simultaneous displacement of the leaving group [167] (Scheme 4.43). The latter proposal would also explain the strong decrease of reactivity of phenacyl derivatives on introduction of *ortho* substituents on the arene. In these *ortho*-substituted phenacyl derivatives addition of nucleophiles to the carbonyl group would lead to significant steric crowding and is therefore more difficult than in unsubstituted aryl ketones [168].

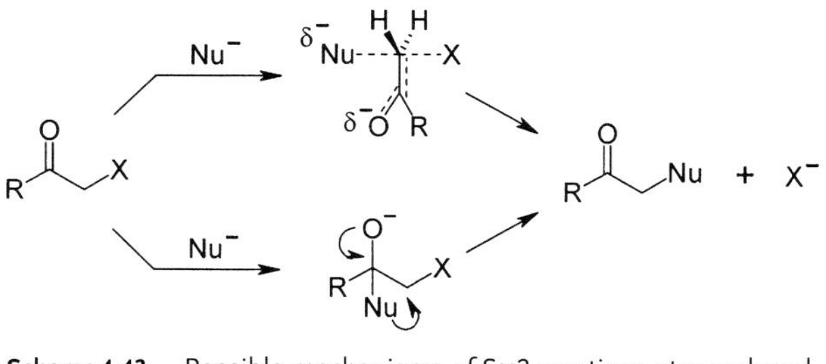

Scheme 4.43. Possible mechanisms of SN2 reactions at α-carbonyl electrophiles.

α-Sulfonyl [165] or α-trifluoromethyl halides [169] are, in contrast, much less reactive in SN2 reactions than comparable, unsubstituted alkyl halides (Schemes 4.44 and 4.45). Fluorinated electrophiles of the type $C_nF_{(2n+1)}CH_2X$ are so unreactive [164, 169–176] that usually only thiolates or similarly powerful nucleophiles will be alkylated by them. If weaker nucleophiles have to be 2,2,2-trifluoroethylated, triflates [177] or phenyliodonium salts [178, 179] must usually be used to achieve acceptable reaction rates. Similarly, 2,2,2-trichloroethyl derivatives Cl_3CCH_2X react only sluggishly with nucleophiles [180] (Scheme 4.44).

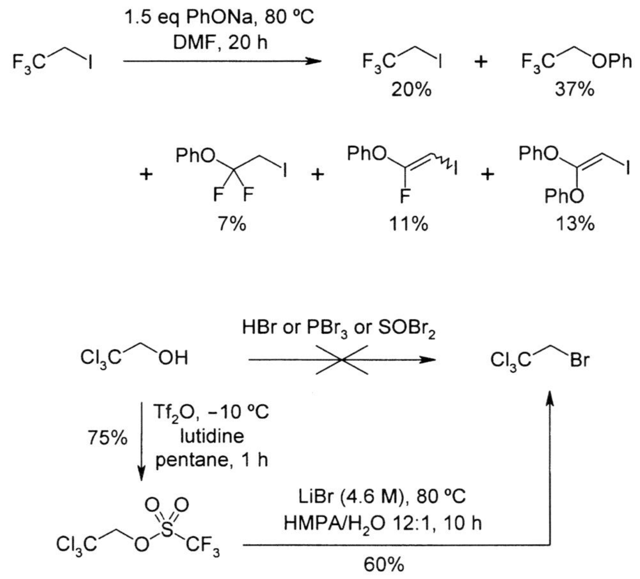

Scheme 4.44. Substitution reactions at 2,2,2-trihaloethyl derivatives [180, 181].

If sulfonyl- or trifluoromethyl groups are separated from the reacting center by a vinyl group, however, a slight increase of the rate of SN2 reactions relative to the unsubstituted allylic substrates is observed [182] (Scheme 4.45). This indicates that the rate-diminishing effect of sulfonyl- and trifluoroalkyl groups is probably due to steric or electrostatic effects [164, 169, 183], which are only active in close proximity to these groups and disappear in the corresponding vinylogous electrophiles. It has also been proposed that the SN2-inhibiting effect of the trifluoromethyl group is because of electrostatic repulsion between the nucleophile and the partial negative charge on the fluorine atoms [184].

The enhanced reactivity of 3-cyano-, 3-sulfonyl-, and 3-(trifluoromethyl)allyl chlorides compared with that of allyl or 2-buten-1-yl chloride can also be explained by a mechanism analogous to that sketched in Scheme 4.43, in which the nucleophile

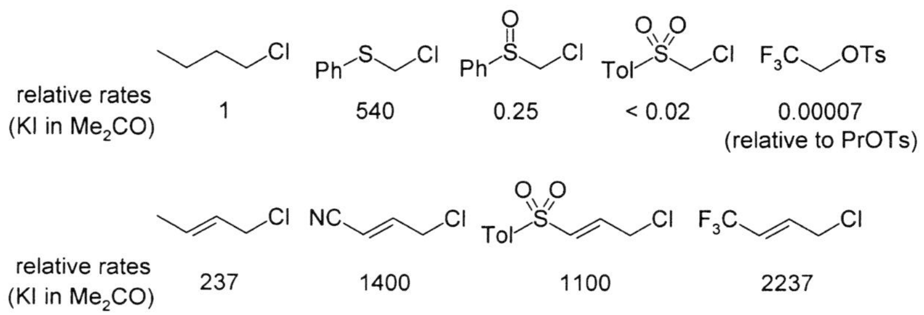

Scheme 4.45. Relative rates of SN2 reactions [164, 165, 182].

first undergoes rapid addition to the electron-poor double bond and then displaces the leaving group by 1,2-migration [182].

A frequent side reaction of electrophiles bearing electron-withdrawing groups is attack of the nucleophile at the leaving group and not at carbon [143]. This reactivity pattern is characteristic of 2,2-dihalo and 2-halomalonic acid derivatives [185] and α-nitroalkyl halides (Section 4.3.3.4), i.e. substrates in which the alkyl group is strongly C,H-acidic and, therefore, is itself a good leaving group. Tetrahalomethanes can also react in this way [154] (Scheme 4.38).

Halomalonic acid derivatives, 1-halo-1-nitroalkanes [186], and related electrophiles can, upon treatment with a base, also dimerize to yield substituted ethylenes, in the same way as nitrobenzyl halides (see above). The reaction conditions required for this dimerization do not differ much from those required for successful nucleophilic substitution (Scheme 4.46), and if substitution is desired a low concentration of the electrophile should be maintained during the reaction to minimize dimerization.

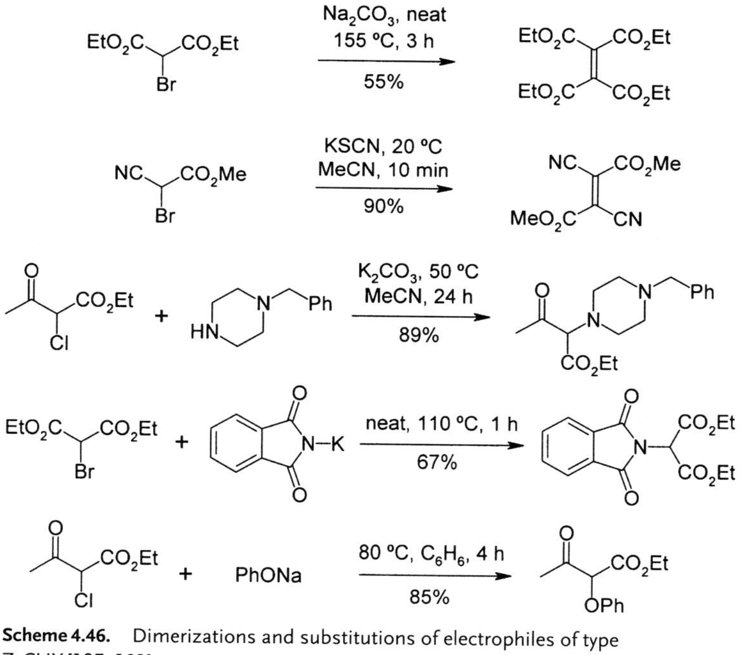

Scheme 4.46. Dimerizations and substitutions of electrophiles of type Z_2CHX [187–191].

Tertiary substrates of the type Z_2CRX can sometimes also undergo nucleophilic substitution (Scheme 4.47), although nucleophilic attack at the leaving group to yield the reduced electrophile (Z_2CHR) will often compete [192]. These reactions probably proceed via (destabilized) carbocations or via SET.

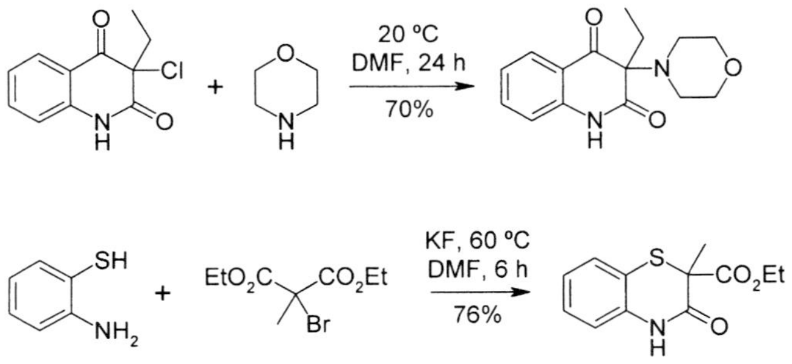

Scheme 4.47. Nucleophilic substitutions at substrates of type Z_2CRX [193–195].

4.3.6
Neighboring-group Participation

Heteroatoms with an unshared electron pair can enhance the SN1-reactivity of electrophiles not only when bound directly to the reacting carbon, but also when positioned further away. In such instances nucleophilic substitutions can occur with retention of configuration and/or rearrangement, because the substitutions are in fact two sequential SN2 reactions proceeding via a cyclic intermediate (Scheme 4.48). The reactivity of such electrophiles will be increased by neighboring-group participation if both the cyclization and the reaction of the resulting intermediate with a nucleophile are faster than the direct SN2 reaction of the starting material with the nucleophile.

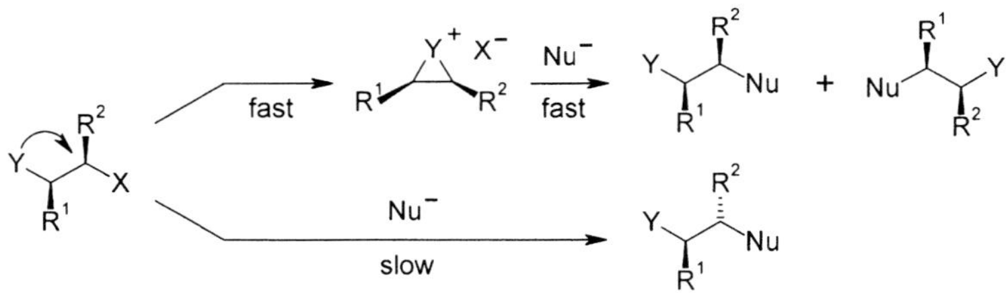

Scheme 4.48. Nucleophilic substitution with neighboring-group participation.

Groups capable of enhancing the reactivity of electrophiles by neighboring-group participation include NR_2, OR, SR, I, Cl, and Br, but phenyl groups, alkenes, alkynes, and C–C σ bonds can also have this effect. The preferred ring sizes formed are three-, five-, and six-membered rings [196–200], although a few examples of four- [201] and seven-membered [202] rings have also been reported. Illustrative examples of substitutions at substrates with neighboring-group participation are sketched in Scheme 4.49. The last example in Scheme 4.49 proceeds via the base-induced formation of an epoxide.

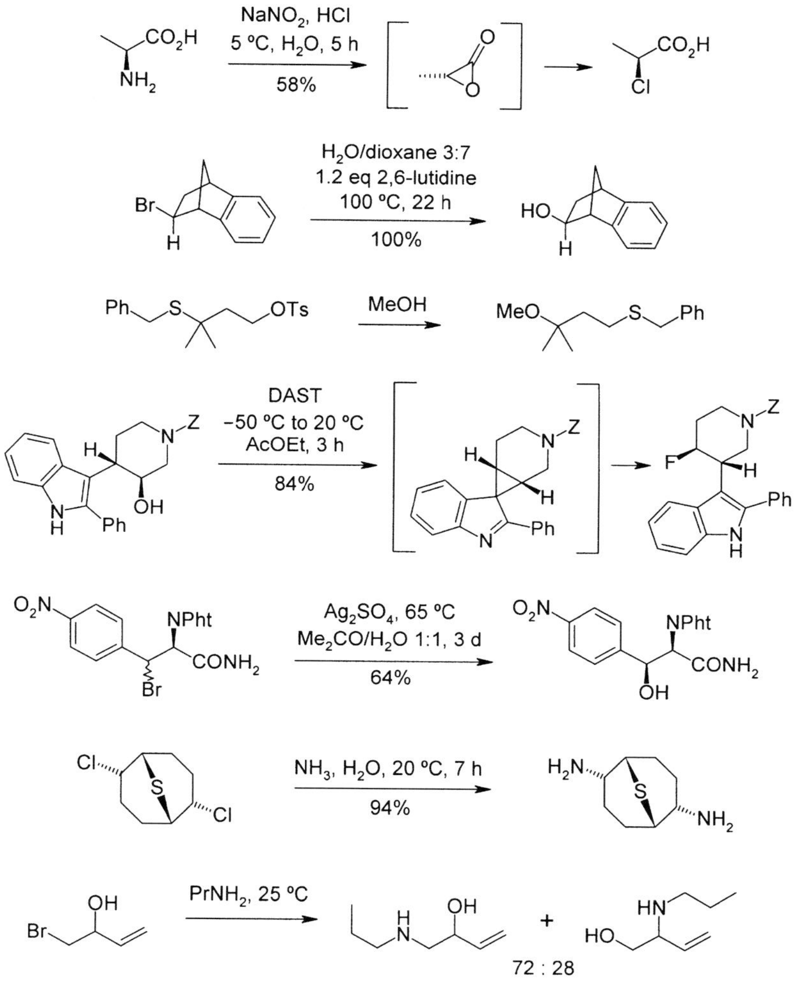

Scheme 4.49. Examples of nucleophilic substitutions with neighboring-group participation [201, 203–208].

With some electrophiles, for example haloalkylamines, irreversible cyclization can occur along with the alkylation of a nucleophile, and mixtures of products can result. Strong nucleophiles can, nevertheless, be alkylated with haloalkylamines in acceptable yields (see, e.g., Scheme 6.11).

As shown by the relative rates of methanolysis of thioethers $PhS(CH_2)_nCl$ (Scheme 4.50), the same substrate can react with or without neighboring-group participation, depending on the nucleophile and on the reaction conditions. Under conditions which favor bimolecular substitution, anchimeric assistance by neighboring groups is observed only rarely.

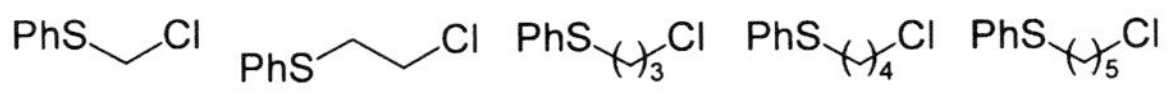

relative rates	PhS–CH2–Cl	PhS–(CH2)2–Cl	PhS–(CH2)3–Cl	PhS–(CH2)4–Cl	PhS–(CH2)5–Cl
solvolysis in MeOH (relative to HexCl)	33 000	150	1.0	130	4.3
KI, Me_2CO (relative to BuCl)	540	0.79	3.1	1.8	1.4

Scheme 4.50. Relative rates of SN1 and SN2 reactions of chloroalkyl thioethers [164].

One example of neighboring-group participation without the formation of cationic intermediates is the aminolysis of 2-bromoethanols (last example, Scheme 4.49). In this instance epoxide formation and opening must be faster or as fast as direct bimolecular substitution of bromide by the amine; otherwise no rearranged product would be observed.

There are also instances in which neighboring groups reduce the reactivity of an electrophile. Thus, as shown in Scheme 4.50, the β-phenylthio group actually reduces the rate of SN2 reactions compared with an ethyl group. As discussed above, electrophiles with one or more fluorine atoms in the β position are strongly deactivated toward nucleophilic substitution [169]. Similarly, the other halogens located β to the reaction center reduce the rate of bimolecular substitution reactions by approximately a factor of 3.6 [209]. Some homoallyl halides or electrophiles containing electron-deficient C–C double bonds can also react more slowly than comparable saturated compounds. Thus, although *anti*-7-tosyloxynorbornene solvolyzes (with retention of configuration) much faster than the corresponding norbornane, because of anchimeric assistance by the double bond [210], 4-bromocyclopentene is less reactive than bromocyclopentane [211] (Scheme 4.51). This surprising effect probably results from the high strain-energy of norbornenes. The energy required to force 4-bromocyclopentene into a suitable conformation for anchimerically assisted

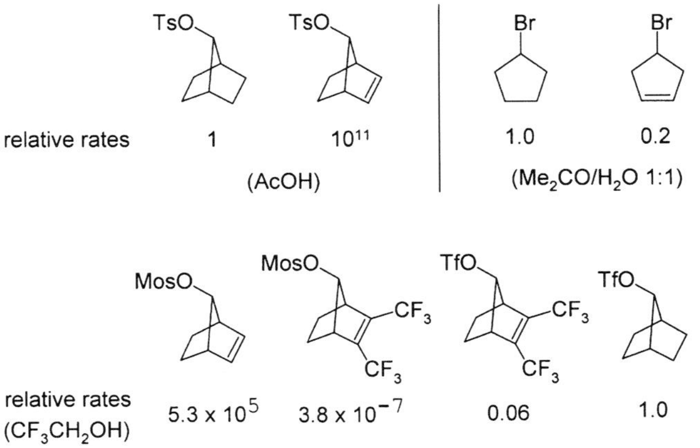

TsO TsO Br Br

relative rates 1 10^{11} (AcOH) | 1.0 0.2 (Me_2CO/H_2O 1:1)

MosO MosO CF_3 CF_3 TfO CF_3 CF_3 TfO

relative rates (CF_3CH_2OH) 5.3×10^5 3.8×10^{-7} 0.06 1.0

Scheme 4.51. Relative solvolysis rates of norbornane and cyclopentane derivatives [210–212].

bromide displacement is larger than the activation energy for an unassisted SN1 or SN2 reaction [211]. Accordingly, neighboring-group participation of double bonds in flexible cyclic or non-cyclic substrates will usually have only a minor effect on reactivity.

The capacity of a homoallylic C–C double bond to assist nucleophilic displacements is, furthermore, a function of its electron density. Electron-withdrawing substituents at the alkene will reduce its electron-donating capability, and electron-deficient alkenes can even reduce the rate of nucleophilic displacements relative to the corresponding saturated compounds (Scheme 4.51).

4.3.7
Allylic and Propargylic Electrophiles

Allylic electrophiles can react with nucleophiles either with or without allylic rearrangement [213]. The outcome of such reactions will depend on whether or not an allylic carbocation is formed as intermediate, and on the steric requirement and hardness of the two electrophilic centers and the nucleophile. Bimolecular substitutions at allylic electrophiles which occur with rearrangement are called SN2′ reactions.

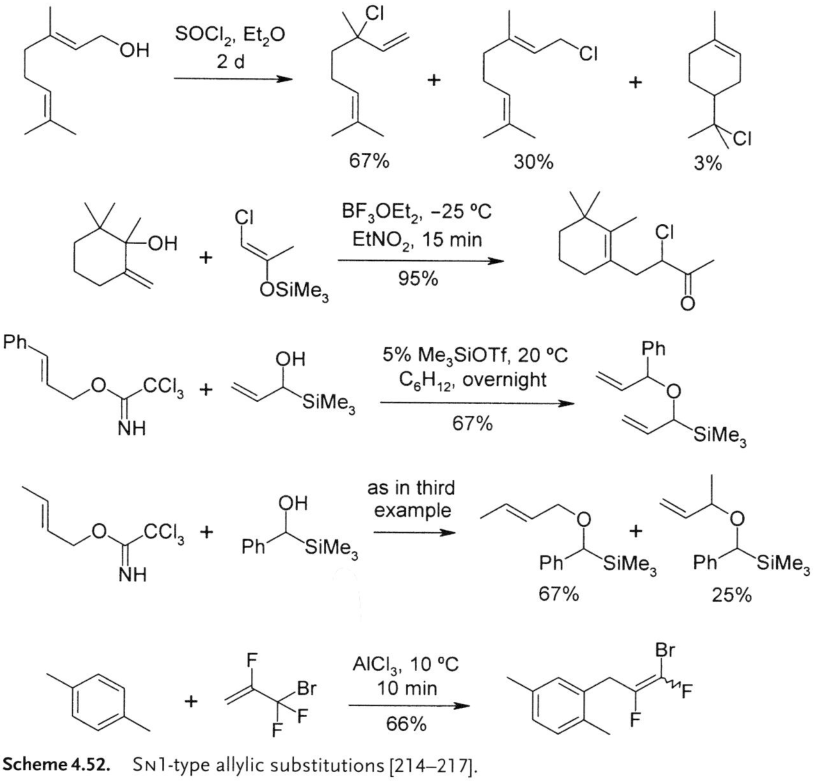

$SOCl_2$, Et_2O, 2 d; 67%, 30%, 3%

BF_3OEt_2, –25 °C, $EtNO_2$, 15 min; 95%

5% Me_3SiOTf, 20 °C, C_6H_{12}, overnight; 67%

as in third example; 67%, 25%

$AlCl_3$, 10 °C, 10 min; 66%

Scheme 4.52. SN1-type allylic substitutions [214–217].

In SN1-type substitutions the nucleophile will generally tend to add to the carbon which forms the better-stabilized carbocation, while avoiding too much steric crowding. The regioselectivity of such reactions is often difficult to predict and to control (Scheme 4.52).

The regioselectivity of bimolecular allylic substitutions, on the other hand, is often easier to control, and can sometimes be reversed by slight modification of the starting materials or reaction conditions. In uncatalyzed, bimolecular substitutions the nucleophile will usually add to the sterically less demanding site of the allylic system (Scheme 4.53).

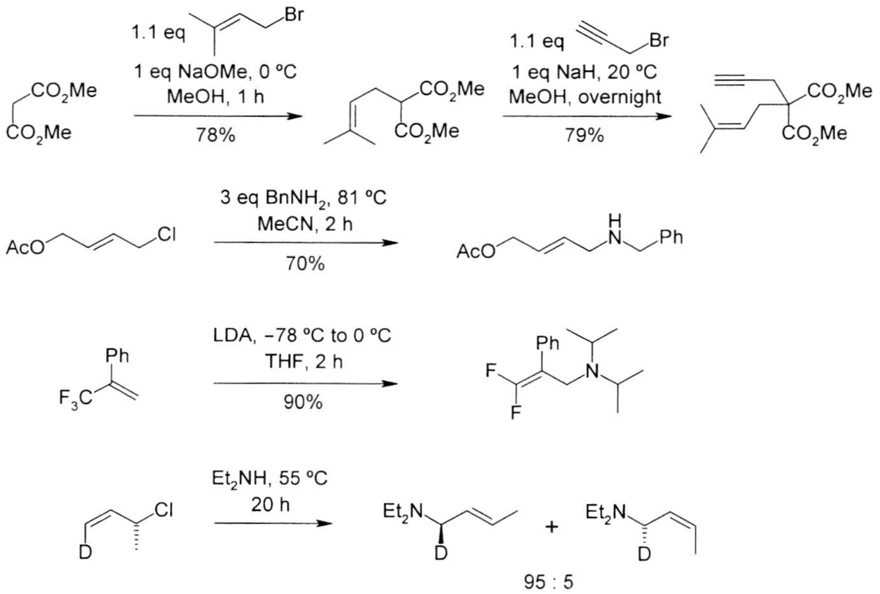

CO_2Me, CO_2Me; 1.1 eq Br; 1 eq NaOMe, 0 °C, MeOH, 1 h; 78%; CO_2Me, CO_2Me; 1.1 eq Br; 1 eq NaH, 20 °C, MeOH, overnight; 79%; CO_2Me, CO_2Me

AcO, Cl; 3 eq $BnNH_2$, 81 °C, MeCN, 2 h; 70%; AcO, H, N, Ph

Ph, F_3C; LDA, −78 °C to 0 °C, THF, 2 h; 90%; Ph, F, F, N

D, Cl; Et_2NH, 55 °C, 20 h; Et_2N, D + Et_2N, D; 95 : 5

Scheme 4.53. SN2 and SN2′ at allylic electrophiles [218–221].

As shown by the last example in Scheme 4.53, SN2′ reactions can proceed with high stereoselectivity. During an SN2′ reaction the plane in which the alkene lies and the C–C–X plane are perpendicular to each other, and the nucleophile attacks the alkene preferentially from the side of the leaving group (first three reactions, Scheme 4.54). This is, however, not always so. Both regio- and stereoselectivity depend on the nucleophile, on the leaving group, and on the precise reaction conditions, and are not always easy to predict (Scheme 4.54).

Allylic electrophiles react readily with Pd(0) complexes to yield η^3-allyl Pd(II) complexes, which retain electrophilic character and react with nucleophiles to yield the product of allylic substitution and Pd(0). Thus, catalytic amounts of Pd(0) provide an additional mechanism (in addition to SN1, SN2, and SN2′) by which an allylic substitution can proceed. Because the metal generally attacks the allylic electrophile

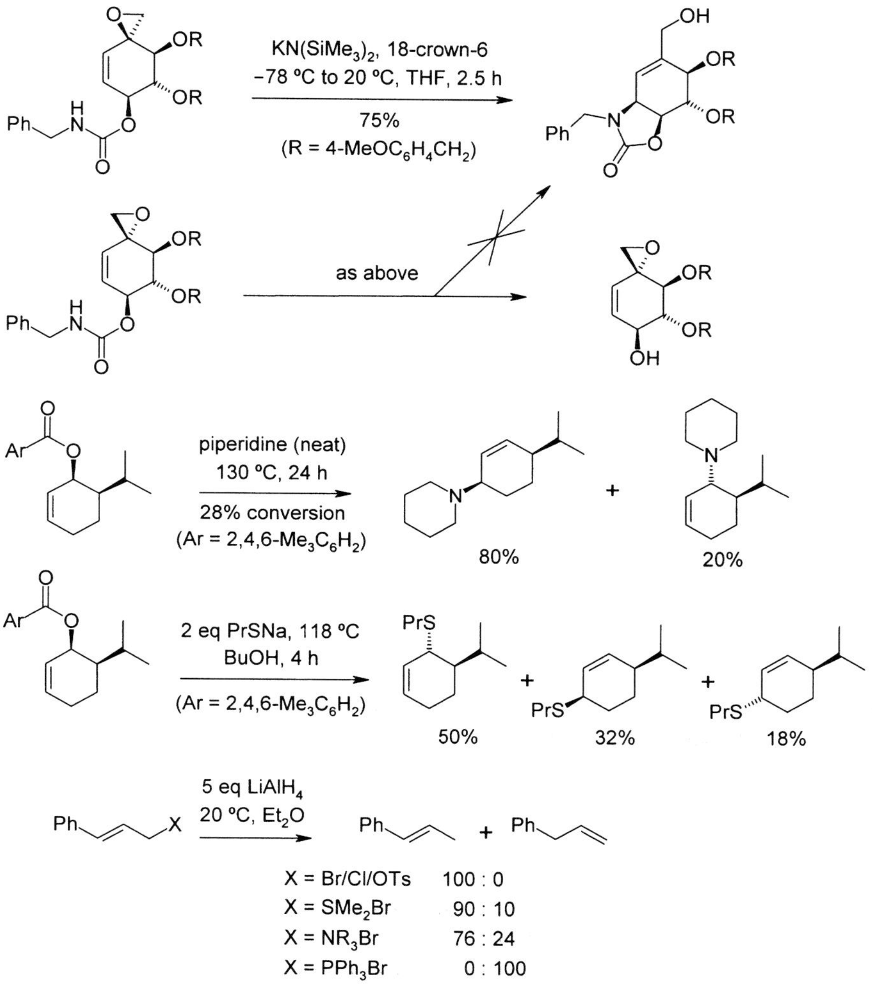

KN(SiMe3)2, 18-crown-6
−78 °C to 20 °C, THF, 2.5 h
75%
(R = 4-MeOC6H4CH2)

as above

piperidine (neat)
130 °C, 24 h
28% conversion
(Ar = 2,4,6-Me3C6H2)
80% 20%

2 eq PrSNa, 118 °C
BuOH, 4 h
(Ar = 2,4,6-Me3C6H2)
50% 32% 18%

5 eq LiAlH4
20 °C, Et2O

X = Br/Cl/OTs 100 : 0
X = SMe2Br 90 : 10
X = NR3Br 76 : 24
X = PPh3Br 0 : 100

Scheme 4.54. Regio- and stereoselectivity of SN2′ reactions [222–224].

from the side opposite to the leaving group, and is itself displaced by nucleophiles in the same fashion (Scheme 4.55) [225], Pd(0)-catalyzed allylic substitutions often proceed with overall retention of configuration. If the starting material or product can be isomerized by the catalyst, the thermodynamically most stable product will usually be isolated from Pd(0)-catalyzed allylic substitutions. Thus, reaction of allylic acetates with sulfinate in the presence of Pd(0) yields exclusively the thermodynamically most stable product (fourth reaction, Scheme 4.55). Isomerization of the product can be suppressed by addition of $NaNO_2$ (which deactivates the catalyst) or by using allylic nitro compounds as electrophiles which liberate nitrite during the course of the reaction [41].

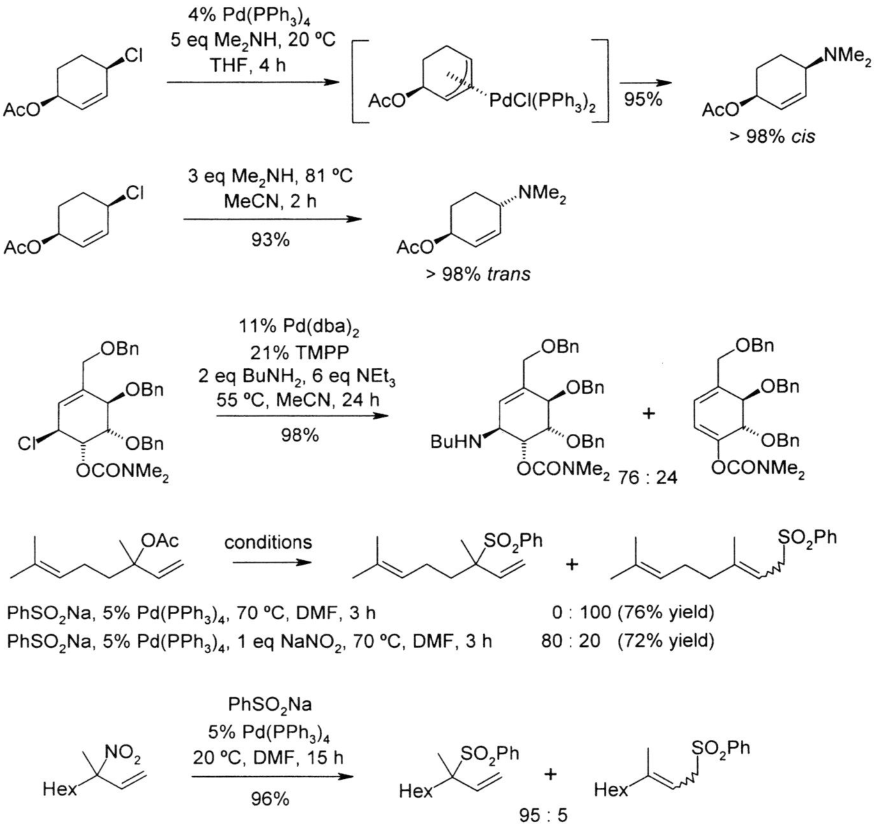

4% $Pd(PPh_3)_4$
5 eq Me_2NH, 20 °C
THF, 4 h
AcO Cl → [AcO $PdCl(PPh_3)_2$] → 95% AcO NMe_2
> 98% *cis*

3 eq Me_2NH, 81 °C
MeCN, 2 h
93%
AcO Cl → AcO NMe_2
> 98% *trans*

11% $Pd(dba)_2$
21% TMPP
2 eq $BuNH_2$, 6 eq NEt_3
55 °C, MeCN, 24 h
98%
OBn, OBn, OBn, Cl, $OCONMe_2$ → OBn, OBn, OBn, BuHN, $OCONMe_2$ + OBn, OBn, OBn, $OCONMe_2$
76 : 24

OAc → conditions → SO_2Ph + SO_2Ph

$PhSO_2Na$, 5% $Pd(PPh_3)_4$, 70 °C, DMF, 3 h — 0 : 100 (76% yield)
$PhSO_2Na$, 5% $Pd(PPh_3)_4$, 1 eq $NaNO_2$, 70 °C, DMF, 3 h — 80 : 20 (72% yield)

$PhSO_2Na$
5% $Pd(PPh_3)_4$
20 °C, DMF, 15 h
96%
Hex NO_2 → Hex SO_2Ph + Hex SO_2Ph
95 : 5

Scheme 4.55. Uncatalyzed and Pd(0)-catalyzed SN2 reactions at a allylic substrates [41, 221, 226].

Organocopper reagents, being soft nucleophiles, often lead to SN2′ reactions when treated with allylic [227–229] or propargylic [230–232] electrophiles. Although highly stereoselective reactions with clean attack from the opposite side of the leaving group can be achieved [233] (Scheme 4.56), these reactions might still require careful optimization of the conditions to achieve high-yielding, stereoselective product formation, because the precise structure of the organocopper reagent and the solvent can have a decisive impact on the selectivity (Scheme 4.56).

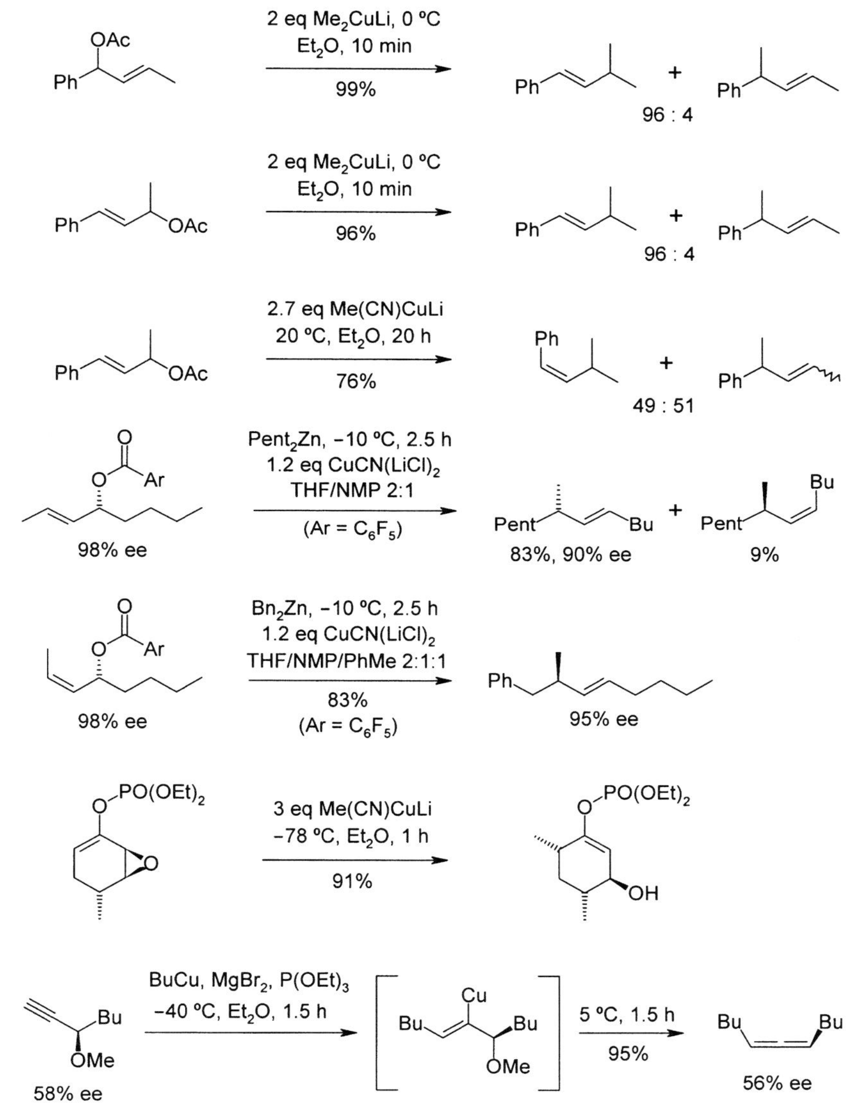

Scheme 4.56. Nucleophilic substitutions at allylic electrophiles with organocopper reagents [230, 234–237].

4.3.8 Epoxides

Epoxides are readily accessible, valuable synthetic intermediates. Because of the high ring strain, epoxides are much more reactive than ethers, and usually react swiftly with nucleophiles either under basic or acidic reaction conditions to yield

β-substituted ethanols. Epoxides are often superior to electrophiles such as 1,2-dihaloethanes or protected 2-haloethanols, because epoxides undergo fewer side reactions than these synthetically equivalent but often unreactive, alternative electrophiles (Scheme 4.57).

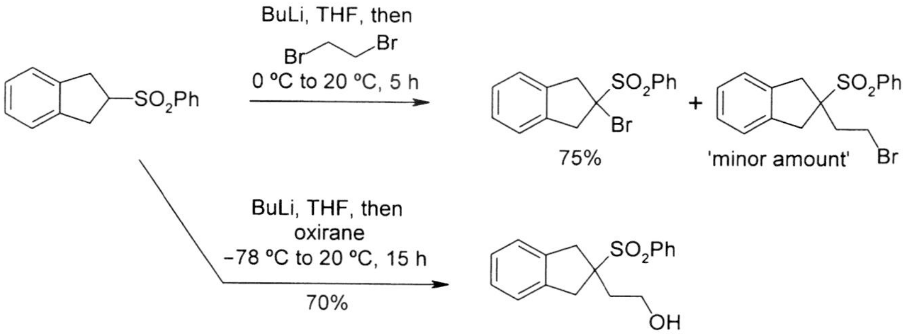

BuLi, THF, then
Br
Br
0 °C to 20 °C, 5 h
SO2Ph
SO2Ph
Br
75%
+
SO2Ph
'minor amount'
Br
BuLi, THF, then
oxirane
-78 °C to 20 °C, 15 h
70%
SO2Ph
OH

Scheme 4.57. Alkylations of a sulfone with 1,2-dibromoethane and oxirane [238].

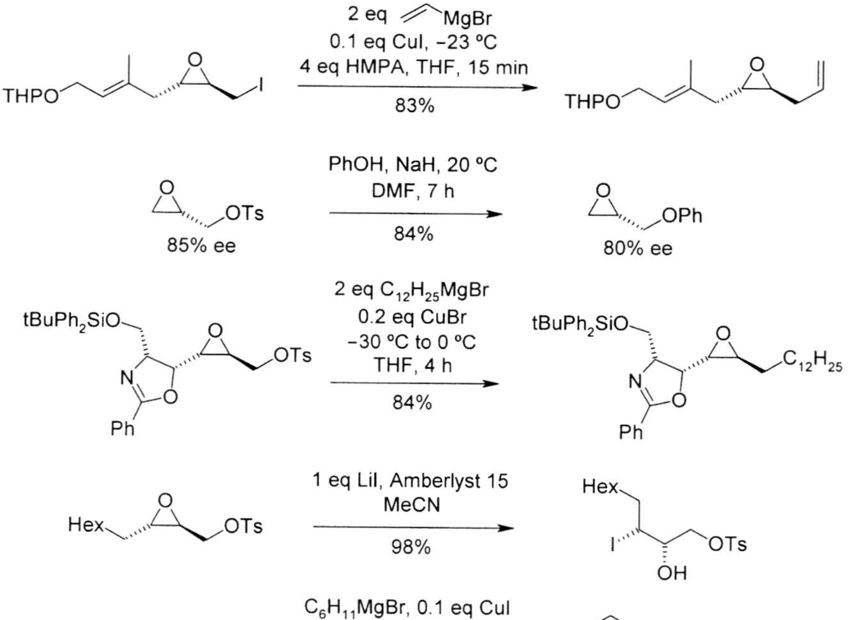

2 eq MgBr
0.1 eq CuI, -23 °C
4 eq HMPA, THF, 15 min
83%
THPO
O
I
THPO
O
PhOH, NaH, 20 °C
DMF, 7 h
84%
O
OTs
85% ee
O
OPh
80% ee
2 eq C12H25MgBr
0.2 eq CuBr
-30 °C to 0 °C
THF, 4 h
84%
tBuPh2SiO
O
OTs
N
O
Ph
tBuPh2SiO
O
C12H25
N
O
Ph
1 eq LiI, Amberlyst 15
MeCN
98%
Hex
O
OTs
Hex
I
OTs
OH
C6H11MgBr, 0.1 eq CuI
-40 °C, Et2O, 3 h
69%
O
OTs
OH
OTs

Scheme 4.58. Reactivity of epoxides vs sulfonates and other electrophiles [243–246]. All reactions required careful monitoring to avoid reaction of both electrophilic sites.

The reactivity of epoxides is similar to that of sulfonates. Compounds containing both an epoxide and a sulfonate can react at either of these functional groups with nucleophiles, depending on the precise structure of the reacting partners (Scheme 4.58) [239]. In most of the reported reactions of carbon nucleophiles with sulfonate-containing epoxides the epoxide is cleaved first [240–243]. As illustrated by the second and third examples in Scheme 4.58, this is, however, not always true (see also Scheme 4.20).

Epoxides with two different substituents at C-2 and C-3 can yield, upon reaction with a nucleophile, two different alcohols. The regioselectivity of this reaction is governed mostly by steric but also by electronic factors [247]. Substituents which generally facilitate nucleophilic substitutions will also enhance the reactivity of the carbon atom of an epoxide to which they are attached. If the nucleophile is sterically demanding the outcome of a reaction can often be predicted, but not so for small nucleophiles. To make matters worse, groups which favor electronically the formation of one product often retard its formation by steric crowding (Scheme 4.59). Instead of reacting with a nucleophile to yield a β-substituted ethanol, epoxides can also undergo acid or base-mediated rearrangements (Scheme 4.59) or act as oxidants while themselves being reduced to alkenes.

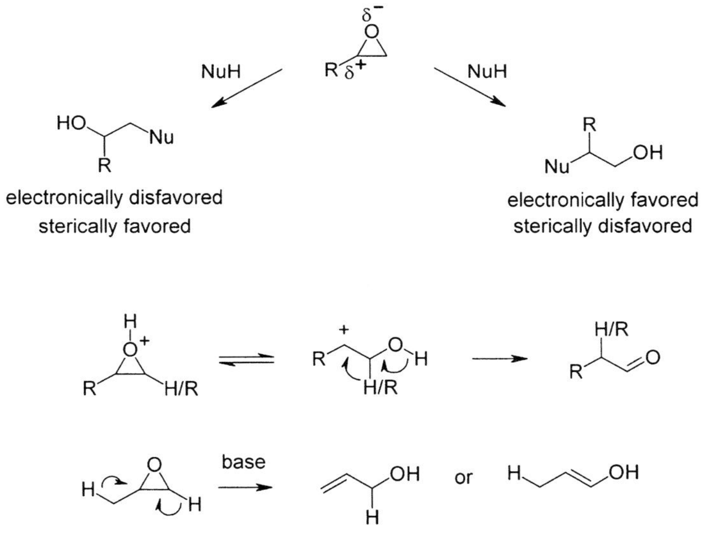

Scheme 4.59. Ring opening of epoxides by nucleophilic attack or acid or base-mediated rearrangement. R = carbocation-stabilizing, sterically demanding group.

Rough guidelines for the prediction of regioselectivity in epoxide ring openings are summarized in Scheme 4.60. Under neutral or basic reaction conditions alkyl- or aryl-substituted epoxides react with most nucleophiles at the less substituted carbon atom [248–253]. Under acidic reaction conditions, however, product mixtures or preferential attack at the most substituted carbon atom can be observed. Acids can usually be used to enhance the reactivity of epoxides and to promote substitution at the site of an epoxide which forms a carbocation more readily.

Epoxides bearing a carbonyl group or vinylogous analogs of such compounds react with nucleophiles under neutral or basic reaction conditions usually at the carbonyl group-bearing carbon atom. This observation is in accord with the increased reactivity of α-halo carbonyl compounds discussed above (Section 4.3.5). Because of the destabilization of carbocations by carbonyl groups, acidic conditions will favor substitution at the non-carbonyl-substituted carbon atom.

Heteroatoms with higher electronegativity than carbon (e.g. nitrogen, oxygen, or the halogens) inductively destabilize carbocations at the β position. Epoxides of the type shown in the last equation of Scheme 4.60 therefore react preferentially at the unsubstituted carbon atom. Only in the presence of certain Lewis acids, capable of chelate formation with simultaneous activation of the substituted carbon atom, is the alternative regiochemistry observed.

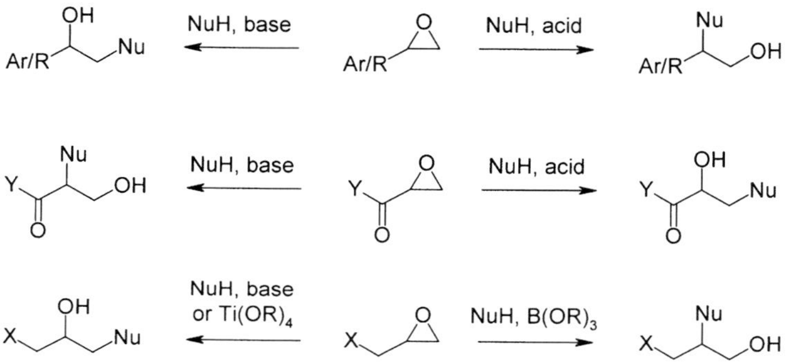

Scheme 4.60. Typical regioselectivity of the ring opening of epoxides. Y = Ar, R, NR_2, OR; X = NR_2, OR, Hal.

4.3.8.1 Epoxide Opening by Hydride

Treatment of epoxides with reducing agents can lead to the formation of either an alkene or an alcohol. Reducing agents such as phosphorus derivatives [254], R_3SiI [255], and some metals [185, 256] can convert epoxides into alkenes, whereas hydrogen in the presence of a catalyst or hydride-donating reagents generally only cleave one C–O bond of epoxides to yield alcohols. If nucleophilic attack of hydride at the least substituted carbon atom is desired, $LiAlH_4$ in ethers can be a suitable reagent (Scheme 4.61).

Styrene oxides can be reduced selectively to either 2-arylethanols or 1-arylethanols. Attack of hydride at the non-benzylic carbon atom can sometimes be achieved with $LiAlH_4$, but most reducing agents, in particular under acidic reaction conditions, will lead to cleavage of the benzylic C–O bond (Scheme 4.62).

Epoxides with an oxymethyl substituent (RO–CR_2) usually react with nucleophiles at the non-oxymethyl-substituted carbon atom, because of the carbocation-destabilizing effect of this group. This is also observed for hydride as nucleophile, as illustrated by the examples in Scheme 4.63.

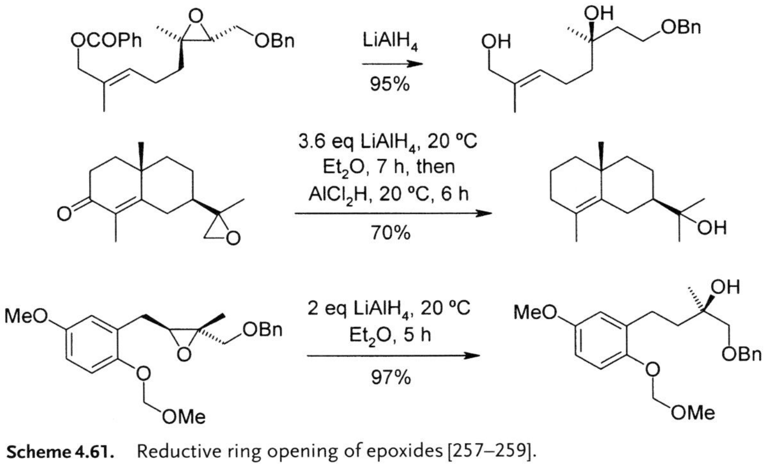

$LiAlH_4$, 95%

3.6 eq $LiAlH_4$, 20 °C, Et_2O, 7 h, then $AlCl_2H$, 20 °C, 6 h; 70%

2 eq $LiAlH_4$, 20 °C, Et_2O, 5 h; 97%

Scheme 4.61. Reductive ring opening of epoxides [257–259].

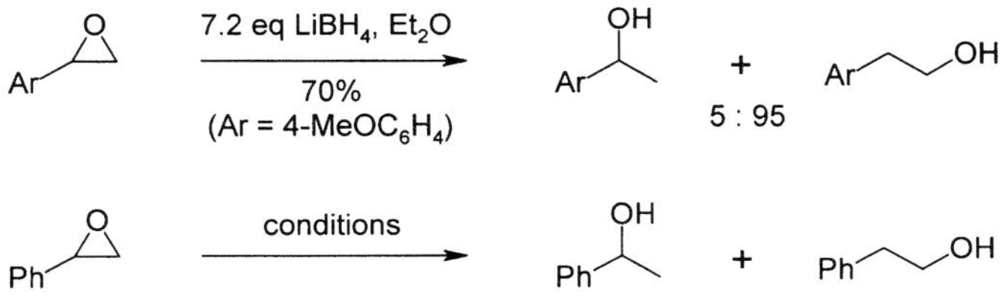

7.2 eq $LiBH_4$, Et_2O; 70% (Ar = 4-$MeOC_6H_4$); 5 : 95

conditions:		
$LiAlH_4$, 25 °C, THF, 1 h	97 : 3	(95% yield)
1.3 eq $LiBH_4$, Et_2O	74 : 26	(100% yield)
iBu_2AlH, 80 °C, heptane, 10 h	32 : 68	(96% yield)
iBu_2AlH, 25 °C, heptane, 20 h	14 : 86	(95% yield)
LiH/Ni/*t*BuOLi, THF, 24 h	5 : 95	(97% yield)
Pd, HCO_2H, NEt_3, 23 °C, AcOEt, 4 h	0 : 100	(84% yield)

Scheme 4.62. Reductive ring opening of styrene oxides [250, 260–262].

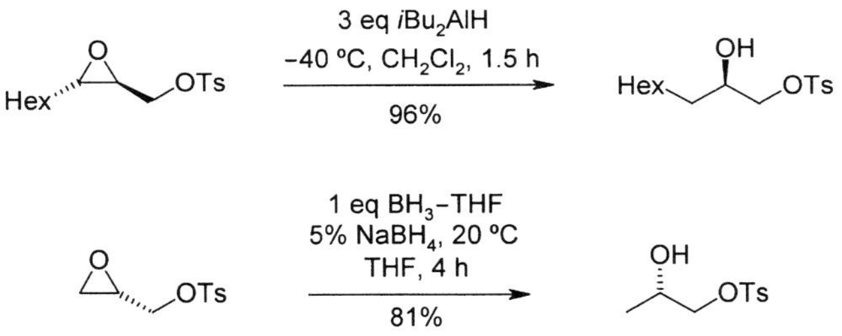

3 eq *i*Bu_2AlH, −40 °C, CH_2Cl_2, 1.5 h; 96%

1 eq BH_3–THF, 5% $NaBH_4$, 20 °C, THF, 4 h; 81%

Scheme 4.63. Reductive ring opening of tosyloxymethyl epoxides [243, 263].

Hydroxymethyl epoxides can be reduced to yield either 1,2- or 1,3-diols. The regioselectivity of this reaction is critically dependent on the structure of the epoxide and on the precise reaction conditions, as illustrated by the examples in Scheme 4.64. Titanium(IV) alcoholates promote attack of hydride at C-3 [264] (to yield 1,2-diols), as does $i\text{Bu}_2\text{AlH}$ in non-coordinating solvents such as benzene, hexane, or dichloromethane [265, 266]. Pure 1,3-diols can be obtained by treatment of hydroxymethyl epoxides with $LiAlH_4$ or Red-Al ($(MeOCH_2CH_2O)_2AlH$) in THF [259, 267, 268], or by single-electron transfer with Cp_2TiCl_2 [269] (Scheme 4.64).

Treatment of hydroxymethyl epoxides with $NaCNBH_3/BF_3OEt_2$ usually leads to attack of hydride at the most highly substituted carbon atom and preferentially to the formation of 1,2-diols (last example, Scheme 4.64). If in a given epoxide cleavage of the C–O bond to the most highly substituted carbon atom leads to formation of a 1,3-diol, mixtures of products can result from use of this reagent [270].

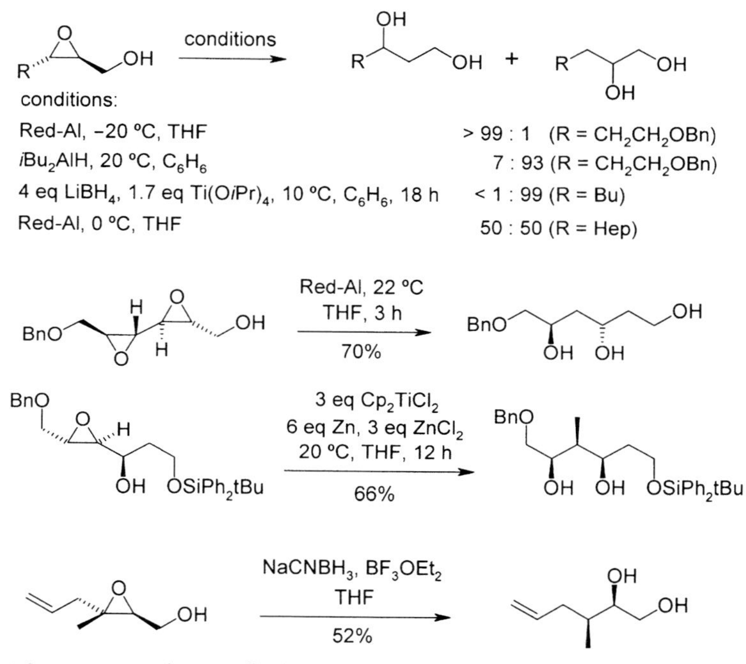

Scheme 4.64. Reduction of hydroxymethyl epoxides [264, 266, 267, 269, 271].

In epoxides substituted with electron-withdrawing groups the C–O bond closest to the electron-withdrawing group will usually be cleaved (Scheme 4.65) [272–274]. This is also observed for reductions proceeding via radicals [273]. Reductions with SmI_2 in methanol, however, can lead to reversal of this regioselectivity with some substrates (last reaction, Scheme 4.65) [275].

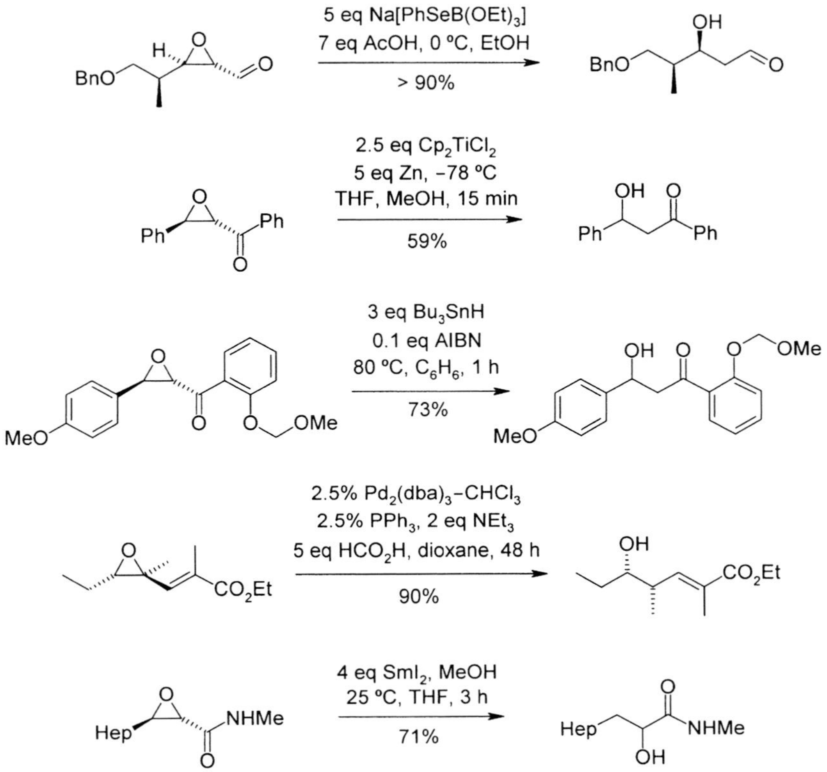

5 eq Na[PhSeB(OEt)$_3$]
7 eq AcOH, 0 °C, EtOH
> 90%

2.5 eq Cp_2TiCl_2
5 eq Zn, −78 °C
THF, MeOH, 15 min
59%

3 eq Bu_3SnH
0.1 eq AIBN
80 °C, C_6H_6, 1 h
73%

2.5% $Pd_2(dba)_3$–$CHCl_3$
2.5% PPh_3, 2 eq NEt_3
5 eq HCO_2H, dioxane, 48 h
90%

4 eq SmI_2, MeOH
25 °C, THF, 3 h
71%

Scheme 4.65. Regioselective reductive ring opening of epoxides substituted with electron-withdrawing groups [275–279].

4.3.8.2 **Epoxide Opening by Carbon Nucleophiles**

Organometallic compounds such as Grignard reagents, organolithium compounds, cuprates, or enolates can react with epoxides to yield 2-substituted ethanols. Because epoxides can rearrange to carbonyl compounds when treated with bases or acids under rather mild conditions [280–288] (e.g. upon treatment with $LiClO_4$ [289], $MgBr_2$ [290], or lithium halides [291]), the type of organometallic reagent used can be of critical importance to the outcome of such reactions. Strongly basic organometallic compounds, such as organolithium or Grignard reagents, usually lead to rapid isomerization of epoxides to enolates [292] or to allyl alcoholates. If isomerization to a ketone or aldehyde is more rapid than addition of the organometallic reagent to the epoxide, the main product can result from addition to the newly formed carbonyl compound (Scheme 4.66). Highly substituted epoxides are particularly prone to such transformations. Rearrangements such as those shown in Scheme 4.66 can sometimes be suppressed by addition of bases (e.g. PPh_3 or Me_2S [293]) to the reaction mixture.

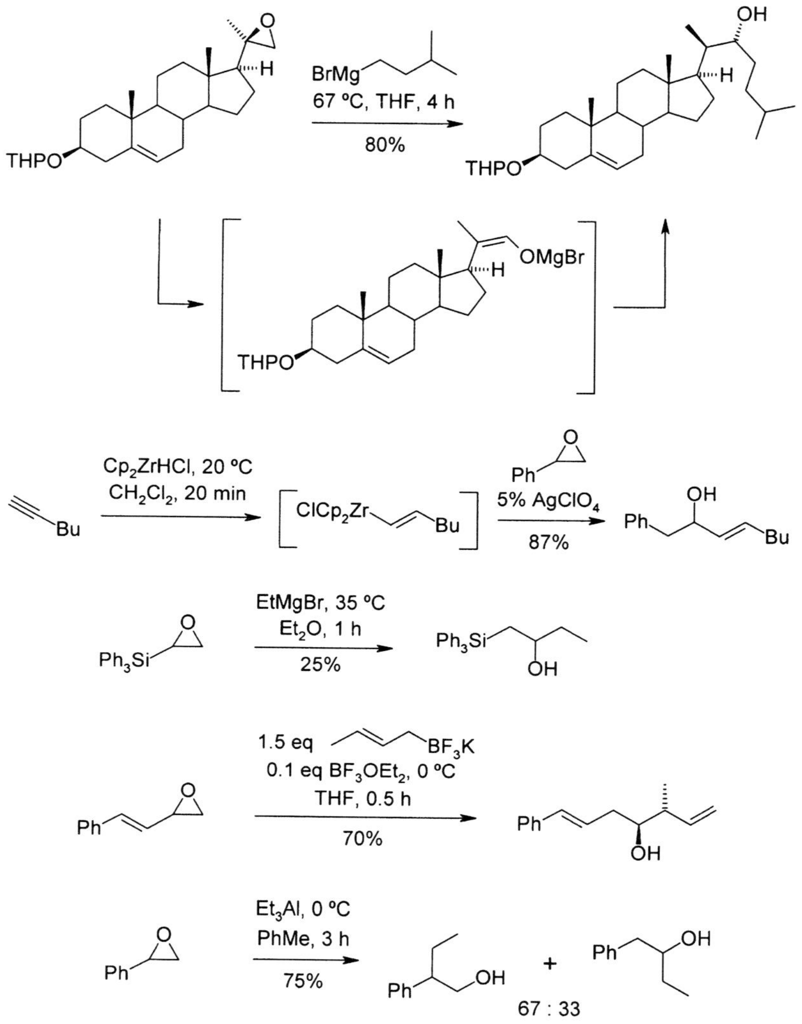

Scheme 4.66. Rearrangement of epoxides followed by addition of organometallic reagents [284, 290, 293–295].

Although the reactivity of epoxides toward organometallic reagents can be enhanced by the addition of Lewis acids [241, 243, 251, 296], this can also lead to cleavage of other ethers present in the reaction mixture, for example THF or Et_2O [238, 240]. Strong Lewis acids, for example trityl cations [297], can also enhance the amount of products resulting from rearranged epoxides.

Few types of epoxide react with Grignard or organolithium reagents in the absence of catalysts without rearrangement. These include oxirane, which has been extensively used to transform these organometallic reagents into 2-substituted ethanols [298, 299], and oxymethyl oxirane derivatives (e.g. glycidol derivatives) [300, 301], which are also quite resistant toward rearrangement. Other epoxides, however, react only slowly with Grignard or related reagents, and tend to give poor yields of 2-substituted ethanols [302, 303].

Carbon nucleophiles which do not readily trigger the rearrangement of epoxides include lithiated dithianes [295, 304], lithiated sulfones [238], lithiated diarylphosphine oxides [240, 305], lithium enolates [306], and allylic organolithium or organomagnesium compounds [298, 307–310] (Scheme 4.67).

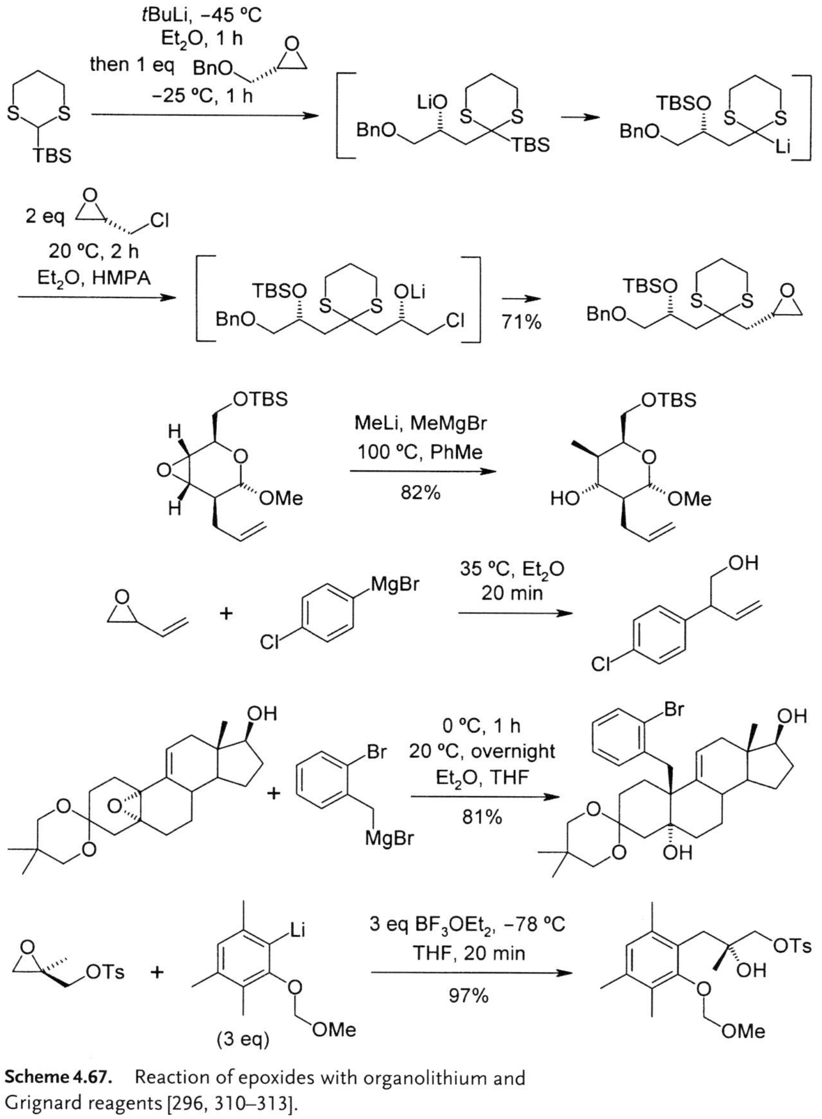

tBuLi, –45 °C, Et_2O, 1 h, then 1 eq BnO-CH2-oxirane, –25 °C, 1 h

2 eq epichlorohydrin, 20 °C, 2 h, Et_2O, HMPA; 71%

MeLi, MeMgBr, 100 °C, PhMe; 82%

35 °C, Et_2O, 20 min

0 °C, 1 h; 20 °C, overnight; Et_2O, THF; 81%

3 eq BF_3OEt_2, –78 °C, THF, 20 min; 97%

(3 eq)

Scheme 4.67. Reaction of epoxides with organolithium and Grignard reagents [296, 310–313].

Consistently better results are obtained when organocopper reagents, rather than Grignard or organolithium reagents, are used for the ring opening of epoxides [314–317]. Cuprates are softer and less basic than the latter, and usually add to epoxides before rearrangement can occur. With vinyl epoxides conjugate addition of the organocopper reagent to the double bond is often (but not always) observed [227, 318]. Other organometallic reagents with a high tendency to yield 2-substituted ethanols on reaction with epoxides are organoaluminum [319], organozinc [320, 321], allyltitanium [322], and allylindium compounds [323]. To illustrate the reactivity of these reagents, the reactions of ethyl oxirane and styrene oxide with a selection of organometallic reagents are shown in Scheme 4.68.

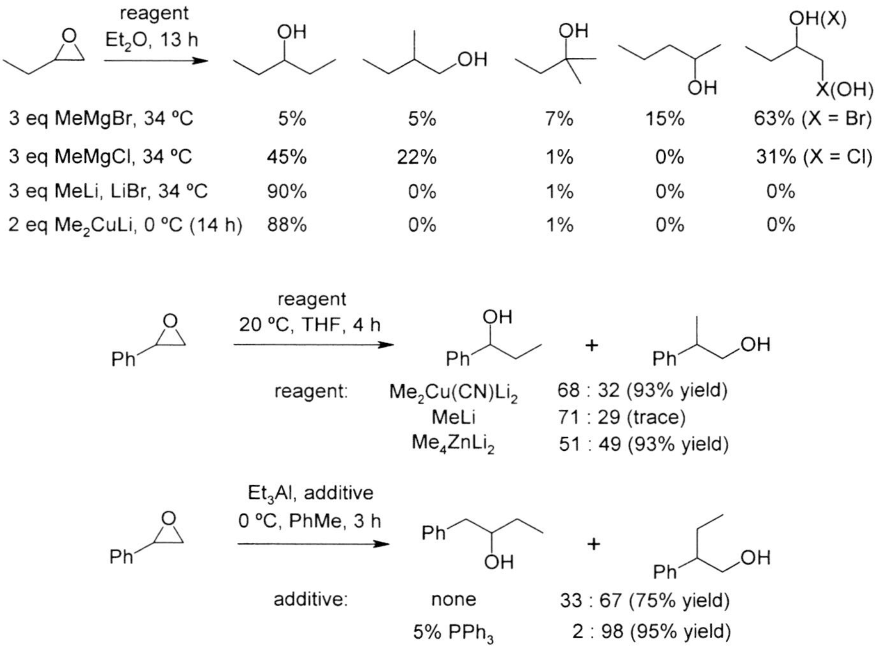

reagent, Et_2O, 13 h

reagent	OH	OH	OH	OH	OH(X) / X(OH)
3 eq MeMgBr, 34 °C	5%	5%	7%	15%	63% (X = Br)
3 eq MeMgCl, 34 °C	45%	22%	1%	0%	31% (X = Cl)
3 eq MeLi, LiBr, 34 °C	90%	0%	1%	0%	0%
2 eq Me_2CuLi, 0 °C (14 h)	88%	0%	1%	0%	0%

reagent, 20 °C, THF, 4 h

reagent:	Ph, OH + Ph, OH
$Me_2Cu(CN)Li_2$	68 : 32 (93% yield)
MeLi	71 : 29 (trace)
Me_4ZnLi_2	51 : 49 (93% yield)

Et_3Al, additive, 0 °C, PhMe, 3 h

additive:	Ph, OH + Ph, OH
none	33 : 67 (75% yield)
5% PPh_3	2 : 98 (95% yield)

Scheme 4.68. Reactions of ethyl oxirane and styrene oxide with a variety of organometallic reagents [293, 302, 320].

The regioselectivity of epoxide ring opening by organometallic reagents depends to a large extent on the type of metal chosen. If the organometallic reagent is soft (cuprates, allylic organometallic reagents, enolates) the regioselectivity is mainly controlled by steric effects. Hard organometallic compounds containing metals with a high affinity for oxygen or which are strongly Lewis-acidic will preferentially attack the most polarized C–O bond of an epoxide. As illustrated by the examples shown in Scheme 4.69, organocopper reagents usually attack epoxides at the sterically less demanding site, or at the position which would undergo SN2 reaction most easily.

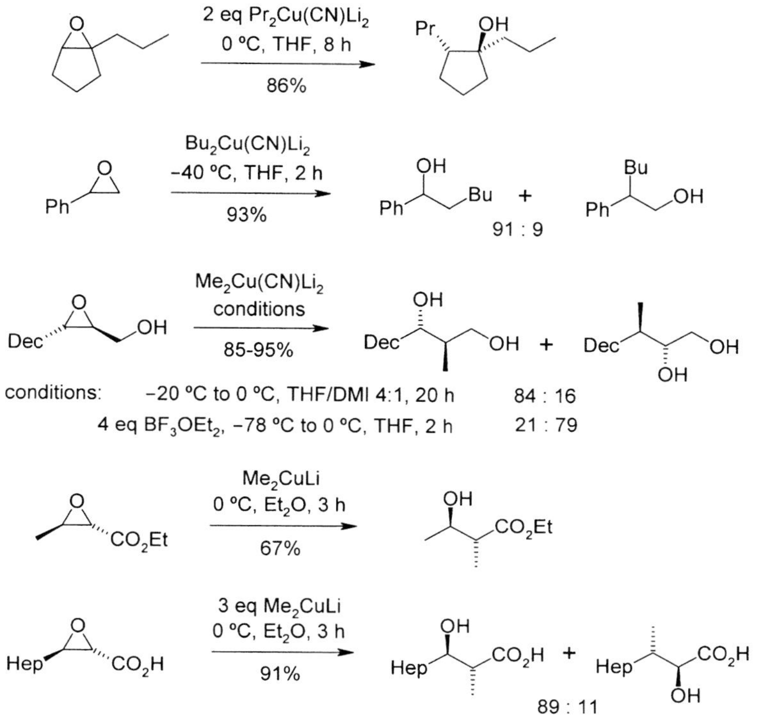

Scheme 4.69. Addition of organocopper reagents to epoxides [303, 314, 316, 317].

Treatment of epoxides with organoaluminum compounds [293] or other Lewis-acidic organometallic reagents can lead to nucleophilic attack at the most highly substituted carbon atom. Thus, as shown in Scheme 4.68, the regioselectivity of styrene oxide opening is reversed when using an organoaluminum reagent instead of a cuprate. Further examples are sketched in Scheme 4.70.

The reactivity of organotitanium compounds is similar to that of organoaluminum compounds, and attack at the most polarized C–O bond is usually observed (Scheme 4.71). The lack of stereoselectivity of the second example in Scheme 4.71 suggests the intermediate formation of carbocations or radicals.

In the presence of Lewis acids allyl silanes and stannanes react with epoxides generally at the sterically less demanding carbon atom. Other electron-rich alkenes, such as ketene acetals, can also be used as nucleophiles. The strong Lewis acids required might, however, also lead to rearrangement of the epoxide before addition of the nucleophile can occur (last reaction, Scheme 4.72).

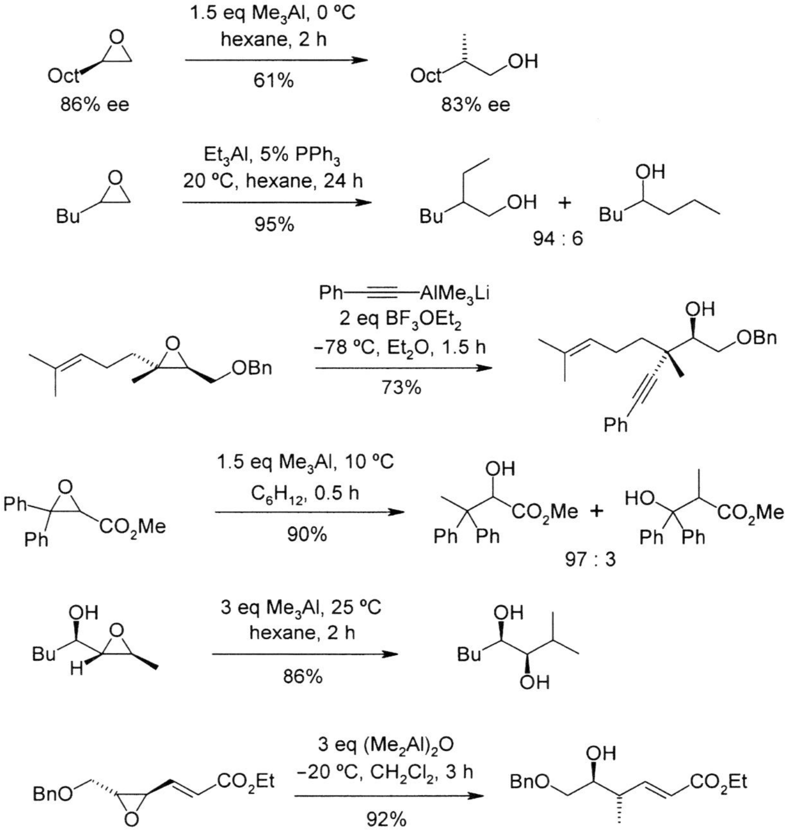

1.5 eq Me_3Al, 0 °C
hexane, 2 h
61%
Oct
86% ee
Oct OH
83% ee

Et_3Al, 5% PPh_3
20 °C, hexane, 24 h
95%
Bu
Bu OH + Bu OH
94 : 6

Ph—≡—$AlMe_3Li$
2 eq BF_3OEt_2
−78 °C, Et_2O, 1.5 h
73%
OBn
OH OBn
Ph

1.5 eq Me_3Al, 10 °C
C_6H_{12}, 0.5 h
90%
Ph Ph CO_2Me
OH CO_2Me Ph Ph + HO CO_2Me Ph Ph
97 : 3

3 eq Me_3Al, 25 °C
hexane, 2 h
86%
OH Bu H
OH Bu OH

3 eq $(Me_2Al)_2O$
−20 °C, CH_2Cl_2, 3 h
92%
BnO CO_2Et O
BnO OH CO_2Et

Scheme 4.70. Addition of organoaluminum reagents to epoxides [293, 319, 324–326].

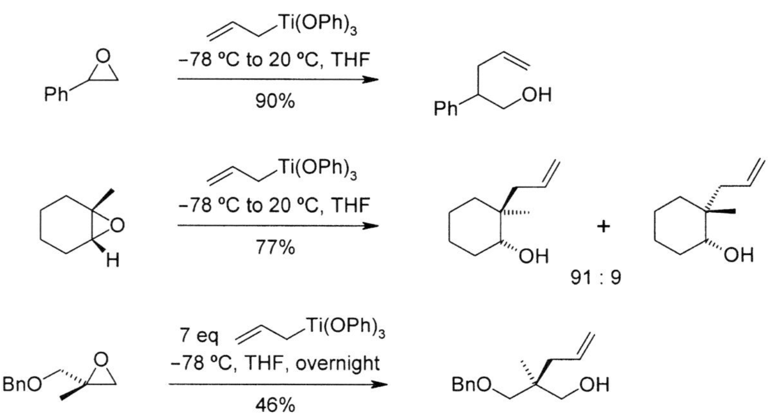

$Ti(OPh)_3$
−78 °C to 20 °C, THF
90%
Ph
Ph OH

$Ti(OPh)_3$
−78 °C to 20 °C, THF
77%
O H
OH + OH
91 : 9

7 eq $Ti(OPh)_3$
−78 °C, THF, overnight
46%
BnO
BnO OH

Scheme 4.71. Addition of organotitanium reagents to epoxides [322, 327].

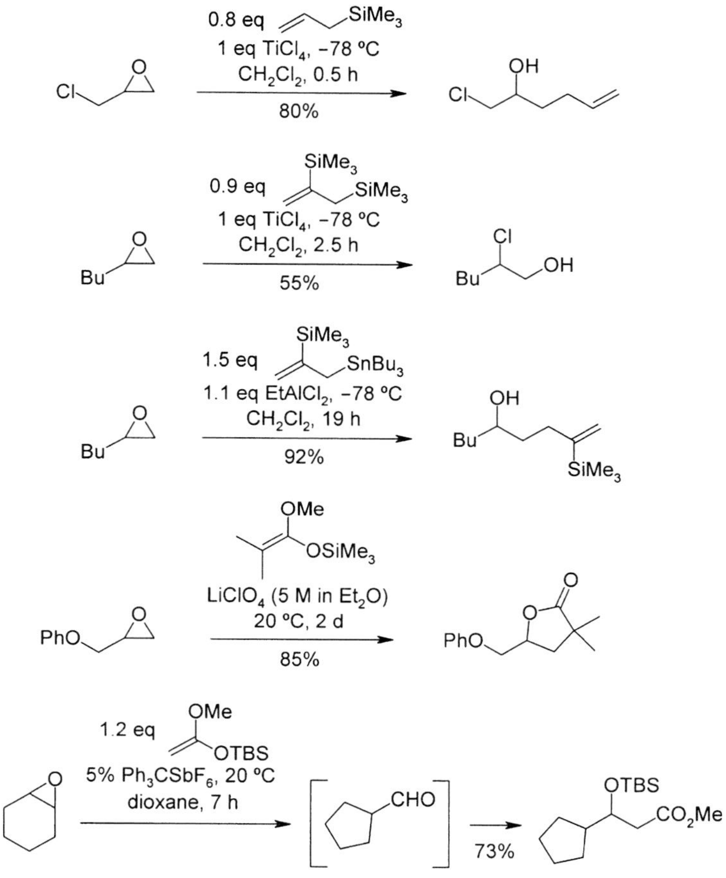

Scheme 4.72. Reaction of epoxides with allyl silanes, allyl stannanes, and silylenol ethers [297, 308, 309, 328].

4.3.8.3 Epoxide Opening by Amines

Amines usually react with epoxides at the less substituted carbon atom (Scheme 4.73) [329, 330]. With sterically demanding reaction partners these reactions will often proceed slowly or, as with tetraalkyl epoxides, not at all [252, 331]. Higher reaction rates can be achieved by increasing the concentration of the reactants, by using lithium amides as nucleophiles [332], or by catalysis with Lewis acids [252, 333] or Brønsted acids [334]. Ammonia can also be alkylated by 2,3-dialkyl epoxides (80 °C, 15–60 h [335]). Hydroxymethyl epoxides (but not alkoxymethyl epoxides) can be activated toward nucleophilic attack by amines by use of stoichiometric amounts of $Ti(OiPr)_4$ [336] (third example, Scheme 4.73).

The regioselectivity of epoxide ring opening by amines can occasionally be modified by using a lithium amide instead of the amine. In the example sketched in

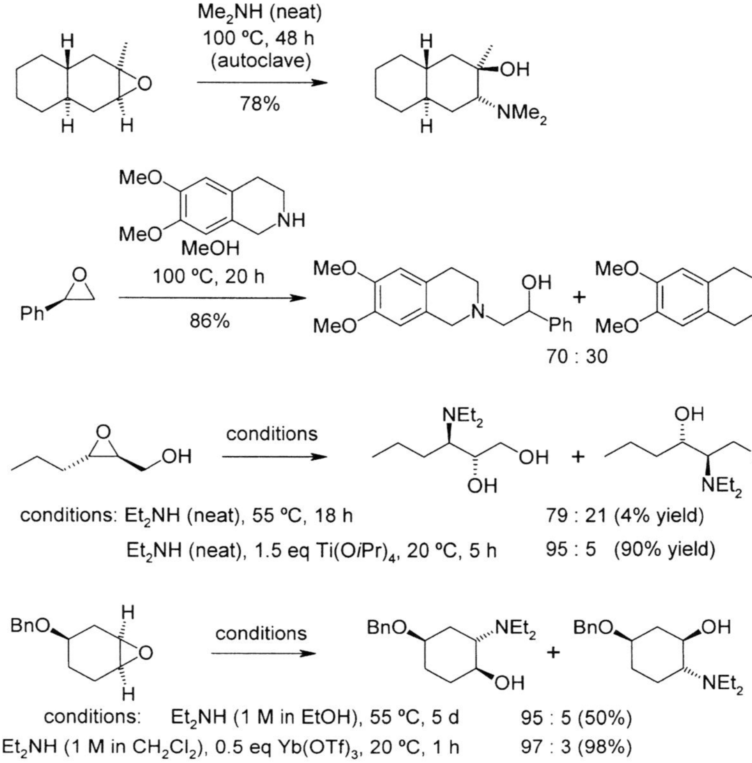

Me$_2$NH (neat)
100 °C, 48 h
(autoclave)
78%

MeOH
100 °C, 20 h
86%
70 : 30

conditions
conditions: Et$_2$NH (neat), 55 °C, 18 h — 79 : 21 (4% yield)
Et$_2$NH (neat), 1.5 eq Ti(O*i*Pr)$_4$, 20 °C, 5 h — 95 : 5 (90% yield)

conditions
conditions: Et$_2$NH (1 M in EtOH), 55 °C, 5 d — 95 : 5 (50%)
Et$_2$NH (1 M in CH$_2$Cl$_2$), 0.5 eq Yb(OTf)$_3$, 20 °C, 1 h — 97 : 3 (98%)

Scheme 4.73. Reactions of epoxides with amines [252, 336–338].

Scheme 4.74 complete reversal of regioselectivity is observed when highly basic lithium amides are used as nucleophiles. Activation of the benzylic C–O bond by chelate formation of Li$^+$ with the oxirane and the carbonyl group has been proposed as a possible reason for this unexpected reversal of regioselectivity [332].

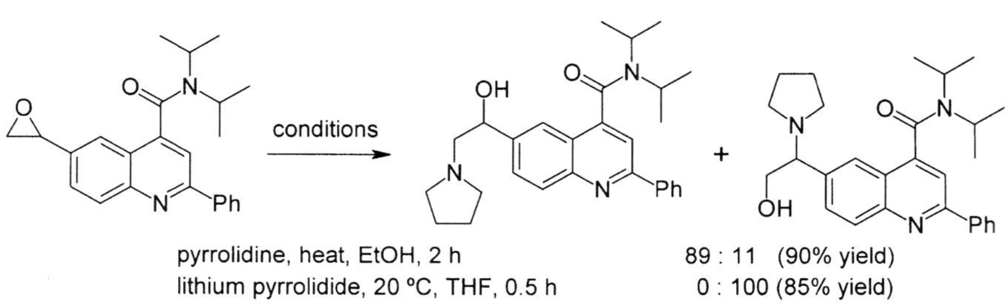

conditions
pyrrolidine, heat, EtOH, 2 h — 89 : 11 (90% yield)
lithium pyrrolidide, 20 °C, THF, 0.5 h — 0 : 100 (85% yield)

Scheme 4.74. Reagent-controlled regioselectivity of epoxide ring opening with amines [332].

Catalysis by acids, which is only rarely effective for aliphatic amines but better suited to the less basic aromatic amines [334], can promote nucleophilic attack at the most strongly polarized C–O bond of the epoxide (Scheme 4.75) [333, 334, 339]. Vinyl epoxides react with amines in the presence of Pd(0) under mild conditions to yield allylamines [340]. If such reactions are performed in the presence of an enantiomerically pure ligand, racemic vinyl epoxides can be converted into enantiomerically enriched products of nucleophilic ring opening (last example, Scheme 4.75).

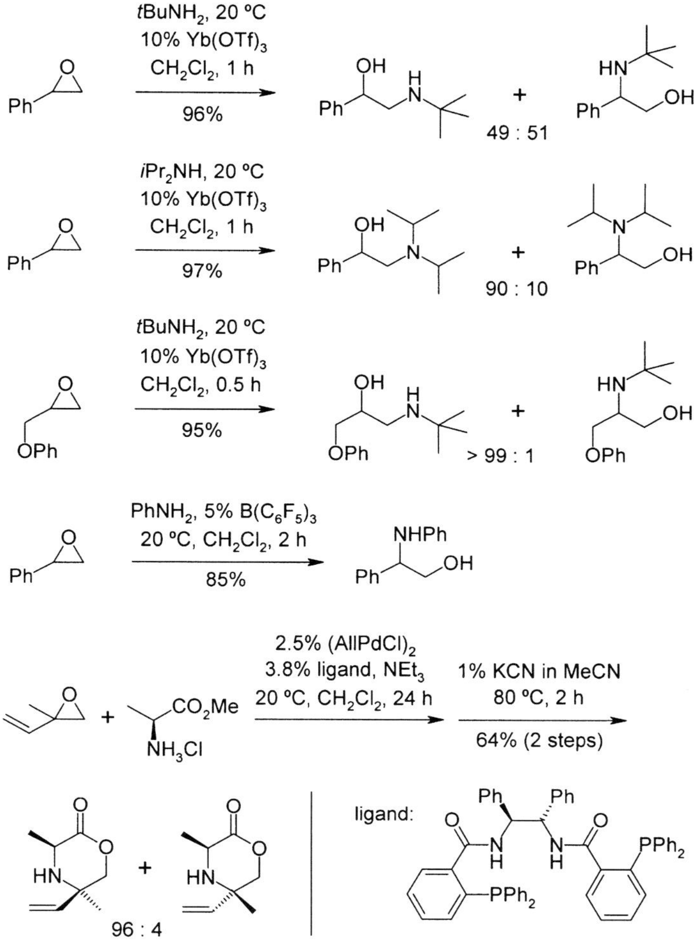

Scheme 4.75. Catalyzed reaction of epoxides with amines [252, 340, 341].

The alkoxycarbonyl group does not have a strong directing effect on the ring opening of epoxides by amines (Scheme 4.76), and steric or electronic effects of the other substituents can be more important to the outcome of these reactions. The regioselectivity of the last reaction in Scheme 4.76 is dominated by the perfluoroalkyl group, which strongly inhibits nucleophilic attack at the carbon atom to which it is attached (see Section 4.3.5).

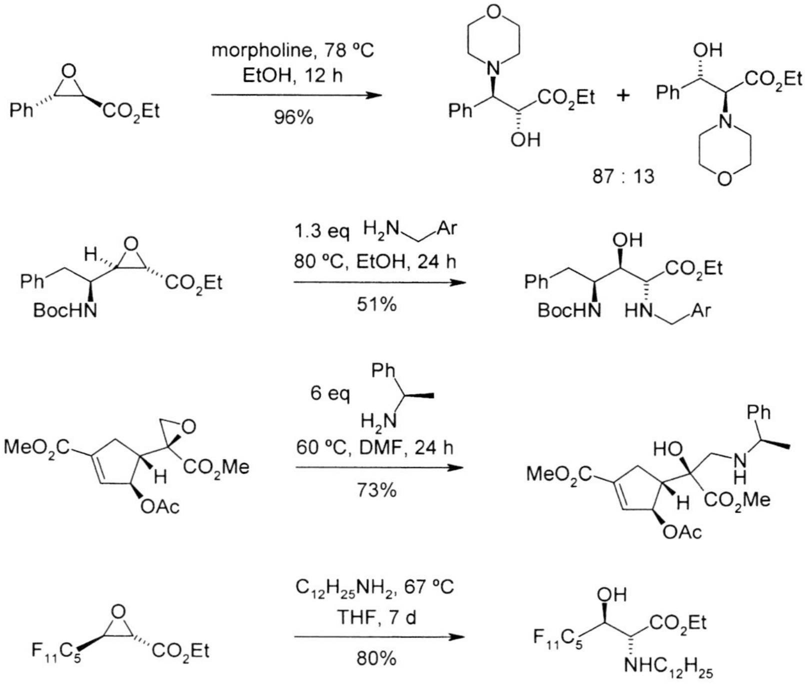

Scheme 4.76. Reactions of amines with alkoxycarbonyl-substituted epoxides [342–345].

The carboxyl group seems to activate epoxides slightly toward nucleophilic attack by amines, and in the absence of catalysts most 2,3-epoxycarboxylic acids react with amines to yield 2-amino-3-hydroxycarboxylic acids [346–348]. This regioselectivity can, however, be overridden by complex formation with Ti(O*i*Pr)$_4$ (Scheme 4.77).

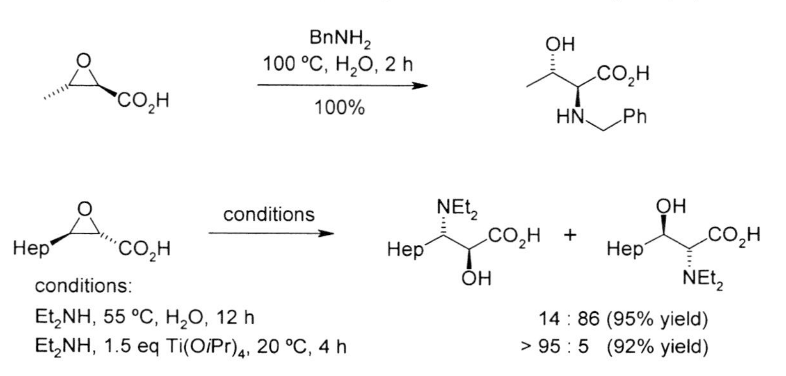

Scheme 4.77. Reactions of amines with 2,3-epoxycarboxylic acids [349, 350].

4.3.8.4 Epoxide Opening by Alcohols

Aliphatic or aromatic alcohols can be alkylated by epoxides under either basic or acidic reaction conditions. Reaction of aliphatic alcoholates with epoxides can be complicated by base-induced rearrangement or oligomerization of the epoxide, because alcoholates are strongly basic and because the product of epoxide ring opening is again an alcoholate. These side reactions can be suppressed by using only catalytic amounts of base (Scheme 4.78). The examples sketched in Scheme 4.78 show that under basic reaction conditions nucleophilic attack occurs preferentially at the sterically most accessible carbon atom.

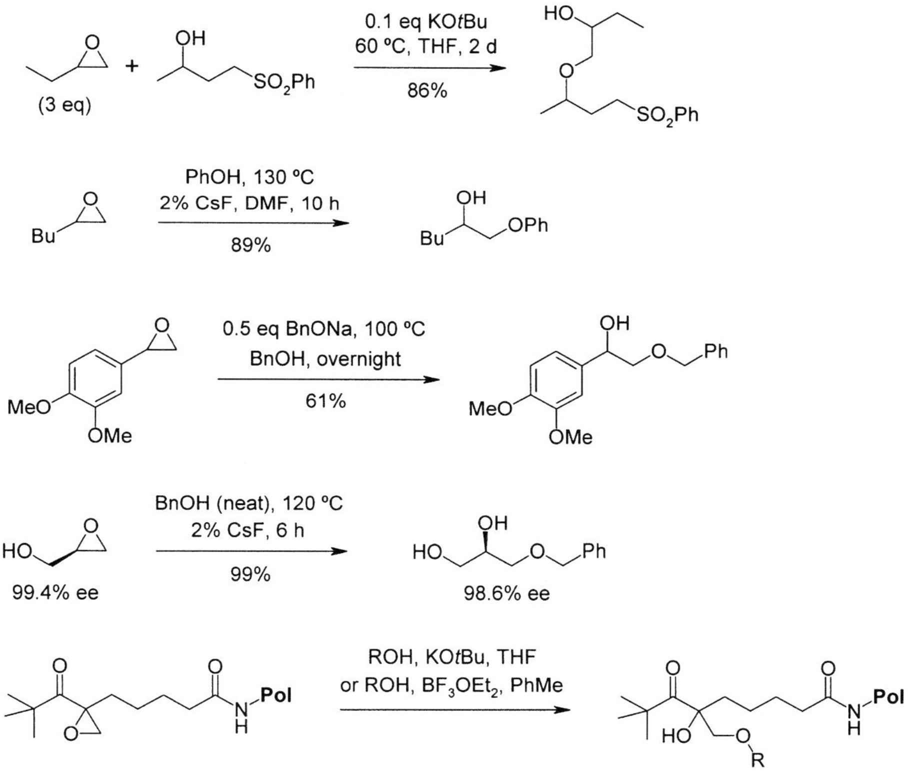

(3 eq) + OH SO$_2$Ph → 0.1 eq KO*t*Bu, 60 °C, THF, 2 d, 86% → HO, O, SO$_2$Ph

Bu (epoxide) → PhOH, 130 °C, 2% CsF, DMF, 10 h, 89% → Bu, OH, OPh

MeO, OMe (aryl epoxide) → 0.5 eq BnONa, 100 °C, BnOH, overnight, 61% → MeO, OMe, OH, O, Ph

HO (epoxide), 99.4% ee → BnOH (neat), 120 °C, 2% CsF, 6 h, 99% → HO, OH, O, Ph, 98.6% ee

O, O, N-H, Pol → ROH, KO*t*Bu, THF or ROH, BF_3OEt_2, PhMe → O, O, HO, O-R, N-H, Pol

Scheme 4.78. Reactions of epoxides with alcohols under basic reaction conditions [329, 351–353]. Pol = polystyrene or TentaGel; R = alkyl.

Less problematic are etherifications with epoxides under acidic reaction conditions. Although we would expect the nucleophile to attack the position which most easily forms a carbocation, this does not always happen and steric effects also seem to be important (Scheme 4.79).

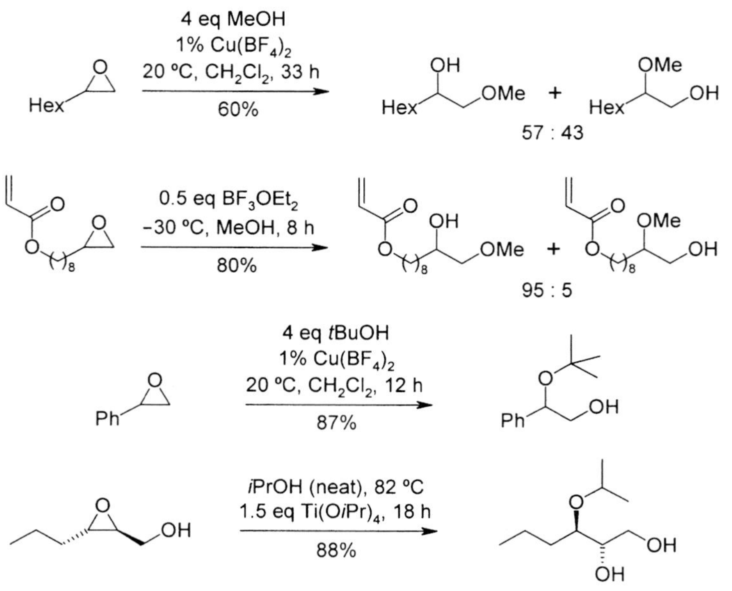

Scheme 4.79. Acid-mediated ring opening of epoxides by alcohols [336, 354, 355].

Vinyl epoxides react with Pd(0) to yield electrophilic allyl complexes which can convert alcohols and phenols into allyl ethers (Scheme 4.80). These alkylations usually yield 2-alkoxy-2-vinylethanols, and if the Pd-mediated etherification is performed in the presence of a chiral, enantiomerically pure diphosphine, enantiomerically enriched ethers may be obtained (Scheme 4.80) [356, 357].

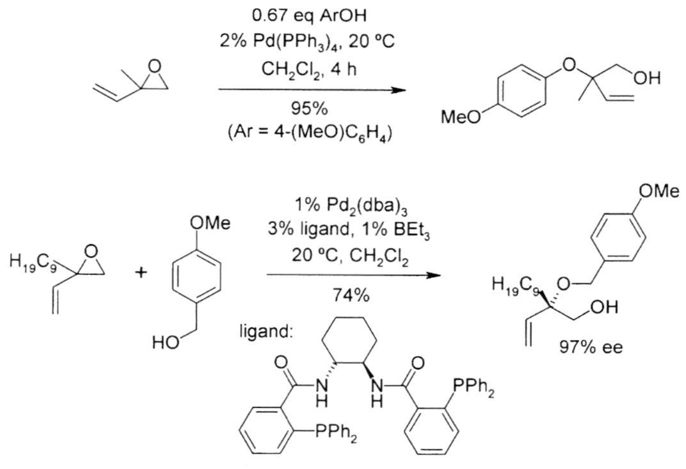

Scheme 4.80. Pd(0)-catalyzed etherification with vinyl epoxides [259, 356].

4.3.8.5 Epoxide Opening by Thiols

Thiols are highly reactive, soft nucleophiles which can be readily alkylated by epoxides under both acidic or basic reaction conditions. Even tetraalkyl epoxides yield the expected 2-mercapto ethanols [358], although these might be rather unstable and undergo fragmentation [359] (last reaction, Scheme 4.82). As with other nucleophiles, steric effects usually control the regioselectivity of epoxide opening under basic reactions, whereas in the presence of acids the stability of the two epoxide-derived carbocations can have a decisive effect on the resulting product ratio (Schemes 4.81 and 4.82). Thus, the regioselectivity of thiol alkylation by styrene oxide (second example, Scheme 4.81) changes significantly on addition of a Lewis acid.

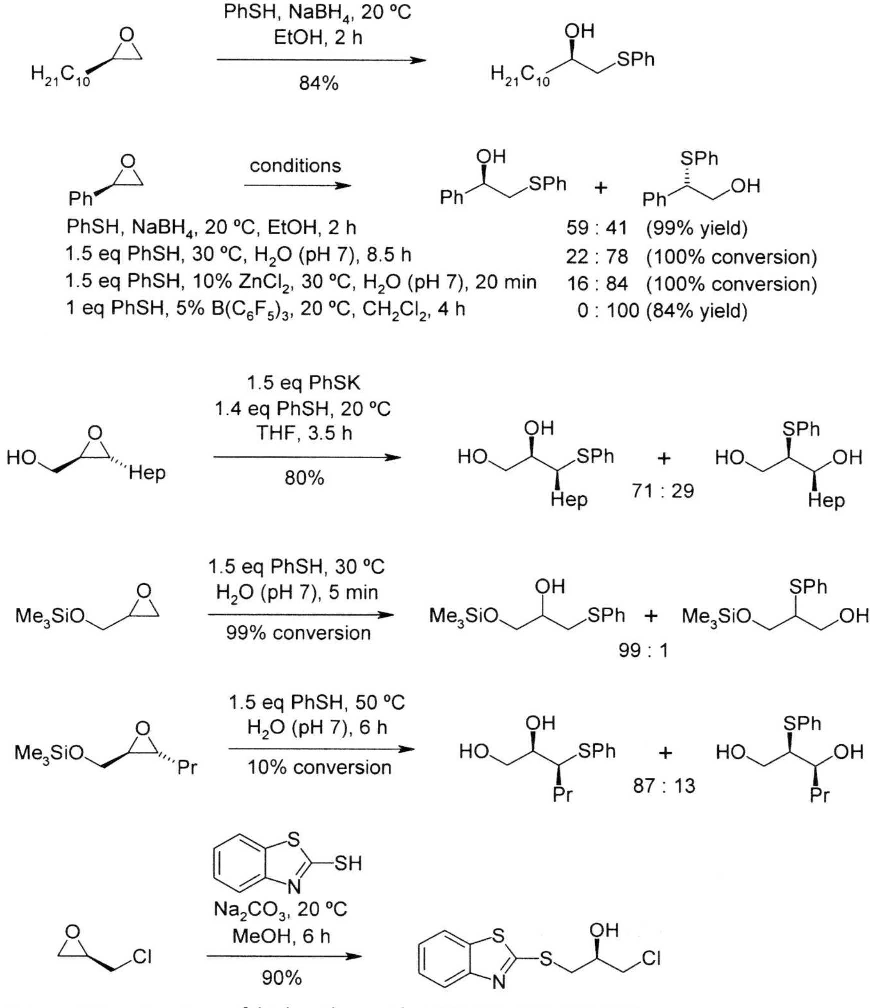

Scheme 4.81. Reactions of thiols with epoxides [247, 249, 341, 360, 361].

Similarly, the outcome of reactions of 2,3-epoxycarboxylic acid derivatives with thiols can be affected by addition of Lewis acids (first reaction, Scheme 4.82). As illustrated by the examples in Scheme 4.82, the regioselectivity of these reactions depends, however, to a large extent on the precise structure of the reacting partners, and can be difficult to predict.

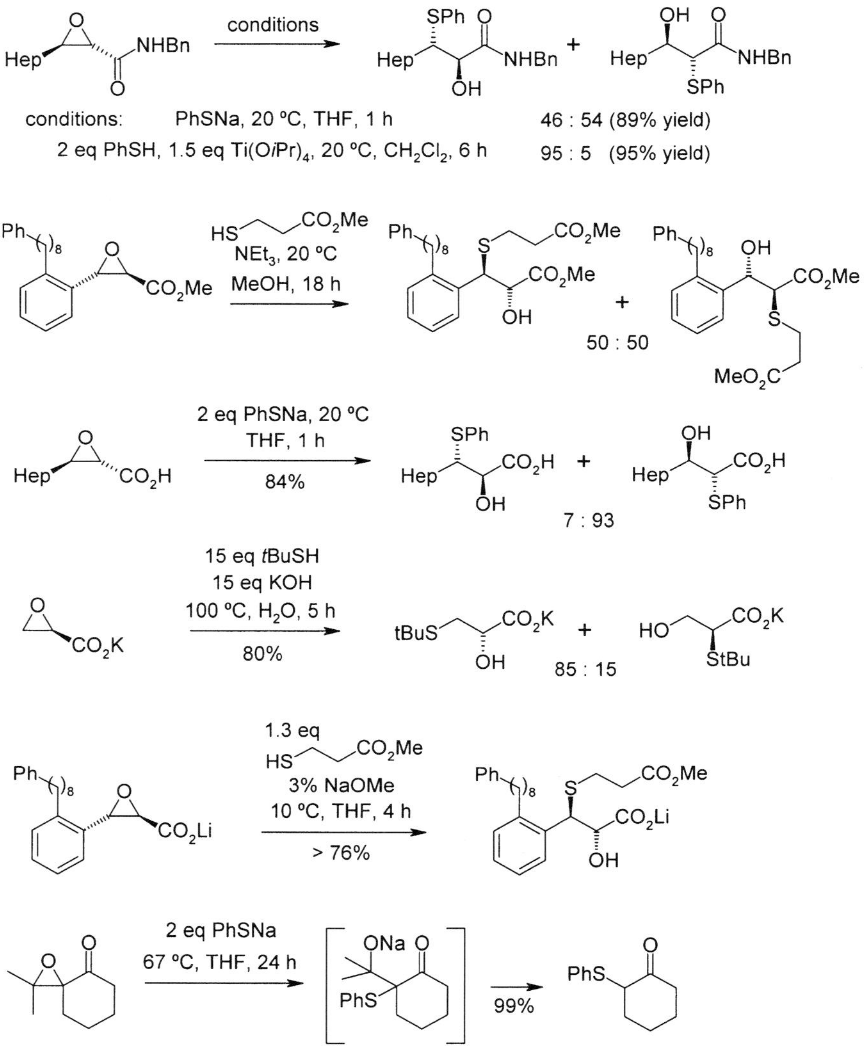

conditions: PhSNa, 20 °C, THF, 1 h — 46 : 54 (89% yield)
2 eq PhSH, 1.5 eq $Ti(OiPr)_4$, 20 °C, CH_2Cl_2, 6 h — 95 : 5 (95% yield)

$HS(CH_2)_2CO_2Me$, NEt_3, 20 °C, MeOH, 18 h — 50 : 50

2 eq PhSNa, 20 °C, THF, 1 h, 84% — 7 : 93

15 eq tBuSH, 15 eq KOH, 100 °C, H_2O, 5 h, 80% — 85 : 15

1.3 eq $HS(CH_2)_2CO_2Me$, 3% NaOMe, 10 °C, THF, 4 h, > 76%

2 eq PhSNa, 67 °C, THF, 24 h; 99%

Scheme 4.82. Reactions of 2,3-epoxycarboxylic acid derivatives and epoxy ketones with thiols [349, 359, 362–364].

2,3-Epoxy-1-propanols with substituents at C-3 can rearrange on treatment with a base to give a terminal epoxide (Payne rearrangement; Scheme 4.83). Because the rearranged epoxide will react with nucleophiles such as amines or thiols more rapidly than the unrearranged epoxide, the main product can result from ring opening of the former (Scheme 4.83).

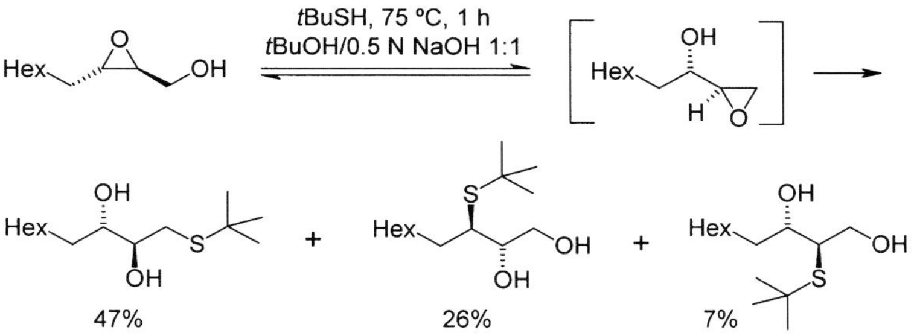

tBuSH, 75 °C, 1 h
tBuOH/0.5 N NaOH 1:1

47% 26% 7%

Scheme 4.83. Payne rearrangement followed by thiol-mediated ring opening of 3-substituted 2,3-epoxy-1-propanols [365].

4.3.8.6 Epoxide Opening by Azide

Either alkali metal azides or Me_3SiN_3 can be used for ring opening of epoxides by the azide ion to yield 2-azido ethanols. With the latter reagent the corresponding trimethylsilyl ethers can be obtained instead of the alcohols.

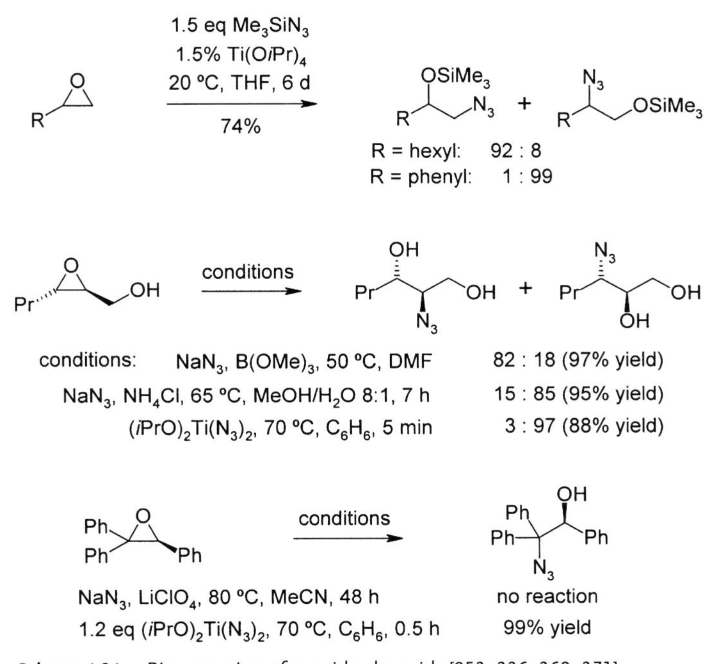

Scheme 4.84. Ring opening of epoxides by azide [253, 336, 369–371].

Most epoxides react with alkali metal azides or Me_3SiN_3 only sluggishly, and different catalysts have been recommended, for example quaternary ammonium salts [248, 366], $AlCl_3$ [367], and copper(II) salts [368]. Other reagents are given in Schemes 4.84 and 4.85.

The regioselectivity of epoxide ring opening by azide is similar to that of other nucleophiles. Monoalkyl epoxides are preferentially attacked at the methylene group whereas in styrene oxides the benzylic C–O bond is usually cleaved, in particular in the presence of acids (Scheme 4.84). By choosing the right reagents and additives, 2,3-epoxypropanols can be converted into either of both possible azido diols.

2,3-Epoxycarboxylic acid derivatives react with azide to yield mainly 2-hydroxy-3-azidoalkanoic acid derivatives [367, 368] (Scheme 4.85). Addition of Lewis acids to the reaction mixture enhances this selectivity further and renders this reaction a valuable strategy for stereoselective preparation of α-hydroxy-β-amino acids from allyl alcohols [368] (last example, Scheme 4.85).

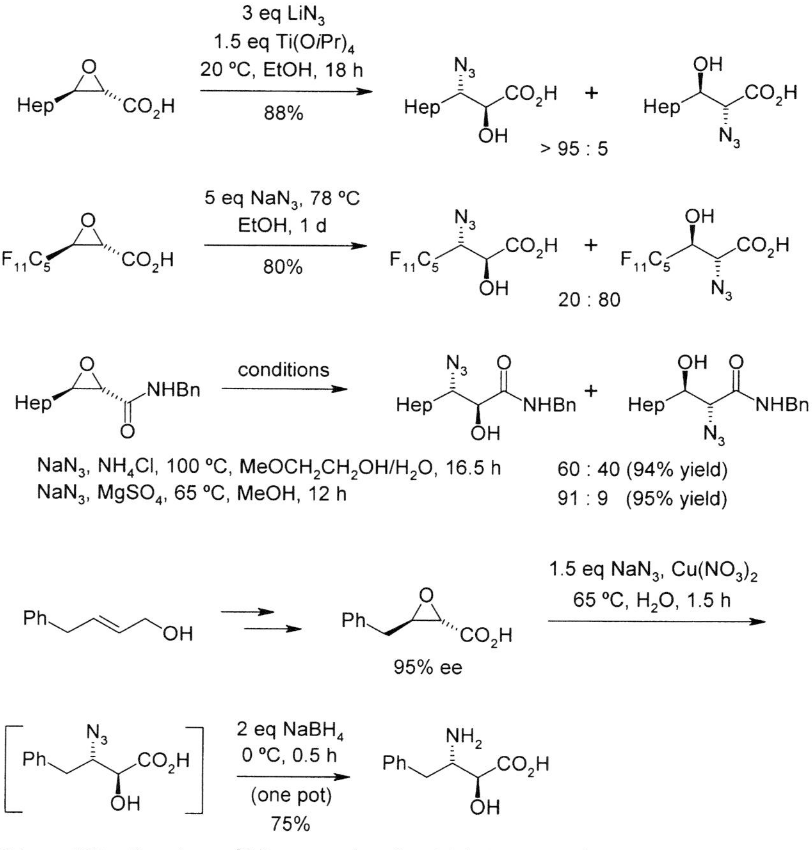

Scheme 4.85. Reactions of 2,3-epoxycarboxylic acid derivatives with azide [247, 345, 349, 368].

4.3.8.7 Epoxide Opening by Cyanide

Epoxides react with cyanide under basic reaction conditions to yield 3-hydroxypropionitriles by nucleophilic attack at the sterically less demanding carbon atom (Scheme 4.86). Me_3SiCN can also be used as reagent, but trimethylsilyl ethers will be the main products. With some types of epoxide (e.g. styrene oxide [372]) the products readily dehydrate to yield α,β-unsaturated nitriles [373] (Scheme 4.86).

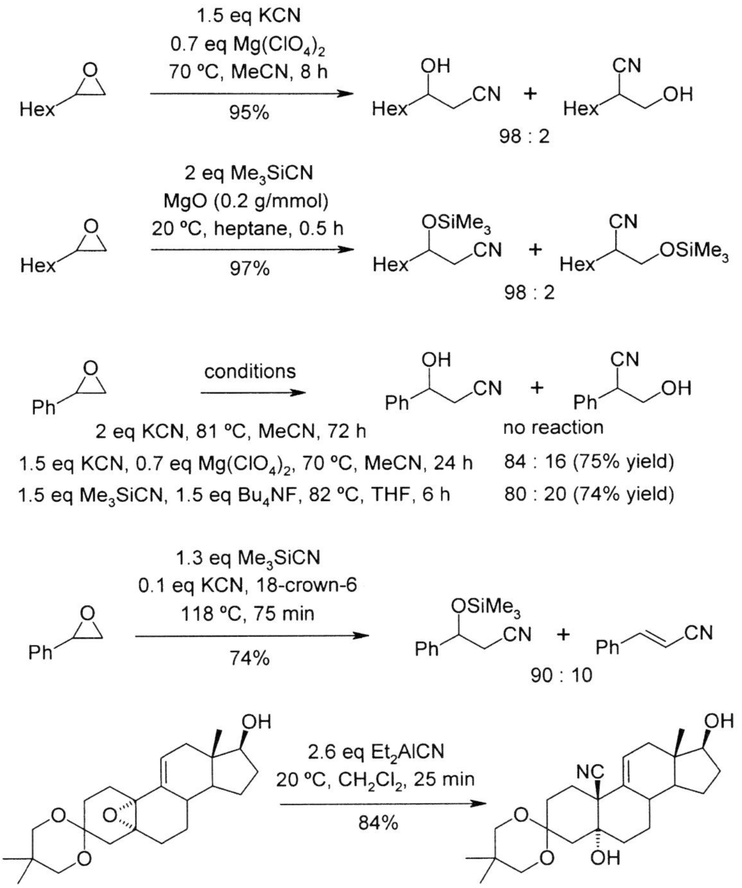

Scheme 4.86. Ring opening of epoxides by cyanide [248, 372, 374–376].

Under acidic reaction conditions the formation of isonitriles can compete efficiently with nitrile formation (Scheme 4.87) [377]. Particularly effective reagents for the formation of isonitriles are mixtures of Me_3SiCN with Lewis acids such as Zn(II), Pd(II), or Sn(II) salts. Aluminum-derived Lewis acids with Me_3SiCN, on the other hand, mediate the conversion of epoxides into nitriles [378, 379].

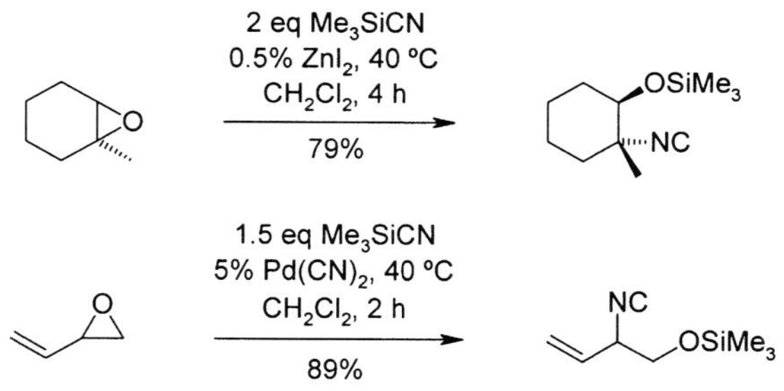

2 eq Me_3SiCN
0.5% ZnI_2, 40 °C
CH_2Cl_2, 4 h
79%
$OSiMe_3$
NC

1.5 eq Me_3SiCN
5% $Pd(CN)_2$, 40 °C
CH_2Cl_2, 2 h
89%
NC
$OSiMe_3$

Scheme 4.87. Formation of isonitriles from epoxides and Me_3SiCN [378, 379].

Epoxides substituted with oxymethyl or other electron-withdrawing groups are usually attacked by cyanide in the presence of acids at the carbon which can form the better stabilized carbocation (Scheme 4.88). As with other nucleophiles, for 2,3-epoxypropanols this regioselectivity can be reversed by addition of boron-derived Lewis acids (first example, Scheme 4.88).

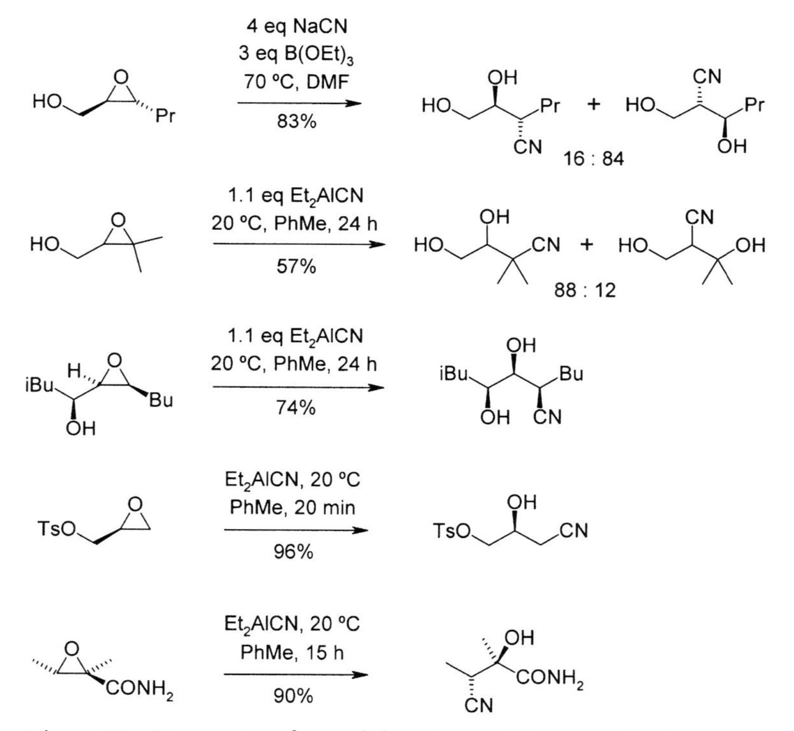

Scheme 4.88. Ring opening of oxymethyl epoxides and 2,3-epoxyamides by cyanide [243, 369, 380, 381].

4.3.8.8 Epoxide Opening by Halides

Many different reagents and reaction conditions have been explored for ring opening of epoxides by halides to prepare 2-haloethanols. As illustrated by the selection of examples given in Scheme 4.89, the regioselectivity of this reaction can be controlled quite effectively.

Styrene oxides are preferentially attacked by halides at the benzylic position, especially so under acidic reaction conditions. It has been claimed, however, that in water in the presence of cyclodextrins, styrene oxides react with halides to yield exclusively benzylic alcohols [382]. Benzylic alcohols can also be obtained from styrene oxides by treatment with halogens in the presence of pyridines [383].

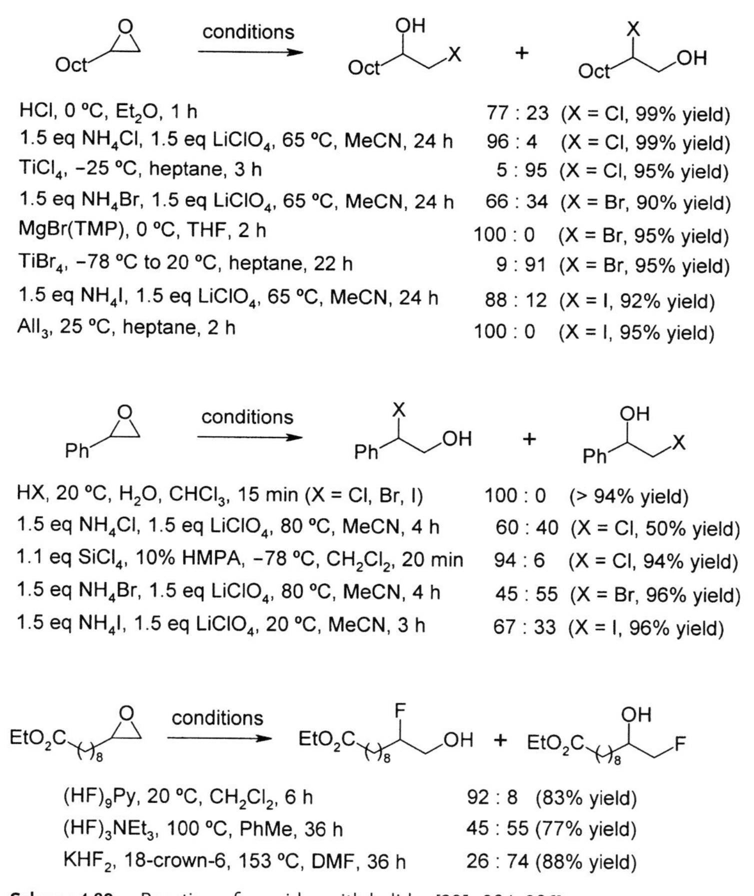

Scheme 4.89. Reaction of epoxides with halides [291, 384–386].

Treatment of epoxides with halides under acidic reaction conditions does not always lead to attack of the halide at the most highly substituted carbon atom. The examples in Schemes 4.89 and 4.90 show that often the opposite regioselectivity is observed, an outcome that is neither readily explained nor predicted. Enantioselective versions of this reaction have been reported [384].

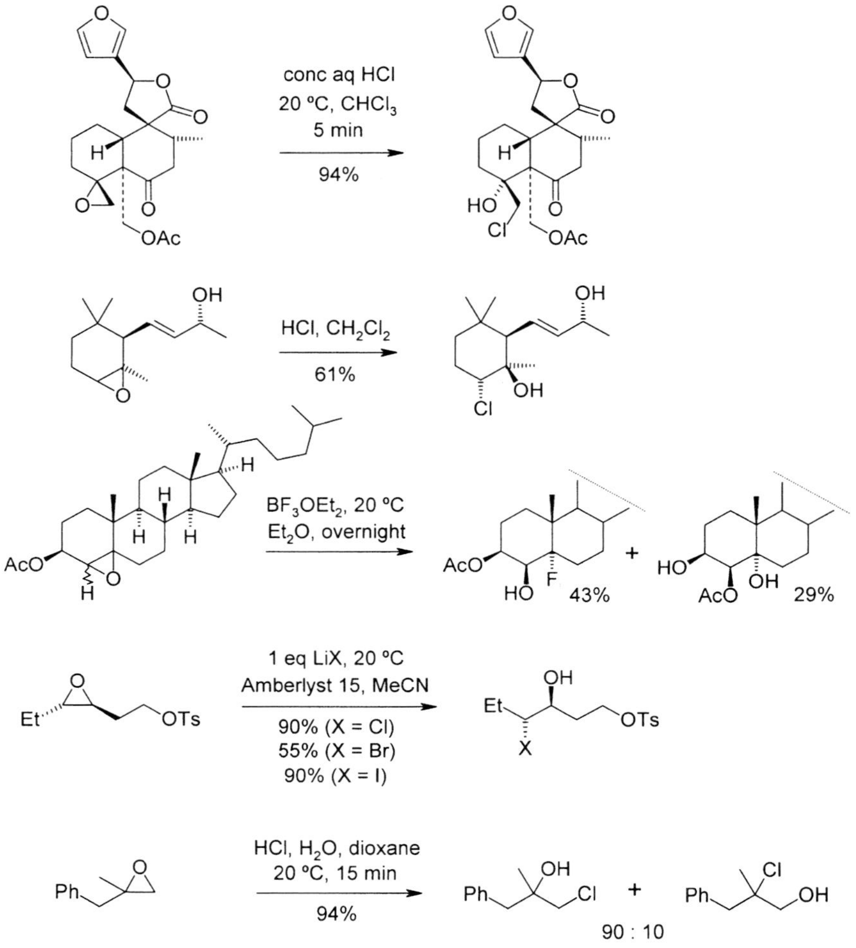

Scheme 4.90. Reaction of epoxides with halides under acidic reaction conditions [246, 255, 387–389].

2,3-Epoxy-1-propanols are preferentially attacked by halides at C-3. This regioselectivity can be enhanced further by acidic reaction conditions or by addition of Ti(IV) salts. Attack at C-2 is promoted by boron-derived Lewis acids (fourth reaction, Scheme 4.91).

Ring opening of epoxides by halides is reversible, and from 2,3-epoxy-1-propanols primary halides can be obtained by a sequence of epoxide openings and epoxide formations (last reaction, Scheme 4.91).

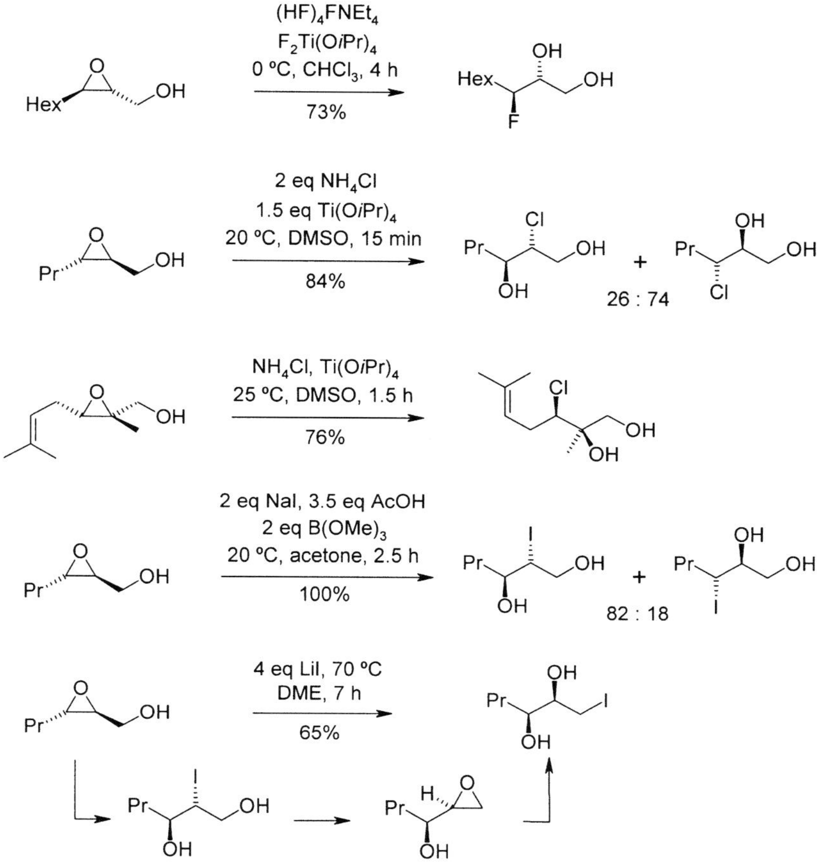

$(HF)_4FNEt_4$, $F_2Ti(OiPr)_4$, 0 °C, $CHCl_3$, 4 h, 73%

2 eq NH_4Cl, 1.5 eq $Ti(OiPr)_4$, 20 °C, DMSO, 15 min, 84%; 26 : 74

NH_4Cl, $Ti(OiPr)_4$, 25 °C, DMSO, 1.5 h, 76%

2 eq NaI, 3.5 eq AcOH, 2 eq $B(OMe)_3$, 20 °C, acetone, 2.5 h, 100%; 82 : 18

4 eq LiI, 70 °C, DME, 7 h, 65%

Scheme 4.91. Reaction of 2,3-epoxypropanols with halides [246, 336, 390–392].

2,3-Epoxycarboxylic acids and derivatives thereof can yield, on reaction with halides, both possible regioisomers (Scheme 4.92). Strongly acidic reaction conditions or Lewis acids can promote the formation of 2-hydroxycarboxylic acids.

Mesyloxymethyl epoxides and related compounds may lead to unexpected products on reaction with iodide, which occasionally acts as reducing agent. As illustrated in Scheme 4.93, halogen-free products can result from such reactions.

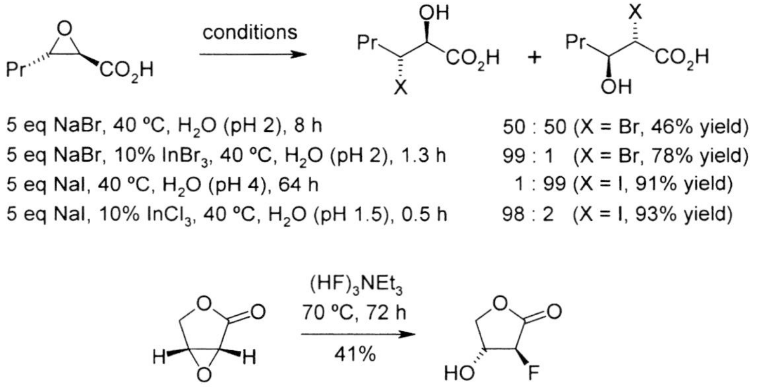

conditions

5 eq NaBr, 40 °C, H_2O (pH 2), 8 h — 50 : 50 (X = Br, 46% yield)
5 eq NaBr, 10% $InBr_3$, 40 °C, H_2O (pH 2), 1.3 h — 99 : 1 (X = Br, 78% yield)
5 eq NaI, 40 °C, H_2O (pH 4), 64 h — 1 : 99 (X = I, 91% yield)
5 eq NaI, 10% $InCl_3$, 40 °C, H_2O (pH 1.5), 0.5 h — 98 : 2 (X = I, 93% yield)

$(HF)_3NEt_3$
70 °C, 72 h
41%

Scheme 4.92. Reaction of 2,3-epoxycarboxylic acids and lactones with halides [393, 394].

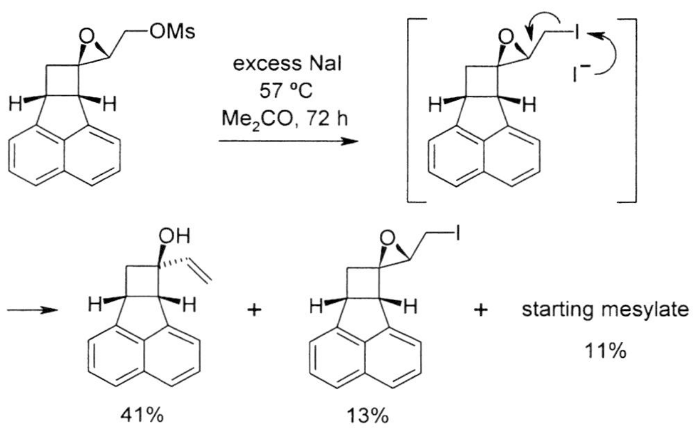

excess NaI
57 °C
Me_2CO, 72 h

41% 13% + starting mesylate 11%

Scheme 4.93. Reductive ring opening of mesyloxy epoxides with iodide [395].

References

1 March, J. *Advanced Organic Chemistry*; John Wiley and Sons: New York, 1992.

2 Fang, Y.-R.; Gao, Y.; Ryberg, P.; Eriksson, J.; Kolodziejska-Huben, M.; Dybala-Defratyka, A.; Madhavan, S.; Danielsson, R.; Paneth, P.; Matsson, O.; Westaway, K. C. Experimental and theoretical multiple kinetic isotope effects for an SN2 reaction. An attempt to determine transition-state structure and the ability of theoretical methods to predict experimental kinetic isotope effects. *Chem. Eur. J.* **2003**, *9*, 2696–2709.

3 Gonzalez, J. M.; Pak, C.; Cox, R. S.; Allen, W. D.; Schaefer, H. F.; Császár, A. G.; Tarczay, G. Definitive ab initio studies of model SN2 reactions $CH_3X + F^-$ (X = F, Cl, CN, OH, SH, NH_2, PH_2). *Chem. Eur. J.* **2003**, *9*, 2173–2192.

4 Reynolds, D. D.; Evans, W. L. β-D-Glucose-1,2,3,4-tetraacetate. *Org. Synth.* **1955**, *Coll. Vol. III*, 432–434.

5 Koppenhoefer, B.; Schurig, V. (*R*)-Alkyloxiranes of high enantiomeric purity from (*S*)-2-chloroalkanoic acids via (*S*)-2-chloro-1-alkanols: (*R*)-methyloxirane. *Org. Synth.* **1993**, *Coll. Vol. VIII*, 434–441.

6 Rossi, R. A.; Pierini, A. B.; Peñéñory, A. B. Nucleophilic substitution reactions by electron transfer. *Chem. Rev.* **2003**, *103*, 71–167.

7 Kornblum, N. Substitution reactions which proceed via radical anion intermediates. *Angew. Chem. Int. Ed.* **1975**, *14*, 734–745.

8 Rossi, R. A.; Pierini, A. B.; Palacios, S. M. Nucleophilic substitution by the SRN1 mechanism on alkyl halides. *Adv. Free Radical Chem.* **1990**, *1*, 193–252.

9 Bowman, W. R. Reactivity of substituted aliphatic nitro compounds with nucleophiles. *Chem. Soc. Rev.* **1988**, *17*, 283–316.

10 Ashby, E. C. Single-electron transfer, a major reaction pathway in organic chemistry. An answer to recent criticisms. *Acc. Chem. Res.* **1988**, *21*, 414–421.

11 Lund, H.; Kristensen, L. H. Electrogenerated anions as electron transfer reagents. *Acta Chem. Scand.* **1979**, *B 33*, 495–498.

12 Gill, N. S.; James, K. B.; Lions, F.; Potts, K. T. β-Acylethylation with ketonic Mannich bases. The synthesis of some diketones, ketonic sulfides, nitroketones and pyridines. *J. Am. Chem. Soc.* **1952**, *74*, 4923–4928.

13 Kornblum, N.; Stuchal, F. W. New and facile substitution reactions at tertiary carbon. The reactions of amines with *p*-nitrocumyl chloride and α,*p*-dinitrocumene. *J. Am. Chem. Soc.* **1970**, *92*, 1804–1806.

14 Zhang, X.-M.; Yang, D.-L.; Liu, Y.-C. Effects of electron acceptors and radical scavengers on nonchain radical nucleophilic substitution reactions. *J. Org. Chem.* **1993**, *58*, 224–227.

15 Lund, T.; Jacobsen, K. B. Complete inversion of configuration in aliphatic nucleophilic substitution reactions with small inner-sphere stabilization. *Acta Chem. Scand.* **1998**, *52*, 778–783.

16 Lund, H.; Daasbjerg, K.; Lund, T.; Pedersen, S. U. On electron transfer in aliphatic nucleophilic substitution. *Acc. Chem. Res.* **1995**, *28*, 313–319.

17 MacTavish, J.; Proctor, G. R.; Redpath, J. Reductive alkylation of pyridinium salts. Part 1. Synthesis of di-, tetra- and hexahydropyridine esters. *J. Chem. Soc. Perkin Trans. 1* **1996**, 2545–2551.

18 Kosower, E. M.; Poziomek, E. J. Stable free radicals. I. Isolation and distillation of 1-ethyl-4-carbomethoxypyridinyl. *J. Am. Chem. Soc.* **1964**, *86*, 5515–5523.

19 Ashby, E. C.; Sun, X.; Duff, J. L. Single-electron transfer in nucleophilic aliphatic substitution. Evidence for single-electron transfer in the reactions of 1-halonorbornanes with various nucleophiles. *J. Org. Chem.* **1994**, *59*, 1270–1278.

20 Kornblum, N.; Boyd, S. D.; Stuchal, F. W. A new reaction of α-nitro esters, ketones, and nitriles and α,α-dinitro compounds. *J. Am. Chem. Soc.* **1970**, *92*, 5783–5784.

21 Benhida, R.; Gharbaoui, T.; Lechevallier, A.; Beugelmans, R. Dérivés *N*-alkylés du 4-nitroimidazole. Synthèse par chimie radicalaire (SRN1) et aménagements fonctionnels. *Bull. Soc. Chim. Fr.* **1994**, *131*, 200–209.

22 Kornblum, N.; Fifolt, M. J. Electron-transfer substitution reactions: facilitation by the cyano group. *Tetrahedron* **1989**, *45*, 1311–1322.

23 Lund, T. Correlation between inner-sphere stabilization and stereochemistry for the aliphatic nucleophilic substitution. *Tetrahedron Lett.* **1991**, *32*, 1595–1598.

24 Al-Khalil, S. I.; Bowman, W. R.; Gaitonde, K.; Marley, M. A.; Richardson, G. D. Radical-nucleophilic substitution (SRN1) reactions. Part 7. Reactions of aliphatic α-substituted nitro compounds. *J. Chem. Soc. Perkin Trans. 2* **2001**, 1557–1565.

25 Beugelmans, R.; Amrollah-Madjdabadi, A.; Frinault, T.; Morris, A. D.; Gharbaoui, T.; Benhida, R.; Lechevallier, A. New and straightforward access to olefins, butenolides and β-nitro γ-butyrolactones from precursors obtained via SRN1 reactions. *Bull. Soc. Chim. Fr.* **1994**, *131*, 1019–1030.

26 Tani, K.; Suwa, K.; Yamagata, T.; Otsuka, S. Fully alkylated chiral diphosphines, RDIOP, and their Rh(I) complexes. *Chem. Lett.* **1982**, 265–268.

27 Herrera, R. P.; Guijarro, A.; Yus, M. Primary alkyl fluorides as regioselective alkylating reagents of lithium arene dianions. Easy prediction of regioselectivity by MO calculations on the dianion. *Tetrahedron Lett.* **2003**, *44*, 1313–1316.

28 Yamauchi, K.; Tanabe, T.; Kinoshita, M. Trimethylsulfonium hydroxide: a new methylating agent. *J. Org. Chem.* **1979**, *44*, 638–639.

29 Metzger, H.; König, H.; Seelert, K. Methylierung mit Dimethyl-oxo-sulfonium methylid. *Tetrahedron Lett.* **1964**, 876–868.

30 Kuhn, R.; Trischmann, H. Trimethyl-sulfoxonium ion. *Liebigs Ann. Chem.* **1958**, *611*, 117–121.

31 Yamauchi, K.; Nakamura, K.; Kinoshita, M. Trimethylselenonium hydroxide: a new methylating agent. *Tetrahedron Lett.* **1979**, 1787–1790.

32 Shieh, W.-C.; Dell, S.; Repic, O. DBU and microwave-accelerated green chemistry in methylation of phenols, indoles, and benzimidazoles with dimethyl carbonate. *Org. Lett.* **2001**, *3*, 4279–4281.

33 Jiang, X.; Tiwari, A.; Thompson, M.; Chen, Z.; Cleary, T. P.; Lee, T. B. K. A practical method for *N*-methylation of indoles using dimethyl carbonate. *Org. Process Res. Dev.* **2001**, *5*, 604–608.

34 Shieh, W.-C.; Dell, S.; Repic, O. Nucleophilic catalysis with DBU for the esterification of carboxylic acids with dimethyl carbonate. *J. Org. Chem.* **2002**, *67*, 2188–2191.

35 Benn, M.; Meesters, A. C. M. Nucleophilic displacement of the nitro group from nitroalkanes. *J. Chem. Soc. Chem. Commun.* **1977**, 597–598.

36 Tamura, R.; Kamimura, A.; Noboru, O. Displacement of aliphatic nitro groups by carbon and heteroatom nucleophiles. *Synthesis* **1991**, 423–434.

37 Kamimura, A.; Sasatani, H.; Hashimoto, T.; Ono, N. Lewis acid induced nucleophilic substitution reaction of β-nitro sulfides. *J. Org. Chem.* **1989**, *54*, 4998–5003.

38 Trost, B. M.; Ghadiri, M. R. Sulfones as chemical chameleons. Cyclization via 1,1-dipole synthons. *J. Am. Chem. Soc.* **1984**, *106*, 7260–7261.

39 Kocienski, P. J.; Pontiroli, A.; Qun, L. Enantiospecific syntheses of pseudopterosin aglycones. Part 2. Synthesis of pseudopterosin K–L aglycone and pseudopterosin A–F aglycone via a B-BA-BAC annulation strategy. *J. Chem. Soc. Perkin Trans. 1* **2001**, 2356–2366.

40 Bosch, H.; Pflieger, P.; Mioskowski, C.; Salaün, J.-P.; Durst, F. Anchimerically assisted substitution of a sulfinyl function in α-sulfinylhydrazones by thiols. *Tetrahedron Lett.* **1991**, *32*, 2237–2240.

41 Ono, N.; Hamamoto, I.; Kawai, T.; Kaji, A.; Tamura, R.; Kakihana, M. Regioselective preparation of allylic sulfones by Pd-catalyzed reactions of allylic nitro compounds with sodium benzenesulfinate. *Bull. Chem. Soc. Jpn.* **1986**, *59*, 405–410.

42 Ono, N.; Hamamoto, I.; Kaji, A. Regio- and stereoselective γ-substitution of allylic nitro compounds with lithium dialkylcuprates. *J. Chem. Soc. Chem. Commun.* **1984**, 274–275.

43 Prakash, G. K. S.; Hu, J.; Olah, G. A. Preparation of tri- and difluoromethylsilanes via an unusual magnesium metal-mediated reductive tri- and difluoromethylation of chlorosilanes using tri- and difluoromethyl sulfides, sulfoxides, and sulfones. *J. Org. Chem.* **2003**, *68*, 4457–4463.

44 Palomo, C.; Oiarbide, M.; González-Rego, M. C.; Sharma, A. K.; García, J. M.; González, A.; Landa, C.; Linden, A. α-Oxymethyl ketone enolates for the asymmetric Mannich reaction. From acetylene and *N*-alkoxycarbonylimines to β-amino acids. *Angew. Chem. Int. Ed.* **2000**, *39*, 1063–1065.

45 Ley, S. V.; Humphries, A. C.; Eick, H.; Downham, R.; Ross, A. R.; Boyce, R. J.; Pavey, J. B. J.; Pietruszka, J. Total synthesis of the protein phosphatase inhibitor okadaic acid. *J. Chem. Soc. Perkin Trans. 1* **1998**, 3907–3911.

46 Schatz, P. F. Synthesis of chrysanthemic acid. *J. Chem. Educ.* **1978**, *55*, 468–470.

47 Khalil, F. Y.; Hanna, M. T.; El-Batouti, M.; Mikhail, A. A. The role of nucleophiles in the kinetics of decomposition of benzyltriphenylphosphonium chloride in different media. *Z. Phys. Chem.* **2002**, *216*, 719–728.

48 Miller, J. A.; Dankwardt, J. W.; Penney, J. M. Nickel-catalyzed cross-coupling and amination reactions of aryl nitriles. *Synthesis* **2003**, 1643–1648.

49 Higgins, R. H.; Faircloth, W. J.; Baughman, R. G.; Eaton, Q. L. Ring opening of azetidinols by phenols: regiochemistry and stereochemistry. *J. Org. Chem.* **1994**, *59*, 2172–2178.

50 von Braun, J. Pentamethylene bromide. *Org. Synth.* **1941**, *Coll. Vol. I*, 428–430.

51 Magnus, P.; Lacour, J.; Evans, P. A.; Rigollier, P.; Tobler, H. Applications of the β-azidonation reaction to organic synthesis. α,β-Enones, conjugate addition, and γ-lactam annulation. *J. Am. Chem. Soc.* **1998**, *120*, 12486–12499.

52 Lee, P. H.; Cho, M.; Han, I.-S.; Kim, S. Facile nucleophilic substitution of 3-*tert*-butyldimethylsilyloxyalk-2-enylphosphonium salts. *Tetrahedron Lett.* **1999**, *40*, 6975–6978.

53 Mazurkiewicz, R.; Grymel, M. *N*-Acyl-α-triphenylphosphonioglycinates: a novel cationic glycine equivalent and its reactions with heteroatom nucleophiles. *Monatsh. Chem.* **1999**, *130*, 597–604.

54 Richard, J. P.; Amyes, T. L.; Rice, D. J. Effects of electronic geminal interactions on the solvolytic reactivity of methoxymethyl derivatives. *J. Am. Chem. Soc.* **1993**, *115*, 2523–2524.

55 Lee, K. C.; Chi, D. Y. An efficient synthesis of (fluoromethyl)pyridylamines for labelling with ^{18}F. *J. Org. Chem.* **1999**, *64*, 8576–8581.

56 Dyer, R. G.; Turnbull, K. D. Hydrolytic stabilization of protected *p*-hydroxybenzyl halides designed as latent quinone methide precursors. *J. Org. Chem.* **1999**, *64*, 7988–7995.

57 Hoffmann, F. W. Aliphatic fluorides. I. ω,ω'-Difluoroalkanes. *J. Org. Chem.* **1949**, *14*, 105–110.

58 Swain, C. G.; Scott, C. B. Rates of solvolysis of some alkyl fluorides and chlorides. *J. Am. Chem. Soc.* **1953**, *75*, 246–248.

59 Bose, A. K.; Das, K. G.; Funke, P. T. Fluoro compounds. II. Reactions and NMR-studies of some fluorobromo esters. *J. Org. Chem.* **1964**, *29*, 1202–1206.

60 Choe, Y. S.; Katzenellenbogen, J. A. TBAF-induced conversion of tresylates to mesylates. *Tetrahedron Lett.* **1993**, *34*, 1579–1580.

61 Hudlicky, M. Elimination of HF from fluorinated succinic acids. (III) Kinetics of dehydrofluorination of *erythro*- and *threo*-α-bromo-α'-fluorosuccinic acids. *J. Fluorine Chem.* **1984**, *25*, 353–361.

62 Hudlicky, M. Reactivity of fluorine in fluorinated succinic esters. *J. Fluorine Chem.* **1973**, *2*, 1–17.

63 King, J. F.; Gill, M. S. Alkyl 2,2,2-trifluoroethanesulfonates (tresylates): elimination–addition vs bimolecular nucleophilic substitution in reactions with nucleophiles in aqueous media. *J. Org. Chem.* **1996**, *61*, 7250–7255.

64 Mikhailopulo, I. A.; Pricota, T. I.; Sivets, G. G.; Altona, C. 2′-Chloro-2′,3′-dideoxy-3′-fluoro-D-ribonucleosides: synthesis, stereospecificity, some chemical transformations, and conformational analysis. *J. Org. Chem.* **2003**, *68*, 5897–5908.

65 Blanksby, S. J.; Ellison, G. B. Bond dissociation energies of organic molecules. *Acc. Chem. Res.* **2003**, *36*, 255–263.

66 Alabugin, I. V.; Zeidan, T. A. Stereoelectronic effects and general trends in hyperconjugative acceptor ability of σ bonds. *J. Am. Chem. Soc.* **2002**, *124*, 3175–3185.

67 Hine, J.; Wiesboeck, R.; Ghirardelli, R. G. The kinetics of the base-catalyzed deuterium exchange of 2,2-dihalo-1,1,1-trifluoroethanes. *J. Am. Chem. Soc.* **1961**, *83*, 1219–1222.

68 Streitwieser, A.; Holtz, D. Acidity of hydrocarbons. XXIII. Base-catalyzed proton exchange of 1*H*-undecafluorobicyclo[2.2.1]heptane and the role of C–F no-bond resonance. *J. Am. Chem. Soc.* **1967**, *89*, 692–693.

69 Shellhamer, D. F.; Allen, J. L.; Allen, R. D.; Gleason, D. C.; Schlosser, C. O.; Powers, B. J.; Probst, J. W.; Rhodes, M. C.; Ryan, A. J.; Titterington, P. K.; Vaughan, G. G.; Heasley, V. L. Ionic reaction of halogens with terminal alkenes: the effect of electron-withdrawing fluorine substituents on the bonding of halonium ions. *J. Org. Chem.* **2003**, *68*, 3932–3937.

70 Takeuchi, Y.; Takagi, K.; Yamaba, T.; Nabetani, M.; Koizumi, T. Synthetic studies for novel structure of α-nitrogenously functionalized α-fluorocarboxylic acids. Part III. Some reactions of α-bromo-α-fluorocarboxylic acids and their ethyl esters with NaN_3. *J. Fluorine Chem.* **1994**, *68*, 149–154.

71 Baklouti, A.; Chaabouni, M. M. Synthese d'ethyleniques monofluores. *J. Fluorine Chem.* **1981**, *19*, 181–190.

72 Eckes, L.; Hanack, M. Herstellung von Vinylfluoriden. *Synthesis* **1978**, 217–219.

73 Anilkumar, R.; Burton, D. J. A highly efficient room temperature non-organometallic route for the synthesis of α,β,β-trifluorostyrenes by dehydrohalogenation. *Tetrahedron Lett.* **2003**, *44*, 6661–6664.

74 Lee, J. G.; Bartsch, R. A. Dehydrohalogenation by complex base ($NaNH_2$/NaO*t*Bu). Preferential loss of "poorer" halogen leaving groups. *J. Am. Chem. Soc.* **1979**, *101*, 228–229.

75 Swain, C. G.; Spalding, R. E. T. Mechanisms of acid catalysis of the hydrolysis of benzyl fluoride. *J. Am. Chem. Soc.* **1960**, *82*, 6104–6107.

76 Namavari, M.; Satyamurthy, N.; Phelps, M. E.; Barrio, J. R. Halogen exchange reactions between alkyl halides and aqueous hydrogen halides. A new method for preparation of alkyl halides. *Tetrahedron Lett.* **1990**, *31*, 4973–4976.

77 Theodoridis, G. Novel applications of alkyl fluorides in organic synthesis: versatile nitrogen protecting groups. *Tetrahedron Lett.* **1998**, *39*, 9365–9368.

78 Olah, G. A.; Kuhn, S. J. Selective Friedel–Crafts reactions. I. Boron halide catalyzed haloalkylation of benzene and alkylbenzenes with fluorohaloalkanes. *J. Org. Chem.* **1964**, *29*, 2317–2320.

79 Bordwell, F. G. Equilibrium acidities in DMSO solution. *Acc. Chem. Res.* **1988**, *21*, 456–463.

80 Postel, D.; Van Nhien, A. N.; Marco, J. L. Chemistry of sulfonate- and sulfonamide-stabilized carbanions. The CSIC reactions. *Eur. J. Org. Chem.* **2003**, 3713–3726.

81 Fraser-Reid, B.; Sun, K. M.; Tsang, R. Y.-K.; Sinaÿ, P.; Pietraszkiewicz, M. Approaches to some carbohydrate-derived sultones. *Can. J. Chem.* **1981**, *59*, 260–263.

82 Chang, F. C. Alkaline hydrolysis of methanesulfonate esters without inversion. *Tetrahedron Lett.* **1964**, 305–309.

83 Abdel-Rahman, A. A. H.; Jonke, S.; El Ashry, E. S. H.; Schmidt, R. R. Stereoselective synthesis of β-D-mannopyranosides with reactive mannopyranosyl donors possessing a neighboring electron-withdrawing group. *Angew. Chem. Int. Ed.* **2002**, *41*, 2972–2974.

84 Cossy, J.; Ranaivosata, J.-L.; Bellosta, V.; Wietzke, R. A selective and efficient demesylation using MeMgBr. *Synth. Commun.* **1995**, *25*, 3109–3112.

85 Creary, X. Reactions of pivaloin derivatives with LiTMP. *J. Org. Chem.* **1980**, *45*, 2419–2425.

86 Creary, X.; Rollin, A. J. Reactions of bicyclic α-keto triflates with bases and nucleophiles. *J. Org. Chem.* **1979**, *44*, 1798–1806.

87 El Nemr, A.; Tsuchiya, T. Reactions of some 2- and 4-*O*-triflylglycopyranosides with MeLi, *t*-BuOK, and pyridine. *Carbohydr. Res.* **2001**, *330*, 205–214.

88 El Nemr, A.; Tsuchiya, T. α-Hydrogen elimination in some carbohydrate triflates. *Tetrahedron Lett.* **1995**, *36*, 7665–7668.

89 Hoffman, R. V.; Jankowski, B. C.; Carr, C. S.; Duesler, E. N. The reactions of α-arylsulfonoxy ketones with nucleophiles. *J. Org. Chem.* **1986**, *51*, 130–135.

90 Wuts, P. G. M.; Northuis, J. M. A cautionary note on the use of *p*-nitrobenzenesulfonamides as protecting groups. *Tetrahedron Lett.* **1998**, *39*, 3889–3890.

91 Patonay, T.; Hegedüs, L.; Patonay-Péli, E. Flavonoids. 43. Deprotonation-initiated aryl migration with sulfur dioxide extrusion: a route to 2,3-dihydro-2,3-diaryl-3-hydroxy-4*H*-1-benzopyran-4-ones. *J. Heterocyclic Chem.* **1993**, *30*, 145–151.

92 Hatch, L. F.; Chiola, V. The preparation and properties of 3-chloro-1-propyne and 1-chloro-2-butyne. *J. Am. Chem. Soc.* **1951**, *73*, 360–362.

93 Jacobs, T. L.; Brill, W. F. Haloallenes. *J. Am. Chem. Soc.* **1953**, *75*, 1314–1317.

94 Fry, J. L.; Engler, E. M.; Schleyer, P. v. R. Steric assistance in the solvolysis of 2-alkyl-2-adamantyl *p*-nitrobenzoates. *J. Am. Chem. Soc.* **1972**, *94*, 4628–4634.

95 Ashby, E. C.; Park, W. S.; Goel, A. B.; Su, W.-Y. Single-electron transfer in reactions of alkyl halides with lithium thiolates. *J. Org. Chem.* **1985**, *50*, 5184–5193.

96 Gharpure, M. M.; Rao, A. S. Enantioconvergent synthesis of (–)-3-methyl-3-phenylcyclopentanone. *J. Chem. Soc. Perkin Trans. 1* **1990**, 2759–2761.

97 Schurink, H. B. Pentaerythrityl bromide and iodide. *Org. Synth.* **1943**, *Coll. Vol. II*, 476–478.

98 Herzog, H. L. Pentaerythrityl tetrabromide. *Org. Synth.* **1963**, *Coll. Vol. IV*, 753–755.

99 Wawzonek, S.; Matar, A.; Issidorides, C. H. Monobromopentaerythritol. *Org. Synth.* **1963**, *Coll. Vol. IV*, 681–683.

100 Metha, G.; Mohal, N. Norbornyl route to cyclopentitols: synthesis of trehazolamine analogues and the purported structure of salpantiol. *Tetrahedron Lett.* **1999**, *40*, 5795–5798.

101 Kitazaki, T.; Tamura, N.; Tasaka, A.; Matsushita, Y.; Hayashi, R.; Okonogi, K.; Itoh, K. Optically active antifungal azoles. VI. Synthesis and antifungal activity of *N*-[(1*R*,2*R*)-2-(2,4-difluorophenyl)-2-hydroxy-1-methyl-3-(1*H*-1,2,4-triazol-1-yl)propyl]-*N*′-(4-substituted phenyl)-3-(2*H*,4*H*)-1,2,4-triazolones and 5(1*H*,4*H*)-tetrazolones. *Chem. Pharm. Bull.* **1996**, *44*, 314–327.

102 Cherney, R. J.; Wang, L. Efficient Mitsunobu reactions with *N*-phenylfluorenyl or *N*-trityl serine esters. *J. Org. Chem.* **1996**, *61*, 2544–2546.

103 Boulton, L. T.; Stock, H. T.; Raphy, J.; Horwell, D. C. Generation of unnatural α,α-disubstituted amino acid derivatives from cyclic sulfamidates. *J. Chem. Soc. Perkin Trans. 1* **1999**, 1421–1429.

104 Hartwig, W.; Mittendorf, J. Enantioselective synthesis of 2,3-diamino acids by the bislactim ether method. *Synthesis* **1991**, 939–941.

105 Atmani, A.; El Hallaoui, A.; El Hajji, S.; Roumestant, M. L.; Viallefont, P. From oxazolines to precursors of amino acids. *Synth. Commun.* **1991**, *21*, 2383–2390.

106 Roberts, J. D.; Chambers, V. C. Small-ring compounds. VIII. Some nucleophilic displacement reactions of cyclopropyl, cyclobutyl, cyclopentyl and cyclohexyl *p*-toluenesulfonates and halides. *J. Am. Chem. Soc.* **1951**, *73*, 5034–5040.

107 Brown, H. C.; Ravindranathan, M.; Peters, E. N.; Rao, C. G.; Rho, M. M. Structural effects in solvolytic reactions. 22. Effects of ring size on the stabilization of developing carbocations as revealed by the tool of increasing electron demand. *J. Am. Chem. Soc.* **1977**, *99*, 5373–5378.

108 Brown, H. C.; Rao, C. G.; Ravindranathan, M. Structural effects in solvolytic reactions. 23. New σ^+ constants for activating substituents. The solvolysis of 1-aryl-1-cyclopropyl 3,5-dinitrobenzoates containing activating substituents in the aryl group. The tool of increasing electron demand and I-strain. *J. Am. Chem. Soc.* **1977**, *99*, 7663–7667.

109 Banert, K. Reaktionen von Cyclopropylsulfonaten mit Nucleophilen: SN2-Reaktionen an Cyclopropanen unter Inversion. *Chem. Ber.* **1985**, *118*, 1564–1574.

110 King, J. F.; Tsang, G. T. Y.; Abdel-Malik, M. M.; Payne, N. C. Nucleophilic substitution factors. 1. Coplanar vs orthogonal bimolecular substitution at a benzylic carbon. X-Ray structure of 2-isobutyl-1,3-dihydrobenzo[*c*]thiophenium perchlorate. *J. Am. Chem. Soc.* **1985**, *107*, 3224–3232.

111 Thorstenson, T.; Eliason, R.; Songstad, J. The effect of 4-methyl and 4-nitro substituents on the reactivity of benzylic compounds. *Acta Chem. Scand.* **1977**, *A 31*, 276–280.

112 Conant, J. B.; Kirner, W. R.; Hussey, R. E. The relation between the structure of organic halides and the speeds of their reaction with inorganic iodides. III. The influence of unsaturated groups. *J. Am. Chem. Soc.* **1925**, *47*, 488–501.

113 Hong, S. W.; Koh, H. J.; Lee, H. W.; Lee, I. Kinetics and mechanism of the pyridinolysis of benzyl bromides in dimethyl sulfoxide. *Bull. Korean Chem. Soc.* **1999**, *20*, 1172–1176.

114 Young, P. R.; Jencks, W. P. Separation of polar and resonance substituent effects in the reactions of acetophenones with bisulfite and of benzyl halides with nucleophiles. *J. Am. Chem. Soc.* **1979**, *101*, 3288–3294.

115 Hudson, R. F.; Klopman, G. Nucleophilic reactivity. Part II. The reaction between substituted thiophenols and benzyl bromides. *J. Chem. Soc.* **1962**, 1062–1067.

116 Fuchs, R.; Nisbet, A. Solvent effects in the reaction of *p*-substituted α-chlorotoluenes with thiosulfate. The relationship of Rho and dielectric constant. *J. Am. Chem. Soc.* **1959**, *81*, 2371–2373.

117 Fuchs, R.; Carlton, D. M. Multiple substituent effects in the solvolysis and thiosulfate reactions of 4-substituted α-chloro-3-nitrotoluenes. *J. Org. Chem.* **1962**, *27*, 1520–1523.

118 Naiki, M.; Shirakawa, S.; Kon-i, K.; Kondo, Y.; Maruoka, K. Tris(2,6-diphenylbenzyl)amine (TDA) and tris(2,6-diphenylbenzyl)phosphine with unique bowl-shaped structures: synthetic application of functionalized TDA to chemoselective silylation of benzylic alcohols. *Tetrahedron Lett.* **2001**, *42*, 5467–5471.

119 Barclay, L. R. C.; Sonawane, H. R.; MacDonald, M. C. Sterically hindered aromatic compounds. III. Acid-catalyzed reactions of 2,4,6-tri-*t*-butyl and 2-methyl-4,6-di-*t*-butylbenzyl alcohols and chlorides. *Can. J. Chem.* **1972**, *50*, 281–290.

120 Smit, K. J. Influence of steric effects on the excited triplet-state lifetime of 2,2′-dinitrostilbene and 2,2′,4,4′,6,6′-hexanitrostilbene in acetonitrile solution. *J. Phys. Chem.* **1992**, *96*, 6555–6558.

121 Hass, H. B.; Bender, M. L. A proposed mechanism of the alkylation of benzyl halides with nitro paraffin salts. *J. Am. Chem. Soc.* **1949**, *71*, 3482–3485.

122 Patey, A. L.; Waldron, N. M. A new indazole ring synthesis. *Tetrahedron Lett.* **1970**, 3375–3376.

123 Hashimoto, H.; Ikemoto, T.; Itoh, T.; Maruyama, H.; Hanaoka, T.; Wakimasu, M.; Mitsudera, H.; Tomimatsu, K. Process development of 4-[*N*-methyl-*N*-(tetrahydropyran-4-yl)-aminomethyl]aniline dihydrochloride: a key intermediate for TAK-779, a small-molecule nonpeptide CCR5 antagonist. *Org. Process Res. Dev.* **2002**, *6*, 70–73.

124 Sollott, G. P. Conversion of 2,4,6-trinitrobenzyl chloride to 2,2′,4,4′,6,6′-hexanitrostilbene by nitrogen bases. *J. Org. Chem.* **1982**, *47*, 2471–2474.

125 Bornstein, J.; Shields, J. E. 2-(*p*-Tolylsulfonyl)-dihydroisoindole. *Org. Synth.* **1973**, *Coll. Vol. V*, 1064–1066.

126 Sardessai, M. S.; Abramson, H. N.; Wormser, H. C. Synthesis of aminated naphthacenetriones: precursors to aminated anthracyclines. *Synth. Commun.* **1993**, *23*, 3223–3229.

127 Hine, J.; Rosscup, R. J. Resonance interactions of substituents attached to the same saturated carbon atom. Reactivity of polychloromethyl ethers. *J. Am. Chem. Soc.* **1960**, *82*, 6115–6118.

128 Hine, J.; Thomas, C. H.; Ehrenson, S. J. The effect of halogen atoms on the reactivity of other halogen atoms in the same molecule. V. The SN2 reactivity of methylene halides. *J. Am. Chem. Soc.* **1955**, *77*, 3886–3889.

129 Meyer, W. P.; Martin, J. C. Substituent effects of alkoxy and amino groups directly bonded to cationic carbon in the perpendicularly twisted geometry. 2-Oxa- and 2-aza-1-adamantyl tosylates. *J. Am. Chem. Soc.* **1976**, *98*, 1231–1241.

130 Apeloig, Y.; Biton, R.; Abu-Freih, A. Importance of electronic geminal interactions in solvolysis reactions. Application to the determination of the α-silyl effect on carbenium ion stability. *J. Am. Chem. Soc.* **1993**, *115*, 2522–2523.

131 Stang, P. J.; Ladika, M.; Apeloig, Y.; Stanger, A.; Schiavelli, M. D.; Hughey, M. R. Bimolecular substitution at carbon in neopentyl-like silylcarbinyl sulfonates. *J. Am. Chem. Soc.* **1982**, *104*, 6852–6854.

132 Apeloig, Y.; Stanger, A. Are carbenium ions stabilized or destabilized by α-silyl substitution? The solvolysis of 2-(trimethylsilyl)-2-adamantyl *p*-nitrobenzoate. *J. Am. Chem. Soc.* **1985**, *107*, 2806–2807.

133 Sato, Y.; Shirai, N.; Machida, Y.; Ito, E.; Yasui, T.; Kurono, Y.; Hatano, K. Rearrangements of 1,6,7-trisubstituted 2-methyl-1,2,3,4-tetrahydroisoquinolinium 2-methylides. *J. Org. Chem.* **1992**, *57*, 6711–6716.

134 Dostrovsky, I.; Hughes, E. D. Mechanism of substitution at a saturated carbon atom. Part XXVI. The role of steric hindrance. (Section A) Introductory remarks, and a kinetic study of the reactions of methyl, ethyl, *n*-propyl, isobutyl, and neopentyl bromides with NaOEt in dry EtOH. *J. Chem. Soc.* **1946**, 157–161.

135 Cook, M. A.; Eaborn, C.; Walton, D. R. M. Organosilicon compounds XLVIII. Effects of α-trimethylsilyl groups on rates of reaction at carbon–halogen bonds. *J. Organomet. Chem.* **1971**, *29*, 389–396.

136 Maleczka, R. E.; Geng, F. Stereospecificity of the 1,2-Wittig rearrangement: how chelation effects influence stereochemical outcome. *J. Am. Chem. Soc.* **1998**, *120*, 8551–8552.

137 Matteson, D. S.; Erdik, E. Epimerization of α-chloro boronic esters by lithium and zinc chlorides. *Organometallics* **1983**, *2*, 1083–1088.

138 Matteson, D. S.; Schaumberg, G. D. Reactions of α-haloalkaneboronic esters. *J. Org. Chem.* **1966**, *31*, 726–731.

139 Matteson, D. S.; Mah, R. W. H. Neighboring boron in nucleophilic displacement. *J. Am. Chem. Soc.* **1963**, *85*, 2599–2603.

140 Matteson, D. S. α-Halo boronic esters: intermediates for stereodirected synthesis. *Chem. Rev.* **1989**, *89*, 1535–1551.

141 Laplante, C.; Hall, D. G. Direct mono-*N*-methylation of solid-supported amino acids: a useful application of the Matteson rearrangement of α-aminoalkylboronic esters. *Org. Lett.* **2001**, *3*, 1487–1490.

142 Jadhav, P. K.; Man, H.-W. Direct synthesis of [α-[(*tert*-butoxycarbonyl)amino]alkyl]boronates from (α-haloalkyl)boronates. *J. Org. Chem.* **1996**, *61*, 7951–7954.

143 Grinblat, J.; Ben-Zion, M.; Hoz, S. Halophilic reactions: anomalies in bromine transfer reactions. *J. Am. Chem. Soc.* **2001**, *123*, 10738–10739.

144 Fishwick, B. R.; Rowles, D. K.; Stirling, C. J. M. Bromonitromethane, a versatile electrophile: reactions with thiolates. *J. Chem. Soc. Chem. Commun.* **1983**, 834–835.

145 Fishwick, B. R.; Rowles, D. K.; Stirling, C. J. M. Bromonitromethane, a versatile electrophile: reactions with feebly basic nucleophiles. *J. Chem. Soc. Chem. Commun.* **1983**, 835–836.

146 Witczak, Z. J.; Boryczewski, D. Thiosugars IV: design and synthesis of *S*-linked fucoside analogs as a new class of α-L-fucosidase inhibitors. *Bioorg. Med. Chem. Lett.* **1998**, *8*, 3265–3268.

147 Kim, K. S.; Hurh, E. Y.; Youn, J. N.; Park, J. I. Dialkyl (1-hydroxyiminoalkyl)phosphonates from 1-bromo-1-nitroalkanes and trialkyl phosphites. *J. Org. Chem.* **1999**, *64*, 9272–9274.

148 Challis, B. C.; Yousaf, T. I. Facile formation of *N*-nitrosamines from bromonitromethane and secondary amines. *J. Chem. Soc. Chem. Commun.* **1990**, 1598–1599.

149 Cerè, V.; De Angelis, S.; Pollicino, S.; Ricci, A.; Reddy, C. K.; Knochel, P.; Cahiez, G. Preparation of 2-zincio-1,3-dithianes and di(1,3-

dithian-2-yl)zinc and their reaction with highly functionalized halides and α,β-unsaturated carbonyl compounds. *Synthesis* **1997**, 1174–1178.

150 DeWolfe, R. H. Synthesis of carboxylic and carbonic ortho esters. *Synthesis* **1974**, 153–172.

151 Kaufmann, W. E.; Dreger, E. E. Ethyl orthoformate. *Org. Synth.* **1941**, *Coll. Vol. I*, 258–261.

152 Ingold, C. K.; Powell, W. J. Experiments on the synthesis of the polyacetic acids of methane. Part II. Some abnormal condensations of malonic and cyanoacetic esters with halogenated methanes. *J. Chem. Soc.* **1921**, *119*, 1222–1229.

153 Hine, J.; Porter, J. J. Methylene derivatives as intermediates in polar reactions. VIII. Difluoromethylene in the reaction of chlorodifluoromethane with NaOMe. *J. Am. Chem. Soc.* **1957**, *79*, 5493–5496.

154 Arnold, R. T.; Kulenovic, S. T. Carbanion halogenations with carbon tetrahalides. α-Halo esters. *J. Org. Chem.* **1978**, *43*, 3687–3689.

155 Grossman, R. B.; Varner, M. A. Selective monoalkylation of diethyl malonate, ethyl cyanoacetate, and malononitrile using a masking group for the second acidic hydrogen. *J. Org. Chem.* **1997**, *62*, 5235–5237.

156 Federsel, H.-J.; Glasare, G.; Högström, C.; Wiestål, J.; Zinko, B.; Ödman, C. A convenient quaternization/rearrangement procedure for conversion of thiazoles to medium- and large-sized *N,S*-heterocycles. *J. Org. Chem.* **1995**, *60*, 2597–2606.

157 Ponticelli, F.; Trendafilova, A.; Valoti, M.; Saponara, S.; Sgaragli, G. Synthesis and antiperoxidant activity of new phenolic *O*-glycosides. *Carbohydr. Res.* **2001**, *330*, 459–468.

158 Salvatore, R. N.; Schmidt, S. E.; Shin, S. I.; Nagle, A. S.; Worrell, J. H.; Jung, K. W. CsOH-promoted chemoselective mono-*N*-alkylation of diamines and polyamines. *Tetrahedron Lett.* **2000**, *41*, 9705–9708.

159 Carocci, A.; Catalano, A.; Corbo, F.; Duranti, A.; Amoroso, R.; Franchini, C.; Lentini, G.; Tortorella, V. Stereospecific synthesis of mexiletine and related compounds: Mitsunobu vs Williamson reaction. *Tetrahedron Asymmetry* **2000**, *11*, 3619–3634.

160 Creary, X.; Geiger, C. C. Properties of α-keto cations. Facile generation under solvolytic conditions. *J. Am. Chem. Soc.* **1982**, *104*, 4151–4162.

161 Creary, X. Carbocationic and related processes in reactions of α-keto mesylates and triflates. *Acc. Chem. Res.* **1985**, *18*, 3–8.

162 Creary, X.; Hopkinson, A. C.; Lee-Ruff, E. α-Carbonyl cations. *Adv. Carbocation Chem.* **1989**, *1*, 45–92.

163 Allen, A. D.; Kanagasabapathy, V. M.; Tidwell, T. T. Doubly destabilized carbocations. Strong aryl delocalization and the attenuation of rate decelerating effects of CF_3 and CN groups. *J. Am. Chem. Soc.* **1986**, *108*, 3470–3474.

164 Bordwell, F. G.; Brannen, W. T. The effect of the carbonyl and related groups on the reactivity of halides in SN2 reactions. *J. Am. Chem. Soc.* **1964**, *86*, 4645–4650.

165 Bordwell, F. G.; Cooper, G. D. The effect of the sulfonyl group on the nucleophilic displacement of halogen in α-halo sulfones and related substances. *J. Am. Chem. Soc.* **1951**, *73*, 5184–5186.

166 McLennan, D. J.; Pross, A. The mechanism for nucleophilic substitution of α-carbonyl derivatives. Application of the valence-bond configuration mixing model. *J. Chem. Soc. Perkin Trans. 2* **1984**, 981–984.

167 Koh, H. J.; Han, K. L.; Lee, H. W.; Lee, I. Kinetics and mechanism of the pyridinolysis of phenacyl bromides in acetonitrile. *J. Org. Chem.* **2000**, *65*, 4706–4711.

168 Kalendra, D. M.; Sickles, B. R. Diminished reactivity of *ortho*-substituted phenacyl bromides toward nucleophilic displacement. *J. Org. Chem.* **2003**, *68*, 1594–1596.

169 Hine, J.; Ghirardelli, R. G. The SN2 reactivity of β-fluoroethyl iodides. *J. Org. Chem.* **1958**, *23*, 1550–1552.

170 Meen, R. H.; Wright, G. F. The nitration of weakly basic secondary amines. *J. Org. Chem.* **1954**, *19*, 391–402.

171 Wu, K.; Chen, Q.-Y. Synthesis of trifluoroethyl ethers from 2,2,2-trifluoroethyl chloride in high temperature aqueous medium. *J. Fluorine Chem.* **2002**, *113*, 79–83.

172 Wu, K.; Chen, Q.-Y. Solvolytically DMSO-promoted reactions of 1,1,1-trifluoroethyl chloride or fluoride with nucleophiles. *Tetrahedron* **2002**, *58*, 4077–4084.

173 McBee, E. T.; Battershell, R. D.; Braendlin, H. P. The kinetics of the reaction between polyfluoroalkyl halides and iodide ion. *J. Am. Chem. Soc.* **1962**, *84*, 3157–3160.

174 Katagiri, T.; Irie, M.; Uneyama, K. Intramolecular SN2 reaction α to a trifluoromethyl

group: preparation of 1-cyano-2-trifluoromethylcyclopropane. *Tetrahedron Asymmetry* **1999**, *10*, 2583–2589.

175 Katagiri, T.; Ihara, H.; Takahashi, M.; Kashino, S.; Furuhashi, K.; Uneyama, K. Intramolecular SN2 reaction at α-carbon of trifluoromethyl group: preparation of optically active 2-trifluoromethylaziridine. *Tetrahedron Asymmetry* **1997**, *8*, 2933–2937.

176 Singh, R. P.; Manandhar, S.; Shreeve, J. M. Mono- and disubstituted polyfluoroalkylimidazolium quaternary salts and ionic liquids. *Synthesis* **2003**, 1579–1585.

177 Hansen, R. L. Perfluoroalkanesulfonate esters as alkylating agents. *J. Org. Chem.* **1965**, *30*, 4322–4324.

178 Umemoto, T.; Gotoh, Y. 1,1-Dihydroperfluoroalkylations of nucleophiles with (1,1-dihydroperfluoroalkyl)phenyliodonium triflates. *J. Fluorine Chem.* **1986**, *31*, 231–236.

179 DesMarteau, D. D.; Montanari, V. Easy preparation of bioactive peptides from the novel N^{α}-trifluoroethyl amino acids. *Chem. Lett.* **2000**, 1052–1053.

180 Chen, K. S.; Tang, D. Y. H.; Montgomery, L. K.; Kochi, J. K. Rearrangements and conformations of chloroalkyl radicals by electron spin resonance. *J. Am. Chem. Soc.* **1974**, *96*, 2201–2208.

181 Nakai, T.; Tanaka, K.; Ishikawa, N. The reaction of 2,2,2-trifluoroethyl iodide with sodium phenolate. A novel competition between substitution and elimination reactions. *J. Fluorine Chem.* **1977**, *9*, 89–93.

182 Pegolotti, J. A.; Young, W. G. Allylic rearrangements. LI. Displacement reactions in trifluoromethylallyl systems. *J. Am. Chem. Soc.* **1961**, *83*, 3258–3262.

183 Nagai, T.; Nishioka, G.; Koyama, M.; Ando, A.; Miki, T.; Kumadaki, I. Reactions of trifluoromethyl ketones. IX. Investigation of the steric effect of a trifluoromethyl group based on the stereochemistry on the dehydration of trifluoromethyl homoallyl alcohols. *J. Fluorine Chem.* **1992**, *57*, 229–237.

184 Katagiri, T.; Uneyama, K. Stereospecific substitution at α-carbon to trifluoromethyl group: application to optically active fluorinated amino acid synthesis. *Chirality* **2003**, *15*, 4–9.

185 Larock, R. C. *Comprehensive organic transformations*; Wiley–VCH: Weinheim, 1989.

186 Bisgrove, D. E.; Brown, J. F.; Clapp, L. B. 3,4-Dinitro-3-hexene. *Org. Synth.* **1963**, *Coll. Vol. IV*, 372–374.

187 Corson, B. B.; Benson, W. L. Ethyl ethylenetetracarboxylate. *Org. Synth.* **1943**, *Coll. Vol. II*, 273–275.

188 Yamada, Y.; Yasuda, H. A convenient synthesis of dialkyl (*E*)-2,3-dicyanobutendioates. *Synthesis* **1990**, 768–770.

189 Wang, H. P.; Bai, T.-S.; Lee, O.; Ker, K.-D. Novel open-ring analogues of N^5,N^{10}-methylenetetrahydrofolic acid with selective activity against brain tumors. *Bioorg. Med. Chem. Lett.* **1995**, *5*, 1909–1912.

190 Osterberg, A. E. Ethyl phthalimidomalonate. *Org. Synth.* **1941**, *Coll. Vol. I*, 271–272.

191 Boehme, W. R. 3-Methylcoumarone. *Org. Synth.* **1963**, *Coll. Vol. IV*, 590–593.

192 Bickel, C. L.; Morris, R. The tautomeric forms of some 1,1-diaroylethanes. *J. Am. Chem. Soc.* **1951**, *73*, 1786–1787.

193 Cizej, V. L.; Urleb, U. The synthesis of 3,4-dihydro-2-methyl-3-oxo-2*H*-benzo-1,4-thiazine-2-carboxylic acids. *J. Heterocyclic Chem.* **1996**, *33*, 97–101.

194 Stadlbauer, W.; Lutschounig, H.; Schindler, G.; Witoszynskyj, T.; Kappe, T. Synthesis of 3-nitro- and 3-aminoquinoline-2,4-diones. An unexpected route to 3-hydroxyquinoline-2,4-diones. *J. Heterocyclic Chem.* **1992**, *29*, 1535–1540.

195 Laschober, R.; Stadlbauer, W. Synthesis of 3-heptyl and 3-nonyl-2,4(1*H*,3*H*)-quinolinediones. *Liebigs Ann. Chem.* **1990**, 1083–1086.

196 Peterson, P. E. Cyclic halonium ions with five-membered rings. *Acc. Chem. Res.* **1971**, *4*, 407–413.

197 Kanoh, S.; Nishimura, T.; Naka, M.; Motoi, M. Unusual cyclodimerization of small cyclic ethers via neighboring carbonyl-group participation and cation transfer. *Tetrahedron* **2002**, *58*, 7065–7074.

198 Winstein, S.; Allred, E.; Heck, R.; Glick, R. Neighboring methoxyl participation in solvolytic nucleophilic substitution. *Tetrahedron* **1958**, *3*, 1–13.

199 Allred, E. L.; Winstein, S. 6-Methoxyl participation and ion pairs in some solvolysis reactions. *J. Am. Chem. Soc.* **1967**, *89*, 4012–4017.

200 Krow, G. R.; Lin, G.; Rapolu, D.; Fang, Y.; Lester, W. S.; Herzon, S. B.; Sonnet, P. E. The rearrangement route to 2-azabicyclo[2.1.1]hexanes. Solvent and electrophile control of neighboring group participation. *J. Org. Chem.* **2003**, *68*, 5292–5299.

201 Eliel, E. L.; Knox, D. E. Neighboring group participation by sulfur involving four-membered-ring intermediates. *J. Am. Chem. Soc.* **1985**, *107*, 2946–2952.

202 Wilen, S. H.; Delguzzo, L.; Saferstein, R. Experimental evidence for AcO-7 neighboring group participation. *Tetrahedron* **1987**, *43*, 5089–5094.

203 Koppenhoefer, B.; Schurig, V. (*S*)-2-Chloroalkanoic acids of high enantiomeric purity from (*S*)-2-amino acids: (*S*)-2-chloropropanoic acid. *Org. Synth.* **1993**, *Coll. Vol. VIII*, 119–123.

204 Wilt, J. W.; Chenier, P. J. Studies of benzonorbornene and derivatives. III. The solvolysis of *syn*- and *anti*-7-bromobenzonorbornadiene and related bromides. *J. Org. Chem.* **1970**, *35*, 1571–1576.

205 Hallett, D. J.; Gerhard, U.; Goodacre, S. C.; Hitzel, L.; Sparey, T. J.; Thomas, S.; Rowley, M.; Ball, R. G. Neighboring group participation of the indole nucleus: an unusual DAST-mediated rearrangement reaction. *J. Org. Chem.* **2000**, *65*, 4984–4993.

206 Easton, C. J.; Hutton, C. A.; Merrett, M. C.; Tiekink, E. R. T. Neighboring group effects promote substitution reactions over elimination and provide a stereocontrolled route to chloramphenicol. *Tetrahedron* **1996**, *52*, 7025–7036.

207 Converso, A.; Burow, K.; Marzinzik, A.; Sharpless, K. B.; Finn, M. G. 2,6-Dichloro-9-thiabicyclo[3.3.1]nonane: a privileged, bivalent scaffold for the display of nucleophilic components. *J. Org. Chem.* **2001**, *66*, 4386–4392.

208 Overman, L. E.; Kakimoto, M.; Okazaki, M. E.; Meier, G. P. C–C Bond formation under mild conditions via tandem cationic aza-Cope rearrangement–Mannich reactions. A convenient synthesis of polysubstituted pyrrolidines. *J. Am. Chem. Soc.* **1983**, *105*, 6622–6629.

209 Hine, J.; Brader, W. H. The effect of halogen atoms on the reactivity of other halogen atoms in the same molecule. III. The SN2 reactivity of ethylene halides. *J. Am. Chem. Soc.* **1953**, *75*, 3964–3966.

210 Winstein, S.; Shatavsky, M. Neighboring carbon and hydrogen. XXI. *Anti*-7 derivatives of norbornene as homoallylic systems. *J. Am. Chem. Soc.* **1956**, *78*, 592–597.

211 Barlett, P. D.; Rice, M. R. 4-Bromocyclopentene by hydride reduction of 3,5-dibromocyclopentene and its unassisted hydrolysis. *J. Org. Chem.* **1963**, *28*, 3351–3353.

212 Gassman, P. G.; Hall, J. B. Testing for symmetry in neighboring group participation in carbocation formation. An insight into double-bond participation via trifluoromethyl group substitution. *J. Am. Chem. Soc.* **1984**, *106*, 4267–4269.

213 Magid, R. M. Nucleophilic and organometallic displacement reactions of allylic compounds: stereo- and regiochemistry. *Tetrahedron* **1980**, *36*, 1901–1930.

214 Cori, O.; Chayet, L.; Perez, L. M.; Bunton, C. A.; Hachey, D. Rearrangement of linalool, geraniol, and nerol and their derivatives. *J. Org. Chem.* **1986**, *51*, 1310–1316.

215 Lambertin, F.; Taran, M.; Delmond, B. Synthesis of ionone and methylionone analogues from δ-pyronene. *J. Chem. Res. (S)* **2000**, 302–303.

216 Maleczka, R. E.; Geng, F. Synthesis and fluoride-promoted Wittig rearrangements of α-alkoxysilanes. *Org. Lett.* **1999**, *1*, 1111–1113.

217 Gyenes, F.; Purrington, S. T.; Liu, Y.-S. Friedel–Crafts reactions of fluorinated allylic compounds. *J. Org. Chem.* **1999**, *64*, 1366–1368.

218 Magid, R. M.; Fruchey, O. S. *Syn* stereospecificity in the SN2′ reaction of an acyclic allylic chloride with secondary amines. *J. Am. Chem. Soc.* **1979**, *101*, 2107–2112.

219 Bégué, J.-P.; Bonnet-Delpon, D.; Rock, M. H. The addition of *N*-lithiated amines to 1,1,1-trifluoromethylstyrene, a concise synthesis of 3-*gem*-difluoro-2-phenyl allylic amines. *Synlett* **1995**, 659–660.

220 Mook, R.; Sher, P. M. Vinyl radical cyclization via addition of tin radicals to triple bonds: 3-methylene-4-isopropyl-1,1-cyclopentanedicarboxylic acid dimethyl ester. *Org. Synth.* **1993**, *Coll. Vol. VIII*, 381–386.

221 Nyström, J. E.; Rein, T.; Bäckvall, J. E. 1,4-Functionalization of 1,3-dienes via Pd-catalyzed chloroacetoxylation and allylic amination: 1-acetoxy-4-diethylamino-2-butene and 1-acetoxy-4-benzylamino-2-butene. *Org. Synth.* **1993**, *Coll. Vol. VIII*, 9–13.

222 Park, T. K.; Danishefsky, S. J. A synthetic route to vallenamine: an interesting observation concerning stereoelectronic preferences in the Sn2′ reaction. *Tetrahedron Lett.* **1994**, *35*, 2667–2670.

223 Stork, G.; Kreft, A. F. Concerning the stereochemistry of the SN2′ reaction in cyclohexenyl systems. *J. Am. Chem. Soc.* **1977**, *99*, 3850–3851.

224 Hirabe, T.; Nojima, M.; Kusabayashi, S. Lithium aluminum hydride reduction of allylic substrates. Notable leaving group effects on the product regiochemistry. *J. Org. Chem.* **1984**, *49*, 4084–4086.

225 Kocovsky, P. Reactivity control in Pd-catalyzed reactions: a personal account. *J. Organomet. Chem.* **2003**, *687*, 256–268.

226 Kok, S. H. L.; Shing, T. K. M. A new synthetic approach towards *N*-alkylated 2-*epi*-valienamines via Pd-catalyzed coupling reaction. *Tetrahedron Lett.* **2000**, *41*, 6865–6868.

227 Marshall, J. A. SN2′ Additions of organocopper reagents to vinyloxiranes. *Chem. Rev.* **1989**, *89*, 1503–1511.

228 Gini, F.; Del Moro, F.; Macchia, F.; Pineschi, M. Regio- and enantioselective Cu-catalyzed addition of dialkylzinc reagents to cyclic 2-alkenyl aziridines. *Tetrahedron Lett.* **2003**, *44*, 8559–8562.

229 Dübner, F.; Knochel, P. Highly enantioselective Cu-catalyzed substitution of allylic chlorides with diorganozincs. *Tetrahedron Lett.* **2000**, *41*, 9233–9237.

230 Alexakis, A.; Marek, I.; Mangeney, P.; Normant, J. F. Mechanistic aspects on the formation of chiral allenes from propargylic ethers and organocopper reagents. *J. Am. Chem. Soc.* **1990**, *112*, 8042–8047.

231 Nantz, M. H.; Bender, D. M.; Janaki, S. A convenient terminal allene synthesis from propargylic acetates. *Synthesis* **1993**, 577–578.

232 Fürstner, A.; Méndez, M. Iron-catalyzed cross-coupling reactions: efficient synthesis of 2,3-allenol derivatives. *Angew. Chem. Int. Ed.* **2003**, *42*, 5355–5357.

233 Calaza, M. I.; Hupe, E.; Knochel, P. Highly *anti*-selective SN2′ substitutions of chiral cyclic 2-iodo-allylic alcohol derivatives with mixed Zn–Cu reagents. *Org. Lett.* **2003**, *5*, 1059–1061.

234 Goering, H. L.; Seitz, E. P.; Tseng, C. C. Alkylation of allylic derivatives. 3. The regiochemistry of alkylation of the isomeric *trans*-α,γ-methylphenylallyl acetates with lithium dialkylcuprates. *J. Org. Chem.* **1981**, *46*, 5304–5308.

235 Harrington-Frost, N.; Leuser, H.; Calaza, M. I.; Kneisel, F. F.; Knochel, P. Highly stereoselective *anti* SN2′ substitutions of (*Z*)-allylic pentafluorobenzoates with polyfunctionalized Zn–Cu reagents. *Org. Lett.* **2003**, *5*, 2111–2114.

236 Marino, J. P.; Abe, H. Stereospecific 1,4-additions of methyl cyanocuprate to enol phosphates of α,β-epoxycyclohexanones: application to the total synthesis of (±)-α-multistriatin. *J. Org. Chem.* **1981**, *46*, 5379–5383.

237 Goering, H. L.; Tseng, C. C. Alkylation of allylic derivatives. 7. Stereochemistry of alkylation of the isomeric *trans*-α,γ-methyl(phenyl)-allyl acetates with lithium dialkylcuprates and alkylcyanocuprates. *J. Org. Chem.* **1983**, *48*, 3986–3990.

238 Kelly, P. A.; Berger, G. O.; Wyatt, J. K.; Nantz, M. H. Synthesis of [ethylene-1-(η^5-4,5,6,7-tetrahydro-1-indenyl)-2-(η^5-4′,5′,6′,7′-tetrahydro-2′-indenyl)]titanium dichloride, the elusive isomer of the Brintzinger-type *ansa*-titanocenes. *J. Org. Chem.* **2003**, *68*, 8447–8452.

239 Fielder, S.; Rowan, D. D.; Sherburn, M. S. Synthesis of α-farnesene hydroperoxides. *Synlett* **1996**, 349–350.

240 Li, Z.; Racha, S.; Dan, L.; El-Subbagh, H.; Abushanab, E. A general and facile synthesis of β- and γ-hydroxy phosphonates from epoxides. *J. Org. Chem.* **1993**, *58*, 5779–5783.

241 Farr, R. N. Synthesis of chiral 3-substituted isoxazolidines. *Tetrahedron Lett.* **1998**, *39*, 195–196.

242 Narjes, F.; Bolte, O.; Icheln, D.; König, W. A.; Schaumann, E. Synthesis of vinylcyclopropanes by intramolecular epoxide ring opening. Application for an enantioselective synthesis of dictyopterene A. *J. Org. Chem.* **1993**, *58*, 626–632.

243 Klunder, J. M.; Onami, T.; Sharpless, K. B. Arenesulfonate derivatives of homochiral glycidol: versatile chiral building blocks for organic synthesis. *J. Org. Chem.* **1989**, *54*, 1295–1304.

244 Nicolaou, K. C.; Duggan, M. E.; Ladduwahetty, T. Reactions of 2,3-epoxyhalides. Synthesis of optically active allylic alcohols and homoallylic epoxides. *Tetrahedron Lett.* **1984**, *25*, 2069–2072.

245 Murakami, T.; Minamikawa, H.; Hato, M. Regio- and stereocontrolled synthesis of D-*erythro*-sphingosine and phytosphingosine from D-glucosamine. *Tetrahedron Lett.* **1994**, *35*, 745–748.

246 Bonini, C.; Federici, C.; Rossi, L.; Righi, G. C-1 Reactivity of 2,3-epoxy alcohols via oxirane opening with metal halides: applications and synthesis of naturally occurring 2,3-octanediol,

muricatacin, 3-octanol, and 4-dodecanolide. *J. Org. Chem.* **1995**, *60*, 4803–4812.
247 Behrens, C. H.; Sharpless, K. B. Selective transformations of 2,3-epoxy alcohols and related derivatives. Strategies for nucleophilic attack at C-3 or C-2. *J. Org. Chem.* **1985**, *50*, 5696–5704.
248 Konno, H.; Toshiro, E.; Hinoda, N. An epoxide ring-opening reaction via a hypervalent silicate intermediate: synthesis of statine. *Synthesis* **2003**, 2161–2164.
249 Fringuelli, F.; Pizzo, F.; Tortoioli, S.; Vaccaro, L. Zn(II)-catalyzed thiolysis of oxiranes in water under neutral conditions. *J. Org. Chem.* **2003**, *68*, 8248–8251.
250 Fuchs, R.; van der Werf, C. A. Direction of ring opening in the reduction of *p*-substituted styrene oxides with $LiBH_4$. *J. Am. Chem. Soc.* **1954**, *76*, 1631–1634.
251 Asao, N.; Kasahara, T.; Yamamoto, Y. σ–π Chelation-controlled chemoselective ring openings of epoxides. *Tetrahedron Lett.* **2001**, *42*, 7903–7905.
252 Chini, M.; Crotti, P.; Favero, L.; Macchia, F.; Pineschi, M. Lanthanide(III) triflates as extraordinarily effective new catalysts for the aminolysis of epoxides. *Tetrahedron Lett.* **1994**, *35*, 433–436.
253 Sutowardoyo, K. I.; Emizane, M.; Lhoste, P.; Sinou, D. Highly regio and chemoselective ring opening of epoxides with Me_3SiN_3 in the presence of $Al(OiPr)_3$ and $Ti(OiPr)_4$. *Tetrahedron* **1991**, *47*, 1435–1446.
254 Bissing, D. E.; Speziale, A. J. Reactions of phosphorus compounds. IX. The opening of epoxides with tertiary phosphines. *J. Am. Chem. Soc.* **1965**, *87*, 2683–2690.
255 Aleu, J.; Brenna, E.; Fuganti, C.; Serra, S. Lipase-mediated synthesis of the enantiomeric forms of 4,5-epoxy-4,5-dihydro-α-ionone and 5,6-epoxy-5,6-dihydro-β-ionone. A new direct access to enantiopure (*R*)- and (*S*)-α-ionone. *J. Chem. Soc. Perkin Trans. 1* **1999**, 271–278.
256 Concellón, J. M.; Bardales, E. Synthesis of (*E*)-α,β-unsaturated esters with total or high diastereoselectivity from α,β-epoxyesters. *Org. Lett.* **2002**, *4*, 189–191.
257 Xiong, Z.; Zhou, G.; Yang, J.; Chen, Y.; Li, Y. An efficient stereocontrolled synthesis of (–)-10-*epi*-5β,11-dihydroxyeudesmane and (–)-4,10-*epi*-5β,11-dihydroxyeudesmane. *Tetrahedron Asymmetry* **1998**, *9*, 1525–1530.
258 Zhang, J.-X.; Wang, G.-X.; Xie, P.; Chen, S.-F.; Liang, X.-T. The total synthesis of chrysotricine. *Tetrahedron Lett.* **2000**, *41*, 2211–2213.
259 Goujon, J.-Y.; Duval, A.; Kirschleger, B. A useful enantioselective synthesis of chroman-2-ylmethanol. *J. Chem. Soc. Perkin Trans. 1* **2002**, 496–499.
260 Eisch, J. J.; Liu, Z.-R.; Singh, M. High regioselectivity in the alternative, reductive cleavages of terminal epoxides with aluminum reagents. *J. Org. Chem.* **1992**, *57*, 1618–1621.
261 Fort, Y. Lithium hydride containing complex reducing agent: a new and simple activation of commercial lithium hydride. *Tetrahedron Lett.* **1995**, *36*, 6051–6054.
262 Ley, S. V.; Mitchell, C.; Pears, D.; Ramarao, C.; Yu, J.-Q.; Zhou, W. Recyclable polyurea-microencapsulated Pd(0) nanoparticles: an efficient catalyst for hydrogenolysis of epoxides. *Org. Lett.* **2003**, *5*, 4665–4668.
263 Chong, J. M.; Johannsen, J. Chemoselective reduction of 2,3-epoxy tosylates with DIBAL-H as a general route to enantiomerically enriched 1-tosyloxy-2-alkanols. *Tetrahedron Lett.* **1994**, *35*, 7197–7200.
264 Dai, L.-X.; Lou, B.-L.; Zhang, Y.-Z.; Guo, G.-Z. Regioselective Ti-mediated reductive opening of 2,3-epoxy alcohols. *Tetrahedron Lett.* **1986**, *27*, 4343–4346.
265 Suzuki, T.; Saimoto, H.; Tomioka, H.; Oshima, K.; Nozaki, H. Regio- and stereoselective ring opening of epoxy alcohols with organoaluminum compounds leading to 1,2-diols. *Tetrahedron Lett.* **1982**, *23*, 3597–3600.
266 Finan, J. M.; Kishi, Y. Reductive ring openings of allyl-alcohol epoxides. *Tetrahedron Lett.* **1982**, *23*, 2719–2722.
267 Ma, P.; Martin, V. S.; Masamune, S.; Sharpless, K. B.; Viti, S. M. Synthesis of saccharides and related polyhydroxylated natural products. 2. Simple deoxyalditols. *J. Org. Chem.* **1982**, *47*, 1378–1380.
268 Pettersson, L.; Frejd, T.; Magnusson, G. Chiral synthons from arabinose. Preparation of 1,3-diols and β-benzyloxy ketones. *J. Org. Chem.* **1984**, *49*, 4540–4541.
269 Chakraborty, T. K.; Das, S. Synthesis of chiral 1,3-diols by radical-mediated regioselective opening of 2,3-epoxy alcohols using Cp_2TiCl_2. *Tetrahedron Lett.* **2002**, *43*, 2313–2315.
270 Nakajima, N.; Tanaka, T.; Hamada, T.; Oikawa, Y.; Yonemitsu, O. Highly stereoselective total synthesis of pikronolide, the aglycon of the

first macrolide antibiotic pikromycin. Crucial role of benzyl-type protecting groups removable by DDQ oxidation. *Chem. Pharm. Bull.* **1987**, *35*, 2228–2237.

271 Chappell, M. D.; Stachel, S. J.; Lee, C. B.; Danishefsky, S. J. En route to a plant scale synthesis of the promising antitumor agent 12,13-desoxyepothilone B. *Org. Lett.* **2000**, *2*, 1633–1636.

272 Miyashita, M.; Suzuki, T.; Hoshino, M.; Yoshikoshi, A. The organoselenium-mediated reduction of α,β-epoxy ketones, α,β-epoxy esters, and their congeners to β-hydroxy carbonyl compounds: novel methodologies for the synthesis of aldols and their analogs. *Tetrahedron* **1997**, *53*, 12469–12486.

273 Díez, D.; Benéitez, M. T.; Marcos, I. S.; Garrido, N. M.; Basabe, P.; Sanz, F.; Broughton, H. B.; Urones, J. G. Chemistry of epoxysulfones: straightforward synthesis of versatile chiral building blocks. *Org. Lett.* **2003**, *5*, 4361–4364.

274 Kakei, H.; Nemoto, T.; Ohshima, T.; Shibasaki, M. Efficient synthesis of chiral α- and β-hydroxy amides: application to the synthesis of (*R*)-fluoxetine. *Angew. Chem. Int. Ed.* **2004**, *43*, 317–320.

275 Concellón, J. M.; Bardales, E. The first transformation of aliphatic α,β-epoxyamides into α-hydroxyamides. *Org. Lett.* **2003**, *5*, 4783–4785.

276 Hayakawa, H.; Miyashita, M. An efficient and stereoselective construction of the C(9)–C(17) dihydropyran segment of swinholides A–C via a novel reductive cleavage of an epoxy aldehyde. *Tetrahedron Lett.* **2000**, *41*, 707–711.

277 Hardouin, C.; Chevallier, F.; Rousseau, B.; Doris, E. Cp_2TiCl-Mediated selective reduction of α,β-epoxy ketones. *J. Org. Chem.* **2001**, *66*, 1046–1048.

278 Nel, R. J. J.; van Rensburg, H.; van Heerden, P. S.; Coetzee, J.; Ferreira, D. Stereoselective synthesis of flavonoids. Part 7. Polyoxygenated β-hydroxydihydrochalcone derivatives. *Tetrahedron* **1999**, *55*, 9727–9736.

279 Shimizu, I.; Hayashi, K.; Ide, N.; Oshima, M. Facile synthesis of (–)-serricornin be means of Pd-catalyzed hydrogenolysis of alkenyloxiranes. *Tetrahedron* **1991**, *47*, 2991–2998.

280 Jensen, B. L.; Slobodzian, S. V. A concise synthesis of 1-substituted 2-tetralones by selective diol dehydration leading to ketone transposition. *Tetrahedron Lett.* **2000**, *41*, 6029–6033.

281 Kita, Y.; Yoshida, Y.; Kitagaki, S.; Mihara, S.; Fang, D.-F.; Furukawa, A.; Higuchi, K.; Fujioka, H. Regio- and stereoselective rearrangement reactions of various α,β-epoxy acylates: suitable combination of acyl groups and Lewis acids. *Tetrahedron* **1999**, *55*, 4979–4998.

282 Ravikumar, K. S.; Bégué, J.-P.; Bonnet-Delpon, D.; Ourévitch, M. Synthesis in fluorous phase: a convenient synthesis of isolongifolene epoxide and its rearrangement to a ketone. *J. Fluorine Chem.* **2000**, *102*, 51–53.

283 Boulin, B.; Miguel, B. A.-S.; Delmond, B. Le γ-pyronène: synthon d'acces au safranal et précurseur d'intermédiaires de synthèse de la forskoline et du strigol. *Tetrahedron* **1998**, *54*, 2753–2762.

284 Lautens, M.; Ouellet, S. G.; Raeppel, S. Amphoteric character of 2-vinyloxiranes: synthetic equivalents of β,γ-unsaturated aldehydes and a vinylogous enolate. *Angew. Chem. Int. Ed.* **2000**, *39*, 4079–4082.

285 Kita, Y.; Higuchi, K.; Yoshida, Y.; Iio, K.; Kitagaki, S.; Akai, S.; Fujioka, H. Asymmetric total synthesis of fredericamycin A. *Angew. Chem. Int. Ed.* **1999**, *38*, 683–686.

286 Reizelman, A.; Zwanenburg, B. Synthesis of the germination stimulants (±)-orobanchol and (±)-strigol via an allylic rearrangement. *Synthesis* **2000**, 1952–1955.

287 Tchilibon, S.; Mechoulam, R. Synthesis of a primary metabolite of cannabidiol. *Org. Lett.* **2000**, *2*, 3301–3303.

288 Morgans, D. J.; Sharpless, K. B.; Traynor, S. G. Epoxy alcohol rearrangements: hydroxyl-mediated delivery of Lewis acid promoters. *J. Am. Chem. Soc.* **1981**, *103*, 462–464.

289 Sudha, R.; Narasimhan, K. M.; Saraswathy, V. G.; Sankararaman, S. Chemo- and regioselective conversion of epoxides to carbonyl compounds in 5 M $LiClO_4$–Et_2O medium. *J. Org. Chem.* **1996**, *61*, 1877–1879.

290 Hudrlik, P. F.; Ahmed, M. E.; Roberts, R. R.; Hudrlik, A. M. Reactions of (triphenylsilyl) ethylene oxide with Grignard reagents (and with $MgBr_2$). A reinvestigation. *J. Org. Chem.* **1996**, *61*, 4395–4399.

291 Eisch, J. J.; Liu, Z.-R.; Ma, X.; Zheng, G.-X. High regioselectivity in the alternative, halogenative cleavages of terminal epoxides with Lewis acid metal halides. *J. Org. Chem.* **1992**, *57*, 5140–5144.

292 Capriati, V.; Florio, S.; Luisi, R.; Russo, V.; Salomone, A. Oxiranyllithium based synthesis of

α-keto-2-oxazolines. *Tetrahedron Lett.* **2000**, *41*, 8835–8838.

293 Schneider, C.; Brauner, J. Lewis base-catalyzed addition of trialkylaluminum compounds to epoxides. *Eur. J. Org. Chem.* **2001**, 4445–4450.

294 Wipf, P.; Xu, W. Organozirconocenes in organic synthesis: tandem epoxide rearrangement–carbonyl addition. *J. Org. Chem.* **1993**, *58*, 825–826.

295 Koreeda, M.; Koizumi, N. Stereochemically controlled synthesis of 20-isocholesterol. *Tetrahedron Lett.* **1978**, 1641–1644.

296 Fukatsu, K.; Fujii, N.; Ohkawa, S. Synthesis of TAK-218 using (*R*)-2-methylglycidyl tosylate as a chiral building block. *Tetrahedron Asymmetry* **1999**, *10*, 1521–1526.

297 Harada, T.; Mukaiyama, T. Trityl antimonate-catalyzed sequential reactions of epoxides with silylated nucleophiles. Rearrangement of epoxides and C–C or C–O bond forming nucleophilic reaction onto the intermediate carbonyl compounds. *Bull. Chem. Soc. Jpn.* **1993**, *66*, 882–891.

298 Linstrumelle, G.; Lorne, R.; Dang, H. P. Cu-Catalysed reactions of allylic Grignard reagents with epoxides. *Tetrahedron Lett.* **1978**, 4069–4072.

299 Dreger, E. E. *n*-Hexyl alcohol. *Org. Synth.* **1941**, *Coll. Vol. I*, 306–308.

300 Hanson, R. M. The synthetic methodology of nonracemic glycidol and related 2,3-epoxy alcohols. *Chem. Rev.* **1991**, *91*, 437–475.

301 Taber, D. F.; Christos, T. E.; Rheingold, A. L.; Guzei, I. A. Synthesis of (–)-fumagillin. *J. Am. Chem. Soc.* **1999**, *121*, 5589–5590.

302 Herr, R. W.; Johnson, C. R. A comparison of the reactions of methylmagnesium, methyllithium, and methylcopper reagents with 1,2-epoxybutane and 3,4-epoxy-1-butene. *J. Am. Chem. Soc.* **1970**, *92*, 4979–4981.

303 Herr, R. W.; Wieland, D. M.; Johnson, C. R. Reactions of organocopper reagents with oxiranes. *J. Am. Chem. Soc.* **1970**, *92*, 3813–3814.

304 Lipshutz, B. H.; Kotsuki, H.; Lew, W. En route to polyene macrolide total synthesis; the key chiral segments of roflamycoin. *Tetrahedron Lett.* **1986**, *27*, 4825–4828.

305 Ergüden, J.-K.; Schaumann, E. Phosphoryl functionalized bishomoallyl alcohols by ring opening of epoxides with lithiated allyldiphenylphosphane oxide. *Synthesis* **1996**, 707–710.

306 Danishefsky, S. J.; Tsai, M.-Y.; Kitahara, T. Specific directing effects in the opening of vicinal hydroxy epoxides. *J. Org. Chem.* **1977**, *42*, 394–396.

307 Taber, D. F.; Green, J. H.; Geremia, J. M. C–C Bond formation with allylmagnesium chloride. *J. Org. Chem.* **1997**, *62*, 9342–9344.

308 Overman, L. E.; Renhowe, P. A. Regioselective opening of terminal epoxides with 3-(trialkylsilyl)allyl organometallic reagents. *J. Org. Chem.* **1994**, *59*, 4138–4142.

309 Imai, T.; Nishida, S. Lewis acid promoted ring-opening allylation of epichlorohydrin with allylic silanes and stannanes to afford 1-chloro-5-alken-2-ols. A short synthesis of (*S*)-(–)-ipsenol. *J. Org. Chem.* **1990**, *55*, 4849–4952.

310 Sunay, U.; Fraser-Reid, B. Synthetic studies relating to the C1-C9 "eastern" half of rosaramicin. *Tetrahedron Lett.* **1986**, *27*, 5335–5338.

311 Smith, A. B.; Pitram, S. M.; Boldi, A. M.; Gaunt, M. J.; Sfouggatakis, C.; Moser, W. H. Multicomponent linchpin couplings. Reaction of dithiane anions with terminal epoxides, epichlorohydrin, and vinyl epoxides: efficient, rapid, and stereocontrolled assembly of advanced fragments for complex molecule synthesis. *J. Am. Chem. Soc.* **2003**, *125*, 14435–14445.

312 Ent, H.; de Koning, H.; Speckamp, W. N. 2-Azonia-Cope rearrangement in *N*-acyliminium cyclizations. *J. Org. Chem.* **1986**, *51*, 1687–1691.

313 Ottow, E.; Neef, G.; Wiechert, R. Stereo- and regiospecific 6-endo-trig-cyclization of aryl radicals, an entry to novel progesterone antagonists of the androstane series. *Angew. Chem. Int. Ed.* **1989**, *28*, 773–776.

314 Lipshutz, B. H.; Kozlowski, J.; Wilhelm, R. S. Chemistry of higher order mixed organocuprates. 2. Reactions of epoxides. *J. Am. Chem. Soc.* **1982**, *104*, 2305–2307.

315 Johnson, C. R.; Herr, R. W.; Wieland, D. M. Reactions of lithium diorganocuprates(I) with oxiranes. *J. Org. Chem.* **1973**, *38*, 4263–4268.

316 Chong, J. M.; Sharpless, K. B. Regioselective openings of 2,3-epoxy acids with organocuprates. *Tetrahedron Lett.* **1985**, *26*, 4683–4686.

317 Chong, J. M.; Cyr, D. R.; Mar, E. K. Regioselective opening of 2,3-epoxy alcohols with organocuprates. Enhanced C-2 selectivity through solvent effects. *Tetrahedron Lett.* **1987**, *28*, 5009–5012.

318 Marshall, J. A.; Trometer, J. D.; Cleary, D. G. Stereoselective SN2′ additions of organocuprates to homochiral acyclic vinyloxiranes. *Tetrahedron* **1989**, *45*, 391–402.

319 Abe, N.; Hanawa, H.; Maruoka, K.; Sasaki, M.; Miyashita, M. Highly efficient alkylation of epoxides with R_3Al/H_2O systems based on the double activation of epoxy oxygens. *Tetrahedron Lett.* **1999**, *40*, 5369–5372.

320 Uchiyama, M.; Koike, M.; Kameda, M.; Kondo, Y.; Sakamoto, T. Unique reactivities of new highly coordinated ate complexes of organozinc derivatives. *J. Am. Chem. Soc.* **1996**, *118*, 8733–8734.

321 Xue, S.; Li, Y.; Han, K.; Yin, W.; Wang, M.; Guo, Q. Addition of organozinc species to cyclic 1,3-diene monoepoxides. *Org. Lett.* **2002**, *4*, 905–907.

322 Tanaka, T.; Inoue, T.; Kamei, K.; Murakami, K.; Iwata, C. Allyltitanium triphenoxide: selective cleavage of oxiranes at the more substituted carbon atom. *J. Chem. Soc. Chem. Commun.* **1990**, 906–908.

323 Yadav, J. S.; Anjaneyulu, S.; Ahmed, M. M.; Reddy, B. V. S. Indium-mediated regioselective allylation of terminal epoxides: a facile synthesis of bishomoallyl alcohols. *Tetrahedron Lett.* **2001**, *42*, 2557–2559.

324 Zhao, H.; Pagenkopf, B. L. Stereospecific and efficient alkynylation at the more hindered carbon of trisubstituted epoxides. *Chem. Commun.* **2003**, 2592–2593.

325 Fukumasa, M.; Furuhashi, K.; Umezawa, J.; Takahashi, O.; Hirai, T. Asymmetric synthesis of α-methyl carboxylic acid derivatives. Stereochemistry in acidic ring opening of epoxides. *Tetrahedron Lett.* **1991**, *32*, 1059–1062.

326 Jansen, R.; Knopp, M.; Amberg, W.; Bernard, H.; Koser, S.; Müller, S.; Münster, I.; Pfeiffer, T.; Riechers, H. Structural similarity and its surprises: endothelin receptor antagonists; process research and development report. *Org. Process Res. Dev.* **2001**, *5*, 16–22.

327 Ohno, H.; Hiramatsu, K.; Tanaka, T. Asymmetric construction of quaternary carbon centers by Ti-mediated stereospecific allylation of 2,3-epoxy alcohol derivatives. *Tetrahedron Lett.* **2004**, *45*, 75–78.

328 Ipaktschi, J.; Heydari, A. $LiClO_4$-katalysierte nucleophile Addition an α-chirale Aldehyde, Aldimine und Oxirane. *Chem. Ber.* **1993**, *126*, 1905–1912.

329 Kitaori, K.; Furukawa, Y.; Yoshimoto, H.; Otera, J. CsF in organic synthesis. Regioselective nucleophilic reactions of phenols with oxiranes leading to enantiopure β-blockers. *Tetrahedron* **1999**, *55*, 14381–14390.

330 Concellón, J. M.; Riego, E. Ring opening of nonactivated 2-(1-aminoalkyl)aziridines: unusual regio- and stereoselective C-2 and C-3 cleavage. *J. Org. Chem.* **2003**, *68*, 6407–6410.

331 Bruns, S.; Haufe, G. Enantioselective introduction of fluoride into organic compounds. First asymmetric ring opening of epoxides by hydrofluorinating reagents. *J. Fluorine Chem.* **2000**, *104*, 247–254.

332 Boa, A. N.; Clark, S.; Hirst, P. R.; Westwood, R. Ring opening reactions of quinoline-substituted epoxides. *Tetrahedron Lett.* **2003**, *44*, 9299–9302.

333 Curini, M.; Epifano, F.; Marcotullio, M. C.; Rosati, O. Zirconium sulfophenyl phosphonate as a heterogeneous catalyst in the preparation of β-amino alcohols from epoxides. *Eur. J. Org. Chem.* **2001**, 4149–4152.

334 Das, U.; Crousse, B.; Kesavan, V.; Bonnet-Delpon, D.; Bégué, J.-P. Facile ring opening of oxiranes with aromatic amines in fluoro alcohols. *J. Org. Chem.* **2000**, *65*, 6749–6751.

335 Pastó, M.; Rodríguez, B.; Riera, A.; Pericàs, M. A. Synthesis of enantiopure amino alcohols by ring-opening of epoxyalcohols and epoxyethers with ammonia. *Tetrahedron Lett.* **2003**, *44*, 8369–8372.

336 Caron, M.; Sharpless, K. B. $Ti(OiPr)_4$-Mediated nucleophilic openings of 2,3-epoxy alcohols. A mild procedure for regioselective ring-opening. *J. Org. Chem.* **1985**, *50*, 1557–1560.

337 Smissman, E. E.; Parker, G. R. Conformational aspects of systems related to acetylcholine. 5. Synthesis of the *dl*-2(e)-methyl-, *dl*-3(e)-methyl, and *dl*-2(e),3(e)-dimethyl-3(a)-trimethylammonium-2(a)-acetoxy-*trans*-decalin halides. *J. Med. Chem.* **1973**, *16*, 23–27.

338 Carbonelle, A.-C.; Gott, V.; Roussi, G. β-Amino alcohol-*N*-oxides as precursors of chiral oxazolidines: synthesis of (*R*)-(–)-cryptostyline I. *Heterocycles* **1993**, *36*, 1763–1769.

339 Blum, S. A.; Walsh, P. J.; Bergman, R. G. Epoxide-opening and group-transfer reactions mediated by monomeric zirconium imido complexes. *J. Am. Chem. Soc.* **2003**, *125*, 14276–14277.

340 Trost, B. M.; Calkins, T. L.; Oertelt, C.; Zambrano, J. Catalyst controlled diastereoselective *N*-alkylations of α-amino esters. *Tetrahedron Lett.* **1998**, *39*, 1713–1716.

341 Chandrasekhar, S.; Reddy, C. R.; Babu, B. N.; Chandrashekar, G. Highly efficient cleavage of epoxides catalyzed by $B(C_6F_5)_3$. *Tetrahedron Lett.* **2002**, *43*, 3801–3803.

342 Chuang, T.-H.; Sharpless, K. B. Applications of aziridinium ions: selective synthesis of pyrazolidin-3-ones and pyrazolo[1,2-*a*]pyrazoles. *Helv. Chim. Acta* **2000**, *83*, 1734–1743.

343 Díaz, M.; Ortuño, R. M. Enantioselective synthesis of novel homochiral α-substituted (*S*)-isoserine derivatives. Incorporation of this amino acid in a highly conformationally constrained dipeptide surrogate. *Tetrahedron Asymmetry* **1996**, *7*, 3465–3478.

344 Ettmayer, P.; Billich, A.; Hecht, P.; Rosenwirth, B.; Gstach, H. Paracyclophanes: a novel class of water-soluble inhibitors of HIV proteinase. *J. Med. Chem.* **1996**, *39*, 3291–3299.

345 Lanier, M.; Le Blanc, M.; Pastor, R. Nucleophilic ring opening of 3-*F*-alkyl 2,3-epoxypropanoates. Access to α,β-difunctional β-*F*-alkylpropanoates. *Tetrahedron* **1996**, *52*, 14631–14640.

346 Corey, E. J.; Lee, D.-H.; Choi, S. An enantioselective synthesis of (2*S*,3*S*)- and (2*R*,3*S*)-3-hydroxyleucine. *Tetrahedron Lett.* **1992**, *33*, 6735–6738.

347 Harada, K.; Oh-hashi, J. Optical resolution and configuration of *trans*-2,3-epoxybutyric acid. *Bull. Chem. Soc. Jpn.* **1966**, *39*, 2311–2312.

348 Caldwell, C. G.; Bondy, S. S. A convenient synthesis of enantiomerically pure (2*S*,3*S*)- or (2*R*,3*R*)-3-hydroxyleucine. *Synthesis* **1990**, 34–36.

349 Chong, J. M.; Sharpless, K. B. Nucleophilic opening of 2,3-epoxy acids and amides mediated by $Ti(OiPr)_4$. Reliable C-3 selectivity. *J. Org. Chem.* **1985**, *50*, 1560–1563.

350 Liwschitz, Y.; Rabinsohn, Y.; Perera, D. Synthesis of α-amino-β-hydroxy acids. Part I. DL-Allothreonine, DL-*erythro*-β-hydroxyleucine, and DL-*erythro*- and *threo*-β-hydroxy-β-methylaspartic acid. *J. Chem. Soc.* **1962**, 1116–1119.

351 Jin, C.; Jacobs, H. K.; Cervantes-Lee, F.; Gopalan, A. S. Intramolecular cyclization reactions of carbonyl derivatives of hydroxysulfones. *Tetrahedron* **2002**, *58*, 3737–3746.

352 Glatthar, R.; Giese, B. A new photocleavable linker in solid-phase chemistry for ether cleavage. *Org. Lett.* **2000**, *2*, 2315–2317.

353 Kaufman, T. S. Extension of the Bobbitt acetal cyclization to the elaboration of 1-hydroxymethyl substituted simple tetrahydroisoquinolines. A new synthesis of calycotomine. *Synth. Commun.* **1993**, *23*, 473–486.

354 Barluenga, J.; Vázquez-Villa, H.; Ballesteros, A.; González, J. M. $Cu(BF_4)_2$-Catalyzed ring-opening reaction of epoxides with alcohols at room temperature. *Org. Lett.* **2002**, *4*, 2817–2819.

355 Olszewski-Ortar, A.; Gros, P.; Fort, Y. Selective ring-opening of ω-epoxyalkyl (meth)acrylates. An efficient access to bifunctional monomers. *Tetrahedron Lett.* **1997**, *38*, 8699–8702.

356 Trost, B. M.; Tang, W.; Schulte, J. L. Asymmetric synthesis of quaternary centers. Total synthesis of (–)-malyngolide. *Org. Lett.* **2000**, *2*, 4013–4015.

357 Trost, B. M.; McEachern, E. J.; Toste, F. D. A two-component catalyst system for asymmetric allylic alkylations with alcohol pronucleophiles. *J. Am. Chem. Soc.* **1998**, *120*, 12702–12703.

358 Bednarski, P. J.; Porubek, D. J.; Nelson, S. D. Thiol-containing androgens as suicide substrates of aromatases. *J. Med. Chem.* **1985**, *28*, 775–779.

359 Avery, M. A.; Chong, W. K. M.; Jennings-White, C. Stereoselective total synthesis of (+)-artemisinin, the antimalarial constituent of *Artemisia annua* L. *J. Am. Chem. Soc.* **1992**, *114*, 974–979.

360 Toshimitsu, A.; Hirosawa, C.; Tamao, K. Retention of configuration in the Ritter-type substitution reaction of chiral β-arylthio alcohols through the anchimeric assistance of the arylthio group. *Tetrahedron* **1994**, *50*, 8997–9008.

361 Di Nunno, L.; Franchini, C.; Scilimati, A.; Sinicropi, M. S.; Tortorella, P. Chemical and hemoenzymatic routes to 1-(benzothiazol-2-ylsulfanyl)-3-chloropropan-2-ol, a precursor of drugs with potential β-blocker activity. *Tetrahedron Asymmetry* **2000**, *11*, 1571–1583.

362 Cavelier, F. First synthesis of the enantiomerically pure α-hydroxy analogue of *S*-*tert*-butyl cysteine. *Tetrahedron Asymmetry* **1997**, *8*, 41–43.

363 Flisak, J. R.; Gombatz, K. J.; Holmes, M. M.; Jarmas, A. A.; Lantos, I. A practical, enantioselective synthesis of SK&F 104353. *J. Org. Chem.* **1993**, *58*, 6247–6254.

364 Gleason, J. G.; Hall, R. F.; Perchonock, C. D.; Erhard, K. F.; Frazee, J. S.; Ku, T. W.; Kondrad, K.; McCarthy, M. E.; Mong, S.; Crooke, S. T.; Chi-Rosso, G.; Wasserman, M. A.; Torphy, T.

J.; Muccitelli, R. M.; Hay, D. W.; Tucker, S. S.; Vickery-Clark, L. High-affinity leukotriene receptor antagonists. Synthesis and pharmacological characterization of 2-hydroxy-3-[(2-carboxyethyl)thio]-3-[2-(8-phenyloctyl)phenyl]propanoic acid. *J. Med. Chem.* **1987**, *30*, 959–961.

365 Behrens, C. H.; Ko, S. Y.; Sharpless, K. B.; Walker, F. J. Selective transformation of 2,3-epoxy alcohols and related derivatives. Strategies for nucleophilic attack at C-1. *J. Org. Chem.* **1985**, *50*, 5687–5696.

366 Schneider, C. Quaternary ammonium salt catalyzed azidolysis of epoxides with Me_3SiN_3. *Synlett* **2000**, 1840–1842.

367 Fringuelli, F.; Pizzo, F.; Vaccaro, L. $AlCl_3$ as an efficient Lewis acid catalyst in water. *Tetrahedron Lett.* **2001**, *42*, 1131–1133.

368 Fringuelli, F.; Pizzo, F.; Rucci, M.; Vaccaro, L. First one-pot Cu-catalyzed synthesis of α-hydroxy-β-amino acids in water. A new protocol for preparation of optically active norstatins. *J. Org. Chem.* **2003**, *68*, 7041–7045.

369 Sasaki, M.; Tanino, K.; Hirai, A.; Miyashita, M. The C-2 selective nucleophilic substitution reactions of 2,3-epoxy alcohols mediated by trialkyl borates: the first endo-mode epoxide-opening reaction through an intramolecular metal chelate. *Org. Lett.* **2003**, *5*, 1789–1791.

370 Caron, M.; Carlier, P. R.; Sharpless, K. B. Regioselective azide opening of 2,3-epoxy alcohols by $[Ti(O\textit{i}Pr)_2(N_3)_2]$: synthesis of α-amino acids. *J. Org. Chem.* **1988**, *53*, 5185–5187.

371 Reddy, K. S.; Solà, L.; Moyano, A.; Pericàs, M. A.; Riera, A. Highly efficient synthesis of enantiomerically pure (*S*)-2-amino-1,2,2-triphenylethanol. Development of a new family of ligands for the highly enantioselective catalytic ethylation of aldehydes. *J. Org. Chem.* **1999**, *64*, 3969–3974.

372 Sassaman, M. B.; Prakash, G. K. S.; Olah, G. A. Regiospecific and chemoselective ring opening of epoxides with Me_3SiCN–KCN/18-crown-6 complex. *J. Org. Chem.* **1990**, *55*, 2016–2018.

373 Bowers, A.; Denot, E.; Sánchez, M. B.; Sánchez-Hidalgo, L. M.; Ringold, H. J. Steroids. CXXIV. Studies in cyano steroids. Part I. The synthesis of a series of C-6-cyano steroid hormones. *J. Am. Chem. Soc.* **1959**, *81*, 5233–5242.

374 Chini, M.; Crotti, P.; Favero, L.; Macchia, F. Easy direct stereo- and regioselective formation of β-hydroxy nitriles by reaction of epoxides with KCN in the presence of metal salts. *Tetrahedron Lett.* **1991**, *32*, 4775–4778.

375 Sugita, K.; Ohta, A.; Onaka, M.; Izumi, Y. Regiospecific ring opening of epoxides with Me_3SiCN on solid bases: reaction features, and role of metal cations of solid bases. *Bull. Chem. Soc. Jpn.* **1991**, *64*, 1792–1799.

376 Neef, G.; Eckle, E.; Müller-Fahrnow, A. A radical approach to the synthesis of 9(10→19)*abeo*-steroids. *Tetrahedron* **1993**, *49*, 833–840.

377 Gassman, P. G.; Okuma, K.; Lindbeck, A.; Allen, R. Mechanistic insights into the opening of epoxides with Me_3SiCN–ZnI_2. *Tetrahedron Lett.* **1986**, *27*, 6307–6310.

378 Imi, K.; Yanagihara, N.; Utimoto, K. Reaction of Me_3SiCN with oxiranes. Effects of catalysts or mediators on regioselectivity and ambident character. *J. Org. Chem.* **1987**, *52*, 1013–1016.

379 Gassman, P. G.; Guggenheim, T. L. Opening of epoxides with Me_3SiCN to produce β-hydroxy isonitriles. A general synthesis of oxazolines and β-amino alcohols. *J. Am. Chem. Soc.* **1982**, *104*, 5849–5850.

380 Benedetti, F.; Berti, F.; Norbedo, S. Ring-opening of epoxyalcohols by Et_2AlCN. Regio- and stereoselective synthesis of 1-cyano-2,3-diols. *Tetrahedron Lett.* **1999**, *40*, 1041–1044.

381 García Ruano, J. L.; Fernández-Ibáñez, M. A.; Martín Castro, A. M.; Rodríguez Ramos, J. H.; Rubio Flamarique, A. C. Regio- and stereocontrolled hydrocyanation of chiral 2-alkylglycidamides with Et_2AlCN: synthesis of enantiomerically pure mono- and disubstituted malic acid derivatives. *Tetrahedron Asymmetry* **2002**, *13*, 1321–1325.

382 Reddy, M. A.; Surendra, K.; Bhanumathi, N.; Rao, K. R. Highly facile biomimetic regioselective ring opening of epoxides to halohydrins in the presence of β-cyclodextrin. *Tetrahedron* **2002**, *58*, 6003–6008.

383 Sharghi, H.; Niknam, K.; Pooyan, M. The halogen-mediated opening of epoxides in the presence of pyridine-containing macrocycles. *Tetrahedron* **2001**, *57*, 6057–6064.

384 Denmark, S. E.; Barsanti, P. A.; Wong, K.-T.; Stavenger, R. A. Enantioselective ring-opening of epoxides with $SiCl_4$ in the presence of a chiral Lewis base. *J. Org. Chem.* **1998**, *63*, 2428–2429.

385 Chini, M.; Crotti, P.; Gardelli, C.; Macchia, F. Regio- and stereoselective synthesis of β-halohydrins from 1,2-epoxides with ammonium halides in the presence of metal salts. *Tetrahedron* **1992**, *48*, 3805–3812.

386 Haufe, G.; Bruns, S.; Runge, M. Enantioselective ring-opening of epoxides by HF-reagents. Asymmetric synthesis of fluoro lactones. *J. Fluorine Chem.* **2001**, *112*, 55–61.

387 Rodríguez, B.; de la Torre, M. C.; Perales, A.; Malakov, P. Y.; Papanov, G. Y.; Simmonds, M. S. J.; Blaney, W. M. Oxirane-opening reactions of some 6,19-oxygenated 4α,18-epoxy-neo-clerodanes isolated from teucrium. Biogenesis and antifeedant activity of their derivatives. *Tetrahedron* **1994**, *50*, 5451–5468.

388 Thomas, M. G.; Suckling, C. J.; Pitt, A. R.; Suckling, K. E. The synthesis of A- and B-ring fluorinated analogues of cholesterol. *J. Chem. Soc. Perkin Trans. 1* **1999**, 3191–3198.

389 Orru, R. V. A.; Mayer, S. F.; Kroutil, W.; Faber, K. Chemoenzymatic deracemization of (±)-2,2-disubstituted oxiranes. *Tetrahedron* **1998**, *54*, 859–874.

390 Hara, S.; Hoshio, T.; Kameoka, M.; Sawaguchi, M.; Fukuhara, T.; Yoneda, N. Regio- and stereoselective fluorinative ring-opening reaction of epoxyalcohols by $(i\mathrm{PrO})_2\mathrm{TiF}_2$–$\mathrm{Et}_4\mathrm{NF}$–nHF. Synthesis of optically active 3-fluoro-1,2-diols. *Tetrahedron* **1999**, *55*, 4947–4954.

391 Jung, M. E.; Fahr, B. T.; D'Amico, D. C. Total syntheses of the cytotoxic marine natural product, aplysiapyranoid C. *J. Org. Chem.* **1998**, *63*, 2982–2987.

392 Tomata, Y.; Sasaki, M.; Tanino, K.; Miyashita, M. The first C-2 selective halide substitution reaction of 2,3-epoxy alcohols by the use of the $(\mathrm{MeO})_3\mathrm{B}$–MX (X = I, Br, Cl) system. *Tetrahedron Lett.* **2003**, *44*, 8975–8977.

393 Amantini, D.; Fringuelli, F.; Pizzo, F.; Vaccaro, L. Bromolysis and iodolysis of α,β-epoxycarboxylic acids in water catalyzed by indium halides. *J. Org. Chem.* **2001**, *66*, 4463–4467.

394 Kvícala, J.; Plocar, J.; Vlasáková, R.; Paleta, O.; Pelter, A. 2-Fluoro-2-buten-4-olide, a new fluorinated synthon. Preparation; 1,2-, 1,4- and tandem additions. *Synlett* **1997**, 986–988.

395 Rehder, K. S.; Reusch, W. Thermal reactions of a 2-aryl-1-vinylcyclobutanol. *Tetrahedron* **1991**, *47*, 7551–7562.

5 The Alkylation of Carbanions

5.1 Introduction

The deprotonation of organic compounds by strong, non-nucleophilic bases followed by *C*-alkylation with a suitable electrophile has become one of the most predictable and straightforward methods for formation of C–C bonds. The popularity of this chemistry is also a consequence of the often-seen close resemblance of feasible structural modifications by a deprotonation/alkylation strategy to an unsophisticated retrosynthetic analysis of a given target compound. Other reactions, which involve, for instance, substantial rearrangement of the carbon framework (e.g. Cope rearrangement, fragmentations, ring contractions or enlargements) are usually more difficult to take into account during retrosynthetic analysis.

LDA and related, sterically hindered, lithium dialkylamides, first investigated by Levine [1], have completely replaced the more nucleophilic sodium amide, which had been the base of choice for many years [2]. Because the reactivity of an organometallic compound depends to a large extent on its state of aggregation (i.e. on the solvent and on additives) and on the metal, transmetalation of the lithiated intermediates and the choice of different solvents and additives emerged as powerful strategies for fine-tuning the reactivity of these valuable nucleophiles.

In this chapter the alkylation of carbanions with simple carbon electrophiles will be discussed, with special emphasis on the structure–reactivity relationship of the carbanion and on side reactions. In this context the term "carbanion" refers to an intermediate prepared by in-situ deprotonation of an organic compound, followed by optional cation exchange (transmetalation), which tends to be alkylated at carbon by soft electrophiles such as MeI or PhCHO. This type of reactivity is characteristic of enolates with an ionic M–O bond or for organometallic compounds with a strongly polarized or ionic M–C bond, M typically being an alkali metal, Mg, Cu, or Zn. Carbanions can, alternatively, also be prepared by halogen–metal exchange [3–8], sulfoxide- or sulfone–metal exchange [9–13], or by transmetalation of stannanes. The most suitable reagents for performing such metalating exchange reactions are organometallic compounds with a metal-bound secondary or tertiary alkyl group, such as *t*BuLi, *s*BuLi, *i*Pr_2Zn [14, 15], or *i*PrMgHal [6]. These reagents are thermodynamically less stable than organometallic compounds with primary alkyl

Side Reactions in Organic Synthesis. Florencio Zaragoza Dörwald

ISBN: 3-527-31021-5

groups, and will not be readily regenerated from other organometallic reagents under the conditions of reversible transmetalation. Formally metal-free carbanions can be generated by treatment of silanes [16, 17] or silylenol ethers [18] with TBAF. For studies on the mechanism of enolate formation, see Refs [19–23].

5.2 The Kinetics of Deprotonations

For discussion of deprotonations of organic compounds both the rate of deprotonation and the equilibrium acidity must be considered. Equilibrium acidities are expressed by the pK_a value, which is defined as follows for an acid AH:

$$AH \underset{k_{-a}}{\overset{k_a}{\rightleftharpoons}} A^- + H^+ \qquad K_a = \frac{[A^-][H^+]}{[AH]} = \frac{k_a}{k_{-a}}$$

$$pK_a = -\lg K_a$$

where [X] = concentration of X, K = equilibrium constant, and k = rate constant.

Equilibrium acidities of representative organic compounds have been determined [24–26], and depend both on the temperature and on the solvent (Table 5.1) [27–29]. Solvents which, because of their high dielectric constant or by hydrogen bonding, can stabilize ions will generally promote ionization, as will higher temperatures. The effect of the solvent on pK_a is particularly strong for acids which, on deprotonation, give anions of high charge density, but becomes smaller for highly delocalized anions of low charge density (Table 5.1).

Table 5.1. Equilibrium acidities in water and in DMSO at 25 °C [24, 30–33].

Acid	pK_a (H_2O)	pK_a (DMSO)	ΔpK_a
H_2O	15.75	32	16.25
MeOH	15.5	29.0	13.5
HF	3.2	15	11.8
PhC≡CH	20.0	28.7	8.7
PhOH	10.0	18.0	8.0
Me_2CO	19.3	26.5	7.2
$MeNO_2$	10.0	17.2	7.2
$PhCO_2H$	4.25	11.1	6.85
MeCOPh	18.3	24.7	6.4
$MeCONMe_2$	29.4	35	5.6
AcOEt	25.6	30	4.4
Ac–CH_2–Ac	8.9	13.3	4.4
HN_3	4.7	7.9	3.2
MeCN	28.9	31.3	2.4
NC–CH_2–CN	11.0	11.0	0.0
Picric acid	0	0	0

From the discussion above follows that the weaker an acid, the larger its pK_a. Acids and bases with a pK_a outside the range given by the pK_a values of OH^- ($pK_a = 15.8$) and H_3O^+ ($pK_a = -1.75$) [24] cannot be deprotonated/protonated to a large extent in water. This does not mean, however, that acids with a $pK_a > 15.8$ cannot be deprotonated at all in water [32, 34]. Propionitrile ($pK_a = 30.9$ in water) and *N,N*-dimethylacetamide ($pK_a = 29.4$ in water) undergo KOD-catalyzed H–D exchange in water at substantial rates ($t_{1/2}$ [EtCN in 0.5 M KOD or $MeCONMe_2$ in 0.25 M KOD] = 138 h [30, 31]); this shows that the corresponding carbanions are indeed formed, although to a small extent only.

The degree of dissociation of an acid in water can be easily estimated with the Henderson–Hasselbach equation. When the pH of the solution equals the pK_a of an acid, then the acid is 50% dissociated ([AH] = [A^-]); for each integer by which the pH differs from the pK_a the ratio [A^-]/[AH] will increase or decrease by a factor of 10:

$$pK_a = -\lg K_a = -\lg \frac{[A^-]}{[AH]} + pH$$

$$pK_a - pH = -\lg \frac{[A^-]}{[AH]}$$

Equilibrium acidities are usually determined by titration, either with indicators of known pK_a [24] or by following the extent of dissociation spectroscopically (e.g. by UV-spectroscopy [35]). Kinetic acidities, i.e. the rates of deprotonation of acids, can be determined by measuring rates of racemization [36] or rates of H–D or H–T exchange [37, 38], by chemical relaxation experiments (e.g. T-jump method [35, 39]), or by ^{1}H NMR (line broadening and saturation recovery [40]).

The rate of deprotonation of an acid by a base depends on their structures [41], on the solvent and temperature, and on the difference (ΔpK_a) between the pK_a of the acid and that of the base. When acid and base have the same pK_a ($\Delta pK_a = 0$) the change of free energy for proton transfer becomes zero and the reaction becomes thermoneutral. Under these conditions the rate of proton transfer is limited only by the so-called intrinsic barrier [34], which is particularly sensitive to structural changes in the reaction partners [39]. When ΔpK_a increases, the rate of proton transfer also increases and approaches a limiting value, which depends on the structures of the acid and base and on the experimental conditions. For normal acids (O–H, N–H) in water the rate of proton transfer becomes diffusion-controlled ($k_a \approx 10^{10}$ L mol^{-1} s^{-1}) when $\Delta pK_a > 2$, but in aprotic solvents the limiting proton transfer rate can be substantially lower [42].

The equilibrium acidities of organic acids do not necessarily correlate with their kinetic acidities [39, 43–46]. Deprotonations of sparsely polarized X–H groups which are accompanied by a large change in hybridization and geometry proceed more slowly than deprotonations with little rehybridization ("principle of least nuclear motion" [45]). In Scheme 5.1 the approximate relative rates of proton transfer to a base from some classes of acid are given, assuming that $\Delta pK_a = 0$ (thermoneutral conditions). Proton transfer from carboxylic acids, phenols, or ammonium salts to bases are usually very fast ($k_a \approx 10^{10}$ L mol^{-1} s^{-1}), the main rate-determining factors under thermodynamically favorable conditions ($\Delta pK_a > 2$) being diffusion and solvent reorganization. Under thermoneutral conditions proton transfer constants of these acids are approximately $5 \times 10^8\ L\ mol^{-1}\ s^{-1}$ [35]. The fact that carboxylic acids and phenols are deprotonated at similar rates as ammonium salts shows that the stabilization of carboxylates or phenolates by charge delocalization and resonance is *small*, and that their high equilibrium acidity is mainly because of the sp^2-hybridization of the oxygen-bound carbon atom and to electrostatic stabilization of O^- by the carbonyl group (in carboxylic acids) [47, 48].

Carbon acids which do not undergo significant rehybridization or structural reorganization upon proton loss, for example HCN [40], alkynes [33], chloroform [33], or thiazolium salts [49], also undergo fast deprotonation [50]. Much slower is, though, the deprotonation of sulfones, ketones, and nitroalkanes (Scheme 5.1). For most C,H-acidic compounds, rates of deprotonation *decrease* with increasing delocalization of the negative charge on to oxygen atoms. The stronger the stabilization of a carbanion by resonance and solvation, the more will this stabilization lag behind proton transfer in the transition state, leading to a lower deprotonation rate [51]. In non-aqueous solvents, where the energy gain by solvation is lower, the rate of deprotonation of, for instance, nitroalkanes is, therefore, larger than in water [52].

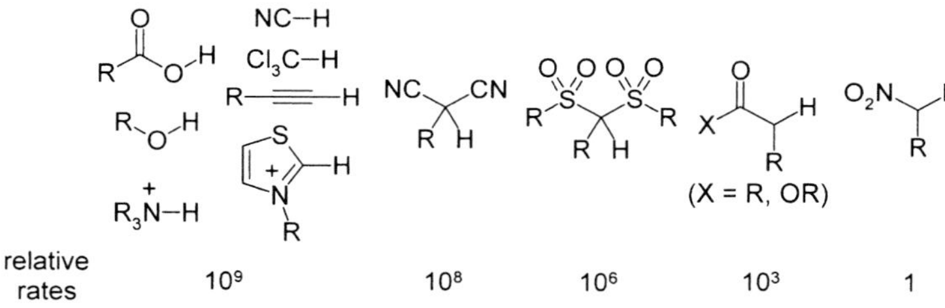

(X = R, OR)

relative rates 10^9 10^8 10^6 10^3 1

Scheme 5.1. Approximate relative rates of proton transfer in water at thermoneutrality ($\Delta pK_a = 0$) [35, 39, 51].

5.3 Regioselectivity of Deprotonations and Alkylations

5.3.1 Introduction

On treatment with strong bases (ΔpK_a acid–base $\gg 2$), compounds containing more than one acidic C,H group are usually deprotonated first at the most acidic position and then at the less acidic sites if the difference between the equilibrium acidities of these C,H groups is large. The equilibrium acidity of C,H-acidic compounds usually (but not always) increases strongly with the number of carbanion stabilizing groups attached to the acidic C,H group ($CH_3Z < CH_2Z_2 < CHZ_3$; Z = carbanion stabilizing group; see, however, Scheme 5.2). Functional groups which stabilize carbanions do this either by charge delocalization to more electronegative atoms (carbonyl compounds, nitroalkanes), by polar effects (nitriles, sulfones [29]), by field effects (ammonium salts), or by chelation of the metal ("dipole stabilization"; e.g. MC–CCONR$_2$, MC–NCO$_2$R, MC–OCONR$_2$ [53–59]). An approximate ranking of functional groups according to their ability to increase the equilibrium acidity of an aliphatic C,H group would be: $NO_2 > SOMe_2^+ \approx SMe_2^+ \approx SO_2CF_3 > C_6H_4\text{-4-}NO_2 > PPh_3^+ > COPh > COR > CN > SO_2Ph \approx CO_2R > CONR_2 \approx SOPh >$ 2-pyridyl $> SPh \approx SePh \approx NMe_3^+ \approx PPh_2 > Ph \approx C{\equiv}CPh \approx CH{=}CH_2 > OPh \approx F > H > CH_3$.

The effect on acidity of carbanion-stabilizing functional groups is not necessarily additive. Certain groups may have a carbanion-stabilizing effect in one compound, but reduce the acidity of another compound. Fluorine, for instance, can increase the acidity of alkanes by electron-withdrawal through σ bonds, but can destabilize a carbanion by electron-donation from its lone pairs into π-type orbitals. Thus, when bound to an sp^3-hybridized carbon atom fluorine will generally enhance acidity, but when bound to an sp^2 carbon it can have a carbanion-destabilizing effect. Fluoroform which, on deprotonation, gives a pyramidalized carbanion with sp^3-hybridized carbon [60], is much more acidic than methane (Scheme 5.2). Fluoroacetophenone or fluoromethyl phenyl sulfone, on the other hand, in which stabilization of the carbanion by delocalization of the negative charge to oxygen is important, are only slightly more acidic than acetophenone and methyl phenyl sulfone, respectively. 9-Fluorofluorene [61] or fluorodinitromethane, which yield on deprotonation planar, sp^2-hybridized carbanions, are even less acidic than the corresponding non-fluorinated compounds (Scheme 5.2).

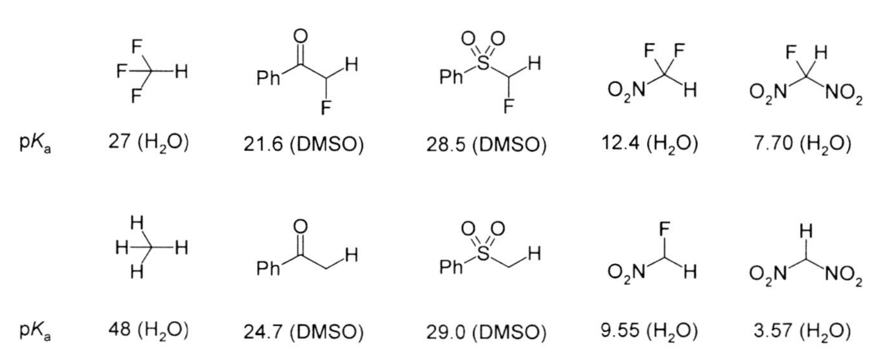

Scheme 5.2. The effect of fluorine on the acidity of organic compounds [24, 62–65].

There is no perfectly linear correlation between the basicity and nucleophilicity of carbanions [66], but higher basicity usually also implies higher nucleophilicity. Carbanions in which the negative charge is highly delocalized (e.g. diethylmalonate) will usually react more slowly with electrophiles than less extensively delocalized carbanions of similar basicity (e.g. malodinitrile) [66].

The more acidic a C,H group, the less basic and nucleophilic will the corresponding carbanion usually be. Consequently, carbanions generated by deprotonation of strongly acidic C,H groups will react slowly with electrophiles, and are usually difficult to alkylate, as illustrated by the examples in Scheme 5.3. Monodeprotonated nitroalkanes (first reaction, Scheme 5.3) are difficult to alkylate at carbon, and are preferentially *O*-alkylated (unless a path via SET is accessible; see Section 4.1). The resulting *O*-benzylated nitroalkanes can decompose during work-up to yield benzaldehydes and oximes [67–69]. The dianions of nitroalkanes, on the other hand, can be readily alkylated at carbon [70, 71].

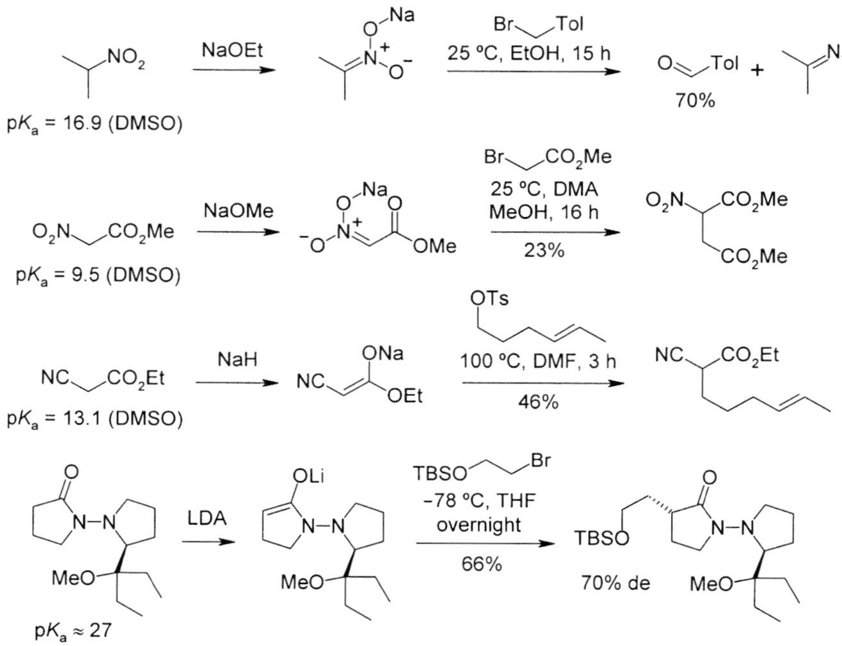

Scheme 5.3. Dependence of the reactivity of carbanions on their basicity [68, 72–74].

5.3.2
Kinetic/Thermodynamic Enolate Formation

Stoichiometric, irreversible formation of enolates from ketones or aldehydes is usually performed by addition of the carbonyl compound to a cold solution of LDA. Additives and the solvent can strongly influence the rate of enolate formation [23]. The use of organolithium compounds as bases for enolate formation is usually not a good idea, because these reagents will add to ketones quickly, even at low temperatures. Slightly less electrophilic carbonyl compounds, for example some methyl esters [75], can, however, be deprotonated by BuLi if the reactants are mixed at low temperatures (typically –78 °C), at which more metalation than addition is usually observed. A powerful lithiating reagent, which can sometimes be used to deprotonate ketones at low temperatures, is *t*BuLi [76].

The organic chemist is occasionally confronted with the problem that a compound has several differrent X–H groups of similar pK_a. If only one of these is to be alkylated, one option would be to perform a regioselective deprotonation. This can sometimes be achieved by exploiting small differences between the kinetic or equilibrium acidities of the acidic sites. If strong bases such as LDA (pK_a = 35.7) or LiTMP (pK_a = 37.3) [77] are used deprotonation of many classes of compound can be conducted irreversibly and at low temperatures, at which interconversion of the possible, different carbanions is slow. If the carbanion formed by deprotonation of the C,H group of highest kinetic acidity (i.e. the initially formed carbanion) is not the thermodynamically most stable carbanion, equilibration of these carbanions at low temperatures might be sufficiently slow to enable the generation of the essentially pure, less stable carbanion ("kinetic control"). As shown in Scheme 5.4, 2-methylcyclohexanone (pK_a ≈ 26.4) will be deprotonated by LDA more rapidly at the α-methylene group, because this group is sterically more accessible and because this position is statistically twice as likely as the methine group to be deprotonated. The enolate resulting from deprotonation of the sterically more demanding methine group is, however, more stable, because it is the more highly substituted alkene. Thus, if the products can equilibrate under the reaction conditions chosen the main product will usually result from the most highly substituted enolate [78, 79]. Acidic reaction conditions are also conducive to the formation of the most highly substituted enol derivative [80, 81]. As shown in Scheme 5.4, positions at which the equilibrium acidity is significantly lower than for other C,H groups might be kinetically sufficiently more acidic to enable their selective deprotonation.

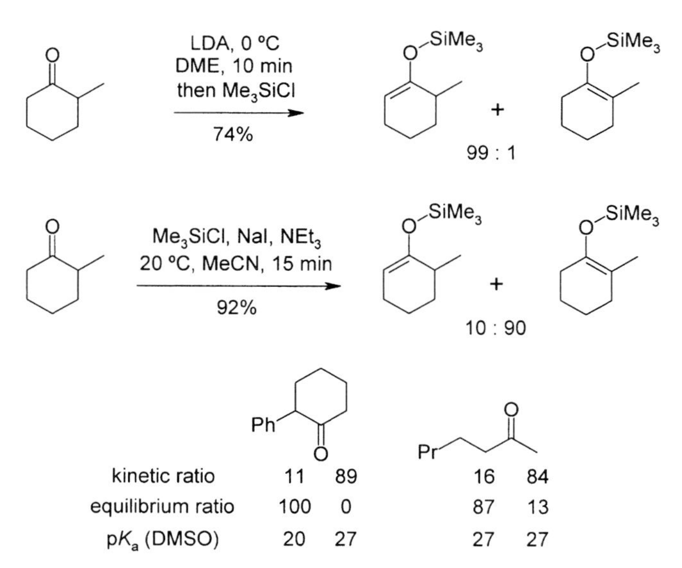

Scheme 5.4. Enolate formation under kinetic and thermodynamic control [79, 82–84].

5.3.3
Allylic and Propargylic Carbanions

Allylic and propargylic carbanions are ambident nucleophiles which can in principle yield unrearranged or rearranged products when treated with an electrophile. Most nucleophilic allyl metals react with electrophiles at the γ position, and only occasionally (in particular with organopotassium and organosodium compounds [85, 86]) are mixtures of γ- and α-alkylated products obtained. Thus, 2-butenyllithium [87], magnesium [88, 89], titanium [90], zinc [91, 92], or copper reagents attack alkyl halides, aldehydes, ketones, imines, or electron-deficient alkenes mainly at the γ carbon to yield 1-methylallyl derivatives (Scheme 5.5). Some of these reactions can be conducted in the presence of enantiomerically pure catalysts with high diastereo-

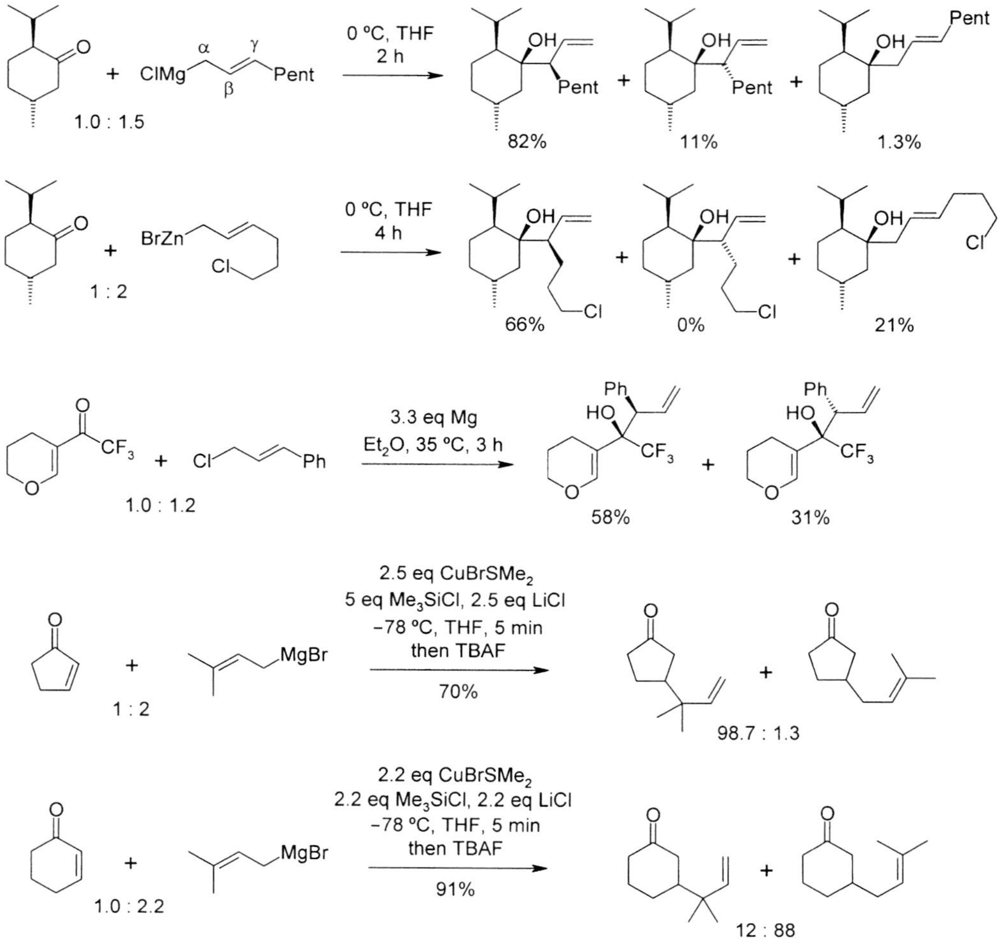

O + ClMg (α, β, γ) Pent; 1.0 : 1.5; 0 °C, THF, 2 h; OH, Pent 82% + OH, Pent 11% + OH, Pent 1.3%

O + BrZn, Cl; 1 : 2; 0 °C, THF, 4 h; OH, Cl 66% + OH, Cl 0% + OH, Cl 21%

O, CF_3 + Cl, Ph; 1.0 : 1.2; 3.3 eq Mg, Et_2O, 35 °C, 3 h; Ph, HO, CF_3, O 58% + Ph, HO, CF_3, O 31%

O + MgBr; 1 : 2; 2.5 eq $CuBrSMe_2$, 5 eq Me_3SiCl, 2.5 eq LiCl, −78 °C, THF, 5 min, then TBAF; 70%; O + O; 98.7 : 1.3

O + MgBr; 1.0 : 2.2; 2.2 eq $CuBrSMe_2$, 2.2 eq Me_3SiCl, 2.2 eq LiCl, −78 °C, THF, 5 min, then TBAF; 91%; O + O; 12 : 88

Scheme 5.5. Reaction of allylic organometallic reagents with different electrophiles [94–96].

and enantioselectivity [93]. As shown by the two last reactions in Scheme 5.5, the α/γ-selectivity of prenyl copper reagents depends on the precise structure of the electrophile.

Substrates in which γ attack implies significant steric congestion can, however, predominantly yield products of α attack. A trick for enhancing the steric demand of aldehydes consists in using bulky acylsilanes instead. These react with organomagnesium or -zinc reagents to yield α-silyl alcoholates which undergo Brook rearrangement to silylethers. The latter can be readily desilylated to the corresponding secondary alcohols [86]. As illustrated by the examples in Scheme 5.6, the triphenylsilyl group is not well suited to enhance the α selectivity, but acyltrimethylsilanes combined with organozinc reagents yield mainly products of α attack. Bulkier electrophiles, such as acyltriisopropylsilanes, for instance, yield almost exclusively products of α attack on treatment with crotyl or prenyl zinc reagents. The corresponding Grignard reagents, however, still react mainly with the γ carbon, even with these sterically demanding silanes [86].

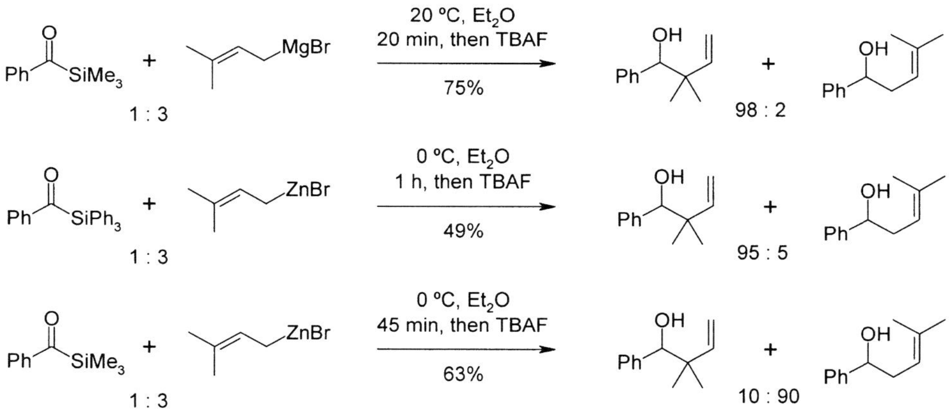

Ph(C=O)SiMe$_3$ + prenyl-MgBr (1 : 3) → 20 °C, Et_2O, 20 min, then TBAF; 75% → 98 : 2

Ph(C=O)SiPh$_3$ + prenyl-ZnBr (1 : 3) → 0 °C, Et_2O, 1 h, then TBAF; 49% → 95 : 5

Ph(C=O)SiMe$_3$ + prenyl-ZnBr (1 : 3) → 0 °C, Et_2O, 45 min, then TBAF; 63% → 10 : 90

Scheme 5.6. Reaction of acylsilanes with prenylmagnesium and zinc bromide [86].

Selective α attack with crotyl metals can be achieved by transmetalation, i.e. by a sequence of two sequential γ derivatizations of the allylic carbanion (Scheme 5.7). This can, for instance, be achieved by treating crotyl magnesium chloride with $AlCl_3$, followed by treatment with the electrophile. Not all transmetalations do, however, lead to an allylic rearrangement. Crotyllithium, for instance, reacts with stannanes with its α carbon to yield pentavalent stannates. In the example shown in Scheme 5.7 (last reaction) after allylation of tin (α attack) a highly diastereoselective intramolecular transfer of the crotyl group from tin to the carbonyl group occurs with allylic rearrangement [97].

Benzylic organometallic compounds may react with electrophiles at either the benzylic or *ortho* position. Benzylic Grignard or organozinc reagents react with elec-

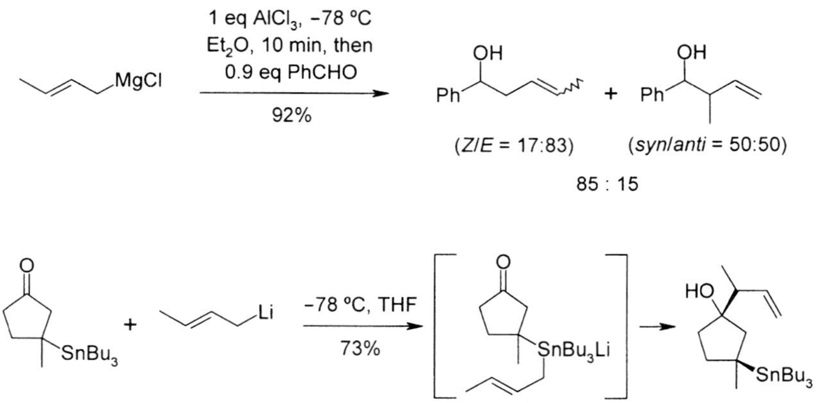

Scheme 5.7. Transmetalation of crotylmagnesium and crotyllithium compounds [97, 98].

trophiles preferentially at the *ortho* position, whereas α attack is observed with the corresponding benzylpotassium or copper derivatives (Scheme 5.8).

Non-stabilized propargylic carbanions can react either without rearrangement to yield alkynes or rearrange to yield allenes. Propargyl bromide reacts with magnesium to yield allenylmagnesium bromide [100], which reacts with ketones [101], aldehydes [102, 103], benzyl halides [104], or tin halides [105] in Et_2O to yield alkynes (Scheme 5.9). Similarly, treatment of aldehydes with propargyl bromide and tin,

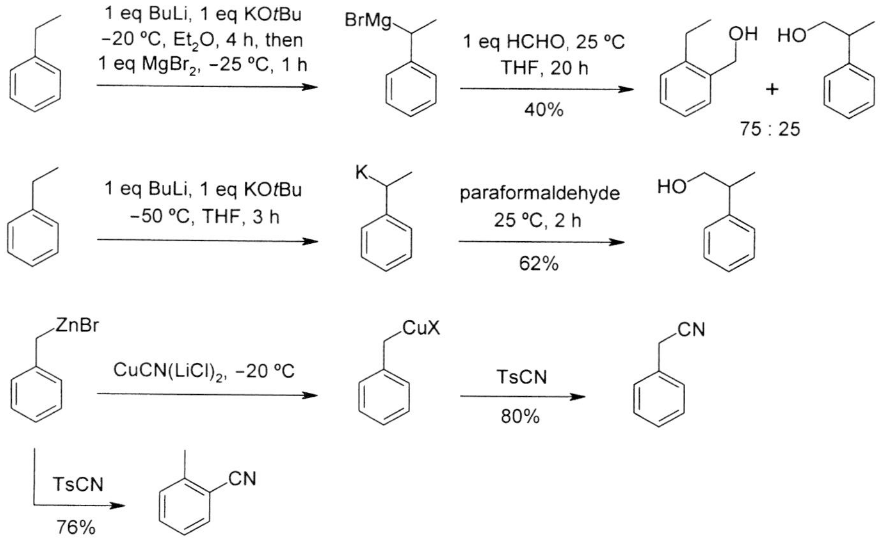

Scheme 5.8. Metal-dependent regioselectivity of the hydroxymethylation and cyanation of benzylic carbanions [85, 99].

zinc, indium, or bismuth in water leads mainly to alkynes [106]. In CH_2Cl_2 under acidic reaction conditions mixtures of allenes and alkynes often result [107, 108]. Propargyl derivatives of the type $RC{\equiv}CCH_2X$ ($R \neq H$) mainly yield allenes on halogen–metal exchange followed by reaction with an electrophile [105–108].

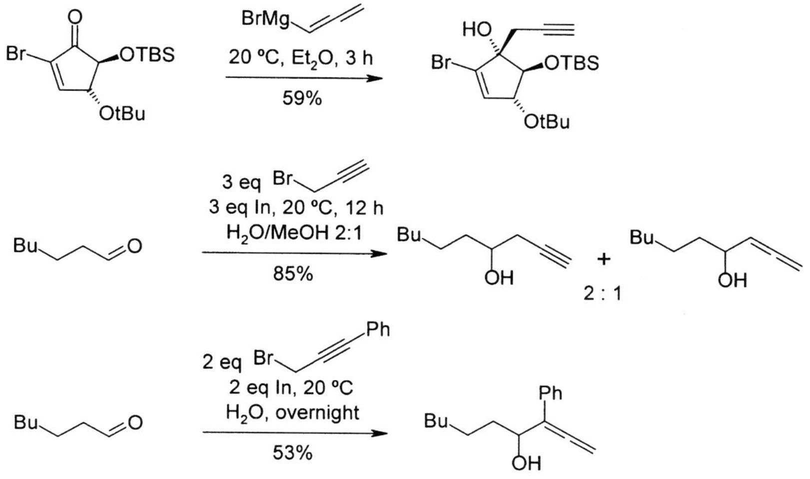

Scheme 5.9. Reactions of propargylic carbanions with electrophiles [101, 106].

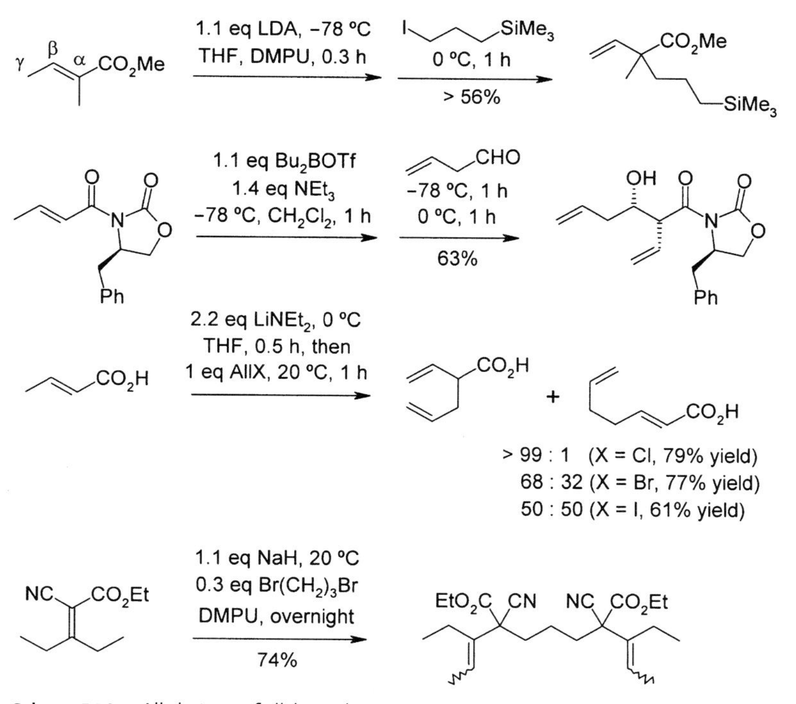

Scheme 5.10. Alkylations of allylic carbanions [109, 111, 122–124].

Metalated enones and related compounds are usually alkylated at the position α to the electron-withdrawing group [109–113], but the precise structure of the electrophile can also have an impact on regioselectivity (Scheme 5.10). Certain substrates, such as crotonic acid dianions [111], crotonamide dianions [114, 115], or certain enones [116] can also give mixtures of α- and γ-alkylated products, whereas metalated β-oxy [117–119] or β-amino [120, 121] acrylic acid derivatives often yield pure products of γ-alkylation.

Selective alkylation of the γ position of α,β-unsaturated esters, aldehydes, or ketones can be achieved if a sterically demanding Lewis acid is used to coordinate to the carbonyl group and inhibit α-alkylation by steric shielding [123, 125, 126] (Scheme 5.11). This method not only results in high regioselectivity but also enables highly stereoselective aldol-type additions to be performed in good yields, even with sensitive substrates, such as α,β-unsaturated aldehydes [126]. Thus, when two diastereotopic γ positions are available, the addition of a bulky aluminum phenolate leads to the clean formation of the Z-alkene (second and third examples, Scheme 5.11).

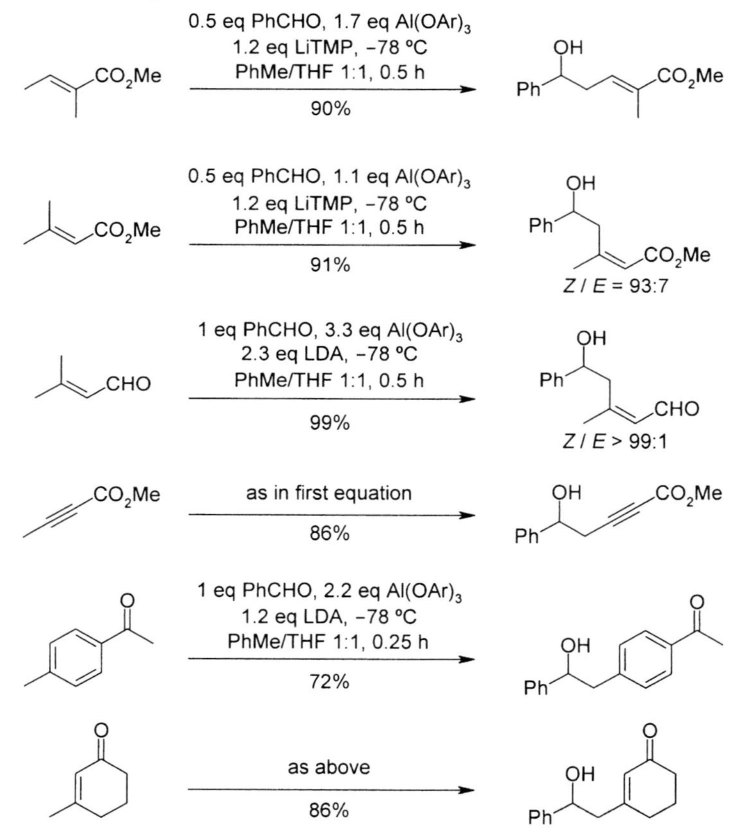

Scheme 5.11. γ-Alkylations of allylic and propargylic carbanions [123, 125, 126]. Ar = 2,6-$Ph_2C_6H_3$.

One potential problem in the reactions of stabilized allylic or propargylic carbanions is the dimerization of the starting material if the carbanions are not formed stoichiometrically. Alkenes substituted with electron-withdrawing groups are good Michael acceptors, to which nucleophiles will undergo conjugate addition. For instance, the Baylis–Hillman reaction of allyl cyanide with benzaldehyde requires careful optimization of the reaction conditions to avoid dimerization of the nitrile (Scheme 5.12). This problem is related to a common side reaction of Michael additions: reaction of the product with the Michael acceptor (Scheme 10.21).

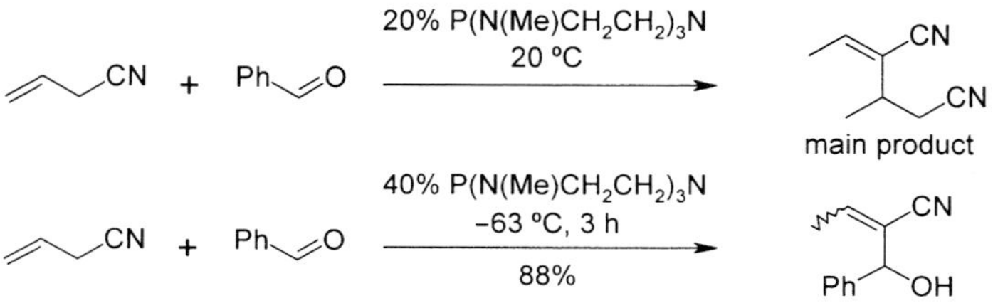

CN + Ph O → 20% $P(N(Me)CH_2CH_2)_3N$, 20 °C → CN, CN (main product)

CN + Ph O → 40% $P(N(Me)CH_2CH_2)_3N$, −63 °C, 3 h, 88% → CN, Ph, OH

Scheme 5.12. Baylis–Hillman reaction of allyl cyanide [110].

5.3.4
Succinic Acid Derivatives and Amide-derived Carbanions

The regioselective deprotonation of heteroatom-substituted succinic acid derivatives has been thoroughly investigated. The choice of protective group for the heteroatom is critical to avoid β-elimination. Because negatively charged atoms or groups destabilize a vicinal carbanion, derivatives of 2-hydroxy or 2-aminosuccinic acid (malic and aspartic acid) can be selectively converted to dianions and alkylated at the methylene group [127] (Scheme 5.13). Succinic acid derivatives in which the two carboxyl groups have been differentiated (e.g. amide/ester or ester/acid) can also be alkylated regioselectively [128–130] (Scheme 5.13).

The regioselectivity of the last reaction in Scheme 5.13 is not only because of the greater acidity of the methylene group, but also because some secondary and tertiary amides (e.g. β-arylamides, β-vinylamides, or β-(phenylthio)amides, or borane complexes of β-phosphino propionamides [132, 133]) are deprotonated at the β position under kinetic control to yield chelate-stabilized carbanions [58, 134]. Illustrative examples of such remarkable metalations are shown in Scheme 5.14.

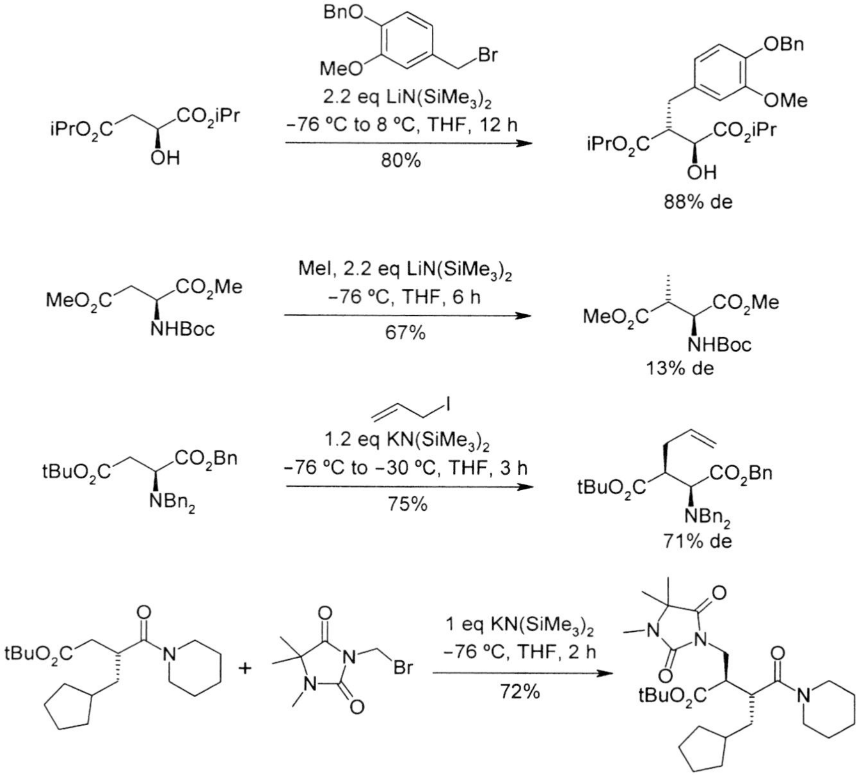

2.2 eq $LiN(SiMe_3)_2$, −76 °C to 8 °C, THF, 12 h, 80%, 88% de

MeI, 2.2 eq $LiN(SiMe_3)_2$, −76 °C, THF, 6 h, 67%, 13% de

1.2 eq $KN(SiMe_3)_2$, −76 °C to −30 °C, THF, 3 h, 75%, 71% de

1 eq $KN(SiMe_3)_2$, −76 °C, THF, 2 h, 72%

Scheme 5.13. Alkylation of substituted succinic acid derivatives [128–131].

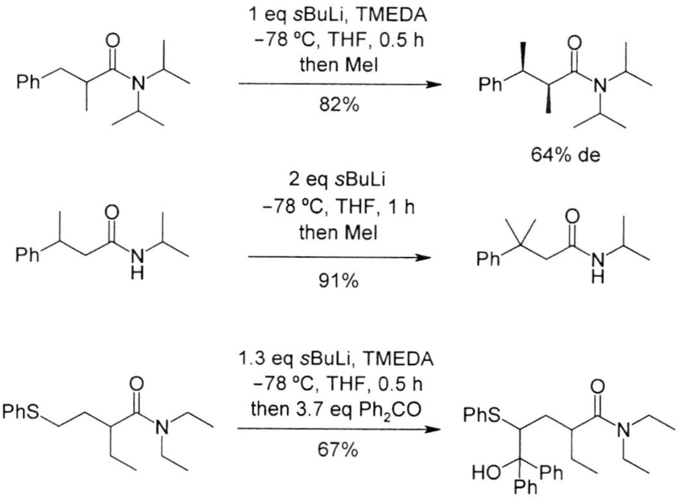

1 eq *s*BuLi, TMEDA, −78 °C, THF, 0.5 h, then MeI, 82%, 64% de

2 eq *s*BuLi, −78 °C, THF, 1 h, then MeI, 91%

1.3 eq *s*BuLi, TMEDA, −78 °C, THF, 0.5 h, then 3.7 eq Ph_2CO, 67%

Scheme 5.14. Alkylation of amides at the β and γ positions [53, 58, 134].

5.3.5
Bridgehead Carbanions

Small bicyclic ketones or lactams with a bridgehead C,H group are much less acidic than comparable acyclic compounds, because delocalization of the negative charge to the carbonyl oxygen atom would imply the formation of a highly strained bridgehead alkene. Such bridgehead enolates usually oligomerize quickly by intermolecular addition of the carbanion to the non-enolized carbonyl groups, sometimes even in the presence of other trapping reagents [135]. Even cyclic 1,3-diketones are no longer C,H-acidic if C-2 is located at a bridgehead position (second example, Scheme 5.15).

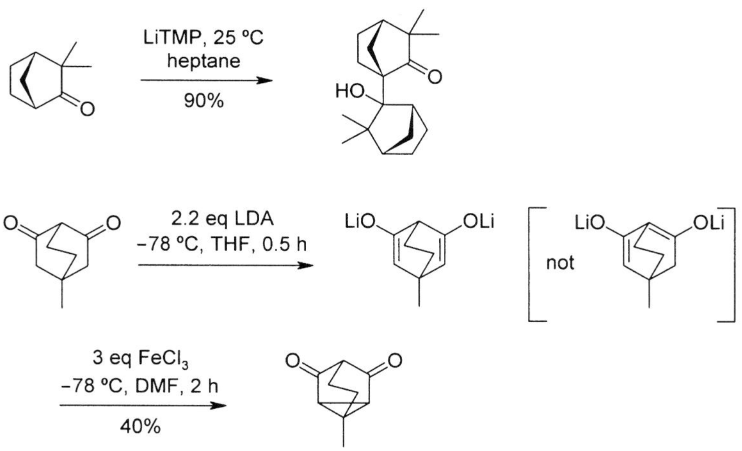

LiTMP, 25 °C, heptane, 90%

2.2 eq LDA, −78 °C, THF, 0.5 h

not

3 eq $FeCl_3$, −78 °C, DMF, 2 h, 40%

Scheme 5.15. Deprotonation and reactions of bicyclic ketones [135, 136].

If bridgehead enolates are, however, generated in the presence of a strong, base-resistant electrophile, under optimized conditions no oligomers but the expected, bridgehead-derivatized products can sometimes be isolated (Scheme 5.16). In the examples in Scheme 5.16 interestingly no *ortho*-metalation of the phenyl group is observed.

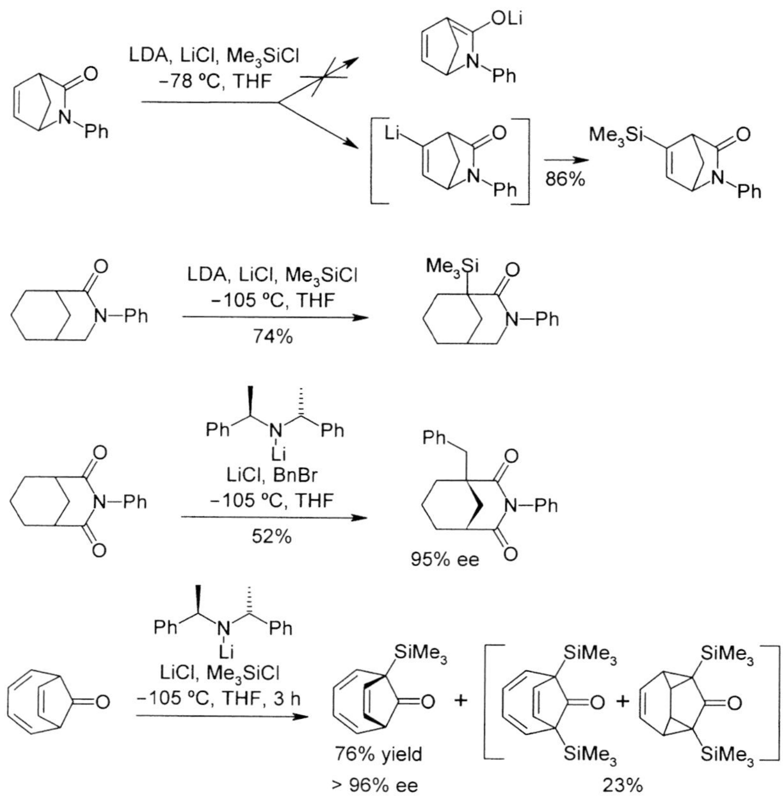

Scheme 5.16. Generation and derivatization of bridgehead enolates [135, 137].

5.3.6
Dianions

A valuable alternative to regioselective deprotonations of substrates with more than one acidic group is the formation of a di- or polyanion by addition of an excess of a strong base. If the nucleophilic sites in the polyanion have different chemical hardness, regioselective alkylation can be achieved by selecting a soft or a hard electrophile (second reaction, Scheme 5.13). If two or more C,H groups are deprotonated, and if the resulting nucleophilic sites are similarly hard, the most nucleophilic site (often that formed last) will be the most reactive, and will be alkylated first. This strategy has been extensively used to alkylate 3-oxobutyrates or related 1,3-dicarbonyl compounds at C-4 (first reaction, Scheme 5.17). Further useful dianions can be generated from ketones (C=C(OM)CM or C=C(OM)CCM) [138–142], carboxylic acids ($C{=}C(OM)_2$) [143–145], succinic acid derivatives ($(C{=}C(OM)OR)_2$) [146], alkynes (MC≡CCM, MC≡CC–OM) [147–155], imidazoles [156], thiophenes [157, 158], β-alanine derivatives (MN–CC=C(OM)X) [159], 3-nitropropanoates [160, 161],

3-hydroxypropanoates [162, 163], 2-hydroxyethylsulfones [164, 165], arenes [166–168], allenes [169], thioamides [170], and sulfonamides ($MC–SO_2NM$ [171]; R_2NSO_2-CM_2 [172]). Twofold metalation of nitroalkanes leads to intermediates which undergo clean *C*-alkylation [70, 71], in contrast to monodeprotonated nitroalkanes, which are quite unreactive and are mainly alkylated at oxygen [67–69, 173] (Scheme 5.3). Illustrative examples of the regioselective alkylation of dianions are illustrated in Schemes 5.17 and 5.18.

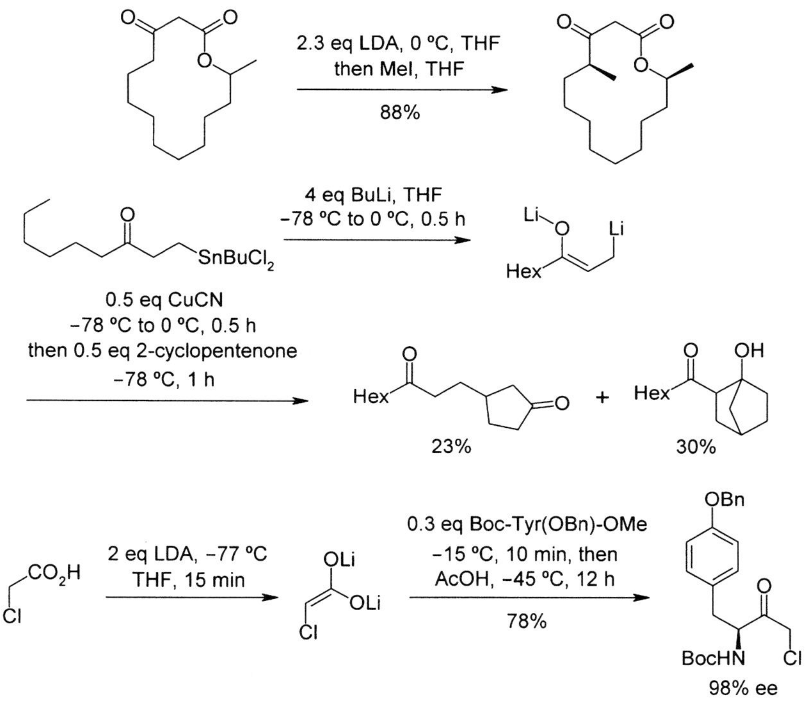

2.3 eq LDA, 0 °C, THF
then MeI, THF
88%

4 eq BuLi, THF
-78 °C to 0 °C, 0.5 h

SnBuCl2

Li
O
Li
Hex

0.5 eq CuCN
-78 °C to 0 °C, 0.5 h
then 0.5 eq 2-cyclopentenone
-78 °C, 1 h

Hex
O
23%
+
Hex
O
OH
30%

CO2H
Cl
2 eq LDA, -77 °C
THF, 15 min
OLi
OLi
Cl
0.3 eq Boc-Tyr(OBn)-OMe
-15 °C, 10 min, then
AcOH, -45 °C, 12 h
78%
OBn
O
BocHN
Cl
98% ee

Scheme 5.17. Alkylation of dianions [141, 144, 174].

Some dianions, in particular Grignard dianions, can be poorly soluble [151], and addition of HMPA or DMPU [175] to enhance their solubility might be advisable. Propargyl derivatives usually yield, on treatment with a base, the acetylide first, because the acetylenic C–H bond is usually more acidic and leads to less rehybridization on deprotonation. Phenyl propargyl sulfide, for instance, can be lithiated and alkylated selectively at the alkynyl group by using one equivalent of BuLi as base [176]. Alkylation of the propargylic position requires either protection of the alkyne (e.g. as trimethylsilane [177]) or the formation of a dianion (Scheme 5.18). The dianions from phenyl propargyl sulfoxide or sulfone have been reported to be significantly less reactive than that derived from phenyl propargyl sulfide [147].

1,3-Oxazoles can be readily deprotonated at C-2, but the resulting carbanions are unstable and can undergo reversible ring opening [179]. An interesting example of

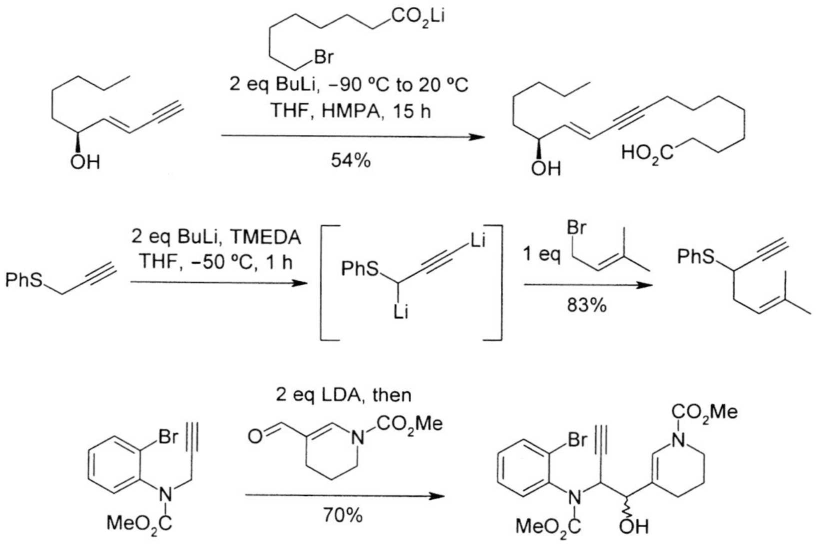

Scheme 5.18. Alkylation of alkyne-derived dianions [147, 148, 178].

electrophile- and solvent-dependent regioselectivity in the derivatization of an oxazole-based dianion is illustrated in Scheme 5.19. When alkylating reagents are used as electrophiles, the dianion undergoes rearrangement-free alkylation at the dithiane. Acylating reagents, however, react faster with the ring-opened dianion to yield an intermediate α,α-diacylisonitrile which, on ring closure, leads to the formation of 4-acyloxazoles.

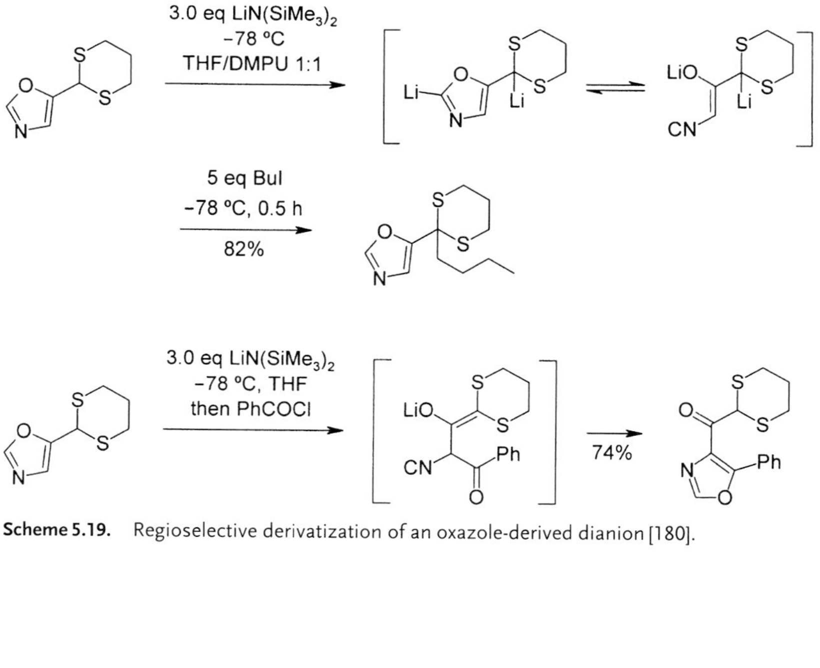

Scheme 5.19. Regioselective derivatization of an oxazole-derived dianion [180].

5.3.7
α-Heteroatom Carbanions

Carbanions in which delocalization of the negative charge to a more electronegative atom than carbon is not possible usually require strong bases for their formation. Typical pK_a values of hydrocarbons and some compounds with carbon–heteroatom single bonds are given in Table 5.2. Because of their high basicity, formation of such "non-stabilized" carbanions rarely competes with the formation of, e.g., enolates or other, better stabilized carbanions. Instead of forming a bond with an electrophile, these strongly basic carbanions can also just deprotonate it if the electrophile can act as an acid. This happens frequently when highly enolized aldehydes or ketones (e.g. acetophenones or tetralones) are chosen as electrophiles, but can also occur with allyl or alkyl halides if the carbanion is very basic [181]. In these circumstances transmetalation (e.g. Li, Na, K to Mg, Mn, Cu, Zn, Cd, Ce, In, Ti, or Zr) will be required to reduce the basicity of the carbanion [171, 182–188]. Organozirconium reagents can even add to ketones containing a CH–NO_2 group without being protonated [183].

Table 5.2. Equilibrium acidities of hydrocarbons in DMSO (H_2O) at 25 °C [24, 25, 33, 189].

Acid	pK_a	Acid	pK_a
Cyclopentadiene	18.0 (16)	Propene	44 (43)
Indene	20.1 (20)	$H_2C{=}CH_2$	(44)
Fluorene	22.6 (23)	Cyclopropane	(46)
PhC≡CH	28.7 (20.0)	CH_4	56 (48)
Ph_3CH	30.6	Me_4N^+	42
Ph_2CH_2	32.2	$MePPh_3^+$	22
$PhCH_3$	43 (40)	MeOPh	49
PhH	(43)	Me_2S	45

Carbanions generated by deprotonation of compounds of type $R'CH_2X$ (X = SiR_3, NR_3^+, NR_2, PR_2, OR, SR, Hal) cannot be stabilized by conjugative charge delocalization, but only by inductive effects, field effects, negative hyperconjugation, or chelate formation [59]. If no possibility of chelation of the metal is available these carbanions will tend to be highly reactive and undergo "forbidden" (i.e. non-concerted) rearrangements via homolytic bond cleavage (e.g. [1,2]-Wittig (X = O) [190–192], Stevens (X = NR_2^+) [193–195], or related rearrangements [195, 196]) or α-elimination to yield carbenes (Scheme 5.20).

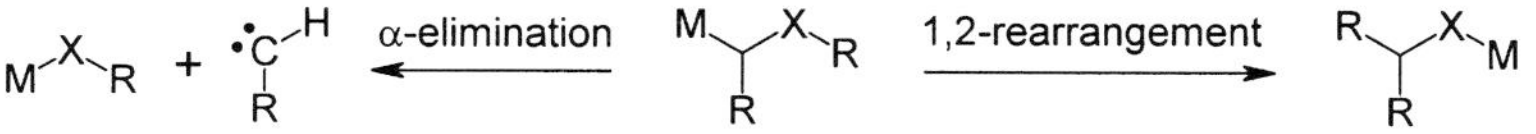

α-elimination ← M–C(R)–X–R → 1,2-rearrangement

Scheme 5.20. Possible decomposition reactions of α-heteroatom substituted carbanions. X = O, NR, NR_2^+.

Such reactive carbanions can also act as reducing agents or as strong bases, and can lead to numerous unexpected reactions [197]. Few solvents are sufficiently inert to withstand them [198]. Chelate-stabilized, α-heteroatom-substituted carbanions can, on the other hand, be quite stable, and in recent years numerous useful transformations involving these intermediates have been developed.

Allylic and propargylic heteroatom-substituted carbanions can yield rearranged or unrearranged products on treatment with an electrophile. The regio- and stereoselectivity of these reactions depends on the precise structure of the carbanion, on the metal and solvent chosen [199], and on the structure of the electrophile [150, 200–203], and can be difficult to predict.

In addition to deprotonation with strong bases, halogen–metal exchange, or transmetalation, α-heteroatom-substituted carbanions can also be prepared by 1,5-hydrogen transfer to vinyl radicals, followed by reduction [204, 205] (Scheme 5.21) or by a related intramolecular 1,4-proton transfer [206].

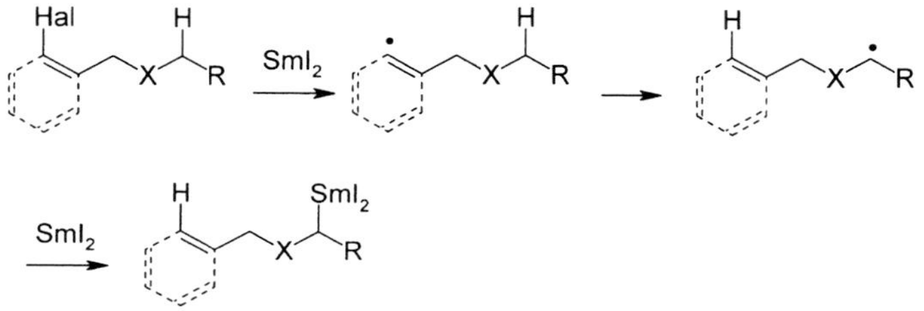

Scheme 5.21. Generation of carbanions by 1,5-hydrogen transfer. X = NR [205], O [204].

5.3.7.1 α-Nitrogen Carbanions

Simple tertiary amines are difficult to deprotonate selectively [195, 196, 200]. To increase the acidity of the α-C,H-groups the amine can be quaternized [207], treated with a Lewis acid [208–211], oxidized to an amine *N*-oxide [161], or, for secondary amines, derivatized with a functional group capable of forming a chelate with the metal (Scheme 5.22).

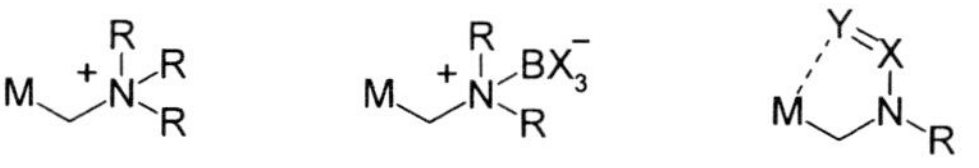

Scheme 5.22. Strategies for the enhancement of the α-acidity of amines.

Non-enolizable amides, for example *N*,*N*-dialkyl pivalamides [212], benzamides, thiobenzamides [213], or phosphinamides ($Ph_2P(O)NR_2$ [214]), can be lithiated α to the amino group by treatment with *s*BuLi [54, 213, 215] or *t*BuLi [216], without further additives, in THF at –78 °C. *N*,*N*-Dimethylbenzamides can be attacked at the carbonyl group by these organolithium reagents to yield ketones [217] or alcohols, but with sterically more demanding amides metalation is usually faster than addition.

N-Benzyl thioamides can be readily lithiated twice with BuLi, to yield benzylic organolithium derivatives. As illustrated by the examples (Scheme 5.23), the regioselectivity of this metalation is quite different than that of comparable amides.

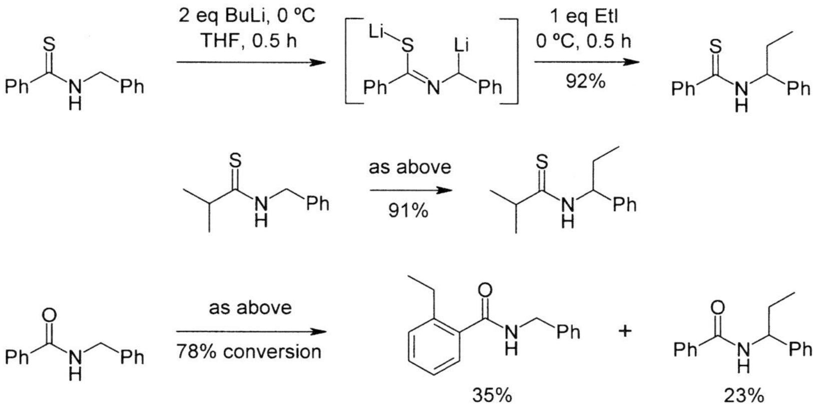

Scheme 5.23. Alkylation of thioamide-derived dianions [170].

As shown by the last reaction in Scheme 5.23, the metalation of benzamides is complicated by several potential side reactions (Scheme 5.24). Thus, benzamides can also undergo *ortho*-metalation [181, 217–222] or metalation at benzylic positions [223–225]. *Ortho*-metalation seems to be promoted by additives such as TMEDA, and benzylic metalation can be performed selectively with lithium amide bases [217, 224], which are often not sufficiently basic to mediate *ortho*- or α-amino metalation. If deprotonation of the CH–N group succeeds, the resulting product might also undergo cyclization by intramolecular attack at the arene [214, 216] (see also Ref. [226] and Scheme 5.27) instead of reacting intermolecularly with an electrophile. That this cyclization occurs, despite the loss of aromaticity, shows how reactive these intermediates are.

Because amides are often difficult to hydrolyze, their utility as transient activating groups for amines is limited. For this reason more readily hydrolyzable groups were

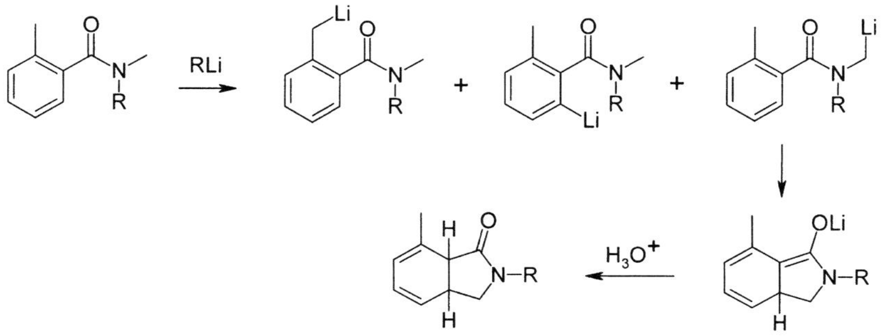

Scheme 5.24. Potential products of the lithiation of benzamides.

evaluated. Formamides and thioformamides are usually not deprotonated α to nitrogen but at the formyl group [59, 227–231] (Section 5.3.9). Formamidines (R_2N–CH=NR), however, have proven to be suitable activating groups for α-metalation of amines [232–237] (Scheme 5.25). The corresponding carbanions are highly reactive, and can act either as nucleophiles or as reducing agents [233, 235].

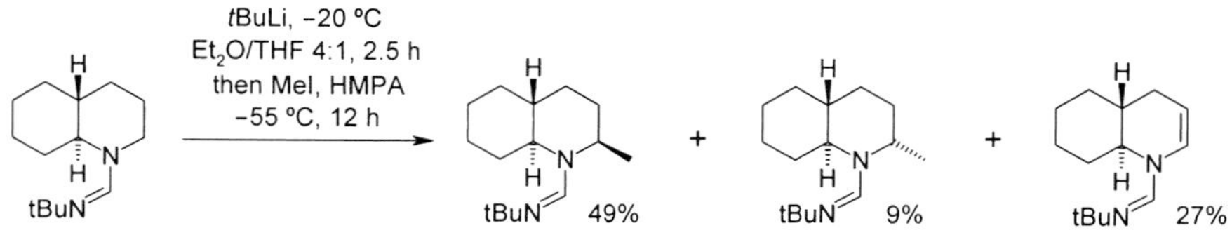

Scheme 5.25. Metalation and alkylation of formamidines [235].

Further derivatives of amines in which the α-C,H groups are sufficiently acidic to enable metalation are carbamates [235, 238–240] (Scheme 5.26), imides [241], *N*-nitroso amines [59, 242], ureas [201, 243], some *N*-phosphorus derivatives [212, 214, 226, 244, 245], *N*-(2-pyridyl)amines (Scheme 5.26), and isonitriles [59]. A potential side reaction in the examples in Scheme 5.26 is the lithiation of the arene; this is, in fact, observed with an isomeric dipyridopyrazine (last reaction, Scheme 5.26).

Sulfonamides also can be metalated at the nitrogen-bound carbon [247] (Scheme 5.27). The sulfonyl group is, however, usually not well suited for the stabilization of α-metalated amines because sulfinate is a good leaving group, and imines

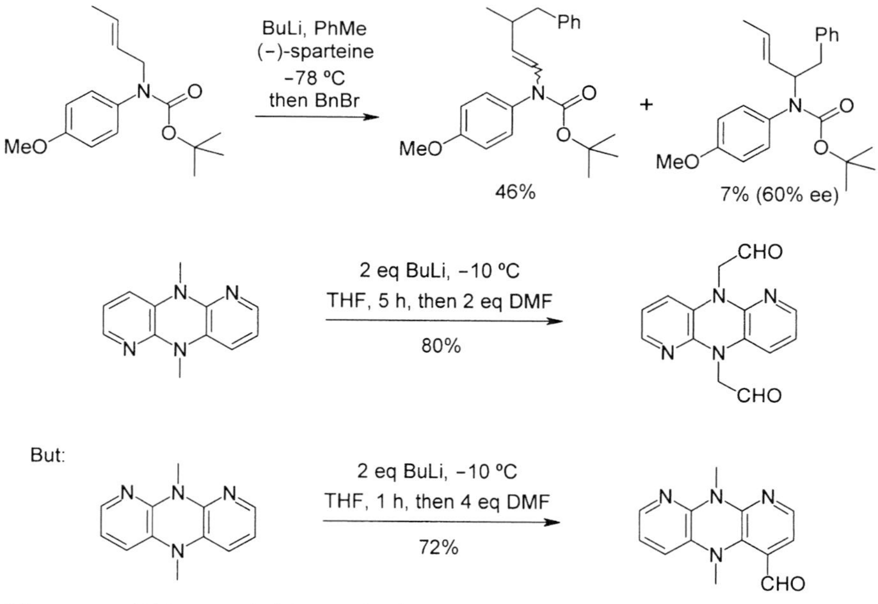

Scheme 5.26. Lithiation and alkylation of carbamates [56] and pyridylamines [246].

usually result when sulfonamides of secondary amines are treated with strong bases [248] (Scheme 5.27). Similarly, some *N*-phosphorus derivatives undergo elimination to yield imines when treated with a strong base [244]. A further side reaction of chelate-stabilized α-amino carbanions is the intramolecular migration of the chelating group to the carbanion (Section 5.4.5).

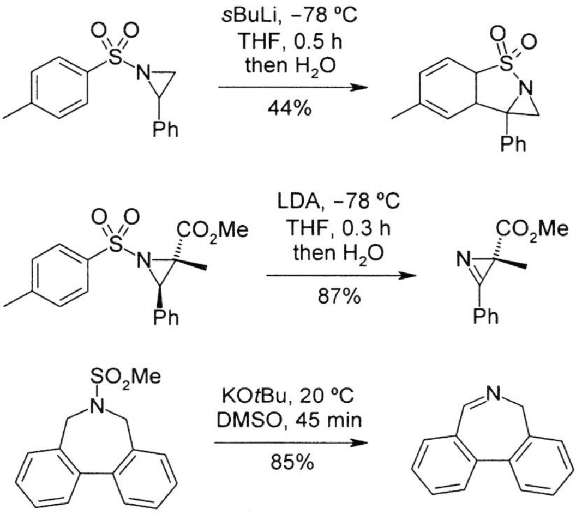

sBuLi, −78 °C
THF, 0.5 h
then H_2O
44%

LDA, −78 °C
THF, 0.3 h
then H_2O
87%

KO*t*Bu, 20 °C
DMSO, 45 min
85%

Scheme 5.27. The sulfonyl group as carbanion-stabilizing or leaving group [247, 249–251].

5.3.7.2 α-Oxygen and α-Sulfur Carbanions

α-Metalated, non-chelate stabilized ethers are highly reactive intermediates which tend to rearrange (Wittig rearrangement) or undergo other decomposition reactions such as eliminations or intermolecular nucleophilic displacement of alkoxide (Scheme 5.28). Because of their high energy, the reaction of metalated ethers with electrophiles can readily lead to product mixtures.

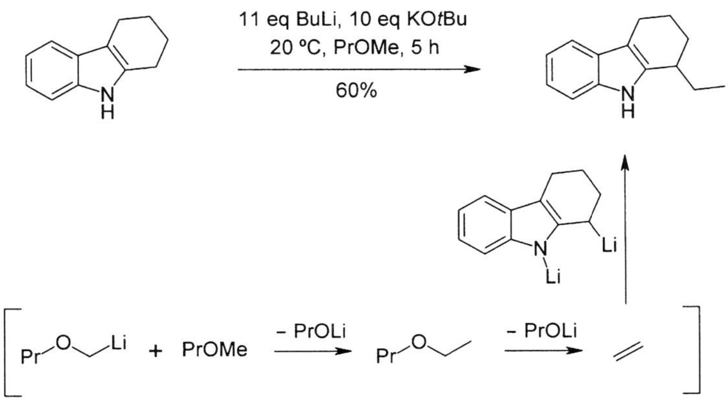

11 eq BuLi, 10 eq KO*t*Bu
20 °C, PrOMe, 5 h
60%

Pr–O–CH$_2$Li + PrOMe —(– PrOLi)→ Pr–O–CH$_2$CH$_2$... —(– PrOLi)→ ethylene

Scheme 5.28. Cleavage of ethers by strong bases [197].

Benzyl methyl ether or allyl methyl ethers can be selectively metalated at the benzylic/allylic position by treatment with BuLi or *s*BuLi in THF at −40 °C to −80 °C, and the resulting organolithium compounds react with primary and secondary alkyl halides, epoxides, aldehydes, or other electrophiles to yield the expected products [187, 252, 253]. With allyl ethers mixtures of α- and γ-alkylated products can result [254], but transmetalation of the lithiated allyl ethers with indium yields γ-metalated enol ethers, which are attacked by electrophiles at the α position (Scheme 5.29). Ethers with β hydrogen usually undergo rapid elimination when treated with strong bases, and cannot be readily *C*-alkylated (last reaction, Scheme 5.29). Metalation of benzyl ethers at room temperature can also lead to metalation of the arene [255] (Section 5.3.11) or to Wittig rearrangement [256]. Epoxides have been lithiated and silylated by treatment with *s*BuLi at −90 °C in the presence of a diamine and a silyl chloride [257].

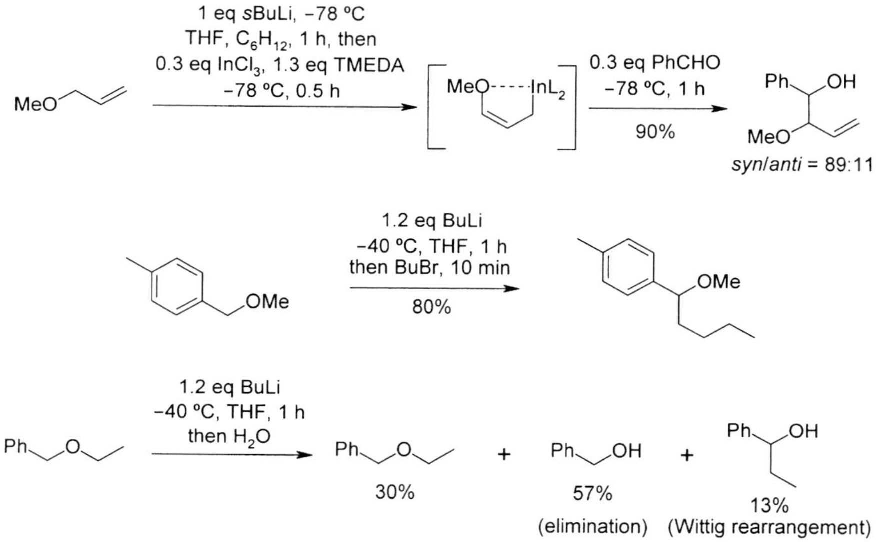

1 eq *s*BuLi, −78 °C
THF, C_6H_{12}, 1 h, then
0.3 eq $InCl_3$, 1.3 eq TMEDA
−78 °C, 0.5 h

0.3 eq PhCHO
−78 °C, 1 h
90%

syn/*anti* = 89:11

1.2 eq BuLi
−40 °C, THF, 1 h
then BuBr, 10 min
80%

1.2 eq BuLi
−40 °C, THF, 1 h
then H_2O

30% + 57% (elimination) + 13% (Wittig rearrangement)

Scheme 5.29. Reactions of metalated ethers [187, 252].

Convenient alternatives to direct deprotonation of ethers are tin–lithium exchange [199, 258–261], halogen–magnesium exchange [262], or reductive cleavage of O,Se-acetals [263, 264]. Another synthetic equivalent of α-metalated ethers are (alkoxymethyl)phosphonium salts [265].

α-Sulfur carbanions are usually more stable than metalated ethers, and lithiated 1,3-dithianes, in particular, have found widespread application in organic synthesis as synthetic equivalents of acyl anions (Scheme 5.30). Lithiated dithianes are, however, in the same way as most other organolithium reagents, highly oxygen- and oxidant-sensitive, and numerous byproducts can be formed if oxygen is not rigorously excluded from the reaction or if the electrophile can act as an oxidant. Oxidation by

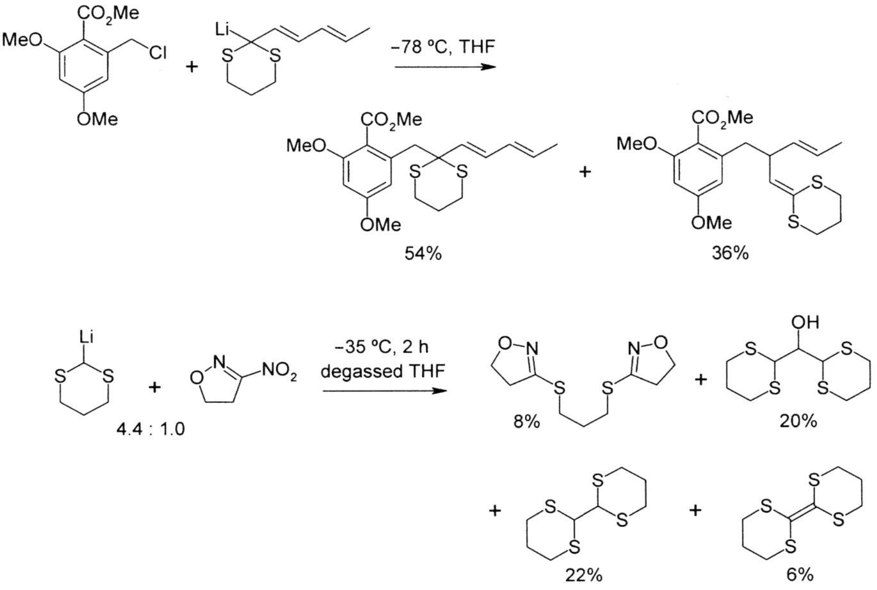

Scheme 5.30. Reactions of lithiated 1,3-dithianes [266, 267].

SET to the nitro group is probably the reason for the product mixture formed in the second example in Scheme 5.30.

α-Oxygen or α-sulfur carbanions can be stabilized by intramolecular chelation. Thus, carbamates such as those developed by Hoppe (Scheme 5.31) yield, on deprotonation, stable carbanions, which undergo clean, highly regio- and stereoselective, and often high-yielding reactions with a variety of electrophiles [202, 203, 254, 268–271]. In the presence of stoichiometric amounts of enantiomerically pure diamines enantioselective deprotonations can be performed (Scheme 5.31). Because the carbamate group can be displaced by nucleophiles, addition of α-metalated carbamates to alkenes can lead to the formation of cyclopropanes (last reaction, Scheme 5.31) [272, 273].

Other compounds which can be deprotonated to yield chelate-stabilized α-oxygen carbanions include non-enolizable esters, for example 2,6-dialkylbenzoates [59, 275–277], 2-alkoxybenzimidazoles [188], and some oxazolines [278]. Pivalates can also be lithiated α to oxygen, but the resulting carbanions are quickly transformed into ketones either by reaction with the starting ester (Scheme 5.32) or by intramolecular rearrangement [276]. Similarly, phosphates, on metalation, undergo rapid rearrangement to α-hydroxyphosphonates [279, 280]. Ester-derived Grignard reagents of the type $RC(=O)OCR_2MgX$ can be prepared by halogen–magnesium or sulfoxide–magnesium exchange, and do not rearrange to ketones at –78 °C (at least within 15 min; Scheme 5.32) [281]. An exceptionally facile deprotonation reported by Christie and

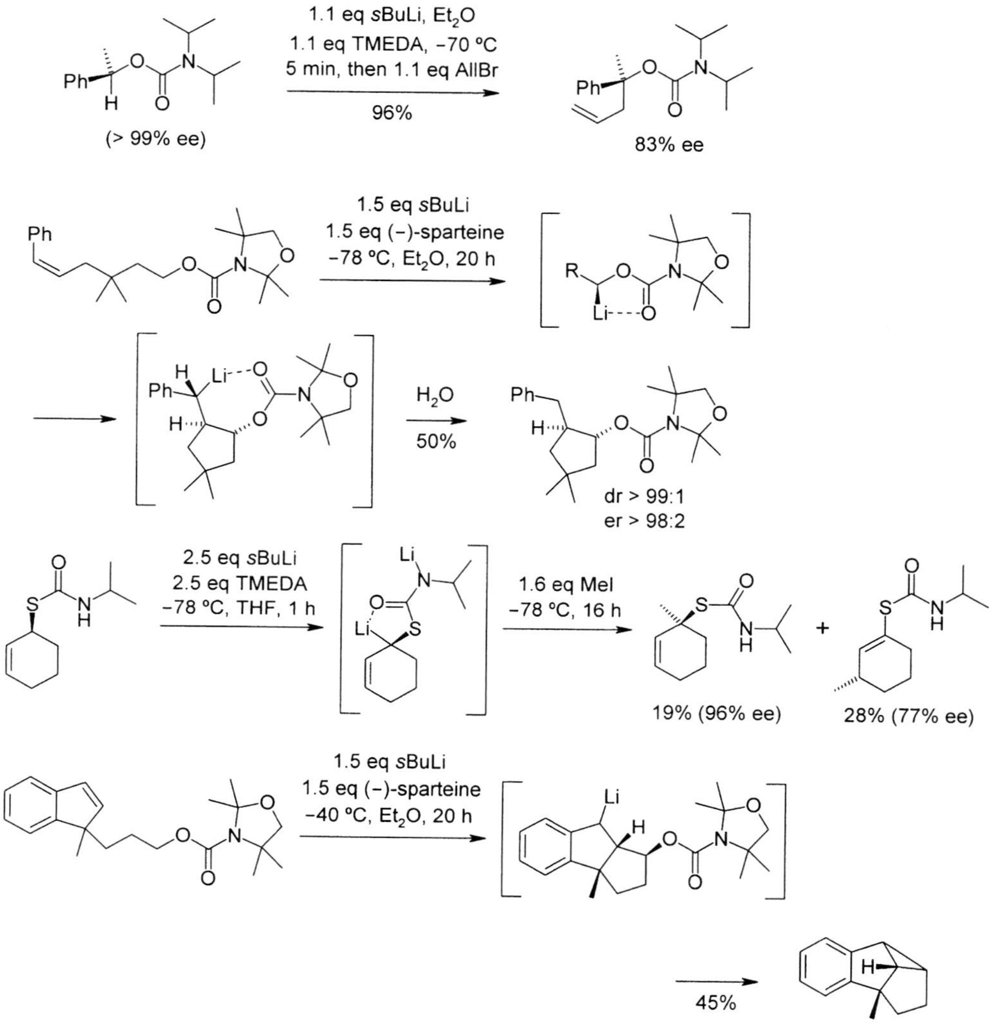

Scheme 5.31. Formation and alkylation of chelate-stabilized α-oxygen and α-sulfur carbanions [55, 272–274].

Rapoport is the lithiation and alkylation of the enolizable α-amino ester shown in Scheme 5.32 (first example).

α-Sulfur chelate-stabilized carbanions have been prepared by metalation of 2-(alkylthio)thiazolines, 2-(alkylthio)oxazolines, 2-(alkylthio)pyridines [283], thiocarbonates, dithiocarbamates [284–286], thiol esters, and related compounds [59] (see also Schemes 5.31 and 5.76). Reaction of ketones or aldehydes with these nucleophiles can yield thiiranes, presumably via intramolecular transfer of the sulfur-

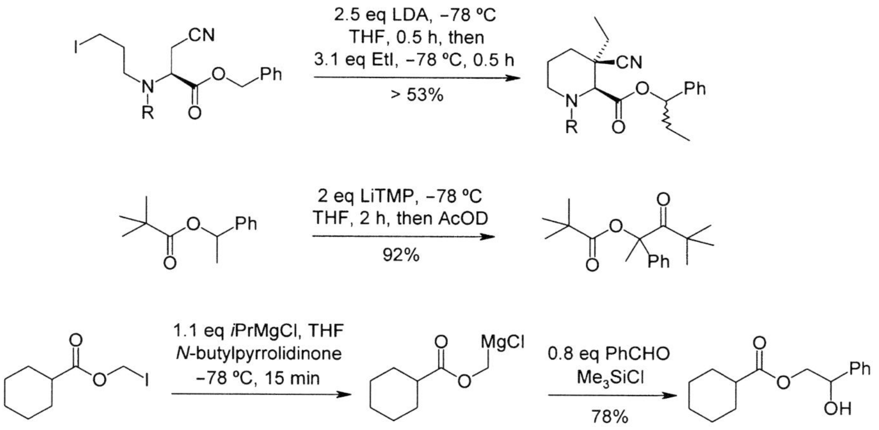

2.5 eq LDA, −78 °C THF, 0.5 h, then 3.1 eq EtI, −78 °C, 0.5 h; > 53%

2 eq LiTMP, −78 °C THF, 2 h, then AcOD; 92%

1.1 eq *i*PrMgCl, THF *N*-butylpyrrolidinone −78 °C, 15 min; 0.8 eq PhCHO Me_3SiCl; 78%

Scheme 5.32. Metalation and reaction of esters [276, 281, 282]. R = 9-phenylfluoren-9-yl.

bound carbanion-stabilizing group to the newly formed alcoholate followed by intramolecular nucleophilic displacement by sulfur.

Unsubstituted aliphatic alcohols cannot usually be α-metalated by treatment with strong bases (to yield a dianion). Dilithiated methanol has been prepared by treatment of Bu_3SnCH_2OH with two equivalents of BuLi, and can be alkylated at carbon [287]. Treatment of allyl alcohol with excess BuLi/TMEDA in pentane at room temperature does not lead to formation of the dianion of allyl alcohol but to addition of BuLi to the C–C double bond [288] (Scheme 5.33). Benzylic alcohols, on the other hand, can be deprotonated twice and, depending on the substitution pattern at the

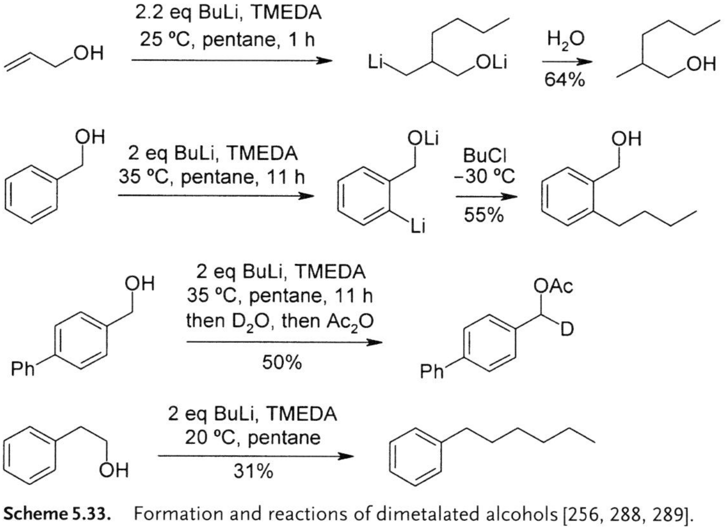

2.2 eq BuLi, TMEDA 25 °C, pentane, 1 h; H_2O 64%

2 eq BuLi, TMEDA 35 °C, pentane, 11 h; BuCl −30 °C 55%

2 eq BuLi, TMEDA 35 °C, pentane, 11 h then D_2O, then Ac_2O; 50%

2 eq BuLi, TMEDA 20 °C, pentane; 31%

Scheme 5.33. Formation and reactions of dimetalated alcohols [256, 288, 289].

arene, metalation can occur either at the benzylic methylene group [256] or at the arene [289, 290] (Scheme 5.33). Attempts to dilithiate 2-phenylethanol at room temperature led to immediate formation of polymers and hexylbenzene, presumably via elimination of lithium oxide to yield styrene, followed by addition of BuLi (Scheme 5.33) [289].

5.3.7.3 α-Halogen Carbanions

The main competing pathways available to metalated alkyl halides are α- and β-elimination [291] and alkylation of the base or carbanion by the starting halide [292]. Metalation and alkylation of alkyl halides at the α position can, therefore, usually only be performed when this position is activated by an additional electron-withdrawing group, and β-elimination is either not possible or difficult. Non-nucleophilic bases must, furthermore, be used to avoid alkylation of the base by the halide. 2-Haloacetic esters [293, 294], 2-haloacetic thiolesters [295], 4-halocrotonic acid esters [123], 2-(1-haloalkyl)oxazolidines [292, 296–298], α-halo ketones [299], α-halo sulfoxides [300, 301], and α-halo imines [302, 303] can all be metalated and alkylated at low temperatures without loss of halide from the nucleophile. Examples of the α-alkylation of less acidic alkyl halides have also been reported (Scheme 5.34). The metalation of allylic [187, 304–308] or propargylic [309] chlorides or bromides can be performed at low temperatures, and addition of Lewis acids can enable highly stereoselective alkylations with these carbanions (Scheme 5.34). Transmetalation may lead to improved yields or to a modified stereochemical outcome [304, 310]. Non-stabilized α-halocarbanions have also been prepared by exchange of a sulfinyl group by lithium or magnesium [300, 311, 312] and by partial halogen–metal exchange of 1,1-dihaloalkanes [310, 313–315] (Scheme 5.34).

The addition of α-deprotonated alkyl halides to alkenes or carbonyl compounds can, because of the good leaving-group properties of halides, also lead to formation of cyclopropanes [292] or epoxides [187, 304, 306, 310], respectively. Because of the inherent instability of α-halo organometallic compounds, these intermediates should be handled carefully and on a small scale only. The ketone produced by the last reaction in Scheme 5.34 is probably formed by Oppenauer oxidation of the intermediate alcohol by the excess benzaldehyde [310].

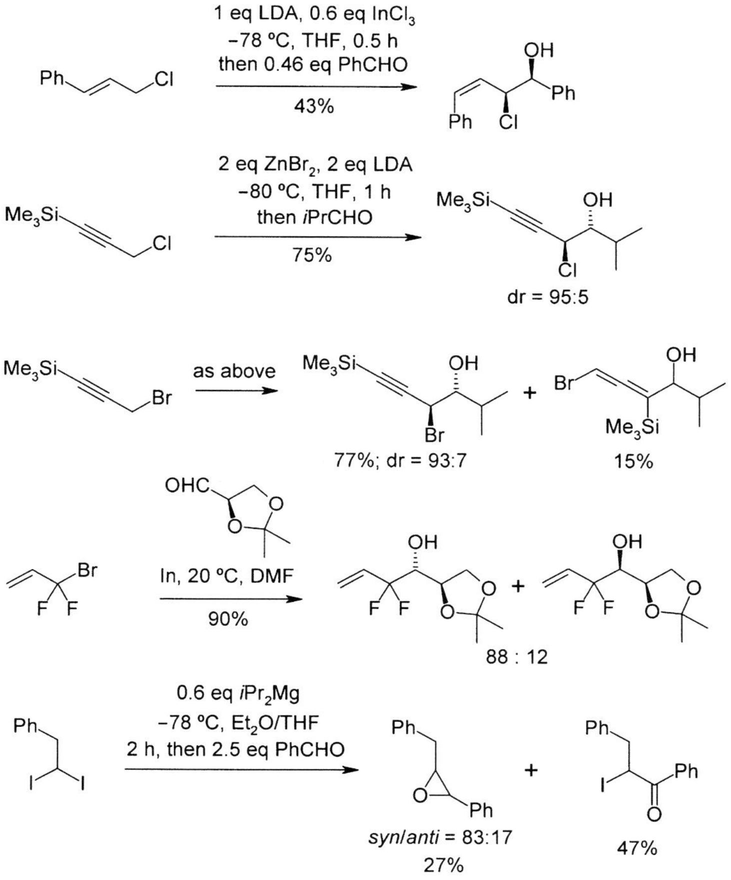

Scheme 5.34. Generation and alkylation of α-halo organometallic compounds [187, 309, 310, 316].

5.3.8
Vinylic Carbanions

Vinylic carbanions are closely related to metalated arenes or heteroarenes, the main difference being that the formation of alkynes from vinylic carbanions with a vicinal leaving group proceeds more readily than the formation of arynes [317, 318]. The most straightforward methods for regioselective preparation of vinyllithium or related organometallic compounds are halogen– or tin–metal exchange [319] and treatment of tosylhydrazones with organolithium compounds (Shapiro reaction) or cuprates [320]. Direct vinylic deprotonations will usually proceed regioselectively only when the substrate contains functional groups which exert a strong directing or carbanion-stabilizing effect and when no allylic protons are available (these are usually removed more rapidly [321, 322]; benzylic protons are, however, often removed more slowly than aromatic protons). Cyclopropenes and allenes are particularly acidic alkenes [323, 324]; the latter occasionally even enable twofold deprotonation (Scheme 5.35).

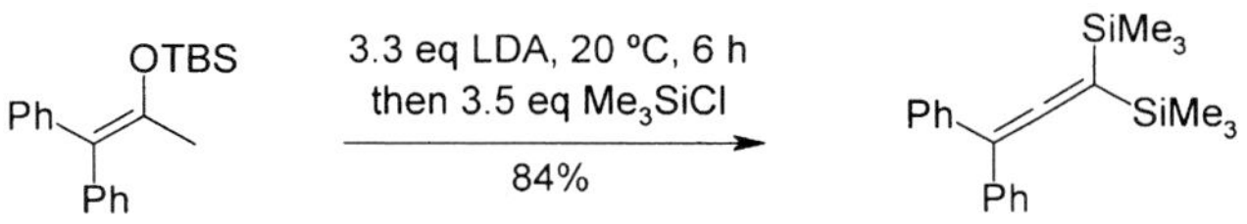

OTBS, Ph, Ph — 3.3 eq LDA, 20 °C, 6 h then 3.5 eq Me_3SiCl, 84% → $SiMe_3$, $SiMe_3$, Ph, Ph

Scheme 5.35. Formation and silylation of an allene-derived dianion [169].

Vinyl halides containing vinylic protons can undergo either halogen–metal exchange or hydrogen–metal exchange when treated with an alkyllithium compound [317, 325–327]. Scheme 5.36 shows some substrates which can be selectively deprotonated at a vinylic position (arrow) by BuLi or a related base. The anion resulting from 1,1-difluoroethene can be generated and alkylated only at temperatures below −100 °C, because it undergoes rapid β-elimination at higher temperatures [328] (Scheme 5.59).

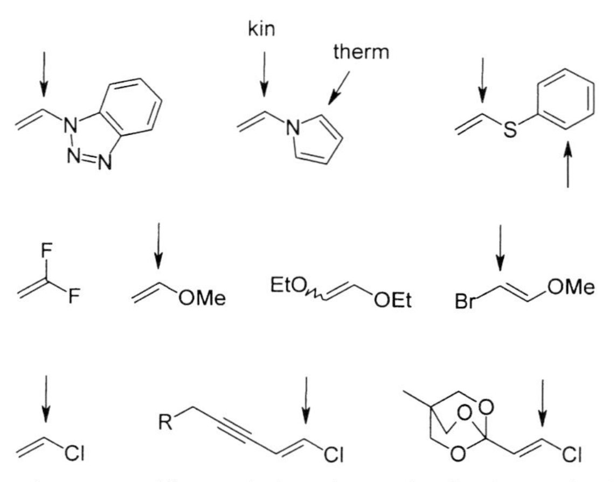

Scheme 5.36. Alkenes which can be metalated and trapped with electrophiles, and preferred sites of vinylic deprotonation [317, 326, 328–332]. kin: kinetically favored; therm: thermodynamically favored.

Acrylic acid derivatives with a heteroatom (N, O, S, Hal) at C-3 can be cleanly deprotonated at this position with BuLi or LDA at low temperatures [329, 333–338] (Scheme 5.37). Some of these anions rearrange to the α-metalated acrylates on warming [329], but can also decompose (see Section 5.4.7). Non-heteroatom-substituted lithium β-lithioacrylate [329] or β-magnesioacrylic acid derivatives [339, 340] have been prepared by bromine–lithium or halogen–magnesium exchange.

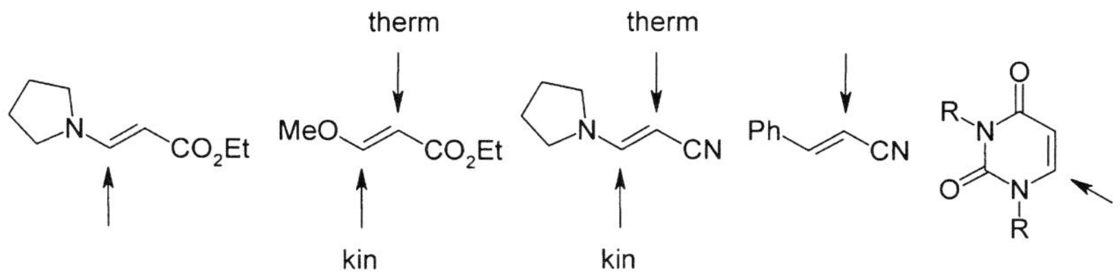

Scheme 5.37. Preferred sites of vinylic deprotonation of acrylic acid derivatives [329, 337] and pyrimidinediones [341–343]. kin: kinetically favored; therm: thermodynamically favored.

5.3.9
Acyl, Imidoyl, and Related Carbanions

Acyl anions (RC(=O)M) are unstable, and quickly dimerize at temperatures > –100 °C (Section 5.4.7). These intermediates are best generated by reaction of organolithium compounds or cuprates with carbon monoxide at –110 °C and should be trapped immediately by an electrophile [344–347]. Metalated formic acid esters (ROC(=O)M) have been generated as intermediates by treatment of alcoholates with carbon monoxide, and can either be protonated to yield formic acid esters, or left to rearrange to carboxylates (ROC(=O)M → RCO_2M) (Scheme 5.38) [348]. Related intermediates are presumably also formed by treatment of alcohols with formamide acetals (Scheme 5.38) [349]. More stable than acyl lithium compounds are acyl silanes or transition metal acyl complexes, which can also be used to perform nucleophilic acylations [350].

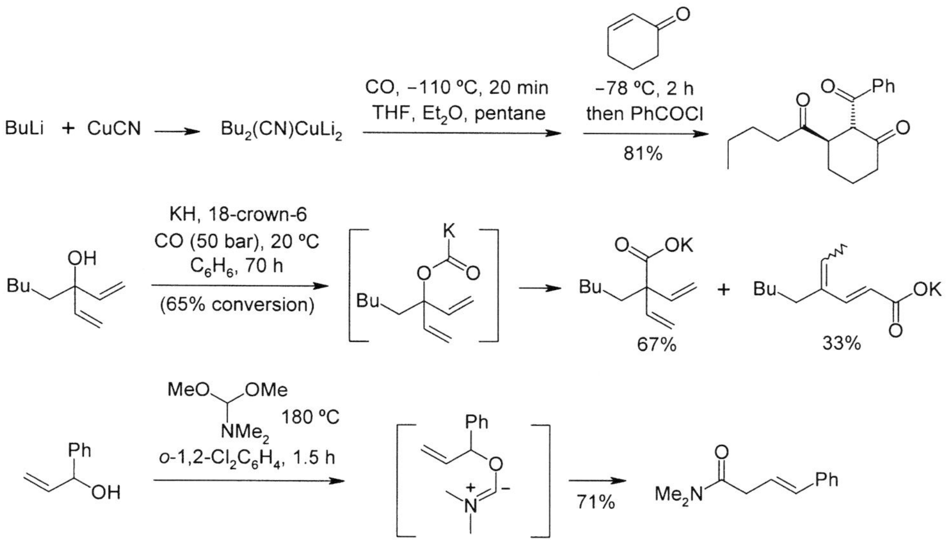

Scheme 5.38. Generation and reactions of acyl anions and related intermediates [347–349].

Formamides are usually not deprotonated α to nitrogen but at the formyl group [227–231]. The resulting carbamoyl lithium derivatives (R_2NCOLi), which can also be generated from deprotonated amines [351] or amides [352] and carbon monoxide, react with electrophiles E^+ to yield the expected products (R_2NCOE), despite the carbene character and consequent low stability of these intermediates [179, 351] (Scheme 5.39, see Section 5.4.7). Palladium-catalyzed versions of the reaction have been reported [353, 354].

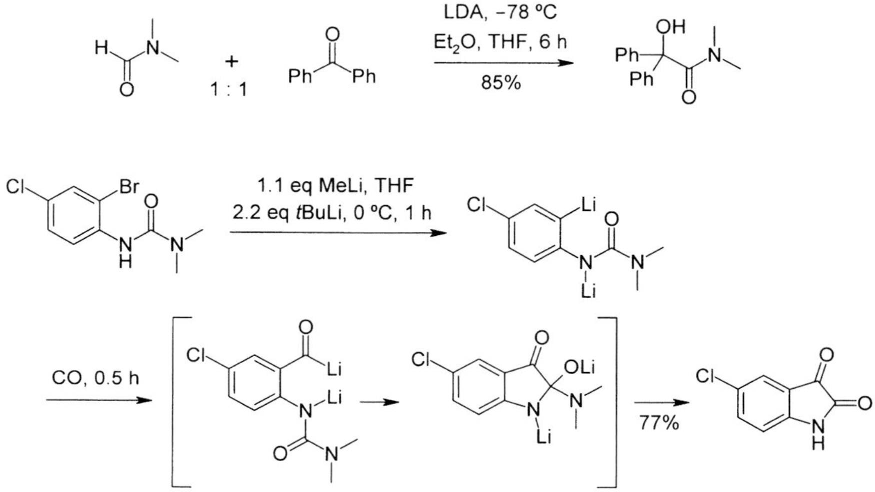

LDA, –78 °C
Et_2O, THF, 6 h
85%
1 : 1
1.1 eq MeLi, THF
2.2 eq *t*BuLi, 0 °C, 1 h
CO, 0.5 h
77%

Scheme 5.39. Formation and reactions of acyllithiums [230, 352].

Metalated imines (RN=C(R)M) have usually been prepared not by deprotonation of imines but by addition of Grignard reagents or organolithium compounds to isonitriles devoid of α hydrogen [355] (to avoid α-metalation of the isonitrile; Scheme 5.40). Alternatively, halogen–lithium exchange at imidoyl iodides [356], tin–lithium exchange at imidoyl stannanes [357], Brook rearrangement of imidoyl silanes [357], or reduction of isonitriles with samarium iodide in the presence of an alkylating agent also lead to the formation of metalated imines, which react with alkylating reagents in the expected way [358, 359] (Scheme 5.40). Dimerization of isonitriles under these reaction conditions has, however, also been observed [360].

Related to acyl and imidoyl anions are carbanions such as ($R_3CC(Li)=N^+(R)O^-$) obtained by deprotonation with *s*BuLi of nitrones devoid of α hydrogen, at –78 °C [362, 363]. These intermediates can react with a variety of electrophiles to yield the corresponding derivatized nitrones [363], but, as with acyl anions, dimerization can also occur [362, 364].

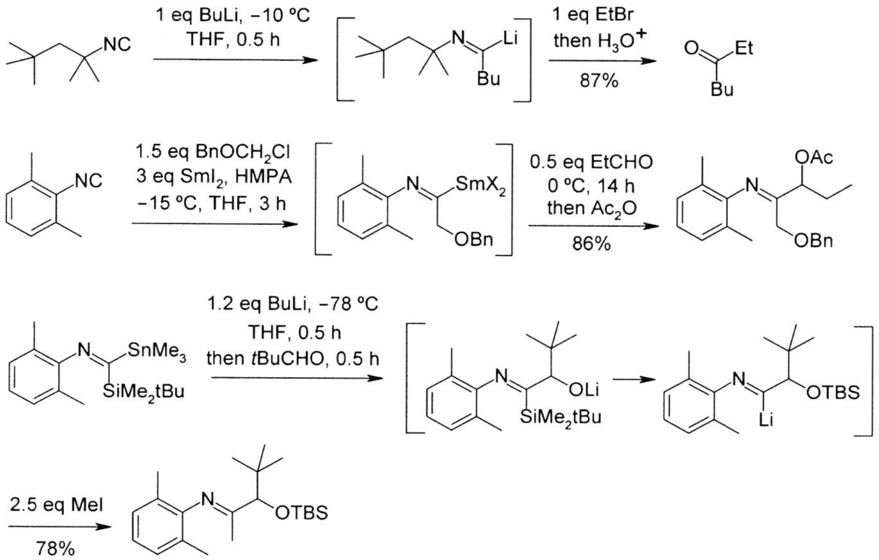

Scheme 5.40. Generation and alkylation of metalated imines [358, 361].

5.3.10
Aromatic Carbanions

Arenes cannot usually be deprotonated with LDA alone, but require mixtures of organosodium [365] or organolithium compounds and tertiary amines [181, 218, 219]. These amines, for instance TMEDA, lead to a partial dissociation of oligomeric BuLi–solvent aggregates and thereby to more powerful metalating reagents [366, 367]. Thus, although benzene cannot be deprotonated with BuLi alone, a mixture of BuLi and TMEDA leads to quantitative lithiation [181].

An important drawback of the use of organolithium compounds as bases is their high nucleophilicity, which limits the scope of suitable substrates. Thus, many carbonyl compounds, benzamides [217], heteroarenes [368], sulfoxides [11, 12], or phosphine oxides [12] will undergo nucleophilic attack instead of deprotonation by organolithium compounds. Organolithium compounds can, furthermore, act as reducing agents by SET. As alternative bases, sterically more demanding lithium [369], zinc [370], or magnesium [371, 372] dialkylamides, or mixtures of LDA and KO*t*Bu or NaO*t*Bu [166, 373] can be used.

The acidity of arenes does not correlate with their electron density [374], because in a metalated arene there is no significant overlap between the aromatic π system and the Ar–M bond. More important for facile aromatic metalation is the presence of substituents able to form bonds with the metal and thereby direct the base into the proximity of an *ortho* Ar–H bond before metalation, and to form a stabilizing chelate with the metal when metalation has occurred [181, 366, 375, 376]. Thus, although metalation of aryl ethers or *N*-(alkoxycarbonyl)anilines (i.e. electron-rich arenes) proceeds readily, and even phenolates undergo metalation to form dianions [377], the metalation of simple alkylbenzenes generally proceeds more sluggishly (Scheme 5.41).

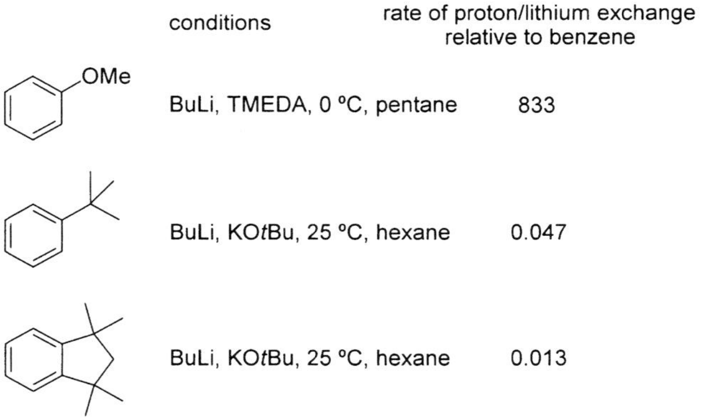

	conditions	rate of proton/lithium exchange relative to benzene
OMe	BuLi, TMEDA, 0 °C, pentane	833
	BuLi, KO*t*Bu, 25 °C, hexane	0.047
	BuLi, KO*t*Bu, 25 °C, hexane	0.013

Scheme 5.41. Relative rates of lithiation of arenes [378, 379].

Arenes and heteroarenes which are particularly easy to metalate are tricarbonyl(η^6-arene)chromium complexes [380, 381], ferrocenes [13, 382, 383], thiophenes [157, 158, 181, 370, 384], furans [370, 385], and most azoles [386–389]. Metalated oxazoles, indoles, or furans can, however, be unstable and undergo ring-opening reactions [179, 181, 388]. Pyridines and other six-membered, nitrogen-containing heterocycles can also be lithiated [59, 370, 390–398] or magnesiated [399], but because nucleophilic organometallic compounds readily add to electron-deficient heteroarenes, dimerization can occur, and alkylations of such metalated heteroarenes often require careful optimization of the reaction conditions [368, 400, 401] (Schemes 5.42 and 5.69).

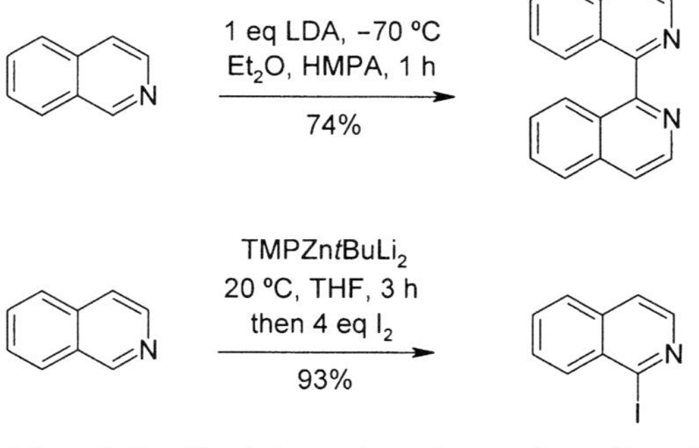

N → 1 eq LDA, –70 °C; Et_2O, HMPA, 1 h; 74% → N, N

N → TMPZn*t*BuLi_2; 20 °C, THF, 3 h; then 4 eq I_2; 93% → N, I

Scheme 5.42. Metalation and ensuing reactions of isoquinoline [370, 402].

Similarly, nitroarenes can also be lithiated (Scheme 5.43), but reactions of metalated nitroarenes with electrophiles only proceed cleanly if the metalation is performed in the presence of the electrophile [403]. Otherwise, the metalated arene can reduce nitro groups to nitroso groups, which quickly react with additional organometallic reagent to yield hydroxylamines [404, 405]. Nitroarylmagnesium halides can be prepared by iodine–magnesium exchange at –80 °C to –40 °C, and react with electrophiles in the expected way [6] (Scheme 5.43).

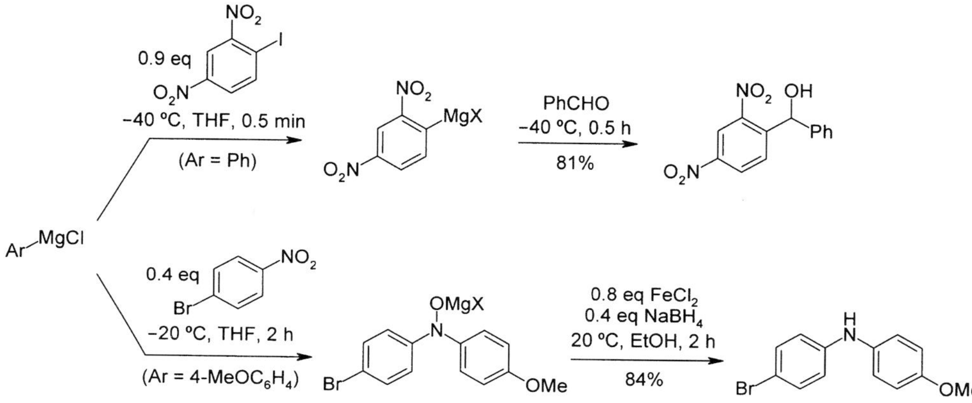

Scheme 5.43. Reactions of arylmagnesium compounds with nitroarenes [404, 406].

As mentioned above, certain functional groups can increase the kinetic and thermodynamic acidity of the aromatic *ortho* protons and thereby control the regioselectivity of the metalation [181, 218]. Typical *ortho*-directing groups (approximately in decreasing order of *ortho*-directing ability) are $OCONR_2$, SO*t*Bu, SO_2tBu, $CONR_2$, OCH_2OMe, CN, SO_2NR_2, NHBoc, $(CH_2)_{1,2}NR_2$, CO_2H, OPh, OMe, $OCSNR_2$, NR_2, CF_3, F, and Cl [181, 374, 377, 407]. Surprisingly, fluoroarenes [377, 394, 403, 408], chloroarenes [369, 409–411], bromoarenes [369, 385, 412, 413], trifluoromethyl arenes [377, 392, 414], and trifluoromethoxy arenes [415] can sometimes be metalated and trapped by different electrophiles without undergoing halogen–metal exchange, elimination (to yield arynes), or other decomposition reactions. Because of their inherent instability, however, metalated halo- or trihalomethyl arenes should be handled with great care, especially when performing reactions on a large scale. Grignard reagents derived from trifluoromethyl arenes, for instance, have led to several explosions, some of them even leading to loss of life [3]. *ortho*-Lithiated bromo- or chlorobenzene undergo β-elimination even at temperatures slightly above –100 °C [412], but some polyhalogenated aryllithium compounds are significantly more stable (Scheme 5.45). Iodoarenes have also been metalated, but migration of iodine ("halogen dance") is often observed [398].

Strong *ortho*-directing groups, for example SOTol, can facilitate aromatic metalation to such an extent that even LDA at low temperatures can lead to metalation. As shown by the example in Scheme 5.44, astonishing selectivity is sometimes observed. That the chlorophenyl group is deprotonated faster than the tolyl group (Scheme 5.44) is in agreement with other studies [243, 410] which show that chlorine strongly enhances the rate of aromatic metalation, in particular at the *meta* position. Chlorobenzene, however, undergoes *ortho*-metalation when treated with *s*BuLi [411].

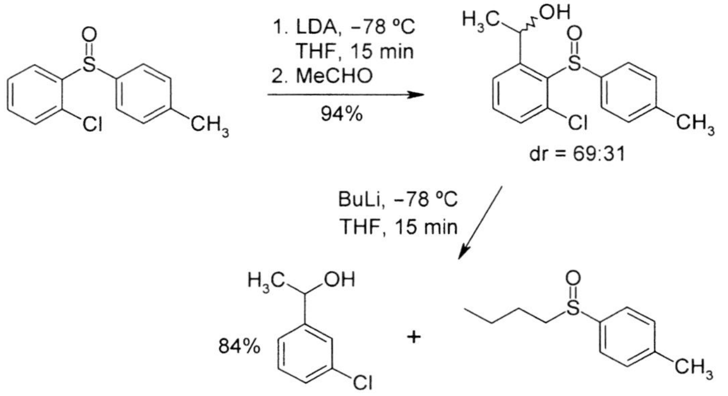

1. LDA, –78 °C THF, 15 min
2. MeCHO
94%

dr = 69:31

BuLi, –78 °C
THF, 15 min

84%

Scheme 5.44. *ortho*-Metalation of diaryl sulfoxides. All reactions proceed without racemization at sulfur [11].

The preferred site of deprotonation of di- or polysubstituted arenes is not easy to predict. In 1,3-disubstituted benzenes in which both substituents facilitate *ortho*-metalation, deprotonation will usually occur between these two groups [181, 365, 408, 416–419] (Scheme 5.45). Dialkylamino groups, however, can sometimes deactivate *ortho* positions (fourth reaction, Scheme 5.45), but this does not always happen [181, 420]. 3-Chloroanisole [411] and 3-fluoroanisole [421] are deprotonated by organolithium compounds between the two functional groups, but the lithiated arenes dimerize readily at –78 °C, presumably via intermediate aryne formation (last example, Scheme 5.45).

For other polysubstituted arenes or heteroarenes, deprotonation at different sites can compete and yield product mixtures. The first reaction in Scheme 5.46 is an example of kinetically controlled carbanion formation, which shows that for some substrates regioselective metalations might be achieved by careful control of the reaction conditions.

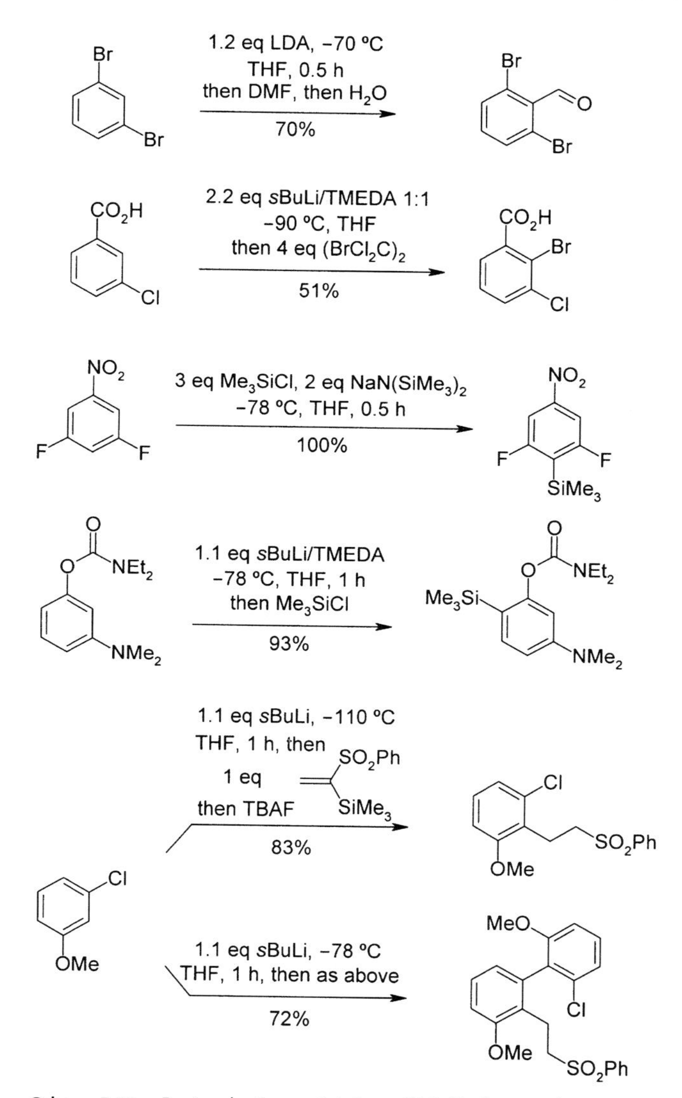

Scheme 5.45. Regioselective metalation of 1,3-disubstituted arenes [403, 411–413, 422].

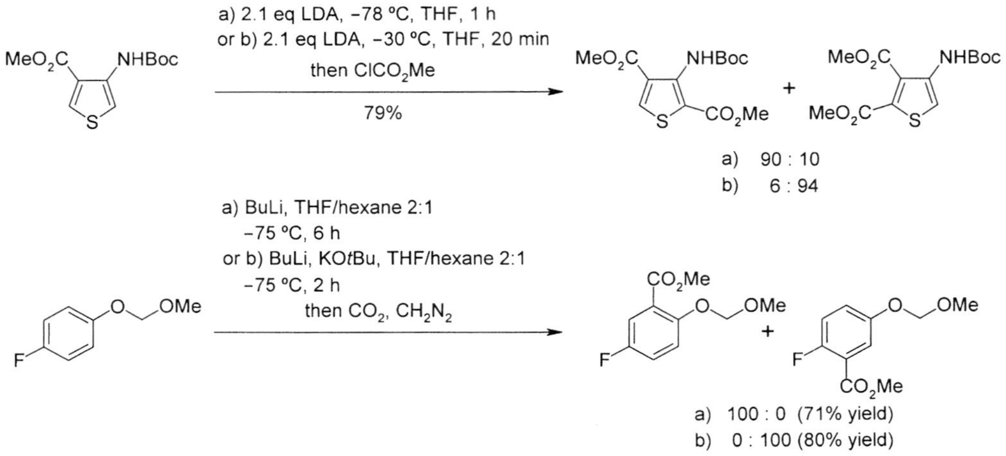

MeO_2C NHBoc (thiophene)

a) 2.1 eq LDA, −78 °C, THF, 1 h
or b) 2.1 eq LDA, −30 °C, THF, 20 min
then $ClCO_2Me$
79%

MeO_2C NHBoc CO_2Me + MeO_2C NHBoc MeO_2C

a) 90 : 10
b) 6 : 94

a) BuLi, THF/hexane 2:1
−75 °C, 6 h
or b) BuLi, KO*t*Bu, THF/hexane 2:1
−75 °C, 2 h
then CO_2, CH_2N_2

CO_2Me OMe + OMe CO_2Me

a) 100 : 0 (71% yield)
b) 0 : 100 (80% yield)

Scheme 5.46. Regioselective metalation of a substituted thiophene [384] and MOM-protected 4-fluorophenol [377].

5.3.11
Aromatic vs Benzylic Deprotonation

According to the "principle of least nuclear motion" [45] aromatic deprotonation should be faster than benzylic metalation, because the benzylic carbanion is expected to rehybridize slightly toward sp^2 to achieve stabilization by conjugation with the aromatic π system. This is, in fact, often observed [217, 401, 423–425], but with some substrates benzylic metalation can effectively compete with aromatic metalation [181, 425, 426] (Scheme 5.47). Thus, treatment of toluene with BuLi/TMEDA or BuLi/DABCO at 80 °C for 0.5 h or with BuLi/KO*t*Bu in Et_2O at −20 °C for 4 h leads to clean formation of benzyllithium [85, 427, 428]. The kinetic preference for aromatic deprotonation, because of the principle of least nuclear motion, thus seems to be too weak to control the regioselectivity of deprotonations in all instances.

The regioselectivity of aromatic metalation can depend on the structure of the base and on the solvent [429], because these will define the structure of the initially formed complex of substrate and base, and thus the site of deprotonation. Similarly, the precise ability of a functional group to be *ortho*-directing will also depend on the solvent and base chosen [430]. This dependence is impressively illustrated by the results obtained by metalation of all the isomers of methoxytoluene (Scheme 5.47).

The metalation of benzylamines is similarly interesting, because slight variations of the reaction conditions can significantly alter the regioselectivity of proton removal. Treatment of *N*,*N*-dimethylbenzylamine with organolithium compounds leads to clean *ortho*-metalation of the phenyl group [418, 431]. If, however, phenylsodium [432] or mixtures of organolithium compounds and KO*t*Bu [193, 209, 433]

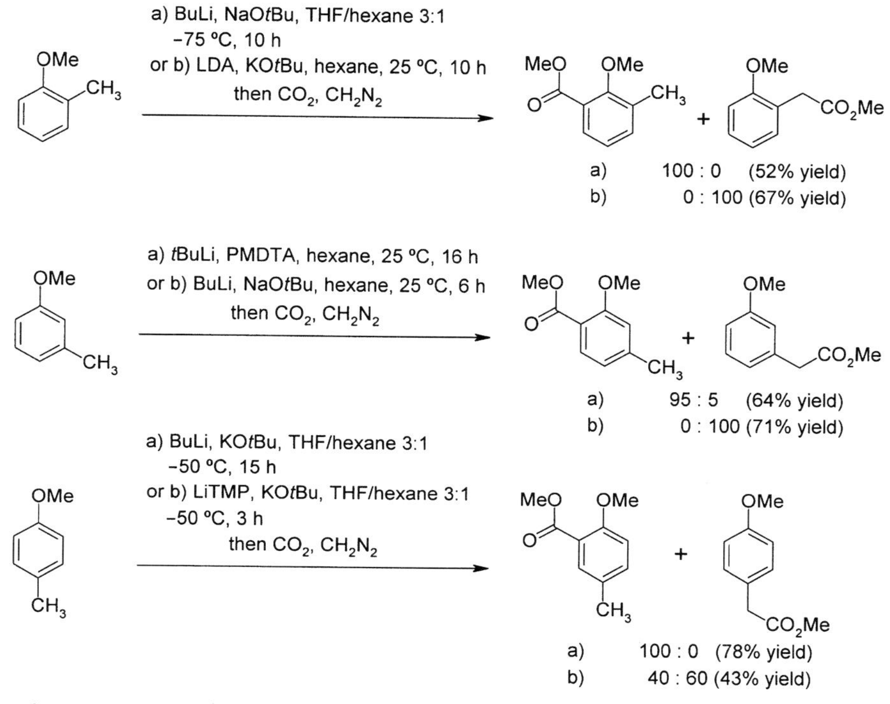

Scheme 5.47. Regioselective metalation of methoxytoluenes [373].

are used as base, exclusive metalation of the benzylic methylene group occurs (Scheme 5.48). The acidity of the benzylic position of benzylamines can also be enhanced by conversion to an amine–borane complex [209].

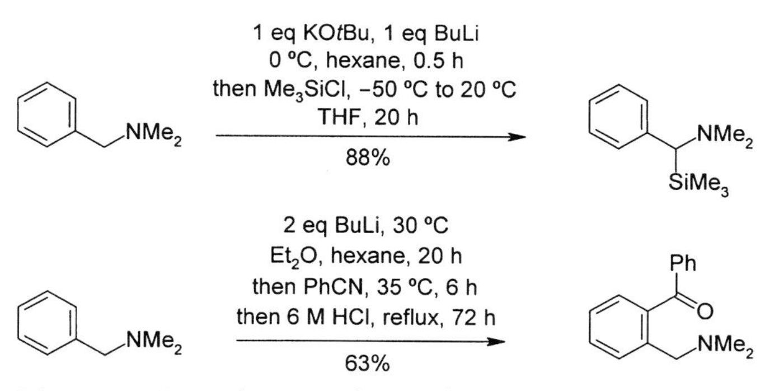

Scheme 5.48. Regioselective metalations of *N*,*N*-dimethylbenzylamine [193, 431].

Similarly, benzyl ethers can be metalated either at the benzylic position or at the arene (Scheme 5.49). As with benzylic amines it seems that benzylic deprotonation is kinetically favored, whereas the metalated arene is the thermodynamically more stable product. The metalation of benzyl alcohols is discussed in Section 5.3.7.2.

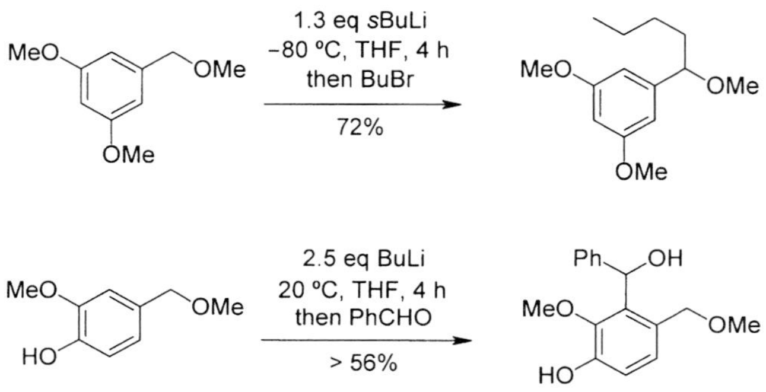

MeO OMe OMe 1.3 eq *s*BuLi −80 °C, THF, 4 h then BuBr 72% MeO OMe OMe

MeO OMe HO 2.5 eq BuLi 20 °C, THF, 4 h then PhCHO > 56% Ph OH MeO OMe HO

Scheme 5.49. Regioselective metalations of benzyl ethers [252, 255].

5.4 The Stability of Carbanions

5.4.1 Introduction

Carbanions can be highly reactive and can therefore undergo not only the expected reaction with an electrophile but also several other, unwanted, transformations. The most common reactions of carbanions in the absence of electrophiles include oxidation, elimination, and rearrangement. The tendency to undergo these reactions usually increases with the basicity and nucleophilicity of a carbanion, which in turn depend on the structure of the organic fragment and on the type of metal chosen. If organolithium compounds are prepared by halogen–lithium exchange with *t*BuLi at temperatures ≥ 0 °C, substantial amounts of *tert*-butylalkanes, alkenes, and homodimers of the alkyl halide can result [8]. These side reactions can effectively be avoided by reducing the reaction temperature or by using mixtures of hydrocarbons and small amounts of ethers as solvent [8].

Simultaneous treatment of a carbonyl compound with a Lewis acid and a tertiary amine or another weak base ("soft enolization") can sometimes be used to generate enolates of sensitive substrates which would have decomposed under strongly basic reaction conditions [434]. Boron enolates, which readily react with aldehydes at low temperatures, can also be prepared in situ from sensitive, base-labile ketones or carboxylic acid derivatives [293, 295, 299]. Unwanted decomposition of a carbanion may also be prevented by generating it in the presence of an electrophile which will not react with the base (e.g. silyl halides or silyl cyanides [435]).

5.4.2
α-Elimination

In Section 5.3.7.3 the formation of α-halogen carbanions and their alkylation was discussed. If these or related intermediates are left to warm, α-elimination will usually occur to yield carbenes, which either react with the solvent, dimerize, or undergo inter- or intramolecular C–H or C–C bond insertion [291, 292, 309, 436]. Because of the electron deficit at the carbene carbon atom (six valence electrons only), these intermediates are highly energetic, and their formation by α-elimination is therefore much slower than the formation of alkenes by β-elimination.

An interesting example of unexpected product formation because of the carbenoid character of α-haloorganometallic reagents is depicted in Scheme 5.50. Depending on the complexing properties of the solvent, treatment of an α-iodo Grignard reagent with *i*PrMgCl either yields the product of nucleophilic substitution of iodide or the product of carbene C–H insertion into the isopropyl group of *i*PrMgCl.

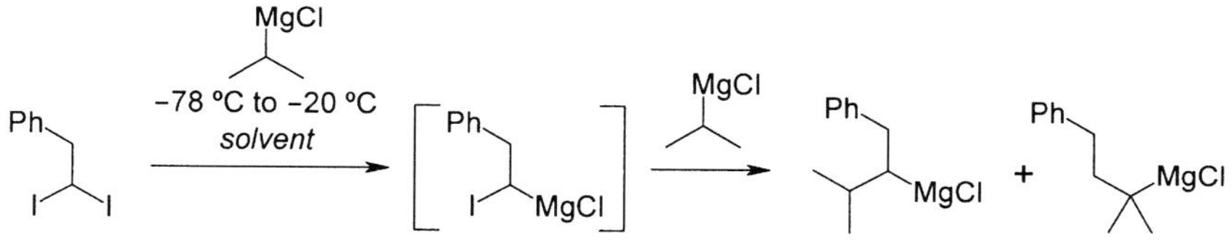

solvent	
THF	90 : 10
*t*BuOMe	31 : 69
Et_2O	25 : 75
iPr_2O	19 : 81

Scheme 5.50. Reactions of an α-iodo Grignard reagent as electrophile or as carbene [437].

Vinylic carbanions with a geminal leaving group can undergo α-elimination to yield vinylidenes and rearrange to yield alkynes (Fritsch–Buttenberg–Wiechell rearrangement). These processes can, however, often be suppressed by keeping the reaction temperature low, and numerous examples of the α-lithiation and alkylation of vinyl halides have been reported [327, 438] (Scheme 5.51). Again, the solvent can have a strong influence on the stability of these intermediates [439]. The examples in Scheme 5.51 illustrate the reaction conditions required for such conversions. Some vinyllithium compounds are sufficiently basic to abstract a proton from THF if allowed to warm (last reaction, Scheme 5.51). The solvent mixture of the first example is required because pure THF (mp –108 °C) becomes too viscous at these low temperatures [440].

Deprotonation of the alkyne group of propargyl halides or sulfonates can also lead to elimination and formation of a vinylidene. Interestingly, these derivatives react with alcoholates, not yielding enol ethers via *O*-alkylation but undergoing C–H bond insertion instead (Scheme 5.52).

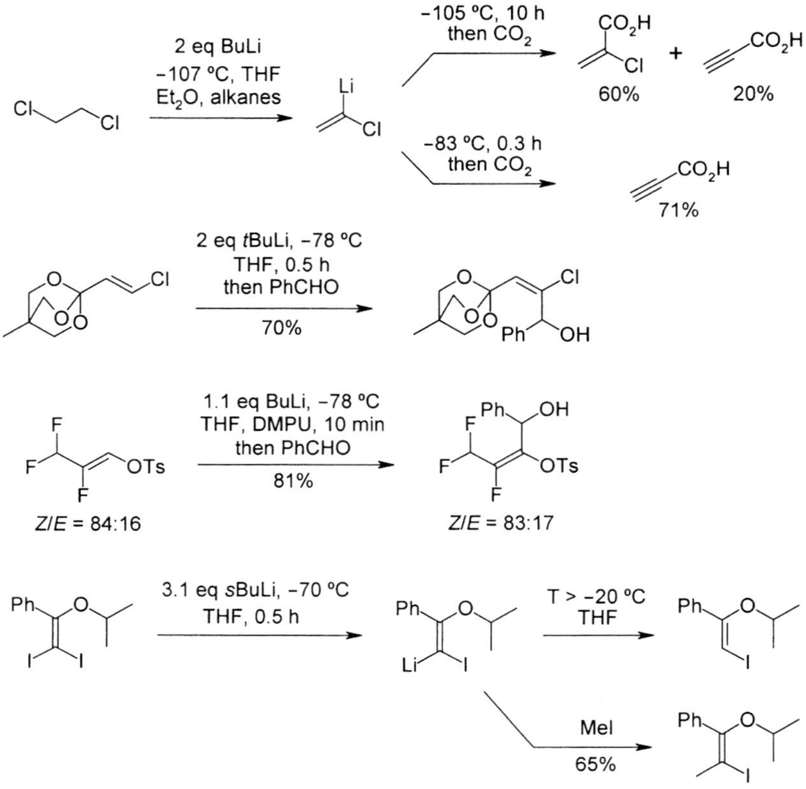

Scheme 5.51. Generation and alkylation of α-halo or α-(tosyloxy)vinyllithium compounds [332, 440–442].

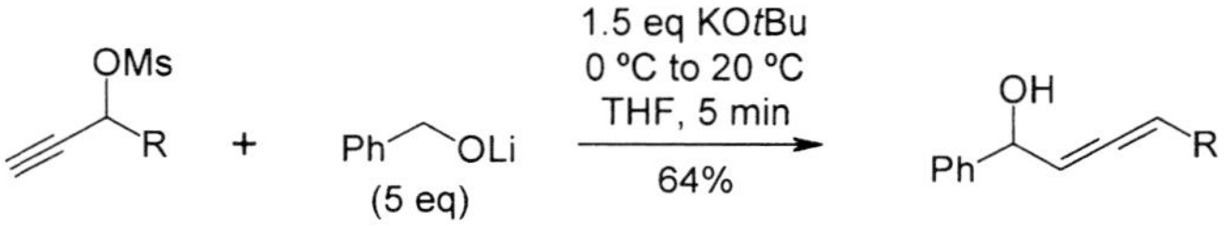

Scheme 5.52. Generation of vinylidenes from propargyl mesylates and their reaction with alcoholates [443].

5.4.3
β-Elimination

In carbanions with a leaving group in the β position β-elimination may compete efficiently with attack by an electrophile. The rate of elimination will increase with increasing reactivity of the carbanion and with increasing ability of the leaving group to act as such [444]. Higher reaction temperatures, moreover, can also promote elimination reactions, which are usually accompanied by a gain of entropy.

Thus, almost no examples have been reported of α-alkylation of β-halo or β-alkoxy ketones, nitriles, sulfones, or propanoic acid derivatives under basic reaction conditions. 1,2-Dihalides are usually converted into alkenes when halogen–metal exchange is attempted, even if highly strained alkenes are thereby formed (Scheme 5.53).

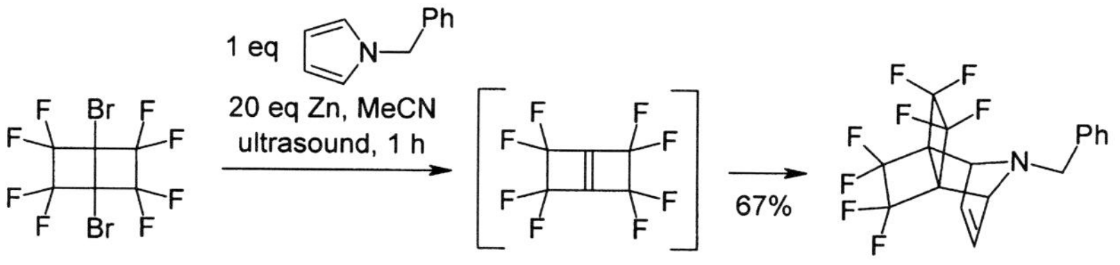

1 eq N-benzylpyrrole; 20 eq Zn, MeCN, ultrasound, 1 h; 67%

Scheme 5.53. Generation of a strained alkene by elimination of $ZnBr_2$ from an intermediate 2-bromoethylzinc compound [445].

β-Elimination can be prevented by choosing a substrate with a worse leaving group in the β position or by reducing the basicity of the carbanion, for instance by selecting a different metal [446]. An alkoxy group, which is a rather good leaving group, might, for instance, be replaced by a hydroxyl group (which can be alkylated at a later stage of the synthesis) and the α-alkylation conducted with the corresponding dianion [162–165] (Scheme 5.54). Because oxide (O^{2-}) is more difficult to generate and a worse leaving group than alkoxide (RO^-), no elimination of oxide usually occurs at low temperatures. *O*-Alkylation will not compete with *C*-alkylation if a soft electrophile and a low reaction temperature are chosen.

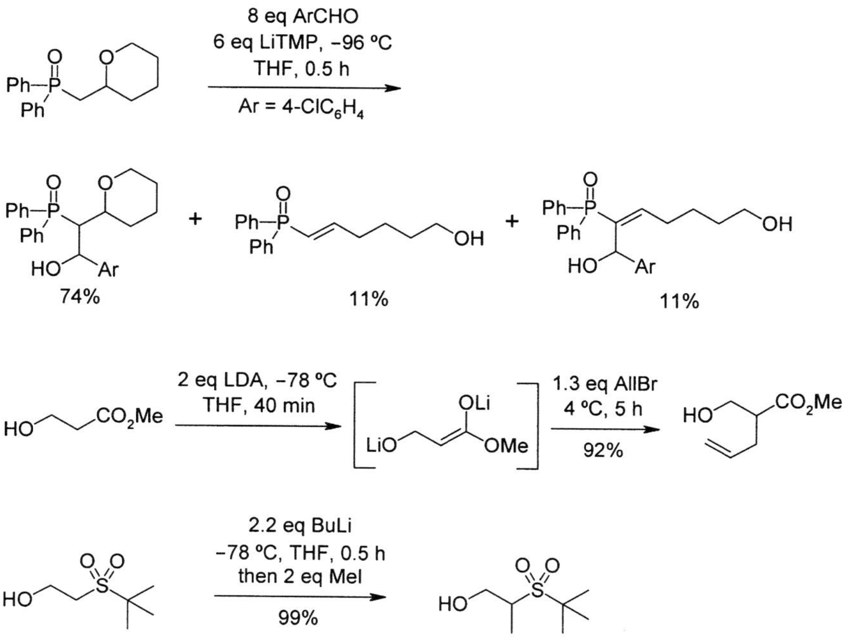

8 eq ArCHO, 6 eq LiTMP, −96 °C, THF, 0.5 h; Ar = 4-ClC$_6$H$_4$ → 74% + 11% + 11%

2 eq LDA, −78 °C, THF, 40 min; 1.3 eq AllBr, 4 °C, 5 h; 92%

2.2 eq BuLi, −78 °C, THF, 0.5 h, then 2 eq MeI; 99%

Scheme 5.54. Alkylation of β-alkoxy and β-hydroxy carbanions [163, 165, 447].

Cyclic substrates, for example 4-pyranones or 4-chromanones, are usually less prone to β-elimination, and can usually be converted into the corresponding enolates and then *C*-alkylated in high yield.

The amino group is a rather poor leaving group [444], and numerous examples of the alkylation of β-amino carbanions are known. Thus, 2-(dialkylamino)ethyl phosphine oxides [448], sulfoxides [449], and sulfones [450, 451], 3-(dialkylamino)propanoic acid esters [452], or 3-aminopropionamides [453] can all be alkylated at the α position of the electron-withdrawing group without loss of the amino group. Even 2-(dialkylamino)ethyl carbamates, which upon lithiation yield highly reactive α-oxy organolithium derivatives, do not eliminate the dialkylamino group [454–456] (first reaction, Scheme 5.55). Similar substrates, in which the amino group is acylated or alkoxycarbonylated can, however, become unstable upon metalation and eliminate $RCONR^-$. In such instances a monoacylated primary amino group, which will also be deprotonated when the carbanion is formed, would be expected to be less likely

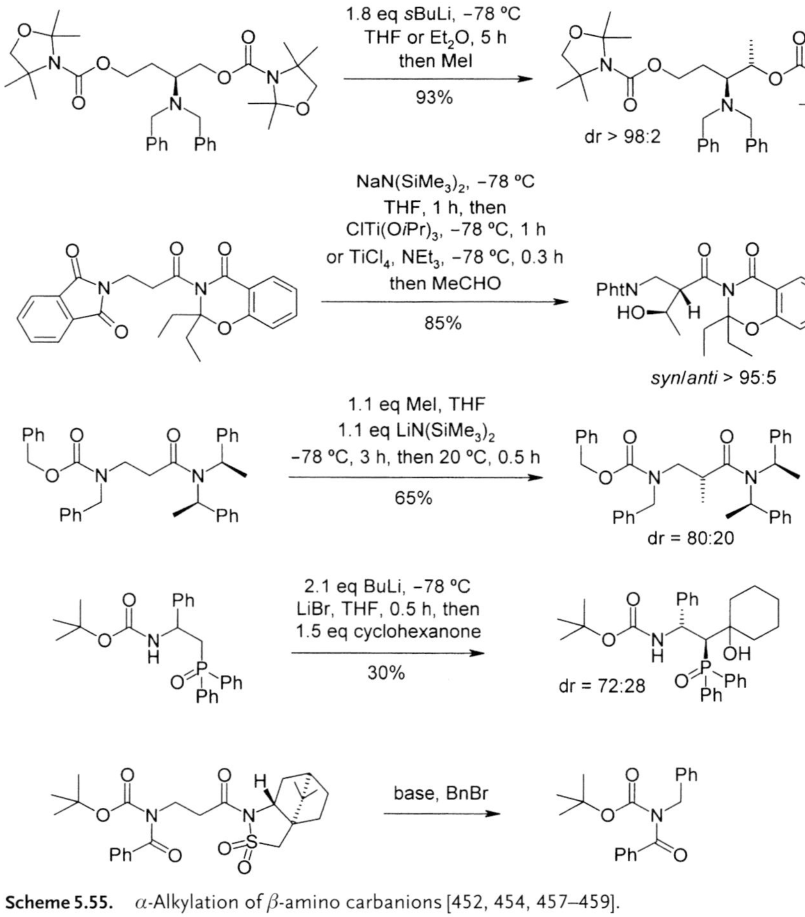

1.8 eq *s*BuLi, −78 °C
THF or Et_2O, 5 h
then MeI
93%
dr > 98:2

$NaN(SiMe_3)_2$, −78 °C
THF, 1 h, then
$ClTi(OiPr)_3$, −78 °C, 1 h
or $TiCl_4$, NEt_3, −78 °C, 0.3 h
then MeCHO
85%
PhtN
HO
syn/*anti* > 95:5

1.1 eq MeI, THF
1.1 eq $LiN(SiMe_3)_2$
−78 °C, 3 h, then 20 °C, 0.5 h
65%
dr = 80:20

2.1 eq BuLi, −78 °C
LiBr, THF, 0.5 h, then
1.5 eq cyclohexanone
30%
dr = 72:28

base, BnBr

Scheme 5.55. α-Alkylation of β-amino carbanions [452, 454, 457–459].

to be eliminated. Substrates of this type (i.e. $ZNHCH_2CH_2Z$), however, often yield only small amounts of *C*-alkylated product [457, 458] (fourth reaction, Scheme 5.55). Transmetalation (second reaction, Scheme 5.55) or a more effective carbanion-stabilizing group can also prevent elimination in critical cases. In the last reaction in Scheme 5.55, which is closely related to the second example in this scheme, the amino group is obviously too strongly activated, and elimination becomes the dominant reaction [457].

Carboxylic acid derivatives or ketones with an alkylthio group at C-3 can be metalated and alkylated at the α position without elimination of thiolate [446] (Scheme 5.56). Amides of 3-(alkylthio)-2-alkylpropanoic acids can, however, also be alkylated at the β position if strong metalating agents are used as base [53, 460].

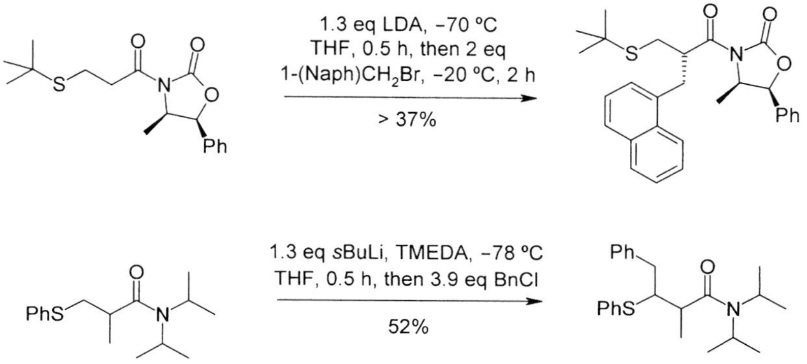

Scheme 5.56. Alkylation of 3-(alkylthio)propanoic acid derivatives [53, 461].

3-Nitropropanoic acid esters can be converted into acrylic acid derivatives by treatment with a slight excess of DBU in THF or DMSO at room temperature. If 3-nitropropanoates are treated with two equivalents of LDA at –78 °C in THF/DMPU or THF/HMPA, however, the resulting dianion can be cleanly *C*-alkylated, and the product isolated without elimination of the nitro group [160, 161] (Scheme 5.57). Without the addition of a cosolvent (DMPU, HMPA, or quinuclidine *N*-oxide [161]) alkylation of the 3-nitropropanoate dianion does not proceed sufficiently quickly [160, 161].

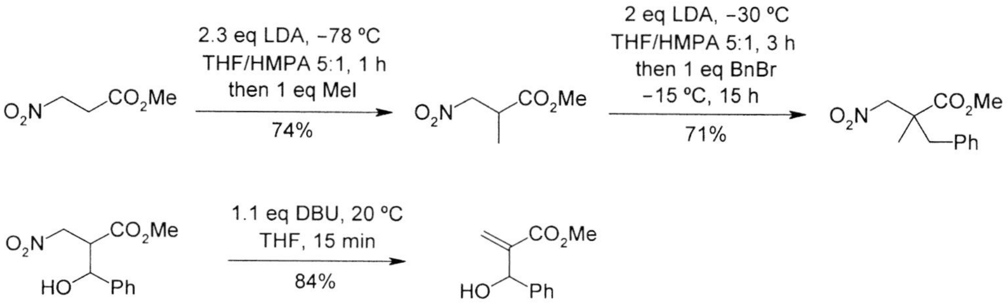

Scheme 5.57. Alkylation of 3-nitropropanoates [160].

Enolates prepared by deprotonation of carboxylic acid derivatives can also undergo elimination to yield ketenes. This is rarely seen with amides, but esters, thiolesters, imides, or *N*-acylsulfonamides can readily decompose to ketenes if left to warm to room temperature (Scheme 5.58). At –78 °C, however, even aryl esters can be converted into enolates stoichiometrically without ketene formation [462, 463].

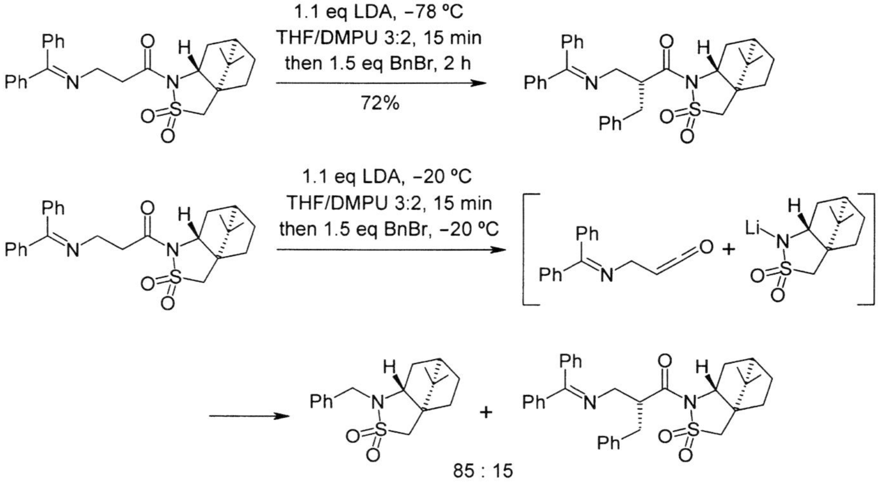

1.1 eq LDA, –78 °C
THF/DMPU 3:2, 15 min
then 1.5 eq BnBr, 2 h
72%

1.1 eq LDA, –20 °C
THF/DMPU 3:2, 15 min
then 1.5 eq BnBr, –20 °C

85 : 15

Scheme 5.58. Temperature-dependent alkylation of *N*-(β-aminopropanoyl)sulfonamides and decomposition via ketene formation [457].

Vinylic carbanions with a vicinal leaving group can undergo β-elimination to yield alkynes. This can sometimes be avoided by keeping reaction temperatures low and by adding the electrophile as soon as possible (Scheme 5.59). The outcome of these reactions often depends not only on the metal and on the substituents at the alkene but also on its configuration. Substrates in which the metal and the leaving group are arranged *anti* undergo elimination more readily than the corresponding *cis* isomers (compare last two reactions in Scheme 5.59 and the reactions in Scheme 5.60). As illustrated by the examples in Scheme 5.59, β-fluorovinyllithiums are sufficiently stable at –100 °C to enable their alkylation before loss of lithium fluoride [464–466]. Similarly, *cis*-2-chlorovinyllithium derivatives require temperatures above –80 °C to undergo elimination. *trans*-2-Chlorovinyllithiums, on the other hand, are extremely unstable, and are transformed into alkynes as soon as they are formed, even at temperatures below –110 °C [440].

Examples have been reported, in which one isomer of a chloroalkene can be α-lithiated whereas the other isomer undergoes β-elimination (Scheme 5.60). Such subtle differences in reactivity will depend on the precise reaction conditions and the substitution pattern of the alkene, and will only rarely be foreseeable.

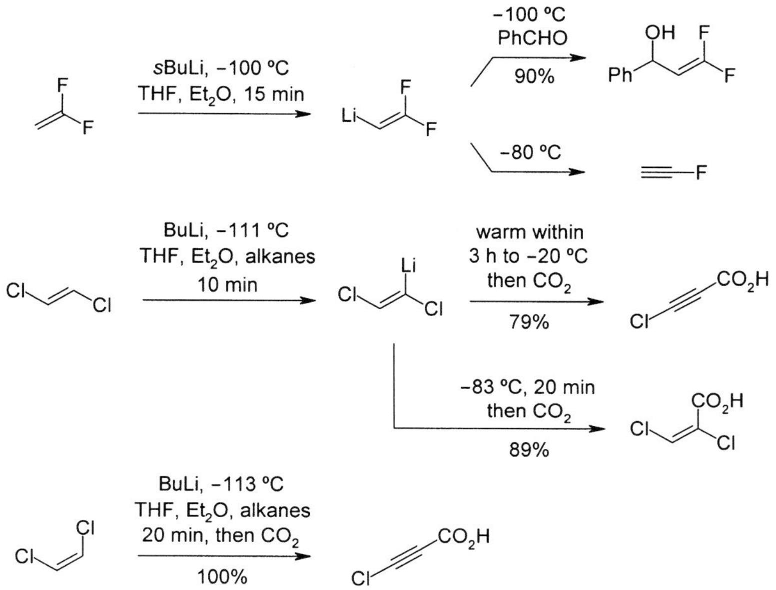

Scheme 5.59. Generation and reactions of lithiated vinyl halides [440, 465, 466].

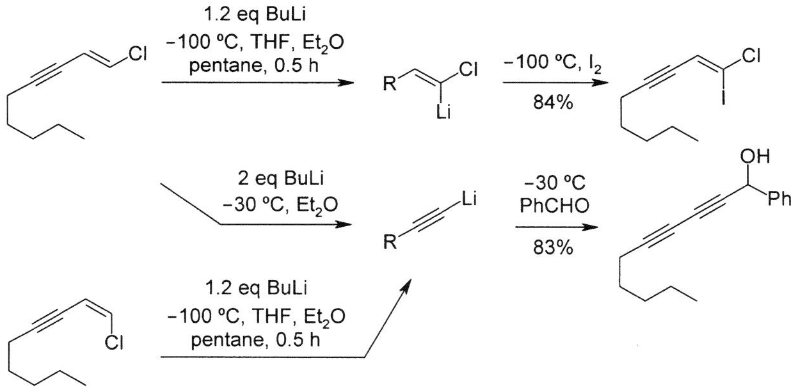

Scheme 5.60. Formation and reactivity of lithiated 1-chloro enynes [331].

Not only halides but also alcoholates or carbamates can act as leaving groups (Scheme 5.61). 2-Lithiovinyl ethers, which cannot be prepared by deprotonation [330] but are accessible by halogen–metal exchange [317], are rather stable when lithium and the alkoxy group are configured *cis*. The corresponding *trans* isomers are, though, less stable, and undergo β-elimination if too much time elapses before the addition of an electrophile.

O-Vinyl carbamates can act as Michael-acceptors toward alkyllithiums (third reaction, Scheme 5.61), because the resulting α-lithiated carbamates are stabilized by chelate formation. A competing process is vinylic metalation, which can lead to elimination of the carbamate and formation of an alkyne. Again, as already mentioned above, the configuration of the alkene can decisively affect the outcome of the reaction (compare third and fourth examples, Scheme 5.61).

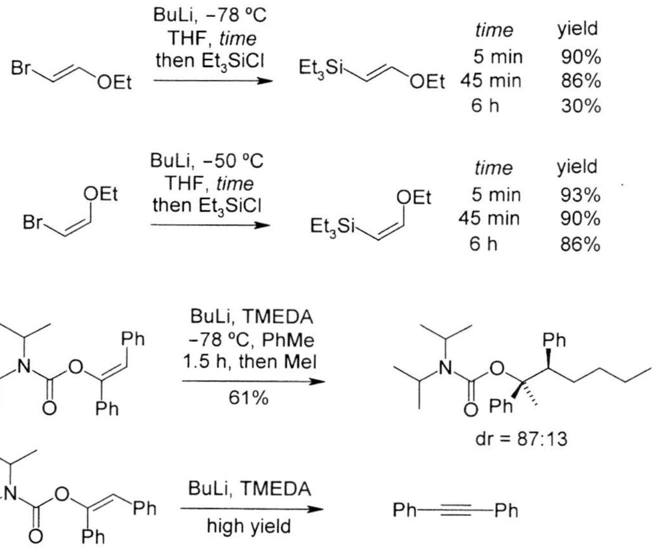

Scheme 5.61. Formation and reactions of β-lithiated vinyl ethers [57, 317].

1-Lithio-2-haloarenes are significantly more stable than 1-lithio-2-haloalkenes. The stability of metalated haloarenes is discussed in Section 5.3.10.

5.4.4 Cyclization

Carbanions with a leaving group or another electrophilic functional group in a suitable position can cyclize or oligomerize instead of reacting intermolecularly with an electrophile, and numerous examples of such cyclizations have been reported [467–470]. Highly reactive carbanions can lead to the formation of strained rings, for example cyclopropanes [272, 292, 471], benzocyclopropenes [472], or benzocyclobutenes [469, 473, 474]. Alternatively, the intermediate carbanion can oligomerize. For the organic chemist it is particularly important to know how readily such cyclizations/oligomerizations will occur, and under which conditions it will be possible to trap the intermediate carbanion intermolecularly with an added electrophile.

Several ω-haloalkyllithiums and Grignard reagents have been described which are sufficiently stable to react intermolecularly with electrophiles before cyclizing (Scheme 5.62). Particularly stable are ω-chloroalkyl derivatives [475–477] whereas, not surprisingly, the ω-bromo- or ω-iodoalkyl carbanions are usually more difficult to prepare and handle (see below). Propargyl chloride has been lithiated at the alkynyl group by treatment with MeLi at low temperatures, and the resulting intermediate could be *C*-acylated with methyl chloroformate in high yield without loss of chloride (Scheme 5.62).

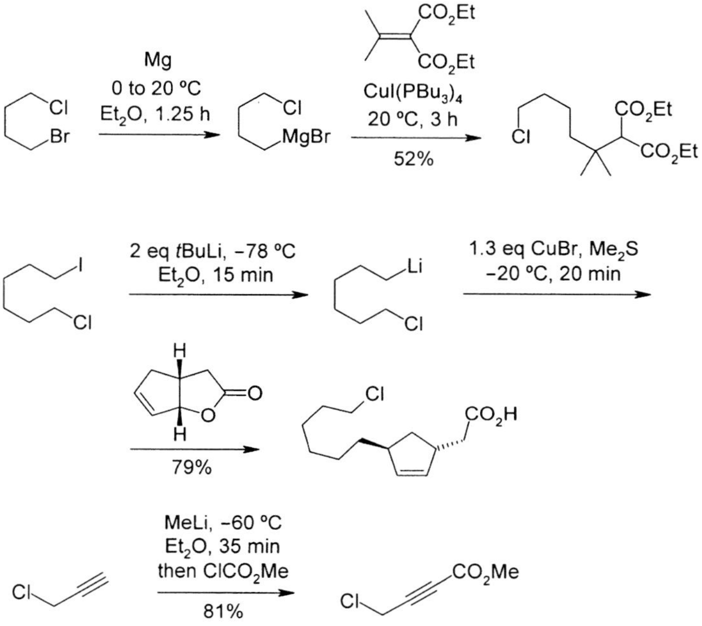

Scheme 5.62. Preparation and reactions of haloalkyl Grignard reagents [478], haloalkyllithium compounds [479], and lithiated propargyl chloride [480].

Dihalides containing an aliphatic and an aromatic halide often undergo selective halogen–metal exchange at the arene [468]. If the aliphatic halogen is chlorine the resulting intermediates are quite stable and can be trapped with external electrophiles (Scheme 5.63). The corresponding bromides are, however, more reactive and cyclize more readily than the chlorides. The stability of these intermediates can also be enhanced by choosing magnesium instead of lithium as the metal.

Enolates with a leaving group in the γ position can cyclize to yield cyclopropanes instead of reacting intermolecularly with an electrophile. 3-Halopropyl ketones or 4-halobutyric acid esters, for instance, are readily converted to cyclopropane derivatives when treated with a base (Scheme 5.64; see also Section 9.4.1).

3-Alkoxypropyl ketones and even 3-acyloxypropyl ketones, however, do generally not cyclize, and can be cleanly metalated and alkylated intermolecularly with an electrophile [485]. Similarly, chloroalkyl enolates with more than two atoms between the nucleophilic and electrophilic carbon atoms do not cyclize readily. Examples of chloroalkyl ketones and nitro compounds, which can be deprotonated and alkylated without undergoing cyclization are shown in Scheme 5.65.

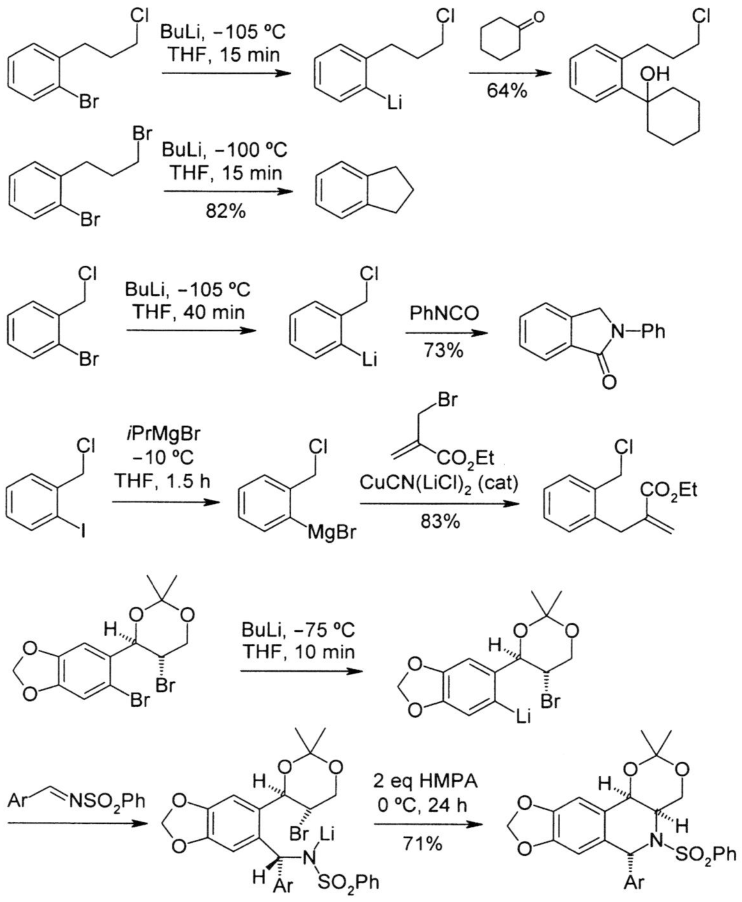

Scheme 5.63. Preparation and reactions of partially metalated arylalkyl dihalides [474, 481, 482]. Ar = 3,4,5-trimethoxyphenyl.

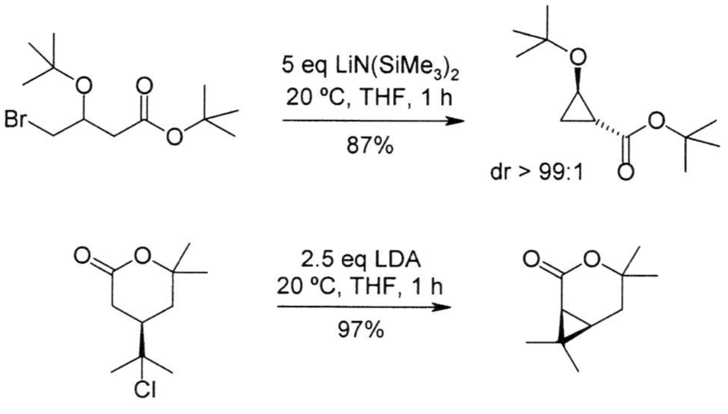

Scheme 5.64. Formation of cyclopropanes from 4-halobutyrates [483, 484].

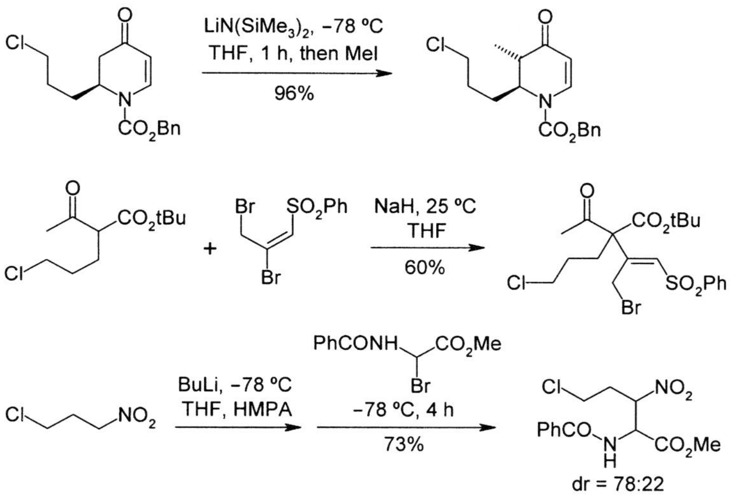

Scheme 5.65. Deprotonation and alkylation/vinylation of chloroalkyl ketones and nitro compounds [486–488].

Ketones usually react quickly with Grignard reagents, even at low temperatures. 2- or 3-Iodoaryl ketones can, however, be converted into Grignard reagents by iodine–magnesium exchange with neopentylmagnesium bromide, which does neither add to nor reduce ketones at significant rates when compared with the rate of halogen–magnesium exchange. These metalations require dipolar aprotic solvents such as NMP or DMA to proceed at acceptable rates. The resulting Grignard reagents are sufficiently stable to enable intermolecular coupling with a variety of electrophiles [489, 490] (Scheme 5.66).

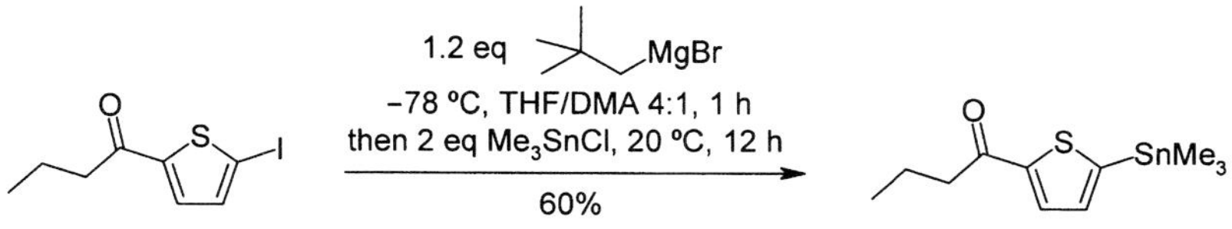

Scheme 5.66. Formation and derivatization of ketone-containing Grignard reagents [489].

5.4.5 Rearrangement

Carbanions which cannot achieve stabilization by charge delocalization to a more electronegative atom than carbon might undergo substantial rearrangement to achieve such stabilization. The Wittig, Stevens, or Grovenstein–Zimmerman rearrangements are examples of such transformations (Scheme 5.67). Such rearrangements are non-concerted when a simple alkyl group migrates, but can become concerted if the migrating groups are allylic or benzylic (2,3-sigmatropic rearrangements).

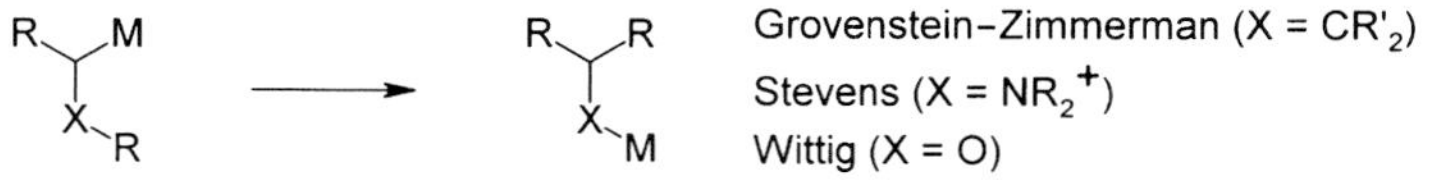

Grovenstein–Zimmerman (X = CR'_2)
Stevens (X = NR_2^+)
Wittig (X = O)

Scheme 5.67. Rearrangements of carbanions (M = alkali metal, R′ = carbanion stabilizing group).

The ease with which these reactions will occur depends to a large extent on the precise structure of the substrate and can be difficult to predict. Low reaction temperatures will generally prolong the half-life of the initially formed carbanions, which might then be trapped intermolecularly with reactive electrophiles before rearrangement can occur [193, 252]. Intramolecular trapping of such carbanions can also be used to prevent their rearrangement [253, 491].

Not only alkyl groups, but also aryl [492, 493], vinyl [494], acyl [276, 495–497], alkoxycarbonyl [498], aminocarbonyl [499–501], silyl [502–504], or phosphoryl groups [279, 280] can migrate to a vicinal carbanion (Scheme 5.68). Because some of these groups can be used to stabilize α-heteroatom-substituted carbanions by chelate formation, migration of these groups to the carbanion is a potential side reaction in the generation and alkylation of chelate-stabilized carbanions.

KO*t*Bu, *i*Pr_2NH, BuLi (1:1:1, 1.25 eq), −78 °C, THF, 0.5 h; 97%

LDA, −78 °C, THF, 0.5 h; 89%

(−)-sparteine, *s*BuLi, PhMe, −78 °C to −40 °C; 88%; 90% ee

BuLi, −78 °C, THF, 3 h; 70%

1.3 eq LDA, −78 °C, THF, 6 h; 86%

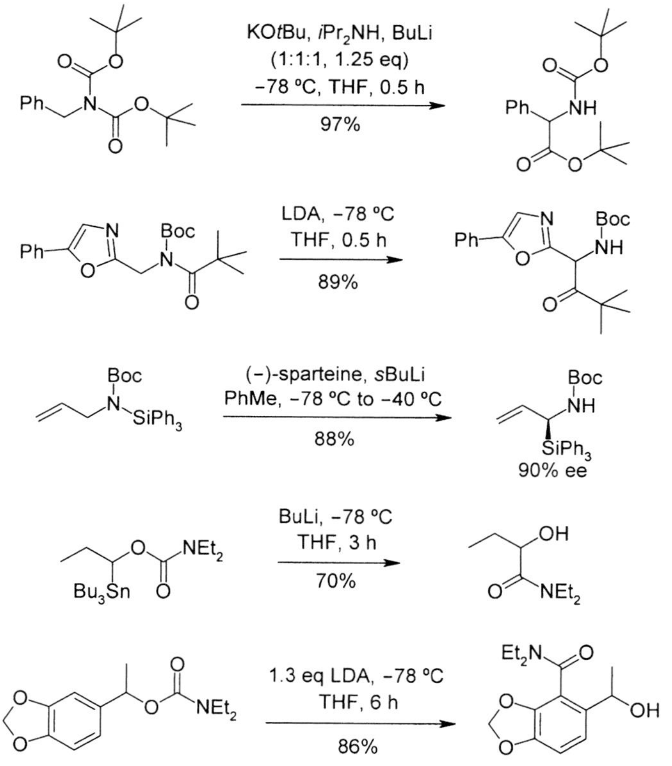

Scheme 5.68. Rearrangements of chelate-stabilized carbanions [496, 498, 500, 502].

5.4.6 Oxidation

Electron-rich organic compounds, for example carbanions, can readily be transformed into radicals by transfer of a single electron to a suitable oxidant. The resulting radicals can dimerize if their half-life is sufficiently long, or undergo fragmentation or other reactions characteristic of radicals. Particularly facile is the formation of radicals substituted with both electron-donating and electron-withdrawing substituents [505, 506]. Potential oxidants can be air, the solvent, the carbanion itself (disproportionation), or an added oxidant or electrophile. Electrophiles with a high tendency to act as single-electron oxidants are alkyl, allyl, and benzyl iodides and bromides, benzophenone, nitro compounds, and electron-poor aldehydes [233, 507]. Carbanions with a high tendency to become oxidized by SET are deprotonated nitroalkanes [508, 509] and carbanions substituted with both electron-donating and electron-withdrawing groups [510–513] (Scheme 5.69). Non-resonance-stabilized carbanions, such as metalated formamidines [233] (Scheme 5.25) or metalated benzyl alcohols [256], can also act as reducing agents. If oxidation of a carbanion by an added electrophile becomes the main reaction pathway, addition of a cosolvent such as DMPU (to accelerate bond formation with the electrophile) or transmetalation

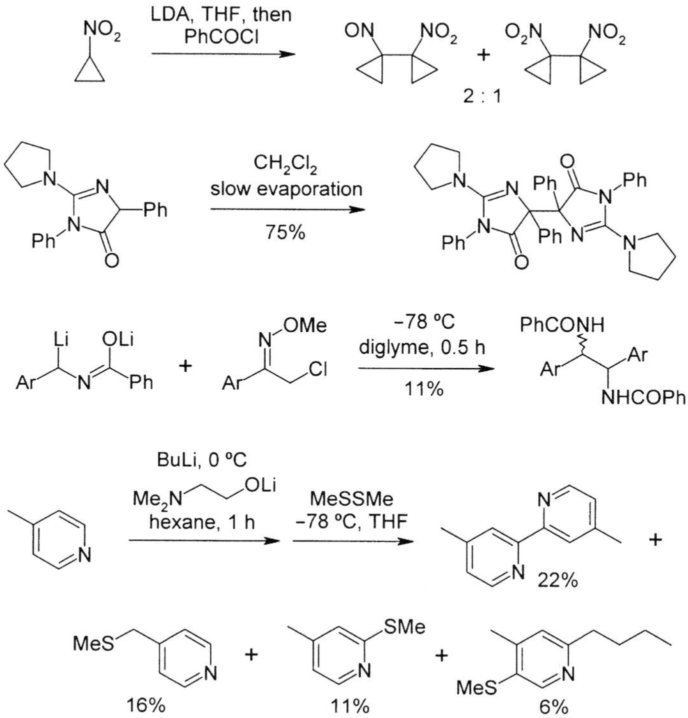

LDA, THF, then PhCOCl
ON NO2 + O2N NO2
2 : 1

CH_2Cl_2 slow evaporation
75%
Ph Ph N Ph O Ph O

-78 °C diglyme, 0.5 h
11%
Li OLi Ar N Ph + N OMe Ar Cl
PhCONH Ar Ar NHCOPh

BuLi, 0 °C
Me_2N OLi
hexane, 1 h
MeSSMe
-78 °C, THF
22%
MeS 16% + SMe 11% + MeS 6%

Scheme 5.69. Dimerization of organolithium compounds and imidazolones [401, 508, 511, 515]. Ar = 4-(MeO)C_6H_4.

can be useful strategies for retarding or suppressing SET processes. Carbanions can also dimerize if treatment with an alkyl halide leads to *C*-halogenation rather than *C*-alkylation [514]. Reaction of the newly formed alkyl halide with the carbanion will then yield symmetric dimers.

5.4.7
Other Factors which Influence the Stability of Carbanions

Carbanions which could, in principle, achieve stabilization by charge delocalization to a more electronegative atom might be rather unstable, anyway, if this tautomer or canonical form is a carbene (Scheme 5.70). This is, for instance, observed for acyl anions, which are difficult to trap with electrophiles and undergo rapid dimerization even at low temperatures [179]. Vinylogous acyl anions, such as metalated propiolic acid esters [75, 516–518], metalated 2-ethynylpyridine [519], or β-metalated acrylic acid derivatives (Section 5.3.8) are more stable than acyl anions, but must still be kept at low temperatures [520] and should be treated with an electrophile immediately after their generation. The stability of these anions can sometimes be enhanced by reducing the electron-withdrawing strength of the group in the β position of the carbanion. The dianions of propiolic acid [175], *N*-alkyl propiolic acid amides [151], or acrylic acid [334], for instance, are more stable than the corresponding metalated esters. The stability can also be enhanced by choosing a metal, for example magnesium, which forms a more covalent bond with carbon than do the alkali metals [340].

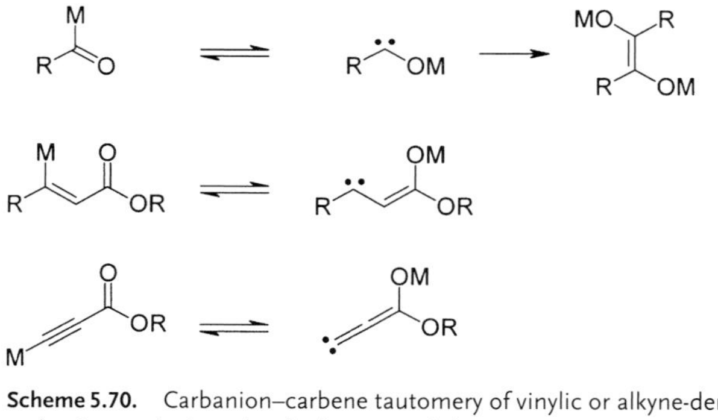

Scheme 5.70. Carbanion–carbene tautomery of vinylic or alkyne-derived carbanions substituted with electron-withdrawing groups.

Another group of unstable carbanions are those with antiaromatic character (Scheme 5.71). Thus, cyclopropenyl anions or oxycyclobutadienes, generated by deprotonation of cyclopropenes or cyclobutenones, respectively, will be highly reactive and will tend to undergo unexpected side reactions. Similarly, cyclopentenediones are difficult to deprotonate and alkylate, because the intermediate enolates are electronically related to cyclopentadienone and thus to the antiaromatic cyclopentadienyl cation.

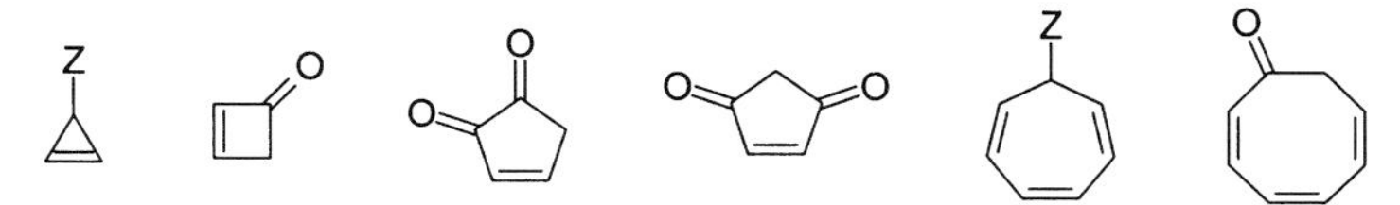

Scheme 5.71. Compounds which upon deprotonation yield antiaromatic carbanions.

5.4.8 Configurational Stability of Carbanions

5.4.8.1 Introduction

A deprotonation is a nucleophilic attack by a base at hydrogen. The formation of aliphatic organolithium compounds (C(sp^3)–Li) by deprotonation of C,H groups with other organolithium compounds or lithium amide bases usually occurs with retention of configuration at carbon [238, 274, 521], as do tin–lithium transmetalations [236, 240, 261, 521, 522] or sulfoxide–magnesium exchanges [9, 311]. Halogen–metal exchange, which can involve radicals [523, 524], can either proceed with retention of configuration [6, 525, 526] or with racemization [527–529].

In carbanions, for instance enolates, in which delocalization of the negative charge to a more electronegative atom (N, O, S) is possible, the cation is usually bound to this atom if the cation is hard (e.g. alkali metal cations, Mg^{2+} [530], Zn^{2+}); in this case the nucleophilic carbon atom becomes a planar sp^2 hybrid. If the cation is soft (e.g. the cations of late transition metals) it will often be bound to the deprotonated carbon atom; these organometallic compounds are, however, usually only weak nucleophiles and will not be treated here.

Electrophilic substitutions at carbon, for example the reaction of an organometallic reagent with an electrophile, can occur either with retention [236, 238, 274, 275, 525, 529] or inversion [234, 471] at the nucleophilic carbon atom [57, 189, 522, 531, 532].

The structures of organometallic compounds prepared by deprotonation of a C(sp^3),H-group range from completely dissociated ion pairs to covalent compounds with strong, kinetically stable C–M bonds. Not only the latter, but also ionic carbanions, can be configurationally stable, for instance if the carbanion is strongly pyramidalized, as some metalated sulfones [533–535], or if inversion is precluded by the structure of the carbanion, as for bridgehead carbanions. Vinylic or cyclopropane-derived carbanions also tend to have high configurational stability, even if the C–M bond is strongly polarized.

The formation of enolates or related compounds with a planar nucleophilic carbon atom by deprotonation of a center of asymmetry does not necessarily mean that the stereochemical information is lost. Examples have been reported in which the enolate remains chiral by assuming a different form of chirality (e.g. axial chirality) and the ensuing reaction with an electrophile proceeds with high enantioselectivity, despite the transient planarization of the stereogenic center (Scheme 5.72).

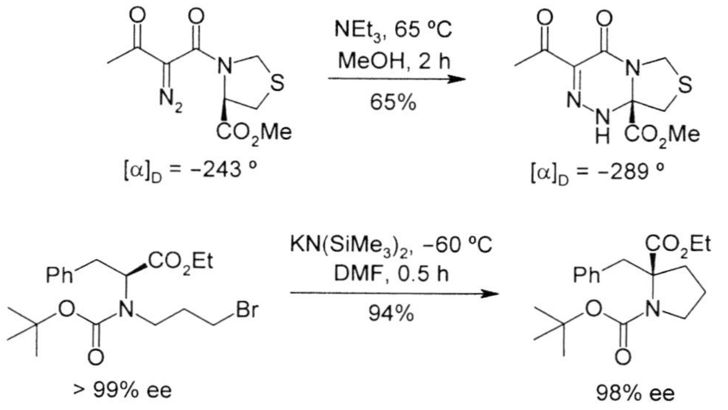

Scheme 5.72. Intramolecular amination and alkylation of enolates with retention of configuration [536, 537].

Configurationally stable carbanions are not a prerequisite for stereoselective reactions with electrophiles. If the organometallic reagent is configurationally labile, as is the case for many organolithium or Grignard reagents at room temperature, highly stereoselective reactions with electrophiles can occur via electrophile- or additive-controlled dynamic kinetic resolution. Examples of such reactions are shown in Scheme 5.73. The intermediate organolithium compound of the second reaction has been shown to be configurationally labile [283], but in the presence of an enantiomerically pure bisoxazoline one of the two enantiomers reacts significantly faster with the electrophile than the other. The first reaction in Scheme 5.73 requires toluene as solvent; in THF no alkylation but mainly elimination of MsOLi occurs [538].

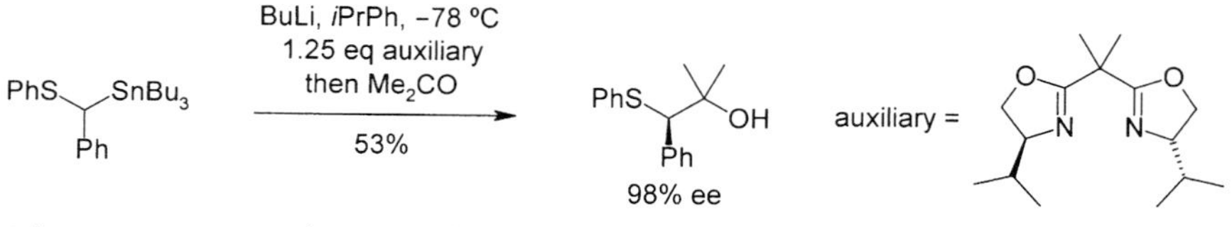

Scheme 5.73. Dynamic kinetic resolution via configurationally labile carbanions [283, 538, 539].

Organometallic compounds with covalent, sparsely polarized C–M bonds do not usually epimerize readily. Thus, organoaluminum or organomercury compounds are configurationally stable up to 150 °C [540, 541] if oxidants or other radical chain initiators are carefully excluded (footnote 10 in Ref. [542]). Similarly, boranes, stannanes, or silanes are configurationally stable at the temperatures usually chosen for most synthetic organic reactions with these intermediates.

5.4.8.2 **Organolithium Compounds**

Organolithium compounds devoid of heteroatoms have little configurational stability [531, 543, 544]. Thus, lithiation of (–)-2-iodooctane with *s*BuLi at –70 °C in Et_2O followed by treatment with CO_2 after 2 min led to 80% racemized 2-methyloctanoic acid [545]. Similarly, lithioalkenes often undergo rapid *cis*/*trans* isomerization, even at low temperatures, in particular if the alkene is substituted with aryl or electron-withdrawing groups [546–548]. For example, *cis*-lithiostilbene isomerized completely to *trans*-lithiostilbene within 0.5 h in THF at –45 °C (Scheme 5.74). Such isomerizations proceed more quickly in ethers as solvent than in pure hydrocarbons [544, 549, 550]. A remarkably stable vinylic organolithium compound is (4-methylcyclohexylidene)methyllithium (second example, Scheme 5.74).

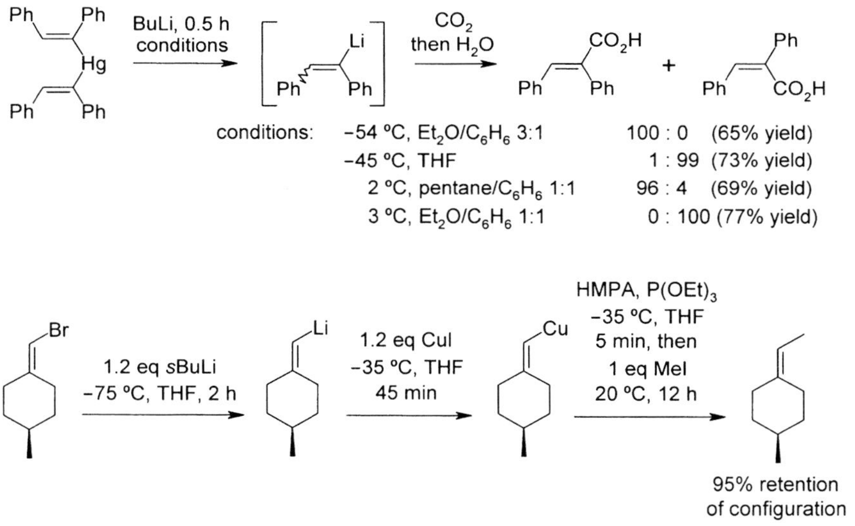

conditions:	
–54 °C, Et_2O/C_6H_6 3:1	100 : 0 (65% yield)
–45 °C, THF	1 : 99 (73% yield)
2 °C, pentane/C_6H_6 1:1	96 : 4 (69% yield)
3 °C, Et_2O/C_6H_6 1:1	0 : 100 (77% yield)

Scheme 5.74. Configurational stability of lithiostilbene [549] and cyclohexylidenemethyllithium and -copper derivatives [551].

Some heteroatom-substituted or chelate-stabilized organolithium compounds, on the other hand, can be sufficiently stable toward racemization to enable their use in stereoselective reactions with electrophiles [223, 225, 271, 531, 543, 552–554] (Scheme 5.75). This increased configurational stability of α-heteroatom-substituted carbanions might be due to the stronger pyramidalization of such carbanions [261, 555] and fixation of the metal by chelate formation.

Enantiomerically pure organolithium compounds have been prepared by tin–lithium exchange [261], mercury–lithium exchange [549], or by lithiation in the presence of a chiral, enantiomerically pure or enriched amine, for example sparteine [271]

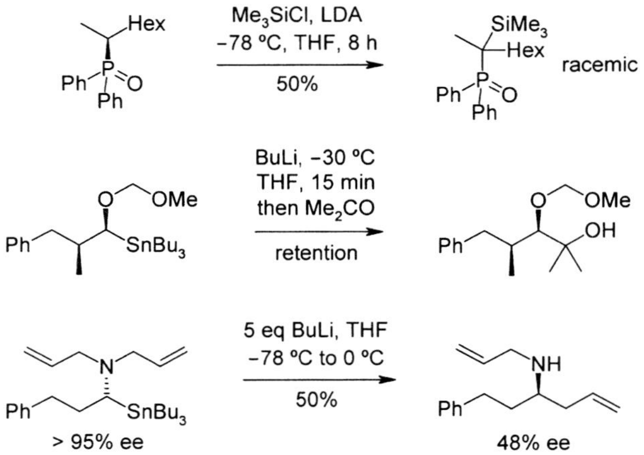

Scheme 5.75. Stereochemical outcome of lithiation and alkylation at stereogenic carbon atoms [261, 556, 557].

(see, e.g., Scheme 5.31). Usually these organometallic reagents are configurationally stable at low temperatures only, but racemize or decompose when left to warm to room temperature [261]. An example of a surprisingly stable α-thio organolithium derivative is shown in Scheme 5.76 [558, 559].

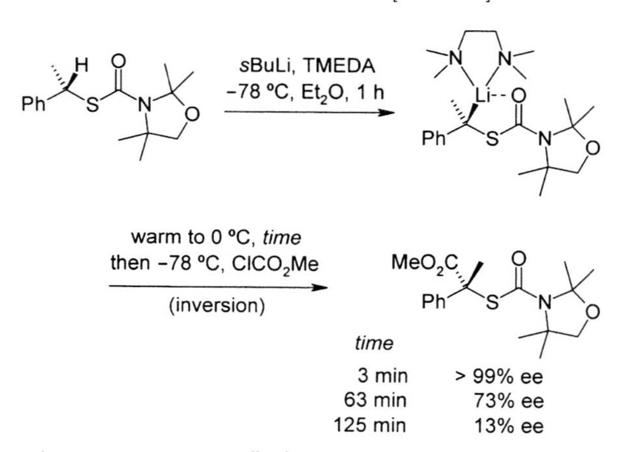

Scheme 5.76. Exceptionally slow racemization of an α-thio organolithium compound [560].

5.4.8.3 Organomagnesium Compounds

The configurational stability of Grignard reagents has been the subject of numerous investigations [312, 540, 561–564]. Unfunctionalized, secondary Grignard reagents isomerize at −10 °C in ethers with a half-life of about 5 h (Scheme 5.77), and are thus significantly more stable toward racemization than the corresponding organolithium compounds.

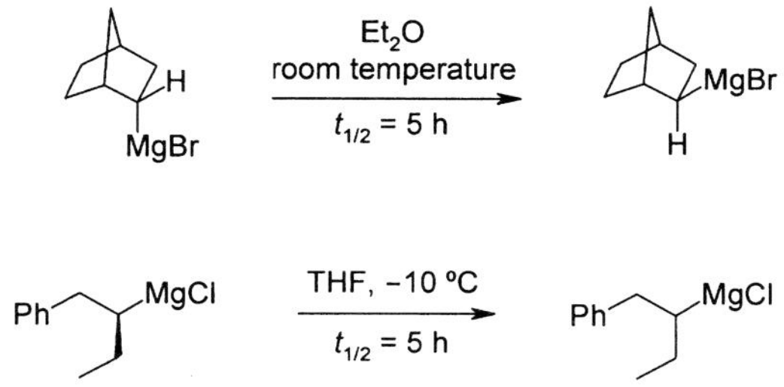

$$\text{(exo-norbornyl-MgBr)} \xrightarrow[t_{1/2} = 5\ \text{h}]{\text{Et}_2\text{O, room temperature}} \text{(endo-norbornyl-MgBr)}$$

$$\text{Ph-CH}_2\text{-CH(Et)-CH}_2\text{MgCl} \xrightarrow[t_{1/2} = 5\ \text{h}]{\text{THF, }-10\ °\text{C}} \text{Ph-CH}_2\text{-CH(Et)-CH}_2\text{MgCl}$$

Scheme 5.77. Isomerization of Grignard reagents [261, 311].

Unfunctionalized, enantiomerically enriched secondary Grignard reagents cannot be prepared by halogen–magnesium exchange with metallic magnesium, because these reactions proceed via radicals [523] and lead to racemates. Halogen–magnesium or sulfoxide–magnesium exchange with other Grignard reagents is not possible either, because Grignard reagents with secondary alkyl groups are too energy-rich and will not be formed in substantial amounts by thermodynamic equilibration with other Grignard reagents [311]. One possible means of access to these compounds is the reaction of enantiomerically pure α-chloro sulfoxides (prepared by chlorination of enantiomerically pure sulfoxides [312, 565]) with excess primary Grignard reagent (Scheme 5.78). First, a sulfoxide–magnesium exchange occurs with retention of configuration, to yield an α-chloro Grignard reagent. The latter undergoes an SN2-like reaction with inversion of configuration with a further equivalent of primary Grignard reagent to yield an enantiomerically enriched secondary organomagnesium compound [9, 564]. These reagents react with a variety of electrophiles (peroxides, PhNCS, PhNCO, CO_2, PhCHO, allyl chloride, aminals [565], vinyl bromide (on catalysis with a transition metal [566]), $ZnCl_2$ [567], boronic esters [568], azides [568]) with retention of configuration; with electrophiles prone to undergo reduction by SET (Ph_2CO, C_6F_5CHO, allyl iodide), however, partial or complete racemization is observed [507, 565].

Vinylmagnesium compounds, which can be prepared by halogen–magnesium or sulfoxide–magnesium exchange, usually have high configurational stability [6]. This is, however, not observed for 1-halo-1-magnesioalkenes. Because of their vinylidene

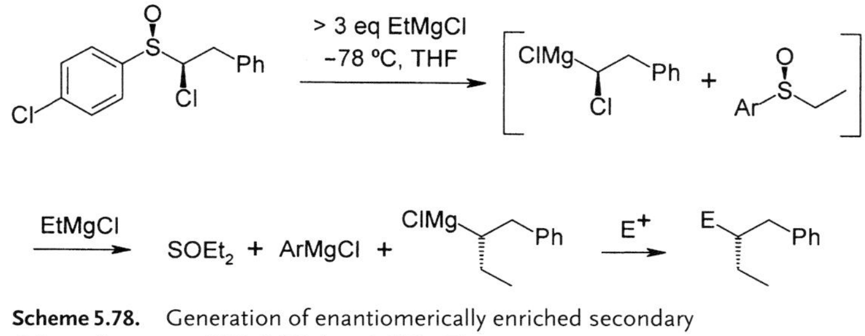

$$\text{4-ClC}_6\text{H}_4\text{S(O)-CHCl-CH}_2\text{Ph} \xrightarrow[-78\ °\text{C, THF}]{>3\ \text{eq EtMgCl}} \left[\text{ClMg-CHCl-CH}_2\text{Ph} + \text{Ar-S(O)-Et}\right] \xrightarrow{\text{EtMgCl}} \text{SOEt}_2 + \text{ArMgCl} + \text{ClMg-CH(Et)-CH}_2\text{Ph} \xrightarrow{\text{E}^+} \text{E-CH(Et)-CH}_2\text{Ph}$$

Scheme 5.78. Generation of enantiomerically enriched secondary Grignard reagents [311, 568]. Ar = 4-ClPh, E^+ = see text.

character, both the C–Mg and C–Hal bonds are weakened [327]; this leads to facile halogen–halogen exchange reactions [569] and rapid loss of stereochemical information (Scheme 5.79).

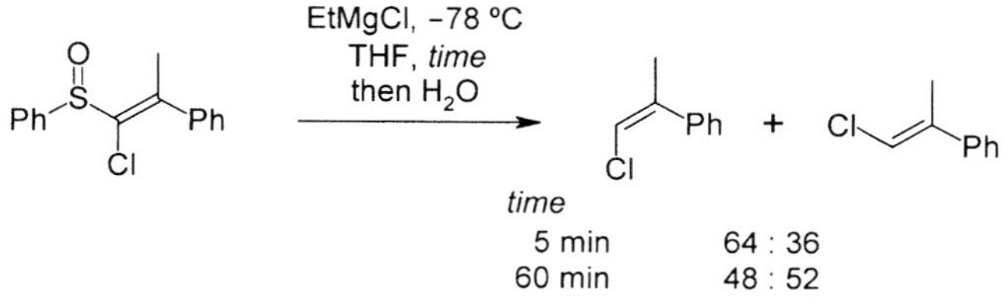

EtMgCl, −78 °C
THF, *time*
then H_2O

time	
5 min	64 : 36
60 min	48 : 52

Scheme 5.79. Conformational lability of 1-chloro-1-magnesioalkenes [569].

5.4.8.4 Organozinc and Organocopper Compounds

The configurational stability of organozinc reagents is greater than that of structurally similar Grignard reagents [540], and configurationally defined organozinc derivatives react with electrophiles at room temperature, usually without epimerization and with retention of configuration [15, 526, 570–572]. An example of the temperature-dependent epimerization of a benzylic organozinc derivative is shown in Scheme 5.80.

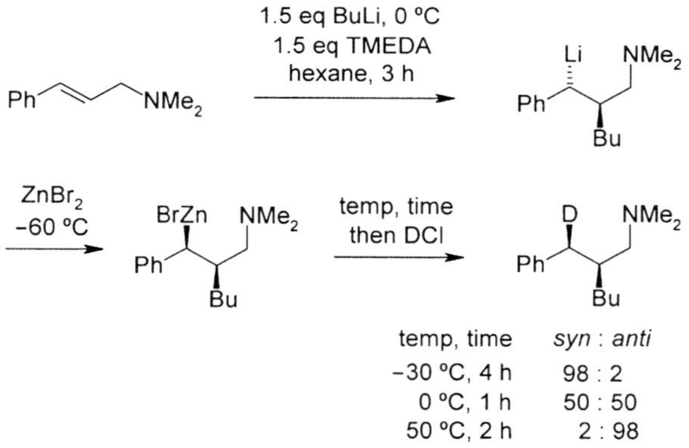

1.5 eq BuLi, 0 °C
1.5 eq TMEDA
hexane, 3 h

$ZnBr_2$
−60 °C

temp, time
then DCl

temp, time	*syn* : *anti*
−30 °C, 4 h	98 : 2
0 °C, 1 h	50 : 50
50 °C, 2 h	2 : 98

Scheme 5.80. Preparation and epimerization of a benzylic organozinc compound [471].

Configurationally defined and stable organozinc reagents have been prepared by reaction of (*sec*-alkyl)dialkylboranes with diisopropylzinc [14, 15, 570, 571] or from secondary Grignard reagents and $ZnCl_2$ [567], giving (*sec*-alkyl)zinc reagents with retention of configuration. These reagents are configurationally stable at room temperature, and undergo transmetalation (Cu(I), Pd(0)) or reaction with electrophiles with retention of configuration (Scheme 5.81). Zinc–copper transmetalation is required when organozinc reagents are too unreactive toward a given carbon electrophile [573]. Other routes to stereochemically defined organocopper reagents are mercury–copper and lithium–copper exchange [542, 551].

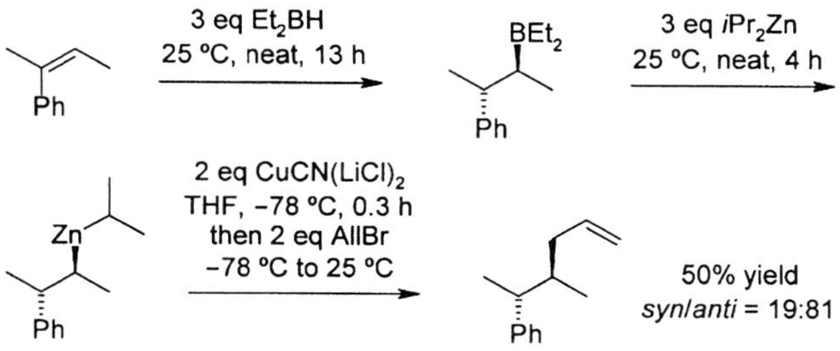

3 eq Et_2BH
25 °C, neat, 13 h
BEt_2
Ph
3 eq iPr_2Zn
25 °C, neat, 4 h
Zn
Ph
2 eq $CuCN(LiCl)_2$
THF, −78 °C, 0.3 h
then 2 eq AllBr
−78 °C to 25 °C
Ph
50% yield
syn/*anti* = 19:81

Scheme 5.81. Stereoselective formation and allylation of organozinc compounds [570, 571].

Direct magnesium–copper transmetalation, however, leads to racemization with most types of copper(I) salts, probably because the large difference between the oxidation potentials of Grignard reagents and copper(I) leads to SET [567]. Similarly, transmetalation of enantiomerically enriched Grignard reagents with iron(III) or cobalt(II) leads to partial racemization [566]. Thus, if enantiomerically enriched Grignard reagents are to be transformed into non-racemic organocopper reagents, a detour via an organozinc intermediate might be required [567].

References

1 Hamell, M.; Levine, R. Condensations effected by the alkali amides. IV. The reactions of esters with lithium amide and certain substituted lithium amides. *J. Org. Chem.* **1950**, *15*, 162–168.
2 Levine, R. Alkali amides as reagents for organic reactions. *Chem. Rev.* **1954**, *54*, 467–573.
3 Leazer, J. L.; Cvetovich, R.; Tsay, F.-R.; Dolling, U.; Vickery, T.; Bachert, D. An improved preparation of 3,5-bis(trifluoromethyl)acetophenone and safety considerations in the preparation of 3,5-bis(trifluoromethyl)phenyl Grignard reagent. *J. Org. Chem.* **2003**, *68*, 3695–3698.
4 Kato, S.; Nonoyama, N.; Tomimoto, K.; Mase, T. Non-cryogenic metalation of aryl bromides bearing proton donating groups: formation of a stable magnesio-intermediate. *Tetrahedron Lett.* **2002**, *43*, 7315–7317.
5 Huo, S. Highly efficient, general procedure for the preparation of alkylzinc reagents from unactivated alkyl bromides and chlorides. *Org. Lett.* **2003**, *5*, 423–425.
6 Knochel, P.; Dohle, W.; Gommermann, N.; Kneisel, F. F.; Kopp, F.; Korn, T.; Sapountzis, I.; Vu, V. A. Highly functionalized organomagnesium reagents prepared through halogen–metal exchange. *Angew. Chem. Int. Ed.* **2003**, *42*, 4302–4320.
7 Keh, C. C. K.; Wei, C.; Li, C.-J. The Barbier–Grignard-type carbonyl alkylation using unactivated alkyl halides in water. *J. Am. Chem. Soc.* **2003**, *125*, 4062–4063.
8 Bailey, W. F.; Brubaker, J. D.; Jordan, K. P. Effect of solvent and temperature on the lithium–iodine exchange of primary alkyl iodides: reaction of *t*BuLi with 1-iodooctane in heptane–ether mixtures. *J. Organomet. Chem.* **2003**, *681*, 210–214.
9 Oestreich, M.; Hoppe, D. Stereospecific preparation of highly enantiomerically enriched organomagnesium reagents. *Chemtracts – Organic Chemistry* **2001**, *14*, 100–105.
10 Prakash, G. K. S.; Hu, J.; Olah, G. A. Preparation of tri- and difluoromethylsilanes via an unusual magnesium metal-mediated reductive tri- and difluoromethylation of chlorosilanes

using tri- and difluoromethyl sulfides, sulfoxides, and sulfones. *J. Org. Chem.* **2003**, *68*, 4457–4463.

11 Ogawa, S.; Furukawa, N. Regiospecific *ortho* lithiation of *o*-halophenyl *p*-tolyl sulfoxides and synthesis of *meta*-substituted optically active aryl alcohols. *J. Org. Chem.* **1991**, *56*, 5723–5726.

12 Furukawa, N.; Ogawa, S.; Matsumura, K.; Fujihara, H. Extremely facile ligand-exchange and disproportionation reactions of diaryl sulfoxides, selenoxides, and triarylphosphine oxides with organolithium and Grignard reagents. *J. Org. Chem.* **1991**, *56*, 6341–6348.

13 Riant, O.; Argouarch, G.; Guillaneux, D.; Samuel, O.; Kagan, H. B. A straightforward asymmetric synthesis of enantiopure 1,2-disubstituted ferrocenes. *J. Org. Chem.* **1998**, *63*, 3511–3514.

14 Hupe, E.; Calaza, M. I.; Knochel, P. Synthesis and reaction of secondary and primary diorganozinc reagents using a B–Zn exchange reaction. A useful method for the stereo- and regioselective formation of new C–C bonds. *J. Organomet. Chem.* **2003**, *680*, 136–142.

15 Hupe, E.; Calaza, M. I.; Knochel, P. Substrate-controlled highly diastereoselective synthesis of primary and secondary diorganozinc reagents by a hydroboration/B–Zn exchange sequence. *Chem. Eur. J.* **2003**, *9*, 2789–2796.

16 Pilcher, A. S.; DeShong, P. Utilization of tetrabutylammonium triphenyldifluorosilicate as a fluoride source for Si–C bond cleavage. *J. Org. Chem.* **1996**, *61*, 6901–6905.

17 Maleczka, R. E.; Geng, F. Synthesis and fluoride-promoted Wittig rearrangements of α-alkoxysilanes. *Org. Lett.* **1999**, *1*, 1111–1113.

18 Kuwajima, I.; Nakamura, E. Reactive enolates from enol silyl ethers. *Acc. Chem. Res.* **1985**, *18*, 181–187.

19 Sun, X.; Collum, D. B. LDA-mediated enolizations: solvent-independent rates, solvent-dependent mechanisms. *J. Am. Chem. Soc.* **2000**, *122*, 2452–2458.

20 Sun, X.; Kenkre, S. L.; Remenar, J. F.; Gilchrist, J. H.; Collum, D. B. Mechanism of LDA-mediated ester deprotonation: the role of disolvated monomers. *J. Am. Chem. Soc.* **1997**, *119*, 4765–4766.

21 Bernstein, M. P.; Collum, D. B. Solvent- and substrate-dependent rates of imine metalations by LDA: understanding the mechanisms underlying 'k_{rel}'. *J. Am. Chem. Soc.* **1993**, *115*, 8008–8018.

22 Pratt, L. M.; Newman, A.; Cyr, J. S.; Johnson, H.; Miles, B.; Lattier, A.; Austin, E.; Henderson, S.; Hershey, B.; Lin, M.; Balamraju, Y.; Sammonds, L.; Cheramie, J.; Karnes, J.; Hymel, E.; Woodford, B.; Carter, C. Ketone enolization with lithium dialkylamides: the effect of structure, solvation, and mixed aggregates with excess BuLi. *J. Org. Chem.* **2003**, *68*, 6387–6391.

23 Zhao, P.; Lucht, B. L.; Kenkre, S. L.; Collum, D. B. $LiN(SiMe_3)_2$-Mediated ketone enolization: the influence of hindered dialkyl ethers and isostructural dialkylamines on reaction rates and mechanisms. *J. Org. Chem.* **2004**, *69*, 242–249.

24 Bordwell, F. G. Equilibrium acidities in DMSO solution. *Acc. Chem. Res.* **1988**, *21*, 456–463.

25 Zhang, X.-M.; Bordwell, F. G. Equilibrium acidities and homolytic bond dissociation energies of the acidic C–H bonds in *P*-substituted triphenylphosphonium cations. *J. Am. Chem. Soc.* **1994**, *116*, 968–972.

26 Bordwell, F. G.; Zhang, S.; Zhang, X.-M.; Liu, W.-Z. Homolytic bond dissociation enthalpies of the acidic H–A bonds caused by proximate substituents in sets of methyl ketones, carboxylic esters, and carboxamides related to changes in ground state energies. *J. Am. Chem. Soc.* **1995**, *117*, 7092–7096.

27 Stanczyk-Dunaj, M.; Galezowski, W.; Jarczewski, A. Potentiometric study of equilibrium acidities of some carbon acids in acetonitrile. *Can. J. Chem.* **2002**, *80*, 1259–1264.

28 Fraser, R. R.; Bresse, M.; Mansour, T. S. pK_a Measurements in THF. *J. Chem. Soc. Chem. Commun.* **1983**, 620–621.

29 Goumont, R.; Magnier, E.; Kizilian, E.; Terrier, F. Acidity inversion of α-NO_2 and α-SO_2CF_3 activated carbon acids as a result of contrasting solvent effects on transfer from water to DMSO solutions. *J. Org. Chem.* **2003**, *68*, 6566–6570.

30 Richard, J. P.; Williams, G.; O'Donoghue, A. C.; Amyes, T. L. Formation and stability of enolates of acetamide and acetate anion: an Eigen plot for proton transfer at γ-carbonyl carbon. *J. Am. Chem. Soc.* **2002**, *124*, 2957–2968.

31 Richard, J. P.; Williams, G.; Gao, J. Experimental and computational determination of

the effects of the cyano group on carbon acidity in water. *J. Am. Chem. Soc.* **1999**, *121*, 715–726.
32 Amyes, T. L.; Richard, J. P. Determination of the pK_a of AcOEt: Brønsted correlation for deprotonation of a simple oxygen ester in aqueous solution. *J. Am. Chem. Soc.* **1996**, *118*, 3129–3141.
33 Lin, A. C.; Chiang, Y.; Dahlberg, D. B.; Kresge, A. J. Base-catalyzed hydrogen exchange of phenylacetylene and chloroform. Brønsted relations and normal acid behavior. *J. Am. Chem. Soc.* **1983**, *105*, 5380–5386.
34 Richard, J. P.; Amyes, T. L.; Toteva, M. M. Formation and stability of carbocations and carbanions in water and intrinsic barriers to their reactions. *Acc. Chem. Res.* **2001**, *34*, 981–988.
35 Aiken, F.; Cox, B. G.; Sørensen, P. E. Proton transfer from carbon. A study of the acid–base-catalyzed relaxation and the bromination of aryl-substituted methanedisulfones. *J. Chem. Soc. Perkin Trans. 2* **1993**, 783–790.
36 Majewski, M.; Nowak, P. Rate of deprotonation of a simple ketone by LDA. *Tetrahedron Lett.* **1998**, *39*, 1661–1664.
37 Lahti, M.; Kankaanperä, A.; Virtanen, P. Kinetic acidity of carbon acids: the hydroxide ion catalyzed ionization of chloroform and acetophenone in aqueous HMPA. *J. Chem. Soc. Perkin Trans. 2* **1993**, 1317–1319.
38 King, J. F.; Rathore, R.; Guo, Z.; Li, M.; Payne, N. C. Experimental evidence for negative hyperconjugation as a component of the polar effect: variation of the ease of α-sulfonyl carbanion formation with the orientation of a β-alkoxy substituent. *J. Am. Chem. Soc.* **2000**, *122*, 10308–10324.
39 Eigen, M. Proton transfer, acid–base catalysis, and enzymatic hydrolysis. Part I. Elementary processes. *Angew. Chem. Int. Ed.* **1964**, *3*, 1–19.
40 Bednar, R. A.; Jencks, W. P. Is HCN a normal acid? Proton transfer from HCN to bases and small inhibition of proton exchange by acid. *J. Am. Chem. Soc.* **1985**, *107*, 7117–7126.
41 Bernasconi, C. F.; Wiersema, D.; Stronach, M. W. Deprotonation of arylnitromethanes. Higher intrinsic rate constants with thiolate ions than with oxyanions or amines as the proton acceptors. Hydrogen bonding in the transition state and desolvation of the base as competing factors in proton transfer at carbon. *J. Org. Chem.* **1993**, *58*, 217–223.
42 Delpuech, J.-J.; Nicole, D. J. Thermodynamic and kinetic acidities in DMSO. Part 3. Alcohols and phenols. *J. Chem. Soc. Perkin Trans. 2* **1977**, 570–576.
43 Bordwell, F. G.; Boyle, W. J.; Hautala, J. A.; Yee, K. C. Brønsted coefficients larger than 1 and less than 0 for proton removal from carbon acids. *J. Am. Chem. Soc.* **1969**, *91*, 4002–4003.
44 Bordwell, F. G.; Matthews, W. S.; Vanier, N. R. Acidities of carbon acids. IV. Kinetic vs equilibrium acidities as measures of carbanion stabilities. The relative effects of phenylthio, diphenylphosphino, and phenyl groups. *J. Am. Chem. Soc.* **1975**, *97*, 442–443.
45 Hine, J. The principle of least nuclear motion. *Adv. Phys. Org. Chem.* **1977**, *15*, 1–61.
46 Stefanidis, D.; Bunting, J. W. Rate–equilibrium relationships for the deprotonation of 4-phenacylpyridines and 4-phenacylpyridinium cations. *J. Am. Chem. Soc.* **1990**, *112*, 3163–3168.
47 Rablen, P. R. Is the acetate anion stabilized by resonance or electrostatics? A systematic structural comparison. *J. Am. Chem. Soc.* **2000**, *122*, 357–368.
48 Wiberg, K. B.; Ochterski, J.; Streitwieser, A. Origin of the acidity of enols and carboxylic acids. *J. Am. Chem. Soc.* **1996**, *118*, 8291–8299.
49 Washabaugh, M. W.; Jencks, W. P. Thiazolium C(2)-proton exchange: general base catalysis, direct proton transfer, and acid inhibition. *J. Am. Chem. Soc.* **1989**, *111*, 674–683.
50 Kresge, A. J. What makes proton transfer fast? *Acc. Chem. Res.* **1975**, *8*, 354–360.
51 Bernasconi, C. F.; Moreira, J. A.; Huang, L. L.; Kittredge, K. W. Intrinsic rate constants for proton transfer from a monoketone to amine bases and electrostatic effects on the intrinsic rate constants for the deprotonation of cationic ketones by OH^-. *J. Am. Chem. Soc.* **1999**, *121*, 1674–1680.
52 Bernasconi, C. F.; Kliner, D. A. V.; Mullin, A. S.; Ni, J. X. Kinetics of ionization of nitromethane and phenylnitromethane by amines and carboxylate ions in DMSO–water mixtures. Evidence of ammonium ion–nitronate ion hydrogen bonded complex formation in DMSO-rich solvent mixtures. *J. Org. Chem.* **1988**, *53*, 3342–3351.
53 Beak, P.; Hunter, J. E.; Jun, Y. M.; Wallin, A. P. Complex-induced proximity effects: remote

lithiation of carboxamides. *J. Am. Chem. Soc.* **1987**, *109*, 5403–5412.

54 Hay, D. R.; Song, Z.; Smith, S. G.; Beak, P. Complex-induced proximity effects and dipole-stabilized carbanions: kinetic evidence for the role of complexes in the α'-lithiations of carboxamides. *J. Am. Chem. Soc.* **1988**, *110*, 8145–8153.

55 Marr, F.; Fröhlich, R.; Wibbeling, B.; Diedrich, C.; Hoppe, D. The synthesis and configurational stability of enantioenriched α-thioallyllithium compounds and the stereochemical course of their electrophilic substitution. *Eur. J. Org. Chem.* **2002**, 2970–2988.

56 Whisler, M. C.; Beak, P. Synthetic applications of lithiated *N*-Boc allylic amines as asymmetric homoenolate equivalents. *J. Org. Chem.* **2003**, *68*, 1207–1215.

57 Peters, J. G.; Seppi, M.; Fröhlich, R.; Wibbeling, B.; Hoppe, D. Stereoselective intermolecular carbolithiation of open-chain and cyclic 1-aryl-1-alkenyl *N,N*-diisopropylcarbamates coupled with electrophilic substitution. Observation of *p*-carboxylation in a benzyllithium derivative. *Synthesis* **2002**, 381–392.

58 Lutz, G. P.; Du, H.; Gallagher, D. J.; Beak, P. Synthetic applications of the β-lithiation of β-aryl secondary amides: diastereoselective and enantioselective substitutions. *J. Org. Chem.* **1996**, *61*, 4542–4554.

59 Beak, P.; Reitz, D. B. Dipole-stabilized carbanions: novel and useful intermediates. *Chem. Rev.* **1978**, *78*, 275–316.

60 Lemal, D. M. Perspective on fluorocarbon chemistry. *J. Org. Chem.* **2004**, *69*, 1–11.

61 Streitwieser, A.; Mares, F. Acidity of hydrocarbons. XXIX. Kinetic acidities of benzal fluoride and 9-fluorofluorene. A pyramidal benzyl anion. *J. Am. Chem. Soc.* **1968**, *90*, 2444–2445.

62 Symons, E. A.; Clermont, M. J. Hydrogen isotope exchange between CF_3H and water. 1. Catalysis by hydroxide ion. *J. Am. Chem. Soc.* **1981**, *103*, 3127–3130.

63 Adolph, H. G.; Kamlet, M. J. Fluoronitroaliphatics. I. The effect of α fluorine on the acidities of substituted nitromethanes. *J. Am. Chem. Soc.* **1966**, *88*, 4761–4763.

64 Lorand, J. P.; Urban, J.; Overs, J.; Ahmed, Q. A. Fluoronitromethane. Synthesis and estimation of acid strength. *J. Org. Chem.* **1969**, *34*, 4176–4178.

65 Rezende, M. C. The acidity of carbon acids in aqueous solutions: correlations with theoretical descriptors. *Tetrahedron* **2001**, *57*, 5923–5930.

66 Bug, T.; Mayr, H. Nucleophilic reactivities of carbanions in water: the unique behavior of the malodinitrile anion. *J. Am. Chem. Soc.* **2003**, *125*, 12980–12986.

67 Hass, H. B.; Bender, M. L. A proposed mechanism of the alkylation of benzyl halides with nitro paraffin salts. *J. Am. Chem. Soc.* **1949**, *71*, 3482–3485.

68 Hass, H. B.; Bender, M. L. The reaction of benzyl halides with the sodium salt of 2-nitropropane. A general synthesis of substituted benzaldehydes. *J. Am. Chem. Soc.* **1949**, *71*, 1767–1769.

69 Weisler, L.; Helmkamp, R. W. The action of some benzyl halides on salts of phenylnitromethane and phenylnitroacetonitrile. *J. Am. Chem. Soc.* **1945**, *67*, 1167–1171.

70 Seebach, D.; Henning, R.; Lehr, F.; Gonnermann, J. Carbon alkylations of α,α- and α,β-doubly deprotonated nitroalkanes. *Tetrahedron Lett.* **1977**, 1161–1164.

71 Eyer, M.; Seebach, D. *l*-2-Nitro-1,3-alkanediols by stereoselective addition of nitroethanol to aldehydes. On the asymmetric electrophilic addition to double bonds. *J. Am. Chem. Soc.* **1985**, *107*, 3601–3606.

72 Enders, D.; Teschner, P.; Raabe, G.; Runsink, J. Asymmetric electrophilic substitutions at the α-position of γ- and δ-lactams. *Eur. J. Org. Chem.* **2001**, 4463–4477.

73 Julia, M.; Maumy, M. Free radical cyclization: ethyl 1-cyano-2-methylcyclohexanecarboxylate. *Org. Synth.* **1988**, *Coll. Vol. VI*, 586–589.

74 Zen, S.; Kaji, E. Dimethyl nitrosuccinate. *Org. Synth.* **1988**, *Coll. Vol. VI*, 503–504.

75 Merino, P.; Castillo, E.; Franco, S.; Merchan, F. L.; Tejero, T. Ready access to enantiopure 5-substituted 3-pyrrolin-2-ones from *N*-benzyl-2,3-*O*-isopropylidene-D-glyceraldehyde nitrone. *Tetrahedron Asymmetry* **1998**, *9*, 1759–1769.

76 Duus, F. Synthesis of β-thioxo ketones by *t*BuLi-promoted Claisen condensation reaction of ketones with thionoesters. *Synthesis* **1985**, 672–674.

77 Fraser, R. R.; Mansour, T. S. Acidity measurements with lithiated amines: steric reduction and electronic enhancement of acidity. *J. Org. Chem.* **1984**, *49*, 3442–3443.

78 Deyine, A.; Dujardin, G.; Mammeri, M.; Poirier, J.-M. A facile isomerization procedure for the access to thermodynamic silyl enol ethers. *Synth. Commun.* **1998**, *28*, 1817–1821.

79 House, H. O.; Czuba, L. J.; Gall, M.; Olmstead, H. D. The chemistry of carbanions. XVIII. Preparation of trimethylsilyl enol ethers. *J. Org. Chem.* **1969**, *34*, 2324–2336.

80 Mahrwald, R.; Gündogan, B. Highly regioselective Lewis acid-mediated aldol additions at the more encumbered α-side of unsymmetrical ketones. *J. Am. Chem. Soc.* **1998**, *120*, 413–414.

81 Gall, M.; House, H. O. The formation and alkylation of specific enolate anions from an unsymmetrical ketone: 2-benzyl-2-methylcyclohexanone and 2-benzyl-6-methylcyclohexanone. *Org. Synth.* **1988**, *Coll. Vol. VI*, 121–130.

82 Bonafoux, D.; Bordeau, M.; Biran, C.; Cazeau, P.; Dunogues, J. Regio- and stereoselective synthesis of silyl enol ethers using a new base electrogenerated from $HN(SiMe_3)_2$. *J. Org. Chem.* **1996**, *61*, 5532–5536.

83 Cazeau, P.; Duboudin, F.; Moulines, F.; Babot, O.; Dunogues, J. A new practical synthesis of silyl enol ethers. Part 1. From simple aldehydes and ketones. *Tetrahedron* **1987**, *43*, 2075–2088.

84 Magnus, P.; Lacour, J.; Coldham, I.; Mugrage, B.; Bauta, W. B. New trialkylsilyl enol ether chemistry: α-*N*-tosylamination of triisopropylsilyl enol ethers. *Tetrahedron* **1995**, *51*, 11087–11110.

85 Guggisberg, Y.; Faigl, F.; Schlosser, M. Optional *ortho* or α-hydroxymethylation of alkylarenes. *J. Organomet. Chem.* **1991**, *415*, 1–6.

86 Yanagisawa, A.; Habaue, S.; Yamamoto, H. Regioselective allylation and propargylation using acylsilanes: facile synthesis of PGE_3 and $F_{3\alpha}$ methyl ester. *Tetrahedron* **1992**, *48*, 1969–1980.

87 Hashimoto, Y.; Takaoki, K.; Sudo, A.; Ogasawara, T.; Saigo, K. Stereoselective addition reaction of organolithium reagents to chiral imines derived from *erythro*-2-amino-1,2-diphenylethanol. *Chem. Lett.* **1995**, 235–236.

88 Zair, T.; Santelli-Rouvier, C.; Santelli, M. Palladium-mediated cyclization of 1,5-hexadien-3-ols to 1-methyl-1,3-cyclopentadienes. *Tetrahedron Lett.* **1991**, *32*, 4501–4502.

89 Cazes, B.; Julia, S. Preparation and utilization of two C_5 conjugated ketene dithioacetals as isoprene synthons. Synthesis of (–)-(*E*)-lanceol. *Tetrahedron Lett.* **1978**, 4065–4068.

90 BouzBouz, S.; Cossy, J. Efficient strategy for the synthesis of stereopentad subunits of scytophycin, rifamycin S, and discodermolide. *Org. Lett.* **2001**, *3*, 3995–3998.

91 Soucy, R. L.; Kozhinov, D.; Behar, V. Remarkable diastereoselectivity in the addition of allylic and unsaturated diorganozinc reagents to β-(*N,N*-dialkylamino)aldehydes. *J. Org. Chem.* **2002**, *67*, 1947–1952.

92 Nakamura, M.; Hara, K.; Hatakeyama, T.; Nakamura, E. Regioselective allylzincation of alkenylboronates. *Org. Lett.* **2001**, *3*, 3137–3140.

93 Denmark, S. E.; Fu, J. Catalytic enantioselective addition of allylic organometallic reagents to aldehydes and ketones. *Chem. Rev.* **2003**, *103*, 2763–2793.

94 Coles, S. J.; Mellor, J. M.; El-Sagheer, A. H.; Salem, E. E.-D. M.; Metwally, R. N. Reaction of benzyl Grignard reagents with trifluoroacetyldihydropyrans and other cyclic β-alkoxy α,β-unsaturated trifluoromethyl ketones. *Tetrahedron* **2000**, *56*, 10057–10066.

95 Nokami, J.; Nomiyama, K.; Matsuda, S.; Imai, N.; Kataoka, K. Highly enantioselective alk-2-enylation of aldehydes through an allyl-transfer reaction. *Angew. Chem. Int. Ed.* **2003**, *42*, 1273–1276.

96 Lipshutz, B. H.; Hackmann, C. Conjugate addition reactions of allylic copper species derived from Grignard reagents: synthetic and spectroscopic aspects. *J. Org. Chem.* **1994**, *59*, 7437–7444.

97 Barbero, A.; Pulido, F. J.; Rincón, J. A.; Cuadrado, P.; Galisteo, D.; Martínez-García, H. Remote stereocontrol in carbonyl additions promoted by vinylstannanes. *Angew. Chem. Int. Ed.* **2001**, *40*, 2101–2103.

98 Yamamoto, Y.; Maruyama, K. Regioreversed addition of crotylmagnesium chloride to carbonyl compounds in the presence of $AlCl_3$. *J. Org. Chem.* **1983**, *48*, 1564–1565.

99 Klement, I.; Lennick, K.; Tucker, C. E.; Knochel, P. Preparation of polyfunctional nitriles by the cyanation of functionalized organozinc halides with TsCN. *Tetrahedron Lett.* **1993**, *34*, 4623–4626.

100 Hopf, H.; Böhm, I.; Kleinschroth, J. Diels–Alder reaction of 1,2,4,5-hexatetraene: tetramethyl[2.2]paracyclophane-4,5,12,13-tetracarboxylate. *Org. Synth.* **1990**, *Coll. Vol. VII*, 485–490.

101 Caddick, S.; Cheung, S.; Doyle, V. E.; Frost, L. M.; Soscia, M. G.; Delisser, V. M.; Williams, M. R. V.; Etheridge, Z. C.; Khan, S.; Hitchcock, P. B.; Pairaudeau, G.; Vile, S. Stereoselective synthesis of polyfunctionalised hydroxylated cyclopentanes from dihydroxylated 2-cyclopentenone derivatives. *Tetrahedron* **2001**, *57*, 6295–6303.

102 Chattopadhyay, A.; Mamdapur, V. R. (*R*)-2,3-*O*-Cyclohexylideneglyceraldehyde, a versatile intermediate for asymmetric synthesis of chiral alcohols. *J. Org. Chem.* **1995**, *60*, 585–587.

103 McDonald, F. E.; Zhu, H. Y. H. Novel strategy for oligosaccharide synthesis featuring reiterative alkynol cycloisomerization. *J. Am. Chem. Soc.* **1998**, *120*, 4246–4247.

104 Irngartinger, H.; Skipinski, M. Synthesis, X-ray structure analysis and topochemical photopolymerization of 2,5-dimethoxyphenyl- and quinone-substituted octa-3,5-diynes. *Tetrahedron* **2000**, *56*, 6781–6794.

105 Kim, E.; Gordon, D. M.; Schmid, W.; Whitesides, G. M. Tin- and indium-mediated allylation in aqueous media: application to unprotected carbohydrates. *J. Org. Chem.* **1993**, *58*, 5500–5507.

106 Yi, X.-H.; Meng, Y.; Hua, X.-G.; Li, C.-J. Regio- and diastereoselective allenylation of aldehydes in aqueous media: total synthesis of (+)-goniofurfurone. *J. Org. Chem.* **1998**, *63*, 7472–7480.

107 Masuyama, Y.; Watabe, A.; Kurusu, Y. Carbonyl allenylations and propargylations by 3-chloro-1-propyne or 2-propynyl mesylates with $SnCl_4$ and Bu_4NI. *Synlett* **2003**, 1713–1715.

108 Lin, M.-J.; Loh, T.-P. Indium-mediated reaction of trialkylsilyl propargyl bromide with aldehydes: highly regioselective synthesis of allenic and homopropargylic alcohols. *J. Am. Chem. Soc.* **2003**, *125*, 13042–13043.

109 Crimmins, M. T.; Tabet, E. A. Formal total synthesis of (+)-trehazolin. Application of an asymmetric aldol–olefin metathesis approach to the synthesis of functionalized cyclopentenes. *J. Org. Chem.* **2001**, *66*, 4012–4018.

110 Kisanga, P. B.; Verkade, J. G. $P(RNCH_2CH_2)_3N$-Catalyzed 1,2-addition reactions of activated allylic synthons. *J. Org. Chem.* **2002**, *67*, 426–430.

111 Aurell, M. J.; Gil, S.; Mestres, R.; Parra, M.; Parra, L. Alkylation of lithium dienediolates of butenoic acids. Regioselectivity effects of structure and leaving group of the alkylating agent. *Tetrahedron* **1998**, *54*, 4357–4366.

112 Lansbury, P. T.; Erwin, R. W.; Jeffrey, D. A. γ-Alkylation of α,β-unsaturated ketones. γ-Arylsulfonyl groups as regioselective control elements. *J. Am. Chem. Soc.* **1980**, *102*, 1602–1608.

113 Abe, T.; Suzuki, T.; Sekiguchi, K.; Hosokawa, S.; Kobayashi, S. Stereoselective construction of a quaternary carbon substituted with multifunctional groups: application to the concise synthesis of (+)-ethosuximide. *Tetrahedron Lett.* **2003**, *44*, 9303–9305.

114 Semple, J. E. A concise synthesis of pellitorine. *Organic Preparations and Procedures International* **1995**, *27*, 582–586.

115 Haynes, R. K.; Starling, S. M.; Vonwiller, S. C. Diastereo- and regioselectivity in the reactions of dilithiated allylic secondary amides with cyclopent-2-enone. *J. Org. Chem.* **1995**, *60*, 4690–4691.

116 Majewski, M.; Irvine, N. M.; Zook, S. E. Reactions of verbenone dienolate with aromatic aldehydes. *Synth. Commun.* **1995**, *25*, 3237–3246.

117 Crimmins, M. T.; Washburn, D. G. Synthesis of the AB spiroketal subunit of spongistatin 1 (altohyrtin A): the pyrone approach. *Tetrahedron Lett.* **1998**, *39*, 7484–7490.

118 Seebach, D.; Mißlitz, U.; Uhlmann, P. Reaktionen von Dienolaten des (*R*)-2-*tert*-Butyl-6-methyl-4*H*-1,3-dioxin-4-ons mit Aldehyden und Ketonen. Ein chirales Acetessigester-d^4-Reagens. *Chem. Ber.* **1991**, *124*, 1845–1852.

119 Crimmins, M. T.; Carroll, C. A.; King, B. W. Synthesis of the C1–C13 fragment of leucascandrolide A. *Org. Lett.* **2000**, *2*, 597–599.

120 Gammill, R. B.; Bryson, T. A. Alkylations of vinylogous amides, vinylogous esters and vinylogous thioesters with potassium hydride. *Synthesis* **1976**, 401–403.

121 Bartoli, G.; Bosco, M.; Cimarelli, C.; Dalpozzo, R.; Palmieri, G. *C*-Alkylation of dianions of β-(isopropylamino)-α,β-enones. *Synlett* **1991**, 229–230.

122 Kuwajima, I.; Urabe, H. Cyclopentanones from carboxylic acids via intramolecular acylation of alkylsilanes: 2-methyl-2-vinylcyclopentanone. *Org. Synth.* **1993**, *Coll. Vol. VIII*, 486–490.

123 Saito, S.; Shiozawa, M.; Yamamoto, H. Mixed crossed aldol condensation between conjugated esters and aldehydes using aluminum

tris(2,6-diphenylphenoxide). *Angew. Chem. Int. Ed.* **1999**, *38*, 1769–1771.

124 Grossman, R. B.; Varner, M. A. Selective monoalkylation of diethyl malonate, ethyl cyanoacetate, and malononitrile using a masking group for the second acidic hydrogen. *J. Org. Chem.* **1997**, *62*, 5235–5237.

125 Saito, S.; Shiozawa, M.; Ito, M.; Yamamoto, H. Conceptually new directed aldol condensation using aluminum tris(2,6-diphenylphenoxide). *J. Am. Chem. Soc.* **1998**, *120*, 813–814.

126 Saito, S.; Shiozawa, M.; Nagahara, T.; Nakadai, M.; Yamamoto, H. Molecular recognition of carbonyl compounds using aluminum tris(2,6-diphenylphenoxide): new regio- and stereoselective alkylation of α,β-unsaturated carbonyl compounds. *J. Am. Chem. Soc.* **2000**, *122*, 7847–7848.

127 Seebach, D.; Aebi, J.; Wasmuth, D. Diastereoselective α-alkylation of β-hydroxycarboxylic esters through alkoxide enolates: diethyl (2*S*,3*R*)-(+)-3-allyl-2-hydroxysuccinate from diethyl (*S*)-(–)-malate. *Org. Synth.* **1990**, *Coll. Vol. VII*, 153–159.

128 Sefkow, M. Enantioselective synthesis of (–)-wikstromol using a new approach via malic acid. *J. Org. Chem.* **2001**, *66*, 2343–2349.

129 Xue, C.-B.; He, X.; Roderick, J.; Corbett, R. L.; Decicco, C. P. Asymmetric synthesis of *trans*-2,3-piperidinedicarboxylic acid and *trans*-3,4-piperidinedicarboxylic acid derivatives. *J. Org. Chem.* **2002**, *67*, 865–870.

130 Wiltshire, H. R.; Prior, K. J.; Dhesi, J.; Maile, G. The synthesis of labelled forms of cipemastat. *J. Labelled Cpd. Radiopharm.* **2001**, *44*, 149–164.

131 Park, J.-I.; Tian, G. R.; Kim, D. H. Synthesis of optically active 2-alkyl-3,4-iminobutanoic acids. β-Amino acids containing an aziridine heterocycle. *J. Org. Chem.* **2001**, *66*, 3696–3703.

132 Léautey, M.; Castelot-Deliencourt, G.; Jubault, P.; Pannecoucke, X.; Quirion, J.-C. Synthesis of α-substituted β-amidophosphines by diastereoselective alkylation. A new access to chiral ligands for asymmetric catalysis. *J. Org. Chem.* **2001**, *66*, 5566–5571.

133 Léautey, M.; Jubault, P.; Pannecoucke, X.; Quirion, J.-C. Synthesis and evaluation of a broad range of new chiral (aminoalkyl)phosphane ligands for asymmetric hydrogen-transfer reduction of prochiral ketones. *Eur. J. Org. Chem.* **2003**, 3761–3768.

134 Lutz, G. P.; Wallin, A. P.; Kerrick, S. T.; Beak, P. Complex induced proximity effects: β-lithiations of carboxamides. *J. Org. Chem.* **1991**, *56*, 4938–4943.

135 Blake, A. J.; Giblin, G. M. P.; Kirk, D. T.; Simpkins, N. S.; Wilson, C. The enantioselective generation of bridgehead enolates. *Chem. Commun.* **2001**, 2668–2669.

136 Paquette, L. A.; Poupart, M.-A. Studies directed toward the total synthesis of cerorubenic acid-III. 1. Expedient construction of the tetracyclic core by oxyanionic sigmatropy. *J. Org. Chem.* **1993**, *58*, 4245–4253.

137 Giblin, G. M. P.; Kirk, D. T.; Mitchell, L.; Simpkins, N. S. Bridgehead enolates: substitution and asymmetric desymmetrization of small bridged carbonyl compounds by lithium amide bases. *Org. Lett.* **2003**, *5*, 1673–1675.

138 Ryu, I.; Ikebe, M.; Sonoda, N.; Yamato, S.-Y.; Yamamura, G.-H.; Komatsu, M. Chemistry of ketone α,β-dianions. Acylation reactions of dianion cuprates by acid chlorides. *Tetrahedron Lett.* **2002**, *43*, 1257–1259.

139 Bates, R. B.; Taylor, S. R. Dialkylation of ketone dianions. *J. Org. Chem.* **1994**, *59*, 245–246.

140 Trimitsis, G. B.; Hinkley, J. M.; TenBrink, R.; Faburada, A. L.; Anderson, R.; Poli, M.; Christian, B.; Gustafson, G.; Erdman, J.; Rop, D. Formation and reactions of 1-phenyl-2-propanone dianion and related systems with electrophilic reagents. *J. Org. Chem.* **1983**, *48*, 2957–2962.

141 Ryu, I.; Nakahira, H.; Ikebe, M.; Sonoda, N.; Yamato, S.-Y.; Komatsu, M. Chelation-aided generation of ketone α,β-dianions and their use as copper ate complexes. Unprecedented enolate intervention in the conjugate addition to enones. *J. Am. Chem. Soc.* **2000**, *122*, 1219–1220.

142 Witt, O.; Mauser, H.; Friedl, T.; Wilhelm, D.; Clark, T. Reactions of the lithium salts of the tribenzylidenemethane dianion, diphenylacetone dianion, and related compounds. *J. Org. Chem.* **1998**, *63*, 959–967.

143 Brun, E. M.; Casades, I.; Gil, S.; Mestres, R.; Parra, M. New conditions for the generation of dianions of carboxylic acids. *Tetrahedron Lett.* **1998**, *39*, 5443–5446.

144 Wang, X.; Thottathil, J. K.; Polniaszek, R. P. Preparation of α-chloro ketones by the chloroacetate Claisen reaction. *Synlett* **2000**, 902–904.

145 Roberts, W. P.; Ebner, C. L. Synthesis of monoalkyl derivatives of 7,7,8,8-tetracyano-*p*-quinodimethane from 2,5-dimethoxybenzoic acid. *J. Org. Chem.* **1987**, *52*, 2297–2299.

146 Pohmakotr, M.; Sampaongoen, L.; Issaree, A.; Tuchinda, P.; Reutrakul, V. Vicinal dianions of diethyl α-aroylsuccinates: a general synthetic route to α-aroyl- and α-arylidene-γ-butyrolactones. *Tetrahedron Lett.* **2003**, *44*, 6717–6720.

147 Negishi, E.-I.; Rand, C. L.; Jadhav, K. P. Highly selective and convenient method for the synthesis of 1,5-enynes and 1,5-dienes by the reaction of 1,3-dilithiopropargyl phenyl sulfide with allylic halides. *J. Org. Chem.* **1981**, *46*, 5041–5044.

148 Jenkinson, J. J.; Parsons, P. J.; Eyley, S. C. A tandem cyclisation for the stereoselective construction of fused tetracyclic systems as precursors of indole alkaloids. *Synlett* **1992**, 679–680.

149 Katritzky, A. R.; Li, J.; Malhotra, N. *N*-Vinyl, *N*-allyl, *N*-propenyl and *N*-propargylbenzotriazoles: reactions of their lithium derivatives. *Liebigs Ann. Chem.* **1992**, 843–853.

150 Stowell, J. C. Three-carbon homologating agents. *Chem. Rev.* **1984**, *84*, 409–435.

151 Coppola, G. M.; Damon, R. E. Acetylenic amides. 2. The generation and reactions of dianions derived from 2-propynamides. *J. Heterocyclic Chem.* **1995**, *32*, 1133–1139.

152 Bury, P.; Hareau, G.; Kocienski, P.; Dhanak, D. Two syntheses of manoalide via heteroatom-assisted alkyne carbometalation. *Tetrahedron* **1994**, *50*, 8793–8808.

153 Hooz, J.; Cabezas, J.; Musmanni, S.; Calzada, J. Propargylation of alkyl halides: (*E*)-6,10-dimethyl-5,9-undecadien-1-yne and (*E*)-7,11-dimethyl-6,10-dodecadien-2-yn-1-ol. *Org. Synth.* **1993**, *Coll. Vol. VIII*, 226–231.

154 Casy, G.; Patterson, J. W.; Taylor, R. J. K. Methyl 7-hydroxyhept-5-ynoate. *Org. Synth.* **1993**, *Coll. Vol. VIII*, 415–420.

155 de Jong, R. L. P.; Brandsma, L. Synthesis of 2,3-disubstituted thiophenes from 1,3-dimetallated acetylenes and non-enolizable thiocarbonyl compounds. *Synth. Commun.* **1990**, *20*, 3427–3431.

156 Butz, R. H.-J.; Lindell, S. D. Synthesis and chemistry of 4,5-dimagnesioimidazole dianions. *J. Org. Chem.* **2002**, *67*, 2699–2701.

157 Doadt, E. G.; Snieckus, V. 3,5-Dilithiated tertiary thiophene 2-carboxamide. Regioselective entries into diversely substituted thiophenes. *Tetrahedron Lett.* **1985**, *26*, 1149–1152.

158 You, Y.; Gibson, S. L.; Hilf, R.; Davies, S. R.; Oseroff, A. R.; Roy, I.; Ohulchanskyy, T. Y.; Bergey, E. J.; Detty, M. R. Water soluble, core-modified porphyrins. 3. Synthesis, photophysical properties, and in vitro studies of photosensitization, uptake, and localization with carboxylic acid-substituted derivatives. *J. Med. Chem.* **2003**, *46*, 3734–3747.

159 Gutiérrez-García, V. M.; Reyes-Rangel, G.; Muñoz-Muñiz, O.; Juaristi, E. Enantioselective synthesis of β-amino acids. Part 13. Diastereoselective alkylation of dianions derived from chiral analogs of β-aminopropanoic acid containing the α-phenylethyl group. *Helv. Chim. Acta* **2002**, *85*, 4189–4199.

160 Seebach, D.; Henning, R.; Mukhopadhyay, T. Doubly deprotonated methyl 3-nitropropanoate, an acrylic ester d^2-reagent. *Chem. Ber.* **1982**, *115*, 1705–1720.

161 O'Neil, I. A.; Lai, J. Y. Q.; Wynn, D. Quinuclidine *N*-oxide: a potential replacement for HMPA. *Chem. Commun.* **1999**, 59–60.

162 Sansbury, F. H.; Warren, S. Stereochemically controlled competitive cyclisation with phenylthio migration in the synthesis of cyclic ethers. *Tetrahedron Lett.* **1992**, *33*, 539–542.

163 Herrmann, J. L.; Schlessinger, R. H. A novel method of preparing α-substituted hydracrylate and acrylate esters. *Tetrahedron Lett.* **1973**, 2429–2432.

164 Jung, J. H.; Lee, J. W.; Oh, D. Y. A facile synthesis of 2,4-disubstituted furans from β-hydroxy sulfones. *Tetrahedron Lett.* **1995**, *36*, 923–926.

165 Gais, H.-J.; von der Weiden, I.; Fleischhauer, J.; Esser, J.; Raabe, G. Lipase catalyzed resolution of α-hydroxymethyl sulfones. Determination of absolute configuration by semiempirical calculation of CD spectra and verification by X-ray structure analysis. *Tetrahedron Asymmetry* **1997**, *8*, 3111–3123.

166 Baston, E.; Maggi, R.; Friedrich, K.; Schlosser, M. Dimetalation: the acidity of monometalated arenes towards superbasic reagents. *Eur. J. Org. Chem.* **2001**, 3985–3989.

167 Cabiddu, M. G.; Cabiddu, S.; Cadoni, E.; Cannas, R.; Fattuoni, C.; Melis, S. Metalation reactions. XXIV. Metalation of (vinylthio)benzene. *Tetrahedron* **1998**, *54*, 14095–14104.

168 Herrera, R. P.; Guijarro, A.; Yus, M. Primary alkyl fluorides as regioselective alkylating

reagents of lithium arene dianions. Easy prediction of regioselectivity by MO calculations on the dianion. *Tetrahedron Lett.* **2003**, *44*, 1313–1316.

169 Langer, P.; Döring, M.; Seyferth, D.; Görls, H. Synthesis of nitrile oligomers through multiple anion capture reactions of allene dianions. *Eur. J. Org. Chem.* **2003**, 1948–1953.

170 Murai, T.; Aso, H.; Tatematsu, Y.; Itoh, Y.; Niwa, H.; Kato, S. Reaction and characterization of thioamide dianions derived from *N*-benzyl thioamides. *J. Org. Chem.* **2003**, *68*, 8514–8519.

171 Johnson, D. C.; Widlanski, T. S. Cerium(III) chloride-mediated reactions of sulfonamide dianions. *J. Org. Chem.* **2003**, *68*, 5300–5309.

172 Postel, D.; Van Nhien, A. N.; Marco, J. L. Chemistry of sulfonate- and sulfonamide-stabilized carbanions. The CSIC reactions. *Eur. J. Org. Chem.* **2003**, 3713–3726.

173 Kornblum, N. Substitution reactions which proceed via radical anion intermediates. *Angew. Chem. Int. Ed.* **1975**, *14*, 734–745.

174 Neeland, E. G.; Sharadendu, A.; Weiler, L. Alkylation of the dianion of 3-oxo-13-tetradecanolide. *Tetrahedron Lett.* **1996**, *37*, 5069–5072.

175 Carlson, R. M.; Oyler, A. R.; Peterson, J. R. Synthesis of substituted 5,6-dihydro-2*H*-pyran-2-ones. Propiolic acid dianion as a reactive three-carbon nucleophile. *J. Org. Chem.* **1975**, *40*, 1610–1616.

176 Dai, W.-M.; Fong, K. C.; Danjo, H.; Nishimoto, S.-I.; Solow, M.; Mak, W. L.; Yeung, M. L. DNA cleavage of novel propargylic sulfones. Enhancement of potency via intercalating interaction. *Bioorg. Med. Chem. Lett.* **1996**, *6*, 1093–1098.

177 Chemla, F.; Bernard, N.; Normant, J.-F. Metalation of 3-trimethylsilyl propargyl chloride: a stereoselective access to *trans*-propargylic oxiranes. *Tetrahedron Lett.* **1999**, *40*, 75–78.

178 Babudri, F.; Fiandanese, V.; Marchese, G.; Punzi, A. Novel synthetic approach to (*S*)-coriolic acid. *Tetrahedron* **2000**, *56*, 327–331.

179 Hilf, C.; Bosold, F.; Harms, K.; Lohrenz, J. C. W.; Marsch, M.; Schimeczek, M.; Boche, G. Carbene structure of stable acyl (formyl) anion equivalents. *Chem. Ber.* **1997**, *130*, 1201–1212.

180 Williams, D. R.; McClymont, E. L. Carbanion methodology for alkylations and acylations in the synthesis of substituted oxazoles. The formation of Cornforth rearrangement products. *Tetrahedron Lett.* **1993**, *34*, 7705–7708.

181 Snieckus, V. Directed *ortho* metalation. Tertiary amide and *O*-carbamate directors in synthetic strategies for polysubstituted aromatics. *Chem. Rev.* **1990**, *90*, 879–933.

182 Weidmann, B.; Widler, L.; Olivero, A. G.; Maycock, C. D.; Seebach, D. Selectivities in the reactions of alkyl-, aryl-, and heterosubstituted organotitanium compounds. *Helv. Chim. Acta* **1981**, *64*, 357–361.

183 Weidmann, B.; Maycock, C. D.; Seebach, D. Alkyl-, aryl-, vinyl-, and heterosubstituted organozirconium compounds. Selective nucleophiles of low basicity. *Helv. Chim. Acta* **1981**, *64*, 1552–1557.

184 Weidmann, B.; Seebach, D. MeTi(O*i*Pr)$_3$, a highly selective nucleophilic methylating reagent. *Helv. Chim. Acta* **1980**, *63*, 2451–2454.

185 Kiefl, C.; Mannschreck, A. 1,8-Disubstituted naphthalenes by directed metalation and subsequent Li–Mn exchange, including Cu catalysis. *Synthesis* **1995**, 1033–1037.

186 Marino, J. P.; Nguyen, H. N. Copper-mediated regioselective allylation and propargylation of 2-(alkylthio)oxazoles. *Tetrahedron Lett.* **2003**, *44*, 7395–7398.

187 Hirashita, T.; Kamei, T.; Horie, T.; Yamamura, H.; Kawai, M.; Araki, S. Preparation of γ-heterosubstituted allylindium and diindium reagents and their reactions with carbonyl compounds. *J. Org. Chem.* **1999**, *64*, 172–177.

188 Yamaguchi, M.; Mukaiyama, T. The stereoselective synthesis of D- and L-ribose. *Chem. Lett.* **1981**, 1005–1008.

189 March, J. *Advanced Organic Chemistry*, John Wiley and Sons: New York, 1992.

190 Antoniotti, P.; Tonachini, G. Mechanism of the anionic Wittig rearrangement. An ab initio theoretical study. *J. Org. Chem.* **1998**, *63*, 9756–9762.

191 Tomooka, K.; Yamamoto, H.; Nakai, T. Recent developments in the [1,2]-Wittig rearrangement. *Liebigs Ann./Recueil* **1997**, 1275–1281.

192 Tomooka, K.; Nakai, T. [1,2]-Wittig rearrangement: stereochemical features and synthetic utility. *J. Synth. Org. Chem. Jpn.* **1996**, *54*, 1000–1007.

193 Maeda, Y.; Sato, Y. Mechanism of the Stevens rearrangement of ammonium ylides. *J. Chem. Soc. Perkin Trans. 1* **1997**, 1491–1493.

194 Heard, G. L.; Yates, B. F. Competing rearrangements of ammonium ylides: a quantum theoretical study. *J. Org. Chem.* **1996**, *61*, 7276–7284.

195 Eisch, J. J.; Kovacs, C. A.; Chobe, P. Carbon-skeletal [1,2] anionic rearrangements and the π-orbital overlap constraint: the question of nucleophilic attack vs electron transfer. *J. Org. Chem.* **1989**, *54*, 1275–1284.

196 Eisch, J. J.; Dua, S. K.; Kovacs, C. A. Carbon-skeletal [1,2] anionic rearrangements of tertiary benzylic amines: geometric and electronic requirements for generating the spiroazacyclopropane intermediate. *J. Org. Chem.* **1987**, *52*, 4437–4444.

197 Naruse, Y.; Kido, K.; Inagaki, S. Ethylation of the indole dianions by alkyl methyl ethers. *J. Org. Chem.* **1995**, *60*, 8334–8335.

198 Fitt, J. J.; Gschwend, H. W. Reaction of *n*-, *sec*-, and *tert*-BuLi with DME: a correction. *J. Org. Chem.* **1984**, *49*, 209–210.

199 Hart, S. A.; Trindle, C. O.; Etzkorn, F. A. Solvent-dependent stereoselectivity in a Still–Wittig rearrangement: an experimental and ab initio study. *Org. Lett.* **2001**, *3*, 1789–1791.

200 Martin, S. F.; DuPriest, M. T. Reactions of allylpyrrolidine carbanions with electrophiles. A new homoenolate equivalent. *Tetrahedron Lett.* **1977**, 3925–3928.

201 Hassel, T.; Seebach, D. A metalated allylurea with sterically protected carbonyl group as novel d^3-reagent. *Angew. Chem. Int. Ed.* **1979**, *18*, 399–400.

202 Zschage, O.; Hoppe, D. 1-(*N,N*-Diisopropylcarbamoyloxy)-1,3-dimethylallyllithium/(–)-sparteine: stereochemistry of the enantioselective carboxylation and methoxycarbonylation. *Tetrahedron* **1992**, *48*, 8389–8392.

203 Zschage, O.; Schwark, J.-R.; Krämer, T.; Hoppe, D. Enantiomerically enriched 1-(*N,N*-diisopropylcarbamoyloxy)-1,3-dimethylallyllithium: stereochemistry of the stannylation, titanation, and the homoaldol reaction. *Tetrahedron* **1992**, *48*, 8377–8388.

204 Kunishima, M.; Hioki, K.; Kono, K.; Kato, A.; Tani, S. SmI_2-Induced 2,3-Wittig rearrangement: regioselective generation of α-allyloxy carbanions via 1,5-hydrogen transfer of vinyl radicals. *J. Org. Chem.* **1997**, *62*, 7542–7543.

205 Murakami, M.; Hayashi, M.; Ito, Y. Generation and alkylation of carbanions α to the nitrogen of amines by a new metalation procedure. *J. Org. Chem.* **1992**, *57*, 793–794.

206 Barluenga, J.; Fañanás, F. J.; Sanz, R.; Marcos, C.; Trabada, M. On the reactivity of *o*-lithioaryl ethers: tandem anion translocation and Wittig rearrangement. *Org. Lett.* **2002**, *4*, 1587–1590.

207 Zhang, L.; Dolbier, W. R. The kinetic acidity of halomethyltrimethylammonium salts. *J. Chem. Soc. Perkin Trans. 2* **2000**, 1431–1434.

208 Kessar, S. V.; Singh, P.; Vohra, R.; Kaur, N. P.; Singh, K. N. Lewis acid complexed heteroatom carbanions; a new concept for α-metalation of tertiary amines. *J. Chem. Soc. Chem. Commun.* **1991**, 568–570.

209 Ebden, M. R.; Simpkins, N. S.; Fox, D. N. A. Activation of benzylic amines towards regioselective metalation by borane complex formation. *Tetrahedron Lett.* **1995**, *36*, 8697–8700.

210 Ariffin, A.; Blake, A. J.; Ebden, M. R.; Li, W.-S.; Simpkins, N. S.; Fox, D. N. A. The diastereoselective and enantioselective substitution reactions of an isoindoline–borane complex. *J. Chem. Soc. Perkin Trans. 1* **1999**, 2439–2447.

211 Blid, J.; Somfai, P. Lewis acid mediated [2,3]-sigmatropic rearrangement of allylic ammonium ylides. *Tetrahedron Lett.* **2003**, *44*, 3159–3162.

212 Seebach, D.; Lohmann, J.-J.; Syfrig, M. A.; Yoshifuji, M. Alkylation of the isoquinoline skeleton in the 1-position. Lithiated 2-pivaloyl and 2-bis(dimethylamino)phosphinoyl-1,2,3,4-tetrahydroisoquinolines. *Tetrahedron* **1983**, *39*, 1963–1974.

213 Ach, D.; Reboul, V.; Metzner, P. Atroposelectivity of reactions of benzylic metalated thiobenzamides and thionaphthamides. *Eur. J. Org. Chem.* **2003**, 3398–3406.

214 Fernández, I.; Forcén-Acebal, A.; García-Granda, S.; López-Ortiz, F. Synthesis of functionalized 1,4-cyclohexadienes through intramolecular anionic dearomatization of *N*-alkyl-*N*-benzyldiphenylphosphinamides. Insight into the reaction mechanism. *J. Org. Chem.* **2003**, *68*, 4472–4485.

215 Beak, P.; Zajdel, W. J. Dipole-stabilized carbanions: the α′ lithiation of piperidines. *J. Am. Chem. Soc.* **1984**, *106*, 1010–1018.

216 Bragg, R. A.; Clayden, J.; Menet, C. J. '*meso*-Selective' functionalization of *N*-benzyl-α-methylbenzylamine derivatives by α-lithiation and alkylation. *Tetrahedron Lett.* **2002**, *43*, 1955–1959.

217 Beak, P.; Brown, R. A. The tertiary amide as an effective director of *ortho* lithiation. *J. Org. Chem.* **1982**, *47*, 34–46.

218 Anctil, E. J.-G.; Snieckus, V. The directed *ortho* metalation–cross coupling symbiosis. Regioselective methodologies for biaryls and heterobiaryls. Deployment in aromatic and hetero-

aromatic natural product synthesis. *J. Organomet. Chem.* **2002**, *653*, 150–160.

219 Chauder, B.; Green, L.; Snieckus, V. The directed *ortho* metalation–transition metal-catalyzed reaction symbiosis in heteroaromatic synthesis. *Pure Appl. Chem.* **1999**, *71*, 1521–1529.

220 Kalinin, A. V.; Bower, J. F.; Riebel, P.; Snieckus, V. The directed *ortho* metalation–Ullmann connection. A new Cu(I)-catalyzed variant for the synthesis of substituted diaryl ethers. *J. Org. Chem.* **1999**, *64*, 2986–2987.

221 Phillion, D. P.; Walker, D. M. Metalation chemistry of *N*-ethyl-*N*-(1-methoxy-2,2-dimethylpropyl)benzamides. A new protective group for secondary amides. *J. Org. Chem.* **1995**, *60*, 8417–8420.

222 Armstrong, D. R.; Boss, S. R.; Clayden, J.; Haigh, R.; Kirmani, B. A.; Linton, D. J.; Schooler, P.; Wheatley, A. E. H. Controlling chemoselectivity in the lithiation of substituted aromatic tertiary amides. *Angew. Chem. Int. Ed.* **2004**, *43*, 2135–2138.

223 Beak, P.; Basu, A.; Gallagher, D. J.; Park, Y. S.; Thayumanavan, S. Regioselective, diastereoselective, and enantioselective lithiation–substitution sequences: reaction pathways and synthetic applications. *Acc. Chem. Res.* **1996**, *29*, 552–560.

224 Derdau, V.; Snieckus, V. Condensation of laterally lithiated *o*-methyl and *o*-ethyl benzamides with imines mediated by (–)-sparteine. Enantioselective synthesis of tetrahydroisoquinolin-1-ones. *J. Org. Chem.* **2001**, *66*, 1992–1998.

225 Clayden, J.; Pink, J. H. Configurational stability and stereospecificity in the reactions of amide-stabilised organolithiums: a non-stereospecific Sn–Li exchange. *Tetrahedron Lett.* **1997**, *38*, 2565–2568.

226 Ruiz Gómez, G.; López Ortiz, F. Nucleophilic naphthalene dearomatization of *N*-alkyl-*N*-benzyl(dinaphthyl)phosphinamides. Application to the synthesis of γ-(*N*-alkylamino)(dihydronaphthalenyl)phosphinic acids. *Synlett* **2002**, 781–783.

227 Fraser, R. R.; Hubert, P. R. Direct formation of the carbonyl anion of diisopropyl formamide. *Can. J. Chem.* **1974**, *52*, 185–187.

228 Fletcher, A. S.; Smith, K.; Swaminathan, K. Generation of diisopropylcarbamoyl-lithium from *N*,*N*-diisopropylformamide and *t*BuLi. Synthesis of α-hydroxy- and α-oxoamides. *J. Chem. Soc. Perkin Trans. 1* **1977**, 1881–1883.

229 Enders, D.; Lotter, H. Synthesis of enantiomerically pure (*R*)- and (*S*)-α-hydroxy ketones and vicinal diols; asymmetric nucleophilic carbamoylation. *Angew. Chem. Int. Ed.* **1981**, *20*, 795–796.

230 Bánhidai, B.; Schöllkopf, U. (Dimethylcarbamoyl)lithium from DMF and LDA; synthesis of α-hydroxy *N*,*N*-dimethylcarboxamides. *Angew. Chem. Int. Ed.* **1973**, *12*, 836–837.

231 Cunico, R. F. Carbamoylsilanes from the in situ metalation–silylation of a formamide. *Tetrahedron Lett.* **2001**, *42*, 1423–1425.

232 Meyers, A. I.; Rieker, W. F.; Fuentes, L. M. Initial complex and the role of solvent in metalations leading to dipole-stabilized anions. *J. Am. Chem. Soc.* **1983**, *105*, 2082–2083.

233 Meyers, A. I.; Edwards, P. D.; Rieker, W. F.; Bailey, T. R. α-Amino carbanions via formamidines. Alkylation of pyrrolidines, piperidines, and related heterocycles. *J. Am. Chem. Soc.* **1984**, *106*, 3270–3276.

234 Meyers, A. I.; Dickman, D. A. Absence of an isotope effect on the metalation of chiral formamidines. The mechanism of asymmetric alkylations leading to chiral amines. *J. Am. Chem. Soc.* **1987**, *109*, 1263–1265.

235 Meyers, A. I.; Milot, G. α-Alkylation and stereochemistry of *cis*- and *trans*-decahydroquinolines mediated by the formamidine and Boc activating groups. Synthesis of pumiliotoxin C. *J. Am. Chem. Soc.* **1993**, *115*, 6652–6660.

236 Elworthy, T. R.; Meyers, A. I. The configurational stability of chiral lithio α-amino carbanions. The effect of Li–O vs Li–N complexation. *Tetrahedron* **1994**, *50*, 6089–6096.

237 Meyers, A. I.; Hellring, S. A novel entry into indole alkaloids. *J. Org. Chem.* **1982**, *47*, 2229–2231.

238 Kopach, M. E.; Meyers, A. I. Metalation–alkylation of *N*-activated pyrrolidines. Rigorous proof of retention for both steps. *J. Org. Chem.* **1996**, *61*, 6764–6765.

239 Kozlowski, M. C.; Xu, Z.; Santos, A. G. Synthesis and conformational analysis of 2,6-dimethyl-1,5-diaza-*cis*-decalins. *Tetrahedron* **2001**, *57*, 4537–4542.

240 Gross, K. M. B.; Jun, Y. M.; Beak, P. Asymmetric deprotonations: lithiation of *N*-(*tert*-butoxycarbonyl)indoline with *sec*-BuLi/(–)-sparteine. *J. Org. Chem.* **1997**, *62*, 7679–7689.

241 Schlecker, R.; Seebach, D. Generation and reactions of tetrasubstituted *N*-lithiomethylsuccinimides. *Helv. Chim. Acta* **1977**, *60*, 1459–1471.
242 Seebach, D.; Enders, D. Umpolung of amine reactivity. Nucleophilic α-(secondary amino)-alkylation via metalated nitrosamines. *Angew. Chem. Int. Ed.* **1975**, *14*, 15–32.
243 Smith, K.; El-Hiti, G. A.; Shukla, A. P. Variation in site of lithiation with ring substituent of *N'*-aryl-*N,N*-dimethylureas: application in synthesis. *J. Chem. Soc. Perkin Trans. 1* **1999**, 2305–2313.
244 Savignac, P.; Leroux, F.; Normant, H. Carbanions-α phosphoramides. II. Formation, stabilité et utilisation en synthèse des carbanions dérivant de *N*-méthyl *N*-benzylphosphoramides. *Tetrahedron* **1975**, *31*, 877–884.
245 Müller, J. F. K.; Spingler, B.; Zehnder, M. Synthesis and reactions of chiral α-lithiophosphoric triamides. *Synlett* **1997**, 1059–1060.
246 Blanchard, S.; Rodriguez, I.; Caubère, P.; Guillaumet, G. Metalation and functionalisation of dihydrodipyridopyrazines. *Synlett* **2002**, 1356–1358.
247 Breternitz, H.-J.; Schaumann, E.; Adiwidjaja, G. Cyclization of deprotonated 2-phenyl-1-tosylaziridine. A surprising example of an intramolecular nucleophilic aromatic addition. *Tetrahedron Lett.* **1991**, *32*, 1299–1302.
248 Silveira, C. C.; Bernardi, C. R.; Braga, A. L.; Kaufman, T. S. Elaboration of 1-benzoyltetrahydroisoquinoline derivatives employing a Pictet–Spengler cyclization with α-chloro α-phenylthio ketones. Synthesis of *O*-methylvelucryptine. *Tetrahedron Lett.* **2001**, *42*, 8947–8950.
249 Davis, F. A.; Liang, C.-H.; Reddy, G. V.; Zhang, Y.; Fang, T.; Titus, D. D. Asymmetric synthesis of 2*H*-azirine 2-carboxylate esters. *J. Org. Chem.* **1999**, *64*, 8929–8935.
250 Kreher, R.; Gerhardt, W. 5*H*-Dibenz[*c,e*]azepin; Synthesen, Eigenschaften, Reaktionen. *Liebigs Ann. Chem.* **1981**, 240–247.
251 Aggarwal, V. K.; Ferrara, M. Highly selective aziridination of imines using trimethylsilyldiazomethane and applications of *C*-silylaziridines in synthesis. *Org. Lett.* **2000**, *2*, 4107–4110.
252 Azzena, U.; Pilo, L.; Sechi, A. Metalation of arylmethyl alkyl ethers. *Tetrahedron* **1998**, *54*, 12389–12398.
253 Mordini, A.; Bindi, S.; Capperucci, A.; Nistri, D.; Reginato, G.; Valacchi, M. Stereoselective access to hydroxy oxetanes and tetrahydrooxepines through isomerization of oxiranyl ethers. *J. Org. Chem.* **2001**, *66*, 3201–3205.
254 Hoppe, D.; Hanko, R.; Brönneke, A.; Lichtenberg, F. Highly alkylated 1-oxyallyl anions from *N,N*-dialkylcarbamic acid allyl esters: γ-hydroxyalkylation (homoaldol reaction). *Angew. Chem. Int. Ed.* **1981**, *20*, 1024–1026.
255 Coll, G.; Costa, A.; Deyá, P. M.; Flexas, F.; Rotger, C.; Saá, J. M. Stereoselective photocyclization of some phenolic, highly congested benzophenones and benzaldehydes. Use of *cis*-2-arylbenzocyclobutenol methyl ethers for the synthesis of lignans. *J. Org. Chem.* **1992**, *57*, 6222–6231.
256 Juaristi, E.; Martínez-Richa, A.; García-Rivera, A.; Cruz-Sánchez, J. S. Use of 4-biphenylylmethanol, 4-biphenylylacetic acid, and 4-biphenylcarboxylic acid/triphenylmethane as indicators in the titration of lithium alkyls. Study of the dianion of 4-biphenylylmethanol. *J. Org. Chem.* **1983**, *48*, 2603–2606.
257 Hodgson, D. M.; Norsikian, S. L. M. First direct deprotonation–electrophile trapping of simple epoxides: synthesis of α,β-epoxysilanes from terminal epoxides. *Org. Lett.* **2001**, *3*, 461–463.
258 Ponzo, V. L.; Kaufman, T. S. Diastereoselective alkoxymethylation of aromatic aldehydes with chiral lithiomethyl ethers. Synthesis of optically active monoprotected glycols. *Can. J. Chem.* **1998**, *76*, 1338–1343.
259 Johnson, C. R.; Medich, J. R. Efficient preparation of [(methoxymethoxy)methyl]tributylstannane, a convenient hydroxymethyl anion equivalent. *J. Org. Chem.* **1988**, *53*, 4131–4133.
260 Kruse, B.; Brückner, R. Does the Wittig–Still rearrangement proceed via a metal free carbanion? *Tetrahedron Lett.* **1990**, *31*, 4425–4428.
261 Still, W. C.; Sreekumar, C. α-Alkoxyorganolithium reagents. A new class of configurationally stable carbanions for organic synthesis. *J. Am. Chem. Soc.* **1980**, *102*, 1201–1202.
262 Pakulski, Z.; Zamojski, A. The synthesis of 6-deoxy-D-galactoheptose. *Polish J. Chem.* **1994**, *68*, 1109–1114.
263 Hoffmann, R.; Rückert, T.; Brückner, R. [1,2]-Wittig rearrangement of a lithioalkyl benzyl ether with inversion of configuration at the carbanion C atom. Diastereoselective reductions of cyclohexyl radicals with Li^+ arene$^-$. *Tetrahedron Lett.* **1993**, *34*, 297–300.
264 Hoffmann, R.; Brückner, R. Ein neuartiger Einstieg in Wittig-Umlagerungen. Eine stereo-

selektive [1,2]-Wittig-Umlagerung mit Konfigurationsumkehr am Carbanion-Zentrum. *Chem. Ber.* **1992**, *125*, 1957–1963.

265 Camuzat-Dedenis, B.; Provot, O.; Moskowitz, H.; Mayrargue, J. Reaction of phosphonium ylides and aromatic nitriles under Lewis acid conditions: and easy access to aryl-substituted α-methoxyacetophenones. *Synthesis* **1999**, 1558–1560.

266 Garbaccio, R. M.; Stachel, S. J.; Baeschlin, D. K.; Danishefsky, S. J. Concise asymmetric synthesis of radicicol and monocillin I. *J. Am. Chem. Soc.* **2001**, *123*, 10903–10908.

267 Wade, P. A.; D'Ambrosio, S. G.; Murray, J. K. Autoxidation of [2-(1,3-dithianyl)]lithium: a cautionary note. *J. Org. Chem.* **1995**, *60*, 4258–4259.

268 Würthwein, E.-U.; Behrens, K.; Hoppe, D. Enantioselective deprotonation of alkyl carbamates by means of (*R,R*)-1,2-bis(*N,N*-dimethylamino)cyclohexane/*sec*-BuLi; theory and experiment. *Chem. Eur. J.* **1999**, *5*, 3459–3463.

269 Hoppe, D.; Brönneke, A. Highly diastereoselective synthesis of di- and trisubstituted 4-butanolides from aldehydes and ketones via three-carbon-extension by allylic homoenolate reagents. *Tetrahedron Lett.* **1983**, *24*, 1687–1690.

270 Hoppe, D.; Zschage, O. Asymmetric homoaldol reaction by enantioselective lithiation of a prochiral 2-butenyl carbamate. *Angew. Chem. Int. Ed.* **1989**, *28*, 69–71.

271 Hoppe, D.; Hense, T. Enantioselective synthesis with lithium/(–)-sparteine carbanion pairs. *Angew. Chem. Int. Ed.* **1997**, *36*, 2282–2316.

272 Laqua, H.; Fröhlich, R.; Wibbeling, B.; Hoppe, D. Synthesis of enantioenriched indene-derived bicyclic alcohols and tricyclic cyclopropanes via (–)-sparteine-mediated lithiation of a racemic precursor and kinetic resolution during the cyclocarbolithiation. *J. Organomet. Chem.* **2001**, *624*, 96–104.

273 Hoppe, D.; Woltering, M. J.; Oestreich, M.; Fröhlich, R. (–)-Sparteine-mediated asymmetric intramolecular carbolithiation of alkenes: synthesis of enantiopure cyclopentanes with three consecutive stereogenic centers. *Helv. Chim. Acta* **1999**, *82*, 1860–1877.

274 Carstens, A.; Hoppe, D. Generation of a configurationally stable, enantioenriched α-oxy-α-methylbenzyllithium: stereodivergence of its electrophilic substitution. *Tetrahedron* **1994**, *50*, 6097–6108.

275 Hammerschmidt, F.; Hanninger, A.; Simov, B. P.; Völlenkle, H.; Werner, A. Configurational stability and stannylation of dipole-stabilized cyclic tertiary benzylic α-oxycarbanions, which occurs with retention or inversion of configuration depending on R and X of R_3SnX used. *Eur. J. Org. Chem.* **1999**, 3511–3518.

276 Hammerschmidt, F.; Hanninger, A. Configurational stability and reactions of α-acyloxy-substituted α-methylbenzyllithium compounds. *Chem. Ber.* **1995**, *128*, 1069–1077.

277 Beak, P.; Carter, L. G. Dipole-stabilized carbanions from esters: α-oxo lithiations of 2,6-substituted benzoates of primary alcohols. *J. Org. Chem.* **1981**, *46*, 2363–2373.

278 Shimano, M.; Meyers, A. I. α-Ethoxyvinyllithium-HMPA. Further studies on its unusual basic properties. *Tetrahedron Lett.* **1997**, *38*, 5415–5418.

279 Hammerschmidt, F.; Schmidt, S. Deprotonation of secondary benzylic phosphates. Configurationally stable benzylic carbanions with a diethoxyphosphoryloxy substituent and their rearrangement to optically active tertiary α-hydroxyphosphonates. *Chem. Ber.* **1996**, *129*, 1503–1508.

280 Hammerschmidt, F.; Hanninger, A. Enantioselective deprotonation of benzyl phosphates by homochiral lithium amide bases. Configurational stability of benzyl carbanions with a dialkoxyphosphoryloxy substituent and their rearrangement to optically active α-hydroxy phosphonates. *Chem. Ber.* **1995**, *128*, 823–830.

281 Avolio, S.; Malan, C.; Marek, I.; Knochel, P. Preparation and reactions of functionalized magnesium carbenoids. *Synlett* **1999**, 1820–1822.

282 Christie, B. D.; Rapoport, H. Synthesis of optically pure pipecolates from L-asparagine. Application to the total synthesis of (+)-apovincamine through amino acid decarbonylation and iminium ion cyclization. *J. Org. Chem.* **1985**, *50*, 1239–1246.

283 Nakamura, S.; Nakagawa, R.; Watanabe, Y.; Toru, T. Highly enantioselective reactions of configurationally labile α-thioorganolithiums using chiral bis(oxazoline)s via two different enantiodetermining steps. *J. Am. Chem. Soc.* **2000**, *122*, 11340–11347.

284 Hayashi, T.; Fujitaka, N.; Oishi, T.; Takeshima, T. Diastereoselection in the reactions of

thioallylic anions with PhCHO. *Tetrahedron Lett.* **1980**, *21*, 303–306.
285 Hoppe, D. Metalated dialkyl *N*-alkyliminocarbonates: new masked α-amino and α-thio carbanions. *Angew. Chem. Int. Ed.* **1975**, *14*, 424–426.
286 Hoppe, D.; Follmann, R. Homologization of aldehydes and ketones to thiiranes or *S*-vinyl thiocarbamates via 2-alkylimino-1,3-oxathiolanes. *Angew. Chem. Int. Ed.* **1977**, *16*, 462–463.
287 Meyer, N.; Seebach, D. Doppelt metalliertes Methanol. Alkohol-d^1- und -d^3-Reagenzien. *Chem. Ber.* **1980**, *113*, 1290–1303.
288 Crandall, J. K.; Clark, A. C. The reaction of organolithium reagents with allylic alcohols. *J. Org. Chem.* **1972**, *37*, 4236–4242.
289 Meyer, N.; Seebach, D. Direkte *ortho*-Metallierung von Benzylalkoholen. Eine neuartige Herstellung von *ortho*-substituierten Benzylalkoholen. *Chem. Ber.* **1980**, *113*, 1304–1319.
290 Uemura, M.; Tokuyama, S.; Sakan, T. Selective nuclear lithiation of aromatic compounds: facile synthesis of methoxyphthalide derivatives by carboxylation of the lithio compounds. *Chem. Lett.* **1975**, 1195–1198.
291 Clayden, J.; Julia, M. Carbenoids from primary alkyl chlorides by heteroatom-assisted metalation. *Synlett* **1995**, 103–104.
292 Capriati, V.; Florio, S.; Luisi, R.; Rocchetti, M. T. Metalation of 2-chloromethyl-2-oxazolines: synthesis of 1,2,3-tris(oxazolinyl)cyclopropanes and derivatives. *J. Org. Chem.* **2002**, *67*, 759–763.
293 Corey, E. J.; Lee, D.-H.; Choi, S. An enantioselective synthesis of (2*S*,3*S*)- and (2*R*,3*S*)-3-hydroxyleucine. *Tetrahedron Lett.* **1992**, *33*, 6735–6738.
294 Corey, E. J.; Choi, S. Highly enantioselective routes to Darzens and acetate aldol products from achiral aldehydes and *t*-butyl bromoacetate. *Tetrahedron Lett.* **1991**, *32*, 2857–2860.
295 Gennari, C.; Vulpetti, A.; Pain, G. Highly enantio- and diastereoselective boron aldol reactions of α-heterosubstituted thioacetates with aldehydes and silyl imines. *Tetrahedron* **1997**, *53*, 5909–5924.
296 Capriati, V.; Degennaro, L.; Florio, S.; Luisi, R.; Tralli, C.; Troisi, L. Lithiation of 2-(1-chloroethyl)-2-oxazolines: synthesis of substituted oxazolinyloxiranes and oxazolinylaziridines. *Synthesis* **2001**, 2299–2306.
297 Meyers, A. I.; Knaus, G.; Kendall, P. M. Synthesis via oxazolines IX. An asymmetric synthesis of 2-methoxy and 2-chloroalkanoic acids. *Tetrahedron Lett.* **1974**, *39*, 3495–3498.
298 Florio, S.; Capriati, V.; Luisi, R. An oxazoline-mediated synthesis of formyl epoxides. *Tetrahedron Lett.* **1996**, *37*, 4781–4784.
299 Brown, H. C.; Zou, M.-F.; Ramachandran, P. V. Efficient diastereoselective synthesis of *anti*-α-bromo β-hydroxy ketones. *Tetrahedron Lett.* **1999**, *40*, 7875–7877.
300 Satoh, T.; Takano, K. A method for generation of α-halo carbanions (carbenoids) from aryl α-haloalkyl sulfoxides with alkylmetals. *Tetrahedron* **1996**, *52*, 2349–2358.
301 Satoh, T.; Oohara, T.; Ueda, Y.; Yamakawa, K. A novel approach to the asymmetric synthesis of epoxides, allylic alcohols, α-amino ketones, and α-amino aldehydes from carbonyl compounds through α,β-epoxy sulfoxides using the optically active *p*-tolylsulfinyl group to induce chirality. *J. Org. Chem.* **1989**, *54*, 3130–3136.
302 Aelterman, W.; De Kimpe, N.; Tyvorskii, V.; Kulinkovich, O. Synthesis of 2,3-disubstituted pyrroles and pyridines from 3-halo-1-azaallylic anions. *J. Org. Chem.* **2001**, *66*, 53–58.
303 De Kimpe, N.; Aelterman, W. An efficient synthesis of (*S*)-(+)-manicone, an alarm pheromone of manica ants. *Tetrahedron* **1996**, *52*, 12815–12820.
304 Julia, M.; Verpeaux, J.-N.; Zahneisen, T. Preparation and alkylation reactions of α-chloroallyl anion with electrophiles. *Synlett* **1990**, 769–770.
305 Macdonald, T. L.; Narayanan, B. A.; O'Dell, D. E. α-Haloallyllithium species. Coupling with alkyl bromides. *J. Org. Chem.* **1981**, *46*, 1504–1506.
306 Hu, S.; Jayaraman, S.; Oehlschlager, A. C. Diastereoselective chloroallylboration of α-chiral aldehydes. *J. Org. Chem.* **1998**, *63*, 8843–8849.
307 Hu, S.; Jayaraman, S.; Oehlschlager, A. C. Diastereo- and enantioselective synthesis of *syn*-α-vinylchlorohydrins and *cis*-vinylepoxides. *J. Org. Chem.* **1996**, *61*, 7513–7520.
308 Hertweck, C.; Boland, W. Asymmetric α-chloroallylboration of amino aldehydes: a novel and highly versatile route to D- and L-*erythro*-sphingoid bases. *J. Org. Chem.* **1999**, *64*, 4426–4430.

309 Chemla, F.; Bernard, N.; Ferreira, F.; Normant, J. F. Preparation of propargylic carbenoids and reactions with carbonyl compounds. A stereoselective synthesis of propargylic halohydrins and oxiranes. *Eur. J. Org. Chem.* **2001**, 3295–3300.

310 Schulze, V.; Nell, P. G.; Burton, A.; Hoffmann, R. W. Simple diastereoselectivity on addition of α-haloalkyl Grignard reagents to PhCHO. *J. Org. Chem.* **2003**, *68*, 4546–4548.

311 Hoffmann, R. W.; Hölzer, B.; Knopff, O.; Harms, K. Asymmetric synthesis of a chiral secondary Grignard reagent. *Angew. Chem. Int. Ed.* **2000**, *39*, 3072–3074.

312 Hoffmann, R. W.; Nell, P. G.; Leo, R.; Harms, K. Highly enantiomerically enriched α-haloalkyl Grignard reagents. *Chem. Eur. J.* **2000**, *6*, 3359–3365.

313 Böhm, V. P. W.; Schulze, V.; Brönstrup, M.; Müller, M.; Hoffmann, R. W. Evidence for an iodine ate complex as an observable intermediate in the I–Mg exchange on a 1,1-diiodoalkane. *Organometallics* **2003**, *22*, 2925–2930.

314 Müller, M.; Brönstrup, M.; Knopff, O.; Schulze, V.; Hoffmann, R. W. Energetics of iodine ate complexes, intermediates in the I–Mg exchange on 1,1-diiodoalkanes. *Organometallics* **2003**, *22*, 2931–2937.

315 Hoffmann, R. W.; Müller, M.; Menzel, K.; Gschwind, R.; Schwerdtfeger, P.; Malkina, O. L.; Malkin, V. G. Reaction of iodoform and isopropyl Grignard reagent revisited. *Organometallics* **2001**, *20*, 5310–5313.

316 Zhang, X.; Xia, H.; Dong, X.; Jin, J.; Meng, W.-D.; Qing, F.-L. 3-Deoxy-3,3-difluoro-D-arabinofuranose: first stereoselective synthesis and application in the preparation of *gem*-difluorinated sugar nucleosides. *J. Org. Chem.* **2003**, *68*, 9026–9033.

317 Schlosser, M.; Wei, H.-X. 2-Ethoxyvinyllithiums and diethoxyvinyllithiums: what makes them stable or fragile? *Tetrahedron* **1997**, *53*, 1735–1742.

318 Martin, S.; Sauvêtre, R.; Normant, J.-F. Reaction de perhalostyrenes avec les organolithiens. Preparation d'aryl-1-alcynes-1 ramifies par l'intermediaire d'aryl fluoro acetylenes. *Tetrahedron Lett.* **1982**, *23*, 4329–4332.

319 Rim, C.; Son, D. Y. Transmetalation as a route to novel styryllithium reagents. *Org. Lett.* **2003**, *5*, 3443–3445.

320 Bertz, S. H. The preparation of hindered cuprates from aldehyde tosylhydrazones. *Tetrahedron Lett.* **1980**, *21*, 3151–3154.

321 Garner, C. M.; Thomas, A. A. Allylic metalation of endo- and exocyclic alkenes: anomalous high reactivity of β-pinene. *J. Org. Chem.* **1995**, *60*, 7051–7054.

322 Revell, J. D.; Ganesan, A. Synthesis of functionalized 1,5-cyclooctadienes by LICKOR metalation. *J. Org. Chem.* **2002**, *67*, 6250–6252.

323 Hooz, J.; Calzada, J. G.; McMaster, D. An efficient coupling reaction of anionic propargyl and organic halides. *Tetrahedron Lett.* **1985**, *26*, 271–274.

324 Johnson, E. P.; Chen, G.-P.; Fales, K. R.; Lenk, B. E.; Szendroi, R. J.; Wang, X.-J.; Carlson, J. A. Macrolactamization via Pd π-allyl alkylation. Preparation of CGS25155: a 10-membered lactam neutral endopeptidase 24.11 inhibitor. *J. Org. Chem.* **1995**, *60*, 6595–6598.

325 Bonnet, B.; Plé, G.; Duhamel, L. Competition between metalation and halogen–lithium exchange in halovinylic acetals. *Synlett* **1996**, 221–224.

326 Lau, K. S. Y.; Schlosser, M. (*Z*)-2-Ethoxyvinyllithium: a remarkably stable and synthetically useful 1,2-counterpolarized species. *J. Org. Chem.* **1978**, *43*, 1595–1598.

327 Duraisamy, M.; Walborsky, H. M. Chiral vinyllithium reagents. Carbenoid reactions. *J. Am. Chem. Soc.* **1984**, *106*, 5035–5037.

328 Normant, J.-F. Synthesis of selectively fluorinated substrates via organometallic reagents derived from $CF_2{=}CFCl$, $CF_2{=}CCl_2$, $CF_2{=}CH_2$. *J. Organomet. Chem.* **1990**, *400*, 19–34.

329 Schmidt, R. R.; Talbiersky, J.; Russegger, P. Generation of functionally substituted vinyllithium compounds. Results and calculations. *Tetrahedron Lett.* **1979**, 4273–4276.

330 Baldwin, J. E.; Höfle, G. A.; Lever, O. W. α-Methoxyvinyllithium and related metalated enol ethers. Practical reagents for nucleophilic acylation. *J. Am. Chem. Soc.* **1974**, *96*, 7125–7127.

331 Alami, M.; Crousse, B.; Linstrumelle, G. A stereoselective route to 1-chloro-1-haloenynes, versatile precursors for the synthesis of chloroenediynes and enetriynes. *Tetrahedron Lett.* **1995**, *36*, 3687–3690.

332 Richardson, S. K.; Jeganathan, A.; Watt, D. S. 1-Vinyl-4-methyl-2,6,7-trioxabicyclo[2.2.2]octanes: unsaturated homoenolate anion equivalents. *Tetrahedron Lett.* **1987**, *28*, 2335–2338.

333 Schmidt, R. R.; Talbiersky, J. Facile synthesis of functionally substituted cyclopentenones and butenolides from functional vinyl carbanions. *Angew. Chem. Int. Ed.* **1978**, *17*, 204–205.

334 Caine, D.; Frobese, A. S. A direct synthesis of α,β-butenolides by reaction of lithium β-lithioacrylates with carbonyl compounds. *Tetrahedron Lett.* **1978**, 5167–5170.

335 Schmidt, R. R.; Hirsenkorn, R. Investigations for synthesizing chlorothricolide. *Tetrahedron Lett.* **1984**, *25*, 4357–4360.

336 Schmidt, R. R.; Betz, R. Synthesis of 3-deoxy-D-manno-2-octulosonic acid (KDO). *Angew. Chem. Int. Ed.* **1984**, *23*, 430–431.

337 Clemo, N. G.; Pattenden, G. Vinylic carbanions in synthesis. Novel synthesis of iso-gregatin B, iso-aspertetronin A and related *O*-methyl tetronic acids. *Tetrahedron Lett.* **1982**, *23*, 585–588.

338 Caine, D.; Ukachukwu, V. C. A new synthesis of 3-butyl-4-bromo-5(*Z*)-(bromomethylidene)-2-(5*H*)furanone, a naturally occurring fimbrolide from *Delisia fimbriata* (Bonnemaisoniaceae). *J. Org. Chem.* **1985**, *50*, 2195–2198.

339 Vu, V. A.; Marek, I.; Knochel, P. Stereoselective preparation of functionalized unsaturated lactones and esters via functionalized magnesium carbenoids. *Synthesis* **2003**, 1797–1802.

340 Sapountzis, I.; Dohle, W.; Knochel, P. Stereoselective preparation of highly functionalized (*Z*)-3-magnesiated enoates by an I–Mg exchange reaction. *Chem. Commun.* **2001**, 2068–2069.

341 Groziak, M. P.; Koohang, A. Facile addition of hydroxylic nucleophiles to the formyl group of uridine-6-carboxaldehydes. *J. Org. Chem.* **1992**, *57*, 940–944.

342 Tanaka, H.; Takashima, H.; Ubasawa, M.; Sekiya, K.; Inouye, N.; Baba, M.; Shigeta, S.; Walker, R. T.; De Clercq, E.; Miyasaka, T. Synthesis and antiviral activity of 6-benzyl analogs of 1-[(2-hydroxyethoxy)methyl]-6-(phenylthio)thymine (HEPT) as potent and selective anti-HIV-1 agents. *J. Med. Chem.* **1995**, *38*, 2860–2865.

343 You, J.; Chen, S.; Chen, Y. A facile method for the synthesis of 6-alkylation products of 1,3-dialkyl-5-fluorouracil. *Synth. Commun.* **2001**, *31*, 1541–1545.

344 Trzupek, L. S.; Newirth, T. L.; Kelly, E. G.; Sbarbati, N. E.; Whitesides, G. M. Mechanism of reaction of CO with PhLi. *J. Am. Chem. Soc.* **1973**, *95*, 8118–8133.

345 Hui, R.; Seyferth, D. Direct nucleophilic acylation by the low-temperature, in situ generation of acyllithium reagents; α-hydroxy ketones from ketones: 3-hydroxy-2,2,3-trimethyloctan-4-one from pinacolone. *Org. Synth.* **1993**, *Coll. Vol. VIII*, 343–346.

346 Seyferth, D.; Weinstein, R. M.; Hui, R. C.; Wang, W.-L.; Archer, C. M. Synthesis of α-hydroxy ketones by direct, low-temperature, in situ nucleophilic acylation of aldehydes and ketones by acyllithium reagents. *J. Org. Chem.* **1992**, *57*, 5620–5629.

347 Li, N.-S.; Yu, S.; Kabalka, G. W. Synthesis of 2-acyl-1,4-diketones via the diacylation of α,β-unsaturated ketones. *Organometallics* **1998**, *17*, 3815–3818.

348 Rautenstrauch, V. Potassium carboxylates by direct carbonylation of potassium alkoxides. *Helv. Chim. Acta* **1987**, *70*, 593–599.

349 Büchi, G.; Cushman, M.; Wüest, H. Conversion of allylic alcohols to homologous amides by DMF acetals. *J. Am. Chem. Soc.* **1974**, *96*, 5563–5565.

350 Chen, J.; Cunico, R. F. α-Aminoamides from a carbamoylsilane and aldehyde imines. *Tetrahedron Lett.* **2003**, *44*, 8025–8027.

351 Viruela-Martin, P.; Viruela-Martin, R.; Tomás, F.; Nudelman, N. S. Theoretical studies of chemical interactions. Ab initio calculations on lithium dialkylamides and their carbonylation reactions. *J. Am. Chem. Soc.* **1994**, *116*, 10110–10116.

352 Smith, K.; El-Hiti, G. A.; Hawes, A. C. Carbonylation of doubly lithiated *N*′-aryl-*N,N*-dimethylureas: a novel approach to isatins via intramolecular trapping of acyllithiums. *Synthesis* **2003**, 2047–2052.

353 Schnyder, A.; Beller, M.; Mehltretter, G.; Nsenda, T.; Studer, M.; Indolese, A. F. Synthesis of primary aromatic amides by aminocarbonylation of aryl halides using formamide as an ammonia synthon. *J. Org. Chem.* **2001**, *66*, 4311–4315.

354 Hosoi, K.; Nozaki, K.; Hiyama, T. Carbon monoxide free aminocarbonylation of aryl and alkenyl iodides using DMF as an amide source. *Org. Lett.* **2002**, *4*, 2849–2851.

355 Sun, L.; Liebeskind, L. S. 2-(Alkylamino)-4-cyclopentene-1,3-diones via addition of imidoyl lithiates to cyclobutenediones. *J. Org. Chem.* **1994**, *59*, 6856–6858.

356 Watanabe, H.; Yan, F.; Sakai, T.; Uneyama, K. (Trifluoroacetimidoyl)lithiums and their reaction with electrophiles. *J. Org. Chem.* **1994**, *59*, 758–761.

357 Ito, Y.; Matsuura, T.; Murakami, M. *N*-Substituted organo(silyliminomethyl)stannanes: synthetic equivalent to organosilyl carbonyl anion and carbonyl dianion. *J. Am. Chem. Soc.* **1987**, *109*, 7888–7890.

358 Murakami, M.; Kawano, T.; Ito, Y. [2-(Benzyloxy)-1-(*N*-2,6-xylylimino)ethyl]samarium as a synthetic equivalent to α-hydroxyacetyl anion. *J. Am. Chem. Soc.* **1990**, *112*, 2437–2439.

359 Murakami, M.; Ito, H.; Ito, Y. Stereoselective synthesis of 2-amino alcohols by use of an isocyanide as an aminomethylene equivalent. *J. Org. Chem.* **1993**, *58*, 6766–6770.

360 Murakami, M.; Masuda, H.; Kawano, T.; Nakamura, H.; Ito, Y. Facile synthesis of vicinal di- and tricarbonyl compounds by SmI_2-mediated double insertion of isocyanides into organic halides. *J. Org. Chem.* **1991**, *56*, 1–2.

361 Niznik, G. E.; Morrison, W. H.; Walborsky, H. M. Metallo aldimines. A masked acyl carbanion. *J. Org. Chem.* **1974**, *39*, 600–604.

362 Voinov, M. A.; Grigor'ev, I. A.; Volodarsky, L. B. Dipole stabilized carbanions in series of cyclic aldonitrones. Aldonitrone metalation and dimerization in LDA and BuLi solutions. *Heterocyclic Comm.* **1998**, *4*, 261–270.

363 Voinov, M. A.; Grigor'ev, I. A. A route to the synthesis of previously unknown α-heteroatom substituted nitrones. *Tetrahedron Lett.* **2002**, *43*, 2445–2447.

364 Cowling, M. P.; Jenkins, P. R.; Lawrence, N. J.; Cooper, K. The base induced dimerisation of α-aryl-*N*-cyclohexylnitrones. *J. Chem. Soc. Chem. Commun.* **1991**, 1581–1582.

365 Gissot, A.; Becht, J.-M.; Desmurs, J. R.; Pévère, V.; Wagner, A.; Mioskowski, C. Directed *ortho* metalation, a new insight into organosodium chemistry. *Angew. Chem. Int. Ed.* **2002**, *41*, 340–343.

366 Stratakis, M. On the mechanism of the *ortho*-directed metalation of anisole by BuLi. *J. Org. Chem.* **1997**, *62*, 3024–3025.

367 Rutherford, J. L.; Hoffmann, D.; Collum, D. B. Consequences of correlated solvation on the structures and reactivities of RLi–diamine complexes: 1,2-addition and α-lithiation reactions of imines by TMEDA-solvated BuLi and PhLi. *J. Am. Chem. Soc.* **2002**, *124*, 264–271.

368 Plé, N.; Turck, A.; Chapoulaud, V.; Quéguiner, G. First functionalization by metalation of the benzene moiety of quinazolines. Diazines XIX. *Tetrahedron* **1997**, *53*, 2871–2890.

369 Gohier, F.; Mortier, J. *ortho*-Metalation of unprotected 3-bromo and 3-chlorobenzoic acids with hindered lithium dialkylamides. *J. Org. Chem.* **2003**, *68*, 2030–2033.

370 Kondo, Y.; Shilai, M.; Uchiyama, M.; Sakamoto, T. TMP–Zincate as highly chemoselective base for directed *ortho* metalation. *J. Am. Chem. Soc.* **1999**, *121*, 3539–3540.

371 Henderson, K. W.; Kerr, W. J. Magnesium bisamides as reagents in synthesis. *Chem. Eur. J.* **2001**, *7*, 3430–3437.

372 Eaton, P. E.; Lee, C.-H.; Xiong, Y. Magnesium amide bases and amido-Grignards. 1. *ortho* Magnesiation. *J. Am. Chem. Soc.* **1989**, *111*, 8016–8018.

373 Schlosser, M.; Maccaroni, P.; Marzi, E. The effect of an alkoxy group on the kinetic and thermodynamic acidity of benzene and toluene. *Tetrahedron* **1998**, *54*, 2763–2770.

374 Fraser, R. R.; Bresse, M.; Mansour, T. S. *ortho* Lithiation of monosubstituted benzenes: a quantitative determination of pK_a values in THF. *J. Am. Chem. Soc.* **1983**, *105*, 7790–7791.

375 Slocum, D. W.; Dumbris, S.; Brown, S.; Jackson, G.; LaMastus, R.; Mullins, E.; Ray, J.; Shelton, P.; Walstrom, A.; Wilcox, J. M.; Holman, R. W. Metalation in hydrocarbon solvents: the mechanistic aspects of substrate-promoted *ortho*-metalations. *Tetrahedron* **2003**, *59*, 8275–8284.

376 Whisler, M. C.; McNeil, S.; Snieckus, V.; Beak, P. Beyond thermodynamic acidity: a perspective on the complex-induced proximity effect in deprotonation reactions. *Angew. Chem. Int. Ed.* **2004**, *43*, 2206–2225.

377 Marzi, E.; Mongin, F.; Spitaleri, A.; Schlosser, M. Fluorophenols and (trifluoromethyl)phenols as substrates of site-selective metalation reactions: to protect or not to protect? *Eur. J. Org. Chem.* **2001**, 2911–2915.

378 Baston, E.; Wang, Q.; Schlosser, M. The rate retarding effect of alkyl groups on arene metalation quantified. *Tetrahedron Lett.* **2000**, *41*, 667–670.

379 Chadwick, S. T.; Rennels, R. A.; Rutherford, J. L.; Collum, D. B. Are BuLi/TMEDA-mediated arene ortholithiations directed? Substituent-dependent rates, substituent-independent

mechanisms. *J. Am. Chem. Soc.* **2000**, *122*, 8640–8647.

380 Price, D.; Simpkins, N. S. Concerning the asymmetric metalation of ferrocenes by chiral lithium amide bases. *Tetrahedron Lett.* **1995**, *36*, 6135–6136.

381 Schmalz, H.-G.; Kiehl, O.; Korell, U.; Lex, J. An enantioselective approach to cytotoxic norcalamenenes via electron-transfer-driven benzylic umpolung of an arene tricarbonyl chromium complex. *Synthesis* **2003**, 1851–1855.

382 Hua, D. H.; Lagneau, N. M.; Chen, Y.; Robben, P. M.; Clapham, G.; Robinson, P. D. Enantioselective synthesis of sulfur-containing 1,2-disubstituted ferrocenes. *J. Org. Chem.* **1996**, *61*, 4508–4509.

383 Slocum, D. W.; Engelmann, T. R.; Ernst, C.; Jennings, C. A.; Jones, W.; Koonsvitsky, B.; Lewis, J.; Shenkin, P. Metalation of metallocenes. *J. Chem. Educ.* **1969**, *46*, 144–150.

384 Carroll, W. A.; Zhang, X. Competitive *ortho* metalation effects: the kinetic and thermodynamic lithiation of 3-(*tert*-butoxycarbonyl)-amino-4-carbomethoxythiophene. *Tetrahedron Lett.* **1997**, *38*, 2637–2640.

385 Alvarez-Ibarra, C.; Quiroga, M. L.; Toledano, E. Synthesis of polysubstituted 3-thiofurans by regiospecific mono-*ipso*-substitution and *ortho*-metalation from 3,4-dibromofuran. *Tetrahedron* **1996**, *52*, 4065–4078.

386 Couture, A.; Grandclaudon, P.; Hoarau, C.; Cornet, J.; Hénichart, J.-P.; Houssin, R. Metalation of 4-oxazolinyloxazole derivatives. A convenient route to 2,4-bifunctionalized oxazoles. *J. Org. Chem.* **2002**, *67*, 3601–3606.

387 Paulson, A. S.; Eskildsen, J.; Vedsø, P.; Begtrup, M. Sequential functionalization of pyrazole 1-oxides via regioselective metalation: synthesis of 3,4,5-trisubstituted 1-hydroxypyrazoles. *J. Org. Chem.* **2002**, *67*, 3904–3907.

388 Vedejs, E.; Monahan, S. D. Metalation of oxazole–borane complexes: a practical solution to the problem of electrocyclic ring opening of 2-lithiooxazoles. *J. Org. Chem.* **1996**, *61*, 5192–5193.

389 Moskalev, N. V.; Gribble, G. W. Synthesis and Diels–Alder reactions of the furo[3,4-*b*]pyrrole ring system. A new indole ring synthesis. *Tetrahedron Lett.* **2002**, *43*, 197–201.

390 Pollet, P.; Turck, A.; Plé, N.; Quéguiner, G. Synthesis of chiral diazine and pyridine sulfoxides. Asymmetric induction by chiral sulfoxides in an "aromatic *ortho* directed metalation–reaction with electrophiles sequence". Diazines. 24. *J. Org. Chem.* **1999**, *64*, 4512–4515.

391 Fruit, C.; Turck, A.; Plé, N.; Mojovic, L.; Quéguiner, G. Synthesis and metalation of pyridazinecarboxamides and thiocarboxamides. Diazenes. Part 32. *Tetrahedron* **2002**, *58*, 2743–2753.

392 Schlosser, M.; Marull, M. The direct metalation and subsequent functionalization of CF_3-substituted pyridines and quinolines. *Eur. J. Org. Chem.* **2003**, 1569–1575.

393 Furukawa, N.; Shibutani, T.; Fujihara, H. Preparation of bipyridyl derivatives via α-lithiation of pyridyl phenyl sulfoxides, substitution with electrophiles and cross-coupling reactions with Grignard reagents. *Tetrahedron Lett.* **1989**, *30*, 7091–7094.

394 Rocca, P.; Marsais, F.; Godard, A.; Quéguiner, G. A short synthesis of the antimicrobial marine sponge pigment fascaplysin. *Tetrahedron Lett.* **1993**, *34*, 7917–7918.

395 Fu, J.-M.; Zhao, B.-P.; Sharp, M. J.; Snieckus, V. Remote aromatic metalation. An anionic Friedel–Crafts equivalent for the regioselective synthesis of condensed fluorenones from biaryl and *m*-teraryl 2-amides. *J. Org. Chem.* **1991**, *56*, 1683–1685.

396 Mongin, O.; Rocca, P.; Thomas-dit-Dumont, L.; Trécourt, F.; Marsais, F.; Godard, A.; Quéguiner, G. Metalation of pyridine *N*-oxides and application to synthesis. *J. Chem. Soc. Perkin Trans. 1* **1995**, 2503–2508.

397 Plé, N.; Turck, A.; Heynderickx, A.; Quéguiner, G. Metalation of diazines. XI. Directed *ortho*-lithiation of fluoropyrimidines and application to synthesis of an azacarboline. *J. Heterocyclic Chem.* **1994**, *31*, 1311–1315.

398 Cochennec, C.; Rocca, P.; Marsais, F.; Godard, A.; Quéguiner, G. Metalation of aryl iodides, part II: directed *ortho*-lithiation of 3-iodo-*N,N*-diisopropyl-2-pyridinecarboxamide: halogen dance and synthesis of an acyclic analogue of meridine. *Synthesis* **1995**, 321–324.

399 Dumouchel, S.; Mongin, F.; Trécourt, F.; Quéguiner, G. Synthesis and reactivity of lithium tri(quinolinyl)magnesates. *Tetrahedron* **2003**, *59*, 8629–8640.

400 Plé, N.; Turck, A.; Couture, K.; Quéguiner, G. Diazines. 13. Metalation without *ortho*-directing group. Functionalization of diazines via direct metalation. *J. Org. Chem.* **1995**, *60*, 3781–3786.

401 Kaminski, T.; Gros, P.; Fort, Y. Side-chain retention during lithiation of 4-picoline and 3,4-lutidine: easy access to molecular diversity in pyridine series. *Eur. J. Org. Chem.* **2003**, 3855–3860.

402 Clarke, A. J.; McNamara, S.; Meth-Cohn, O. Novel aspects of the metalation of heterocycles. Side-chain metalation of thiophene and ring metalation of six-membered nitrogen heterocycles. *Tetrahedron Lett.* **1974**, 2373–2376.

403 Black, W. C.; Guay, B.; Scheuermeyer, F. Metalation of nitroaromatics with in situ electrophiles. *J. Org. Chem.* **1997**, *62*, 758–760.

404 Sapountzis, I.; Knochel, P. A new general preparation of polyfunctional diarylamines by the addition of functionalized arylmagnesium compounds to nitroarenes. *J. Am. Chem. Soc.* **2002**, *124*, 9390–9391.

405 Dohle, W.; Staubitz, A.; Knochel, P. Mild synthesis of polyfunctional benzimidazoles and indoles by the reduction of functionalized nitroarenes with PhMgCl. *Chem. Eur. J.* **2003**, *9*, 5323–5331.

406 Sapountzis, I.; Knochel, P. General preparation of functionalized *o*-nitroarylmagnesium halides through an I–Mg exchange. *Angew. Chem. Int. Ed.* **2002**, *41*, 1610–1611.

407 Ameline, G.; Vaultier, M.; Mortier, J. Directed metalation reactions. Intermolecular competition of the carboxylic acid group and various substituents. *Tetrahedron Lett.* **1996**, *37*, 8175–8176.

408 Grega, K. C.; Barbachyn, M. R.; Brickner, S. J.; Mizsak, S. A. Regioselective metalation of fluoroanilines. An application to the synthesis of fluorinated oxazolidinone antibacterial agents. *J. Org. Chem.* **1995**, *60*, 5255–5261.

409 Slocum, D. W.; Dietzel, P. *ortho*-Metalation of *p*-chloroanisole: a media study. *Tetrahedron Lett.* **1999**, *40*, 1823–1826.

410 Schlosser, M.; Marzi, E.; Cottet, F.; Büker, H. H.; Nibbering, N. M. M. The acidity of chloro-substituted benzenes: a comparison of gas phase, ab initio, and kinetic data. *Chem. Eur. J.* **2001**, *7*, 3511–3516.

411 Iwao, M. Directed lithiation of chlorobenzenes. Regioselectivity and application to a short synthesis of benzocyclobutenes. *J. Org. Chem.* **1990**, *55*, 3622–3627.

412 Lulinski, S.; Serwatowski, J. Regiospecific metalation of oligobromobenzenes. *J. Org. Chem.* **2003**, *68*, 5384–5387.

413 Moyroud, J.; Guesnet, J.-L.; Bennetau, B.; Mortier, J. Regiospecific synthesis of mixed 2,3-dihalobenzoic acids and related acetophenones via *ortho*-metalation reactions. *Tetrahedron Lett.* **1995**, *36*, 881–884.

414 Cabiddu, M. G.; Cabiddu, S.; Cadoni, E.; Corrias, R.; Fattuoni, C.; Floris, C.; Melis, S. Metalation reactions XXII. Regioselective metalation of (trifluoromethyl)(alkylthio)-benzenes. *J. Organomet. Chem.* **1997**, *531*, 125–140.

415 Leroux, F.; Castagnetti, E.; Schlosser, M. Trifluoromethoxy substituted anilines: metalation as the key step for structural elaboration. *J. Org. Chem.* **2003**, *68*, 4693–4699.

416 Pocci, M.; Bertini, V.; Lucchesini, F.; De Munno, A.; Picci, N.; Iemma, F.; Alfei, S. Unexpected behavior of the methoxymethoxy group in the metalation/formylation reactions of 3-methoxymethoxyanisole. *Tetrahedron Lett.* **2001**, *42*, 1351–1354.

417 Sinha, S.; Mandal, B.; Chandrasekaran, S. Selective *para* metalation of unprotected 3-methoxy and 3,5-dimethoxy benzoic acids with BuLi–KO*t*Bu: synthesis of 3,5-dimethoxy-4-methyl benzoic acid. *Tetrahedron Lett.* **2000**, *41*, 3157–3160.

418 Carroll, J. D.; Jones, P. R.; Ball, R. G. Novel ring–chain tautomers derived from (*o*-formylphenyl)ethylamines. *J. Org. Chem.* **1991**, *56*, 4208–4213.

419 Heiss, C.; Marzi, E.; Schlosser, M. Buttressing effects rerouting the deprotonation and functionalization of 1,3-dichloro- and 1,3-dibromobenzene. *Eur. J. Org. Chem.* **2003**, 4625–4629.

420 Murahashi, S.-I.; Naota, T.; Tanigawa, Y. Palladium-phosphine-complex-catalyzed reaction of organometallic compounds and alkenyl halides: (*Z*)-β-[2-(*N*,*N*-dimethylamino)phenyl]-styrene. *Org. Synth.* **1990**, *Coll. Vol. VII*, 172–176.

421 Adejare, A.; Miller, D. D. Synthesis of fluorinated biphenyls via aryne reaction. *Tetrahedron Lett.* **1984**, *25*, 5597–5598.

422 Skowronska-Ptasinska, M.; Verboom, W.; Reinhoudt, D. N. Effect of different dialkylamino groups on the regioselectivity of lithiation of *O*-protected 3-(dialkylamino)phenols. *J. Org. Chem.* **1985**, *50*, 2690–2698.

423 Reed, J. N.; Snieckus, V. *ortho*-Amination of lithiated tertiary benzamides. Short route to polysubstituted anthranilamides. *Tetrahedron Lett.* **1983**, *24*, 3795–3798.

424 Watanabe, M.; Date, M.; Kawanishi, K.; Tsukazaki, M.; Furukawa, S. *ortho*-Lithiation of phenols using the bis(dimethylamino)phosphoryl group as a directing group. *Chem. Pharm. Bull.* **1989**, *37*, 2564–2566.

425 Benkeser, R. A.; Trevillyan, A. E.; Hooz, J. Factors governing orientation in metalation reactions. I. The metalation of ethylbenzene with organosodium and organopotassium compounds. *J. Am. Chem. Soc.* **1962**, *84*, 4971–4975.

426 Faigl, F.; Thurner, A.; Vass, B.; Töke, L. Efficient methods for optional metalation of 1-(methylphenyl)pyrroles in α of benzylic positions. *J. Chem. Res. (S)* **2003**, 132–133.

427 Hargreaves, S. L.; Pilkington, B. L.; Russell, S. E.; Worthington, P. A. The synthesis of substituted pyridylpyrimidine fungicides using Pd-catalyzed cross-coupling reactions. *Tetrahedron Lett.* **2000**, *41*, 1653–1656.

428 Mannekens, E.; Tourwé, D.; Lubell, W. D. Efficient synthesis of 1-benzyloxyphenyl-3-phenylacetones. *Synthesis* **2000**, 1214–1216.

429 Slocum, D. W.; Reed, D.; Jackson, F.; Friesen, C. Media effects in directed *ortho* metalation. *J. Organomet. Chem.* **1996**, *512*, 265–267.

430 Maggi, R.; Schlosser, M. Optional site selectivity in the metalation of *o*- and *p*-anisidine through matching of reagents with neighboring groups. *J. Org. Chem.* **1996**, *61*, 5430–5434.

431 Jones, F. N.; Vaulx, R. L.; Hauser, C. R. *o*-Metalation of benzyldimethylamine and related amines with BuLi. Condensations with carbonyl compounds to form *ortho* derivatives. *J. Org. Chem.* **1963**, *28*, 3461–3465.

432 Puterbaugh, W. H.; Hauser, C. R. Synthesis of functional α-derivatives of benzyldimethylamine through metalation with PhNa and condensations with carbonyl compounds. *J. Org. Chem.* **1963**, *28*, 3465–3467.

433 Ahlbrecht, H.; Harbach, J.; Hauck, T.; Kalinowski, H.-O. Struktur von α-(Dimethylamino)benzyllithium in Lösung: Dynamisches Gleichgewicht zwischen einer η^1- und einer η^3-Spezies. *Chem. Ber.* **1992**, *125*, 1753–1762.

434 Sasai, H.; Suzuki, T.; Arai, S.; Arai, T.; Shibasaki, M. Basic character of rare earth metal alkoxides. Utilization in catalytic C–C bond-forming reactions and catalytic asymmetric nitroaldol reactions. *J. Am. Chem. Soc.* **1992**, *114*, 4418–4420.

435 Cunico, R. F. On the metalation–silylation of 1,3,5-trioxane. *Synth. Commun.* **2000**, *30*, 433–436.

436 Hine, J.; Porter, J. J. The formation of difluoromethylene from difluoromethyl phenyl sulfone and NaOMe. *J. Am. Chem. Soc.* **1960**, *82*, 6178–6181.

437 Hoffmann, R. W.; Knopff, O.; Kusche, A. Formation of rearranged Grignard reagents by carbenoid-C–H insertion. *Angew. Chem. Int. Ed.* **2000**, *39*, 1462–1464.

438 Braun, M.; Mahler, H. Non-chelate-controlled addition of 1-bromo-1-lithio-1-alkenes to *O*-protected lactaldehydes and 3-alkoxybutyraldehydes. *Liebigs Ann. Chem.* **1995**, 29–40.

439 Pelter, A.; Kvicala, J.; Parry, D. E. Preparation and direct observation by ^{19}F NMR spectroscopy of 1-fluoro-1-lithioalkenes. *J. Chem. Soc. Perkin Trans. 1* **1995**, 2681–2682.

440 Köbrich, G.; Flory, K. Chlorsubstituierte Vinyllithium-Verbindungen. *Chem. Ber.* **1966**, *99*, 1773–1781.

441 Funabiki, K.; Ohtsuki, T.; Ishihara, T.; Yamanaka, H. Facile generation of polyfluoro-1-(tosyloxy)prop-1-enyllithiums and their reaction with electrophiles. A new, efficient and convenient access to (*Z*)-1,1-di- and 1,1,1-trifluoro-3-(tosyloxy)alk-3-en-2-ones. *J. Chem. Soc. Perkin Trans. 1* **1998**, 2413–2423.

442 Barluenga, J.; Rodríguez, M. A.; Campos, P. J.; Asensio, G. Synthesis of 2-functionalized 1,1-diiodo-1-alkenes. Generation and reactions of 1-iodo-1-lithio-1-alkenes and 1,1-dilithio-1-alkenes. *J. Am. Chem. Soc.* **1988**, *110*, 5567–5568.

443 Katsuhira, T.; Harada, T.; Oku, A. New method for generation of alkenylidenecarbenes from propargylic methanesulfonates and its use in regioselective C–H insertion reactions. *J. Org. Chem.* **1994**, *59*, 4010–4014.

444 Marshall, D. R.; Thomas, P. J.; Stirling, C. J. M. Leaving group ability in base-promoted alkene-forming 1,2-eliminations. *J. Chem. Soc. Chem. Commun.* **1975**, 940–941.

445 He, Y.; Junk, C. P.; Cawley, J. J.; Lemal, D. M. A remarkable [2.2.2]propellane. *J. Am. Chem. Soc.* **2003**, *125*, 5590–5591.

446 Brocchini, S. J.; Eberle, M.; Lawton, R. G. Molecular yardsticks. Synthesis of extended equilibrium transfer alkylating cross-link reagents and their use in the formation of macrocycles. *J. Am. Chem. Soc.* **1988**, *110*, 5211–5212.

447 Kawashima, T.; Nakamura, M.; Inamoto, N. Tandem Peterson–Michael reaction using α-silylalkylphosphine chalcogenides and Horner–Emmons reaction of in situ generated α-carbanions of its products. *Heterocycles* **1997**, *44*, 487–507.

448 Pietrusiewicz, K. M.; Zablocka, M. Amine-directed lithiation in aliphatic organophosphorus systems. An approach to α-monoalkylation of α,β-unsaturated phosphine oxides. *Tetrahedron Lett.* **1989**, *30*, 477–480.

449 Maignan, C.; Guessous, A.; Rouessac, F. Acces aux (*R*)*S* acyl-1 vinyl *p*-tolylsulfoxydes a partir du (*R*)*S* dimethylamino-2 *p*-tolylsulfinyl-1 ethane. Leurs reactivites en tant que dienophile. *Tetrahedron Lett.* **1986**, *27*, 2693–2606.

450 Alonso, D. A.; Costa, A.; Mancheño, B.; Nájera, C. Lithiated β-aminoalkyl sulfones as mono and dinucleophiles in the preparation of nitrogen heterocycles: application to the synthesis of capsazepine. *Tetrahedron* **1997**, *53*, 4791–4814.

451 Inomata, K.; Tanaka, Y.; Sasaoka, S.-I.; Kinoshita, H.; Kotake, H. A convenient method for the preparation of α-alkylated vinylic sulfones and their conversion to allylic sulfones. *Chem. Lett.* **1986**, 341–344.

452 Gutiérrez-García, V. M.; López-Ruiz, H.; Reyes-Rangel, G.; Juaristi, E. Enantioselective synthesis of β-amino acids. Part 11. Diastereoselective alkylation of chiral derivatives of β-aminopropionic acid containing the β-phenethyl group. *Tetrahedron* **2001**, *57*, 6487–6496.

453 Nagula, G.; Huber, V. J.; Lum, C.; Goodman, B. A. Synthesis of α-substituted β-amino acids using pseudoephedrine as a chiral auxiliary. *Org. Lett.* **2000**, *2*, 3527–3529.

454 Guarnieri, W.; Sendzik, M.; Fröhlich, R.; Hoppe, D. Regio- and stereoselective lithiation and *C*-substitution of (*S*)-2-(dibenzylamino)butane-1,4-diol via dicarbamates. *Synthesis* **1998**, 1274–1286.

455 Sendzik, M.; Guarnieri, W.; Hoppe, D. Monocarbamates derived from (*S*)-2-(dibenzylamino)butane-1,4-diol and the influence of the second *O*-protecting group on the regioselectivity of deprotonation. Application to the synthesis of the Boletus toxin (2*S*,4*S*)-γ-hydroxynorvaline. *Synthesis* **1998**, 1287–1297.

456 Schwerdtfeger, J.; Kolczewski, S.; Weber, B.; Fröhlich, R.; Hoppe, D. Stereoselective deprotonation of chiral and achiral 2-aminoalkyl carbamates: synthesis of optically active β-amino alcohols via 1-oxy-substituted alkyllithium intermediates. *Synthesis* **1999**, 1573–1592.

457 Ponsinet, R.; Chassaing, G.; Vaissermann, J.; Lavielle, S. Diastereoselective synthesis of β^2-amino acids. *Eur. J. Org. Chem.* **2000**, 83–90.

458 Feeder, N.; Fox, D. J.; Medlock, J. A.; Warren, S. Synthesis, X-ray crystal structures and Horner–Wittig addition reactions of some protected β-aminophosphine oxides. *J. Chem. Soc. Perkin Trans. 1* **2002**, 1175–1180.

459 Seki, M.; Furutani, T.; Miyake, T.; Yamanaka, T.; Ohmizu, H. A novel synthesis of a key intermediate for penems and carbapenems utilizing lipase-catalyzed kinetic resolution. *Tetrahedron Asymmetry* **1996**, *7*, 1241–1244.

460 Shinozuka, T.; Kikori, Y.; Asaoka, M.; Takei, H. Stereoselective reaction of arylsulfanyl-stabilized homoenolates with aldehydes. *J. Chem. Soc. Perkin Trans. 1* **1996**, 119–120.

461 Holla, E. W.; Napierski, B.; Rebenstock, H.-P. Short and efficient synthesis of (2*S*)-3-(1′,1′-dimethylethylsulfonyl)-2-(1-naphthylmethyl)-propionic acid, *N*-terminal building block for protease inhibitors. *Synlett* **1994**, 333–334.

462 Eames, J.; Fox, D. J.; de las Heras, M. A.; Warren, S. Stereochemically controlled synthesis of 1,8-dioxaspiro[4.5]decanes and 1-oxa-8-thiaspiro[4.5]decanes by phenylsulfanyl migration. *J. Chem. Soc. Perkin Trans. 1* **2000**, 1903–1914.

463 Saito, S.; Hatanaka, K.; Yamamoto, H. Asymmetric Mannich-type reactions of aldimines with a chiral acetate. *Org. Lett.* **2000**, *2*, 1891–1894.

464 Lu, H.; Burton, D. J. A general method for the preparation of 1,1-bis(trifluoromethyl)-substituted olefins. *Tetrahedron Lett.* **1995**, *36*, 3973–3976.

465 Tellier, F.; Sauvêtre, R. Straightforward synthesis of α,β-unsaturated acids and derivatives. *Tetrahedron Lett.* **1993**, *34*, 5433–5436.

466 Sauvêtre, R.; Normant, J. F. Réaction de CF_2=CH_2 avec les organolithiens. Une nouvelle préparation du difluorovinyllithium. *Tetrahedron Lett.* **1981**, *22*, 957–958.

467 Sibi, M. P.; Shankaran, K.; Alo, B. I.; Hahn, W. R.; Snieckus, V. Overriding normal Friedel–Crafts regiochemistry in cycliacylation. Regiospecific carbodesilylation and Parham cyclization routes to 7-methoxy-1-indanols. *Tetrahedron Lett.* **1987**, *28*, 2933–2936.

468 Parham, W. E.; Bradsher, C. K.; Reames, D. C. Selective halogen–lithium exchange in some secondary and tertiary (bromophenyl)alkyl halides. *J. Org. Chem.* **1981**, *46*, 4804–4806.

469 Dhawan, K. L.; Gowland, B. D.; Durst, T. Preparation of benzocyclobutenols from *o*-halostyrene oxides. *J. Org. Chem.* **1980**, *45*, 922–924.

470 Inoue, A.; Kitagawa, K.; Shinokubo, H.; Oshima, K. Selective halogen–magnesium exchange reaction via organomagnesium ate complexes. *J. Org. Chem.* **2001**, *66*, 4333–4339.

471 Norsikian, S.; Marek, I.; Klein, S.; Poisson, J. F.; Normant, J. F. Enantioselective carbometalation of cinnamyl derivatives: new access to chiral disubstituted cyclopropanes. Configurational stability of benzylic organozinc halides. *Chem. Eur. J.* **1999**, *5*, 2055–2068.

472 Saward, C. J.; Vollhardt, K. P. C. 1,2-Cyclopropa-4,5-cyclobutabenzene. A novel strained benzene derivative. *Tetrahedron Lett.* **1975**, 4539–4542.

473 Lear, Y.; Durst, T. Synthesis of regiospecifically substituted 2-hydroxybenzocyclobutenones. *Can. J. Chem.* **1997**, *75*, 817–824.

474 Parham, W. E.; Jones, L. D.; Sayed, Y. A. Selective halogen–lithium exchange in bromophenylalkyl halides. *J. Org. Chem.* **1976**, *41*, 1184–1186.

475 Molander, G. A.; Köllner, C. Development of a protocol for eight- and nine-membered ring synthesis in the annulation of sp^2,sp^3-hybridized organic dihalides with keto esters. *J. Org. Chem.* **2000**, *65*, 8333–8339.

476 Meyers, A. I.; Licini, G. Intramolecular asymmetric tandem additions to chiral naphthyl oxazolines. *Tetrahedron Lett.* **1989**, *30*, 4049–4052.

477 Fleming, F. F.; Zhang, Z.; Wang, Q.; Steward, O. W. Cyclic alkenenitriles: synthesis, conjugate addition, and stereoselective annulation. *J. Org. Chem.* **2003**, *68*, 7646–7650.

478 Bernady, K. F.; Poletto, J. F.; Nocera, J.; Mirando, P.; Schaub, R. E.; Weiss, M. J. Prostaglandins and congeners. 28. Synthesis of 2-(ω-carbalkoxyalkyl)cyclopent-2-en-1-ones, intermediates for prostaglandin syntheses. *J. Org. Chem.* **1980**, *45*, 4702–4715.

479 Robertson, J.; Burrows, J. N.; Stupple, P. A. Bicyclo[10.2.1]pentadecenone derivatives by free radical macrocyclization. *Tetrahedron* **1997**, *53*, 14807–14820.

480 Olomucki, M.; Le Gall, J. Y. Alkoxycarbonylation of propargyl chloride: methyl 4-chloro-2-butynoate. *Org. Synth.* **1993**, *Coll. Vol. VIII*, 371–373.

481 Delacroix, T.; Bérillon, L.; Cahiez, G.; Knochel, P. Preparation of functionalized arylmagnesium reagents bearing an *o*-chloromethyl group. *J. Org. Chem.* **2000**, *65*, 8108–8110.

482 Pearce, H. L.; Bach, N. J.; Cramer, T. L. Synthesis of 2-azapodophyllotoxin. *Tetrahedron Lett.* **1989**, *30*, 907–910.

483 White, D. A.; Parshall, G. W. Reductive cleavage of aryl–Pd bonds. *Inorg. Chem.* **1970**, *9*, 2358–2361.

484 Henniges, H.; Militzer, H.-C.; de Meijere, A. A new versatile synthesis of *tert*-butyl 2-alkoxycyclopropanecarboxylates. *Synlett* **1992**, 735–737.

485 McGarvey, G. J.; Andersen, M. W. The diastereoselective influences of remote substituents on enolate alkylations. *Tetrahedron Lett.* **1990**, *31*, 4569–4572.

486 Comins, D. L.; LaMunyon, D. H.; Chen, X. Enantiopure *N*-acyldihydropyridones as synthetic intermediates: asymmetric synthesis of indolizine alkaloids (–)-205A, (–)-207A, and (–)-235B. *J. Org. Chem.* **1997**, *62*, 8182–8187.

487 Padwa, A.; Muller, C. L.; Rodriguez, A.; Watterson, S. H. Alkylation reactions of 3-(phenylsulfonyl)methyl substituted cyclopentanones. *Tetrahedron* **1998**, *54*, 9651–9666.

488 Easton, C. J.; Roselt, P. D.; Tiekink, E. R. T. Synthesis of side-chain functionalized amino acid derivatives through reaction of alkyl nitronates with α-bromoglycine derivatives. *Tetrahedron* **1995**, *51*, 7809–7822.

489 Kneisel, F. F.; Knochel, P. Synthesis and reactivity of aryl- and heteroaryl-magnesium reagents bearing keto groups. *Synlett* **2002**, 1799–1802.

490 Yang, X.; Rotter, T.; Piazza, C.; Knochel, P. Successive I–Mg or –Cu exchange reactions for the selective functionalization of polyhalogenated aromatics. *Org. Lett.* **2003**, *5*, 1229–1231.

491 Matsumoto, M.; Watanabe, N.; Ishikawa, A.; Murakami, H. Base-induced cyclization of 1-benzyloxy-2,2,4,4-tetramethylpentan-3-ones: intramolecular nucleophilic addition of an anion of a benzyl ether to the carbonyl moiety without Wittig rearrangement or protophilic decomposition. *Chem. Commun.* **1997**, 2395–2396.

492 Eisch, J. J.; Galle, J. E.; Piotrowski, A.; Tsai, M.-R. Rearrangement and cleavage of [(aryloxy)methyl]silanes by organolithium reagents: conversion of phenols into benzylic alcohols. *J. Org. Chem.* **1982**, *47*, 5051–5056.

493 Kiyooka, S.-I.; Tsutsui, T.; Kira, T. Complete asymmetric induction in the [1,2]-Wittig rearrangement of a system involving a binaphthol moiety. *Tetrahedron Lett.* **1996**, *37*, 8903–8704.

494 Tomooka, K.; Inoue, T.; Nakai, T. Stereochemistry and mechanism of vinyl-migrating [1,2]-Wittig rearrangement of α-lithioalkyl vinyl ethers. *Chem. Lett.* **2000**, 418–419.

495 Hara, O.; Ito, M.; Hamada, Y. Novel N–C acyl migration reaction of acyclic imides: a facile method for α-amino ketones and β-amino alcohols. *Tetrahedron Lett.* **1998**, *39*, 5537–5540.

496 Wipf, P.; Methot, J.-L. Synthetic studies toward diazonamide A. A novel approach for polyoxazole synthesis. *Org. Lett.* **2001**, *3*, 1261–1264.

497 Horne, S.; Rodrigo, R. A short efficient route to acronycine and other acridones. *J. Chem. Soc. Chem. Commun.* **1991**, 1046–1048.

498 Kise, N.; Ozaki, H.; Terui, H.; Ohya, K.; Ueda, N. A convenient synthesis of *N*-Boc-protected *tert*-butyl esters of phenylglycines from benzylamines. *Tetrahedron Lett.* **2001**, *42*, 7637–7639.

499 Focken, T.; Hopf, H.; Snieckus, V.; Dix, I.; Jones, P. G. Stereoselective lateral functionalization of monosubstituted [2.2]paracyclophanes by directed *ortho* metalation–homologous anionic Fries rearrangement. *Eur. J. Org. Chem.* **2001**, 2221–2228.

500 Zhang, P.; Gawley, R. E. Directed metalation/Snieckus rearrangement of *O*-benzylic carbamates. *J. Org. Chem.* **1993**, *58*, 3223–3224.

501 Kells, K. W.; Ncube, A.; Chong, J. M. An unusual 1,2-*N* to *C* acyl migration in urea derivatives of α-aminoorganolithiums. *Tetrahedron* **2004**, *60*, 2247–2257.

502 Sieburth, S. M.; O'Hare, H. K.; Xu, J.; Chen, Y.; Liu, G. Asymmetric synthesis of α-amino allyl, benzyl, and propargyl silanes by metalation and rearrangement. *Org. Lett.* **2003**, *5*, 1859–1861.

503 Sakaguchi, K.; Fujita, M.; Suzuki, H.; Higashino, M.; Ohfune, Y. Reverse Brook rearrangement of 2-alkynyl trialkylsilyl ether. Synthesis of optically active (1-hydroxy-2-alkynyl)-trialkylsilanes. *Tetrahedron Lett.* **2000**, *41*, 6589–6592.

504 Danheiser, R. L.; Fink, D. M.; Okano, K.; Tsai, Y.-M.; Szczepanski, S. W. A practical and efficient synthesis of α,β-unsaturated acylsilanes. *J. Org. Chem.* **1985**, *50*, 5393–5396.

505 Bordwell, F. G.; Lynch, T.-Y. Radical stabilization energies and synergistic (captodative) effects. *J. Am. Chem. Soc.* **1989**, *111*, 7558–7562.

506 Tanaka, H.; Yoshida, S. Kinetic study of the radical homopolymerization of captodative substituted methyl α-(acyloxy)acrylates. *Macromolecules* **1995**, *28*, 8117–8121.

507 Hoffmann, R. W.; Hölzer, B. Concerted and stepwise Grignard additions, probed with a chiral Grignard reagent. *Chem. Commun.* **2001**, 491–492.

508 Kai, Y.; Knochel, P.; Kwiatkowski, S.; Dunitz, J. D.; Oth, J. F. M.; Seebach, D.; Kalinowski, H.-O. Structure, synthesis, and properties of some persubstituted 1,2-dinitroethanes. In quest of nitrocyclopropyl-anion derivatives. *Helv. Chim. Acta* **1982**, *65*, 137–161.

509 Asaro, M. F.; Nakayama, I.; Wilson, R. B. Formation of sterically hindered primary vicinal diamines from vicinal and geminal dinitro compounds. *J. Org. Chem.* **1992**, *57*, 778–782.

510 Rodriguez, H.; Marquez, A.; Chuaqui, C. A.; Gomez, B. Oxidation of mesoionic oxazolones by oxygen. *Tetrahedron* **1991**, *47*, 5681–5688.

511 Heras, M.; Ventura, M.; Linden, A.; Villalgordo, J. M. Reaction of α-iminomethylene amino esters with mono- and bidentate nucleophiles: a straightforward route to 2-amino-1*H*-5-imidazolones. *Tetrahedron* **2001**, *57*, 4371–4388.

512 Dudley, K. H.; Bius, D. L. Diphenylhydantil. *J. Org. Chem.* **1969**, *34*, 1133–1136.

513 Russell, G. A.; Kaupp, G. Oxidation of carbanions. IV. Oxidation of indoxyl to indigo in basic solution. *J. Am. Chem. Soc.* **1969**, *91*, 3851–3859.

514 Kitamura, C.; Maeda, N.; Kamada, N.; Ouchi, M.; Yoneda, A. Synthesis of 2-(substituted methyl)quinolin-8-ols and their complexation with Sn(II). *J. Chem. Soc. Perkin Trans. 1* **2000**, 781–785.

515 Kaiser, A.; Wiegrebe, W. 1,3-Diphenylpropan-1,3-diamines IX. Reaction of α-chlorooxime ethers with α-lithiobenzylamines. *Monatsh. Chem.* **1996**, *127*, 763–774.

516 Su, Y.-L.; Yang, C.-S.; Teng, S.-J.; Zhao, G.; Ding, Y. Total synthesis of four diastereoisomers of goniofufurone from D-(–)- or L-(+)-tartaric acid. *Tetrahedron* **2001**, *57*, 2147–2153.

517 Denis, J.-N.; Tchertchian, S.; Tomassini, A.; Vallée, Y. The reaction of propiolate acetylides with nitrones. Synthesis of γ-amino-α,β-ethylenic acid derivatives. *Tetrahedron Lett.* **1997**, *38*, 5503–5506.

518 Corey, E. J.; Kim, C. U.; Chen, R. H. K.; Takeda, M. A method for stereospecific synthesis of 1,3- and 1,4-dienes via organocopper reagents. *J. Am. Chem. Soc.* **1972**, *94*, 4395–4396.

519 Tucker, T. J.; Lyle, T. A.; Wiscount, C. M.; Britcher, S. F.; Young, S. D.; Sanders, W. M.; Lumma, W. C.; Goldman, M. E.; O'Brien, J. A.; Ball, R. G.; Homnick, C. F.; Schleif, W. A.; Emini, E. A.; Huff, J. R.; Anderson, P. S. Synthesis of a series of 4-(arylethynyl)-6-chloro-4-cyclopropyl-3,4-dihydroquinazolin-2(1*H*)-ones as novel non-nucleoside HIV-1 reverse transcriptase inhibitors. *J. Med. Chem.* **1994**, *37*, 2437–2444.

520 Midland, M. M.; Tramontano, A.; Cable, J. R. Synthesis of alkyl 4-hydroxy-2-alkynoates. *J. Org. Chem.* **1980**, *45*, 28–29.

521 Christoph, G.; Hoppe, D. Asymmetric synthesis of 2-alkenyl-1-cyclopentanols via Sn–Li exchange and intramolecular cycloalkylation. *Org. Lett.* **2002**, *4*, 2189–2192.

522 Gawley, R. E.; Zhang, Q. Alkylation of 2-lithio-*N*-methylpiperidines and -pyrrolidines: scope, limitations, and stereochemistry. *J. Org. Chem.* **1995**, *60*, 5763–5769.

523 Walborsky, H. M. Mechanism of Grignard reagent formation. The surface nature of the reaction. *Acc. Chem. Res.* **1990**, *23*, 286–293.

524 Walborsky, H. M.; Hamdouchi, C. Mechanism of organocalcium reagent formation. *J. Org. Chem.* **1993**, *58*, 1187–1193.

525 Erickson, G. W.; Fry, J. L. 2-Bornyllithium. Preparation, characterization, and use in synthesis. *J. Org. Chem.* **1987**, *52*, 462–464.

526 Duddu, R.; Eckhardt, M.; Furlong, M.; Knoess, H. P.; Berger, S.; Knochel, P. Preparation and reactivity of chiral β-amido-alkylzinc iodides and related configurationally stable zinc organometallics. *Tetrahedron* **1994**, *50*, 2415–2432.

527 Glaze, W. H.; Selman, C. M. Cyclohexylmetal compounds. II. NMR and carbonation results on (+)-menthyllithium. *J. Org. Chem.* **1968**, *33*, 1987–1990.

528 Tarbell, D. S.; Weiss, M. The action of lithium on an optically active aliphatic chloride. *J. Am. Chem. Soc.* **1939**, *61*, 1203–1205.

529 Schlosser, M.; Heinz, G. Die Fluorolyse metallorganischer Bindungen durch Perchlorylfluorid. *Chem. Ber.* **1969**, *102*, 1944–1953.

530 Allan, J. F.; Clegg, W.; Henderson, K. W.; Horsburgh, L.; Kennedy, A. R. Solvent effects and molecular rearrangements during the reaction of Hauser bases with enolizable ketones: structural characterization of [{*t*BuC(=CH_2)OMgBr/HMPA}$_2$] and [$MgBr_2$/(HMPA)$_2$]. *J. Organomet. Chem.* **1998**, *559*, 173–179.

531 Basu, A.; Thayumanavan, S. Configurational stability and transfer of stereochemical information in the reactions of enantioenriched organolithium reagents. *Angew. Chem. Int. Ed.* **2002**, *41*, 716–738.

532 Glaze, W. H.; Selman, C. M.; Ball, A. L.; Bray, L. E. Cyclohexyl metal compounds. III. Stereochemistry of some substitution reactions at the C–Li bond. *J. Org. Chem.* **1969**, *34*, 641–644.

533 Raabe, G.; Gais, H.-J.; Fleischhauer, J. Ab initio study of the effect of fluorination upon the structure and configurational stability of α-sulfonyl carbanions: the role of negative hyperconjugation. *J. Am. Chem. Soc.* **1996**, *118*, 4622–4630.

534 Cram, D. J.; Partos, R. D.; Pine, S. H.; Jäger, H. Configurational stability of carbanions stabilized by *d*-orbitals. *J. Am. Chem. Soc.* **1962**, *84*, 1742–1743.

535 Cram, D. J.; Wingrove, A. S. Configurational stability of sulfonyl carbanions generated by decarboxylation reaction. *J. Am. Chem. Soc.* **1962**, *84*, 1496–1497.

536 Beagley, B.; Betts, M. J.; Pritchard, R. G.; Schofield, A.; Stoodley, R. J.; Vohra, S. 'Hidden' axial chirality as a stereodirecting element in reactions involving enol(ate) intermediates. Part 1. Cyclization reactions of methyl (4*R*)-3-(2-diazo-3-oxobutanoyl)thiazolidine-4-carboxylate and related compounds. *J. Chem. Soc. Perkin Trans. 1* **1993**, 1761–1770.

537 Kawabata, T.; Kawakami, S.; Majumdar, S. Asymmetric cyclization via memory of chirality: a concise access to cyclic amino acids with a quaternary stereocenter. *J. Am. Chem. Soc.* **2003**, *125*, 13012–13013.

538 Kuo, S.-C.; Chen, F.; Hou, D.; Kim-Meade, A.; Bernard, C.; Liu, J.; Levy, S.; Wu, G. G. A novel enantioselective alkylation and its application

to the synthesis of an anticancer agent. *J. Org. Chem.* **2003**, *68*, 4984–4987.
539 Nakamura, S.; Nakagawa, R.; Watanabe, R.; Toru, T. Enantioselective reactions of configurationally unstable α-thiobenzyllithium compounds. *Angew. Chem. Int. Ed.* **2000**, *39*, 353–338.
540 Witanowski, M.; Roberts, J. D. ^{1}H NMR. Configurational stability of neohexyl(3,3-dimethylbutyl) organometallic compounds. *J. Am. Chem. Soc.* **1966**, *88*, 737–741.
541 Jensen, F. R. The stereochemistry of the cleavage of di-*sec*-butylmercury by $HgBr_2$. *J. Am. Chem. Soc.* **1960**, *82*, 2469–2471.
542 Bergbreiter, D. E.; Whitesides, G. M. Generation and utilization of Cu(I) ate complexes from diastereomeric and enantiomeric alkylmercury reagents. *J. Am. Chem. Soc.* **1974**, *96*, 4937–4944.
543 Ahlbrecht, H.; Harbach, J.; Hoffmann, R. W.; Ruhland, T. On the racemization of α-heterosubstituted benzyllithium compounds. *Liebigs Ann. Chem.* **1995**, 211–216.
544 Reich, H. J.; Medina, M. A.; Bowe, M. D. Stereochemistry of a cyclohexyllithium reagent. A case of higher configurational stability in strongly coordinating media. *J. Am. Chem. Soc.* **1992**, *114*, 11003–11004.
545 Letsinger, R. L. Formation of optically active 1-methylheptyllithium. *J. Am. Chem. Soc.* **1950**, *72*, 4842
546 Miller, J. A.; Leong, W.; Zweifel, G. Conformational instability of α-alkenyl and α-alkynyl vinyllithiums. Synthesis of stereodefined 2-alkyl-1-en-3-ynes. *J. Org. Chem.* **1988**, *53*, 1839–1840.
547 Curtin, D. Y.; Crump, J. W. Configurational stabilities of stereoisomeric vinyllithium compounds. *J. Am. Chem. Soc.* **1958**, *80*, 1922–1926.
548 Knorr, R.; Lattke, E. Configurational isomerization of α-arylvinyllithium: kinetics, mechanism, and steric acceleration. *Tetrahedron Lett.* **1977**, 3969–3972.
549 Curtin, D. Y.; Koehl, W. J. Effect of solvent on the steric stability of lithium reagents. *J. Am. Chem. Soc.* **1962**, *84*, 1967–1973.
550 Panek, E. J.; Neff, B. L.; Chu, H.; Panek, M. G. *cis–trans* Isomerizations of 1-lithio-1-phenyl-1-butene. Solvent effects on the rate of isomerization and on NMR spectra. *J. Am. Chem. Soc.* **1975**, *97*, 3996–4000.
551 Walborsky, H. M.; Banks, R. B.; Banks, M. L. A.; Duralsamy, M. Stability and oxidative coupling of chiral vinyl- and cyclopropylcopper reagents. Formation of a novel dissymmetric diene. *Organometallics* **1982**, *1*, 667–674.
552 Burchat, A. F.; Chong, J. M.; Park, S. B. Observations on Sn–Li exchange in α-aminoorganostannanes and the configurational stability of non-stabilized α-aminoorganolithiums. *Tetrahedron Lett.* **1993**, *34*, 51–54.
553 Chong, J. M.; Park, S. B. Enantiomerically enriched Boc-protected α-aminoorganolithiums: preparation and configurational stability. *J. Org. Chem.* **1992**, *57*, 2220–2222.
554 Pearson, W. H.; Lindbeck, A. C.; Kampf, J. W. Configurational stability of chiral, nonconjugated nitrogen-substituted organolithium compounds generated by Sn–Li exchange of *N*-[(1-tributylstannyl)alkyl]imidazolidin-2-ones and -oxazolidin-2-ones. *J. Am. Chem. Soc.* **1993**, *115*, 2622–2636.
555 Niemeyer, H. M. On the inversion barriers of pyramidal carbanions. *Tetrahedron* **1977**, *33*, 2267–2270.
556 Tomoyasu, T.; Tomooka, K.; Nakai, T. Asymmetric synthesis of enantio-enriched acyclic α-amino alkylstannanes and rearrangement behavior of carbanions thereof. *Tetrahedron Lett.* **2003**, *44*, 1239–1242.
557 O'Brien, P.; Warren, S. On the configurational stability of lithiated phosphine oxides. *Tetrahedron Lett.* **1995**, *36*, 8473–8476.
558 Brandt, P.; Haeffner, F. A DFT-derived model predicts solvation-dependent configurational stability of organolithium compounds: a case study of a chiral α-thioallyllithium compound. *J. Am. Chem. Soc.* **2003**, *125*, 48–49.
559 Hoppe, D.; Kaiser, B.; Stratmann, O.; Fröhlich, R. A highly enantiomerically enriched α-thiobenzyl derivative with unusual configurational stability. *Angew. Chem. Int. Ed.* **1997**, *36*, 2784–2786.
560 Stratmann, O.; Kaiser, B.; Fröhlich, R.; Meyer, O.; Hoppe, D. The configurational stability of an enantioenriched α-thiobenzyllithium derivative and the stereochemical course of its electrophilic substitution reactions; synthesis of enantiomerically pure, tertiary benzylic thiols. *Chem. Eur. J.* **2001**, *7*, 423–435.
561 Whitesides, G. M.; Witanowski, M.; Roberts, J. D. Magnetic resonance spectroscopy. The configurational stability of primary Grignard

reagents. 3,3-Dimethylbutylmagnesium chloride. *J. Am. Chem. Soc.* **1965**, *87*, 2854–2862.

562 Fraenkel, G.; Cottrell, C. E.; Dix, D. T. Mechanism for inversion in primary organomagnesium compounds. *J. Am. Chem. Soc.* **1971**, *93*, 1704–1708.

563 Pechhold, E.; Adams, D. G.; Fraenkel, G. On the rigidity to carbanion inversion of four-, five-, and six-membered cyclic organomagnesium compounds. *J. Org. Chem.* **1971**, *36*, 1368–1374.

564 Hoffmann, R. W. The quest for chiral Grignard reagents. *Chem. Soc. Rev.* **2003**, *32*, 225–230.

565 Hoffmann, R. W.; Nell, P. G. α-Chloroalkylmagnesium reagents of > 90% ee by sulfoxide/magnesium exchange. *Angew. Chem. Int. Ed.* **1999**, *38*, 338–340.

566 Hölzer, B.; Hoffmann, R. W. Kumada–Corriu coupling of Grignard reagents, probed with a chiral Grignard reagent. *Chem. Commun.* **2003**, 732–733.

567 Hoffmann, R. W.; Hölzer, B. Stereochemistry of the transmetalation of Grignard reagents to Cu(I) and Mn(II). *J. Am. Chem. Soc.* **2002**, *124*, 4204–4205.

568 Hoffmann, R. W.; Hölzer, B.; Knopff, O. Amination of Grignard reagents with retention of configuration. *Org. Lett.* **2001**, *3*, 1945–1948.

569 Satoh, T.; Takano, K.; Ota, H.; Someya, H.; Matsuda, K.; Koyama, M. Magnesium alkylidene carbenoids: generation from 1-halovinyl sulfoxides with Grignard reagents and studies on the properties, mechanism, and some synthetic uses. *Tetrahedron* **1998**, *54*, 5557–5574.

570 Boudier, A.; Darcel, C.; Flachsmann, F.; Micouin, L.; Oestreich, M.; Knochel, P. Stereoselective preparation and reactions of configurationally defined dialkylzinc compounds. *Chem. Eur. J.* **2000**, *6*, 2748–2761.

571 Boudier, A.; Hupe, E.; Knochel, P. Highly diastereoselective synthesis of monocyclic and bicyclic secondary diorganozinc reagents with defined configuration. *Angew. Chem. Int. Ed.* **2000**, *39*, 2294–2297.

572 Poisson, J.-F.; Chemla, F.; Normant, J. F. Configurational stability of propargyl zinc reagents. *Synlett* **2001**, 305–307.

573 Knochel, P.; Yeh, M. C. P.; Berk, S. C.; Talbert, J. Synthesis and reactivity toward acyl chlorides and enones of the new highly functionalized copper reagents RCu(CN)ZnI. *J. Org. Chem.* **1988**, *53*, 2390–2392.

6
The Alkylation of Heteroatoms

Although carbon–heteroatom bonds are often easier to make than C–C bonds, and are therefore often chosen as retrosynthetic points of disconnection, some non-carbon nucleophiles can be difficult to alkylate. Problems can arise when the nucleophile is highly basic and hard, for these will preferentially attack (hard) protons and cause elimination instead of being alkylated [1]. Similarly, sterically demanding nucleophiles will also tend to attack protons faster than undergo alkylation. Ambident nucleophiles, such as amides, phenols, thiocyanate, or nitrite, can be a further cause of problems, because a variety of different products can be formed, depending on the type of electrophile chosen and on the precise reaction conditions [2]. Substrates with two or more similar nucleophilic groups, for example aminophenols or hydroxybenzoic acids, can often be alkylated with high chemoselectivity, but the outcome of these reactions can be difficult to predict. In the following sections the alkylation of the most common, trouble-causing non-carbon nucleophiles will be discussed.

As for carbanions, the reactivity of anionic non-carbon nucleophiles depends on the cation. The nucleophilicity and basicity of a given anionic nucleophile will usually be enhanced if it does not form strong bonds either with the cation or with the solvent. Hard cations, for example Li^+ or Ti^{4+}, will significantly reduce the reactivity of hard anions (RO^-, R_2N^-, F^-), whereas soft cations (Cs^+, Cu^+, Pd^{2+}) will form strong bonds with soft anions (RS^-, I^-, CN^-, H^-, R^-) and thereby reduce their reactivity.

6.1
Alkylation of Fluoride

Hydrogen fluoride (pK_a 3.2 in water) is the hydrogen halide with the lowest acidity, and fluoride is, therefore, a rather strong base. Nucleophilic substitutions with fluoride do not usually proceed smoothly, despite the strength of the C–F bond, and alkenes are often obtained instead of alkyl fluorides if simple alkyl monohalides are treated with fluoride under basic reaction conditions. Eliminations occur particularly readily from alkyl fluorides with an electron-withdrawing group in the β position. Common reagents for displacing halides or related leaving groups by fluoride include AgF, KF, CsF, Bu_4NF [3], HF, and SbF_5, and suitable solvents are ethylene

Side Reactions in Organic Synthesis. Florencio Zaragoza Dörwald

ISBN: 3-527-31021-5

glycol or ionic liquids such as *N*-butyl-*N'*-methylimidazolium tetrafluoroborate ([bimim][BF_4], Scheme 6.1). The best results are obtained with hard electrophiles, for example sulfonates or chlorides. The selective displacement of iodide by fluoride in the presence of other halogens by treatment with TolIF_2 has been reported [4].

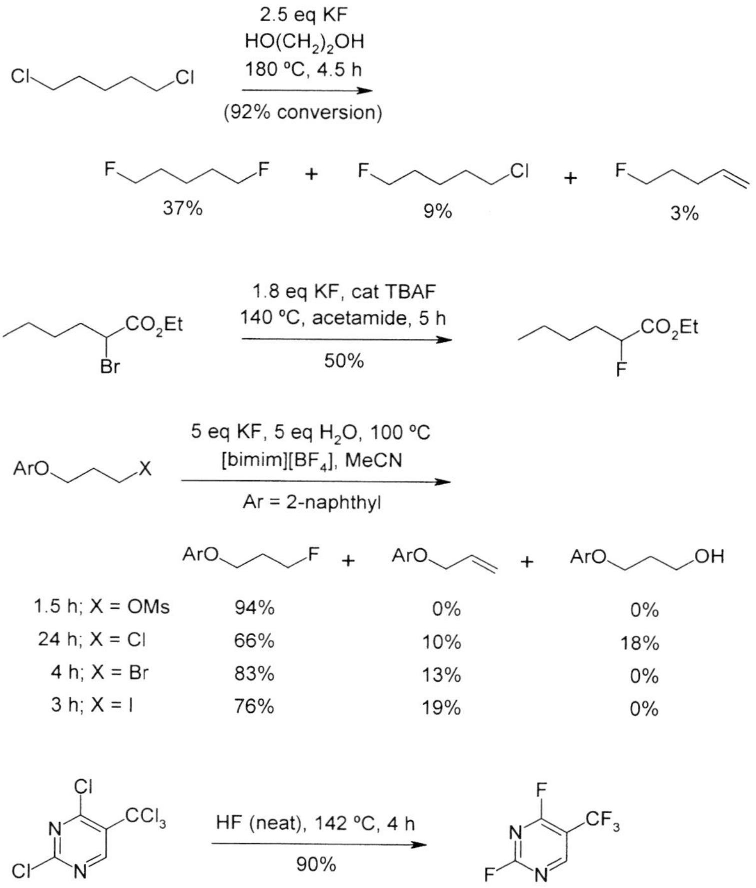

2.5 eq KF, $HO(CH_2)_2OH$, 180 °C, 4.5 h (92% conversion)

37% + 9% + 3%

1.8 eq KF, cat TBAF, 140 °C, acetamide, 5 h; 50%

5 eq KF, 5 eq H_2O, 100 °C, [bimim][BF_4], MeCN; Ar = 2-naphthyl

	ArO–F	ArO–alkene	ArO–OH
1.5 h; X = OMs	94%	0%	0%
24 h; X = Cl	66%	10%	18%
4 h; X = Br	83%	13%	0%
3 h; X = I	76%	19%	0%

HF (neat), 142 °C, 4 h; 90%

Scheme 6.1. Displacement of different leaving groups by fluoride [5–8].

Replacement of the hydroxyl group of alcohols by fluoride using the phosphine-derived reagent Ph_3PF_2 is successful, but requires much higher reaction temperatures (150–170 °C, 5–7 h) than the analogous conversions to the other halides [9], probably because of the low nucleophilicity of fluoride and the strength of the P–F bond. The conversion of alcohols into fluorides or of carbonyl compounds into geminal difluorides under mild conditions can be achieved with diethylaminosulfur trifluoride (DAST).

A further peculiarity of fluoride is its high tendency to add to fluorinated alkenes and to form stable CF_2 and CF_3 groups (Scheme 6.2). The high stability of these groups is because of negative hyperconjugation between the (antibonding) σ^*_{C-F} orbitals and the lone pairs on fluorine (see Section 3.4).

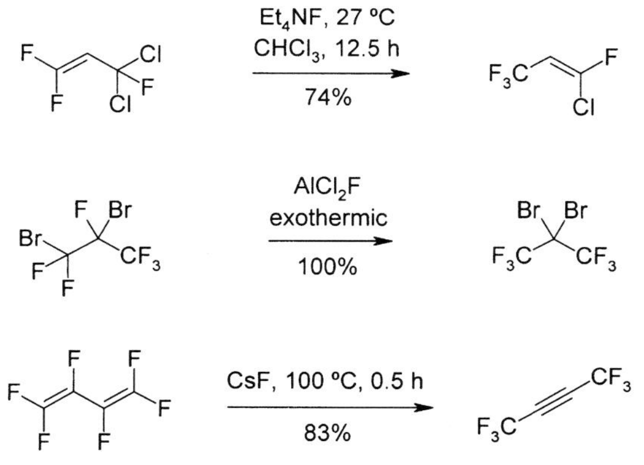

Scheme 6.2. Formation of CF_3 groups by addition of fluoride to fluorinated alkenes [10] and by isomerization [11, 12].

6.2
Alkylation of Aliphatic Amines

The most conspicuous property of aliphatic amines, apart from their fishy smell, is their high basicity, which usually precludes *N*-alkylations under acidic reaction conditions (last reaction, Scheme 6.3). Hence, alkylation of amines with tertiary alkyl groups is not usually possible without the use of highly stabilized carbocations which can be formed under basic reaction conditions. Rare exceptions are *N*-alkylations of amines via radicals (Scheme 4.2), copper-catalyzed propargylations (Scheme 6.3), and the addition of amines to some Michael acceptors and allyl palladium or iridium complexes. Better strategies for the preparation of *tert*-alkylamines include the addition of Grignard reagents to ketone-derived imines [13] or the reduction of *tert*-alkyl nitro compounds.

The direct alkylation of primary or secondary amines with alkyl halides can lead to the formation of secondary or tertiary amines or of quaternary ammonium salts [17]. The conversion of primary aliphatic amines into secondary amines with alkylating agents is difficult, because secondary amines are often more nucleophilic than primary amines. Hence, the product will be alkylated faster than the starting amine, and product mixtures will often result (Scheme 6.4), unless a large excess of primary amine can be used. Only if sterically demanding alkylating agents, for example diarylmethyl halides or alkyl halides with an electron-withdrawing group are used (Hal–CR_2–Z; third reaction in Scheme 6.4; these lead to the formation of electronically deactivated secondary amines), or if the primary amine is bulky, can clean monoalkylations be achieved. Small, reactive alkylating agents, on the other hand, for example methyl iodide or triflate, will usually lead to peralkylation or quaternization of aliphatic amines. It has been found empirically that addition of CsOH

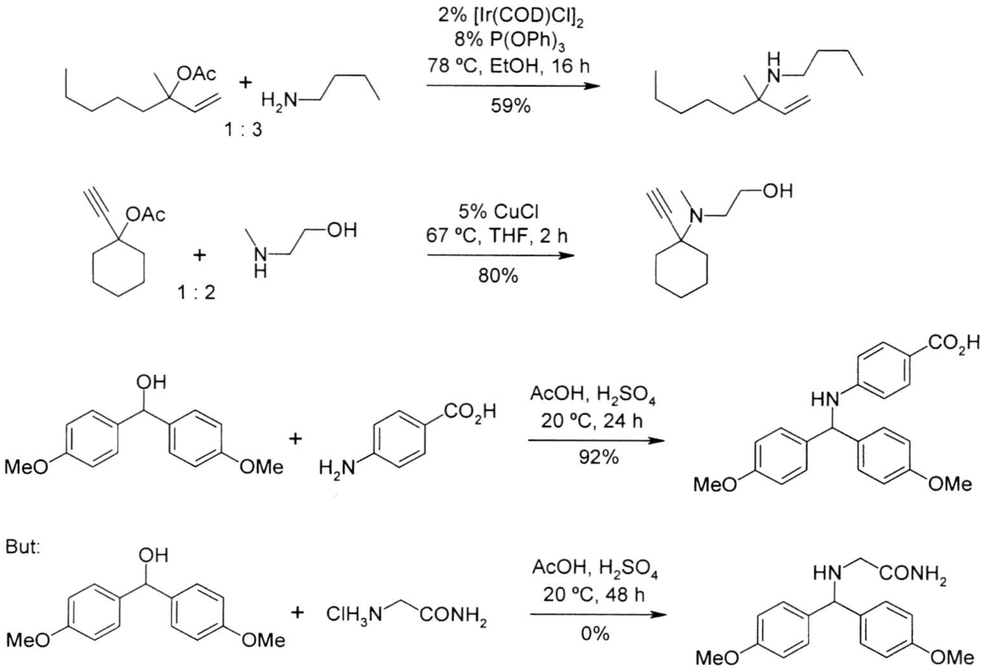

Scheme 6.3. Alkylation of amines with allyl iridium complexes and with carbocations [14–16].

can occasionally promote the formation of secondary amines from primary amines and alkylating agents and inhibit the formation of tertiary amines (Scheme 6.4) [18, 19]. Polyamines may be partially protected from alkylation by complexation with Lewis acids, for example $ZnCl_2$ [20].

A better strategy for monoalkylation of primary aliphatic amines is their condensation with aldehydes or ketones followed by reduction of the resulting imine. Alternatively, primary amines can be converted into sulfonamides, which can be readily *N*-alkylated and then hydrolyzed (less readily) with strong acids [21–23] or cleaved with thiols (Scheme 6.32).

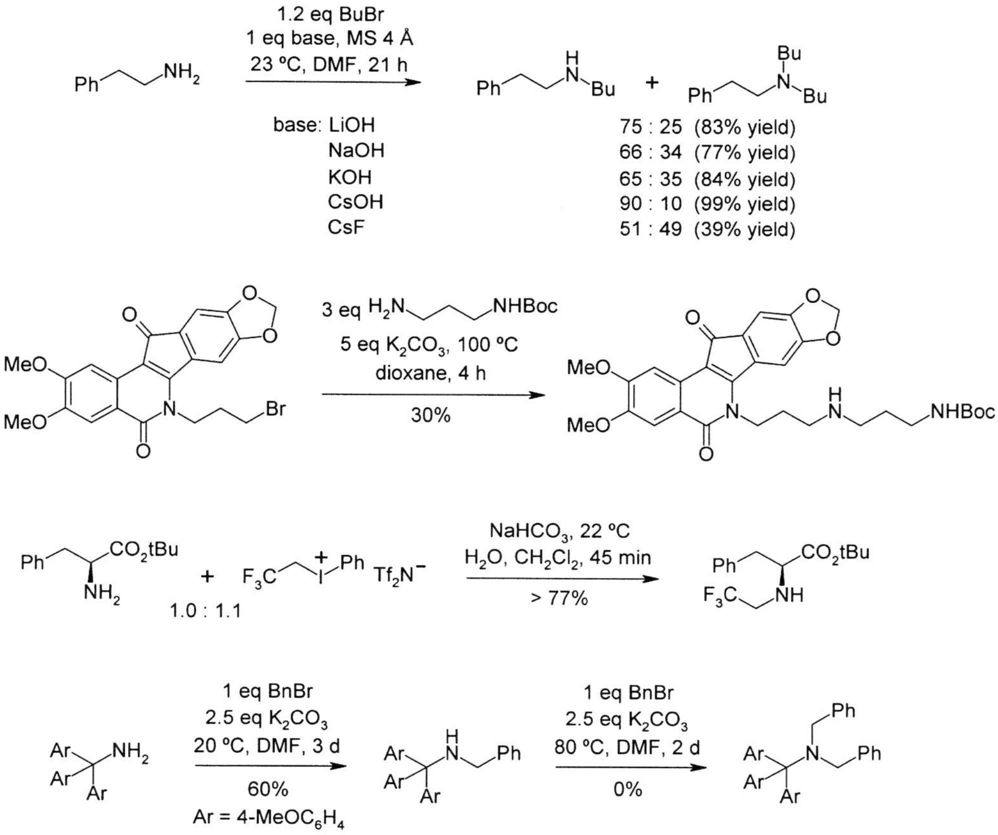

Scheme 6.4. Alkylation of primary aliphatic amines [19, 24–26].

Sterically demanding amines have a high tendency to induce β-elimination when treated with alkyl halides. Their alkylation is, however, feasible with reactive alkylating agents, and even 2,2,6,6-tetramethylpiperidines have been *N*-alkylated (Scheme 6.5). Secondary tritylamines, however, cannot usually be alkylated or acylated intermolecularly (last reaction, Scheme 6.4), but examples of intramolecular alkylations have been reported [27].

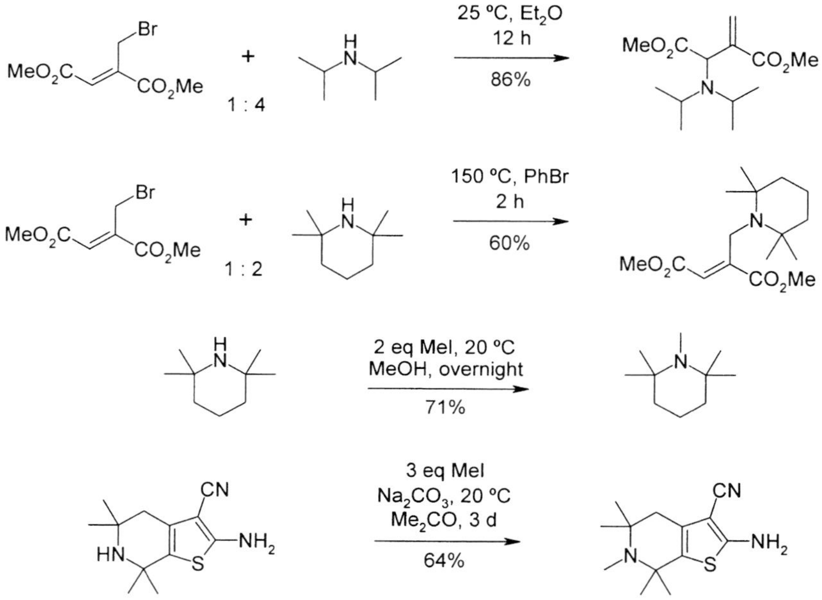

Scheme 6.5. Alkylation of sterically hindered amines [17, 28, 29].

Unsymmetric diamines can be monoalkylated regioselectively if the reactivity of both amino groups is substantially different. If the two groups are similar, however, such reactions are usually difficult to perform, and low yields are usually obtained (Scheme 6.6).

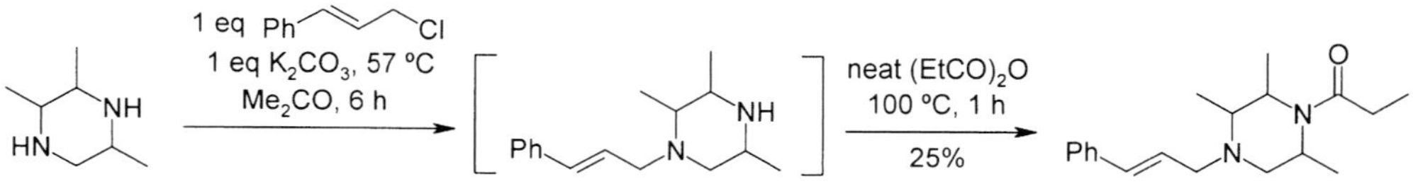

Scheme 6.6. Monoalkylation of an unsymmetric diamine [30].

6.3 Alkylation of Anilines

Anilines are generally less basic and nucleophilic than aliphatic amines, but can still be alkylated with alkyl halides under relatively mild reaction conditions under which, for instance, aliphatic alcohols will not undergo alkylation (Scheme 6.7). Monoalkylations of primary anilines with highly reactive alkylating agents can be difficult, and usually require use of excess of aniline and/or careful optimization of the reaction conditions [31–33].

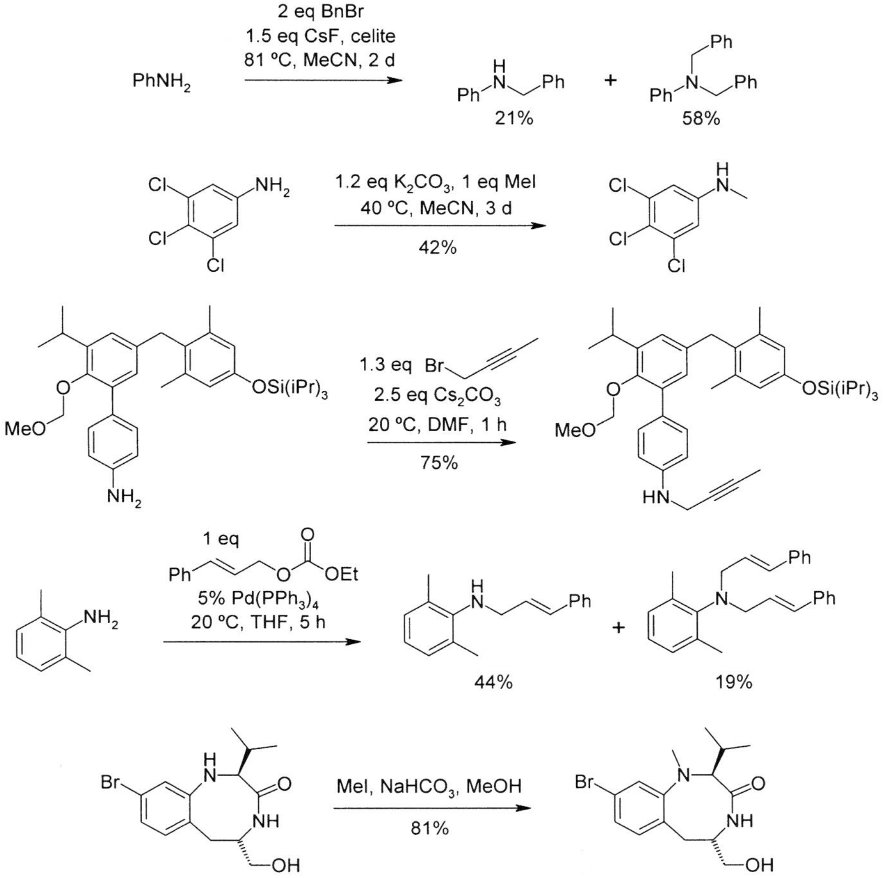

Scheme 6.7. Alkylation of anilines [32, 34–37].

Anilines with strongly electron-withdrawing groups or diarylamines [38] are only weak nucleophiles, and might require deprotonation to react with electrophiles at acceptable rates (Scheme 6.8). These anilines can also be allylated by allyl palladium complexes [34]. Electron-deficient anilines are electrophiles themselves, and can transfer the aryl group to other nucleophiles by aromatic nucleophilic substitution [39].

Anthranilic acid derivatives are also rather poor nucleophiles. Unsubstituted anthranilic acid can be *N*-alkylated selectively with some electrophiles under basic reaction conditions without the formation of esters by *O*-alkylation (Scheme 6.9) [40]. Even use of excess alkylating agent still gives reasonable yields of monoalkylated derivatives; this indicates that the second *N*-alkylation is significantly slower than

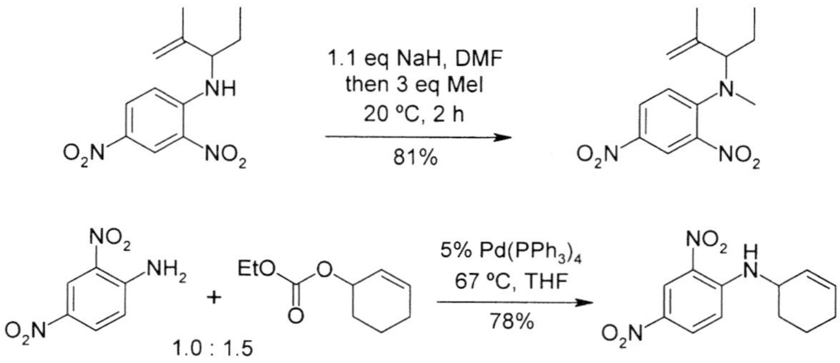

Scheme 6.8. *N*-Alkylation of electron-deficient anilines [34, 39].

the first. Phenacyl halides, however, lead to the clean *O*-alkylation of anthranilic acid [41, 42] (Scheme 6.9).

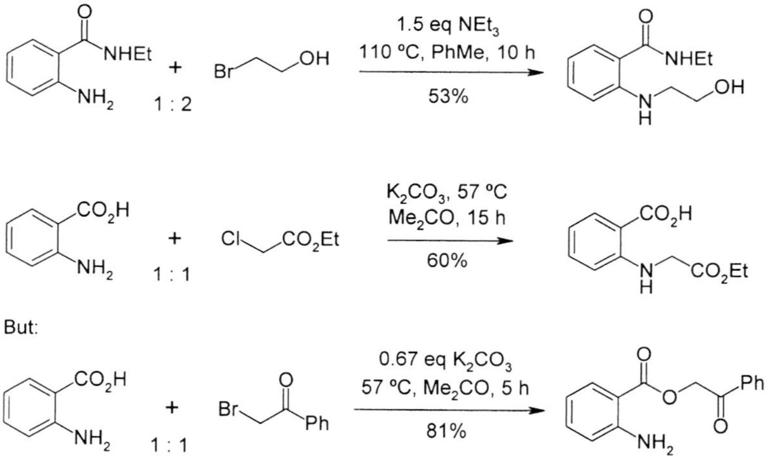

But:

Scheme 6.9. Alkylation of *N*-ethyl anthranilamide and of anthranilic acid [40, 41, 43].

The close proximity of functional groups in 1,2-disubstituted benzenes can sometimes bring about an unexpected reactivity. Attempts to *N*-alkylate *ortho*-nitroanilines under strongly basic reaction conditions, for instance, lead to the formation of *N*-alkoxybenzimidazoles (Scheme 6.10). The main force driving this reaction is the formation of an imidazole ring, a heteroarene with high resonance energy and thermal stability.

Haloalkylamines are problematic reagents which will readily polymerize or cyclize under basic conditions and must always be stored as ammonium salts. Because anilines are not very basic, their direct alkylation with salts of haloalkylamines under neutral or acidic reaction conditions can sometimes be achieved

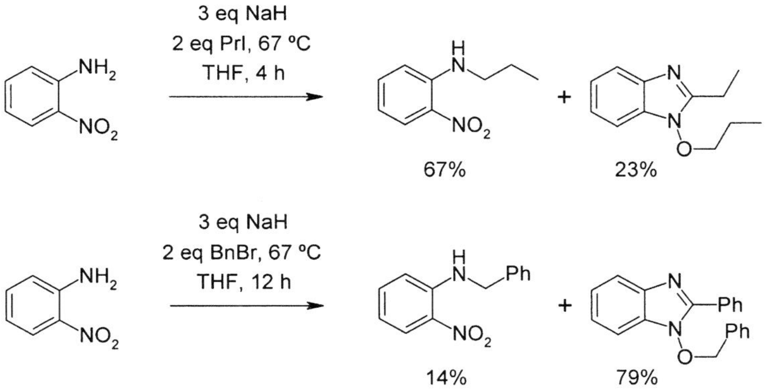

Scheme 6.10. Formation of benzimidazoles during the alkylation of 2-nitroanilines [44].

(Scheme 6.11). Reactions of this type will only proceed well with electron-rich, nucleophilic anilines, but fail with electron-deficient anilines. Yields will usually be low if the reaction is conducted under basic conditions [35].

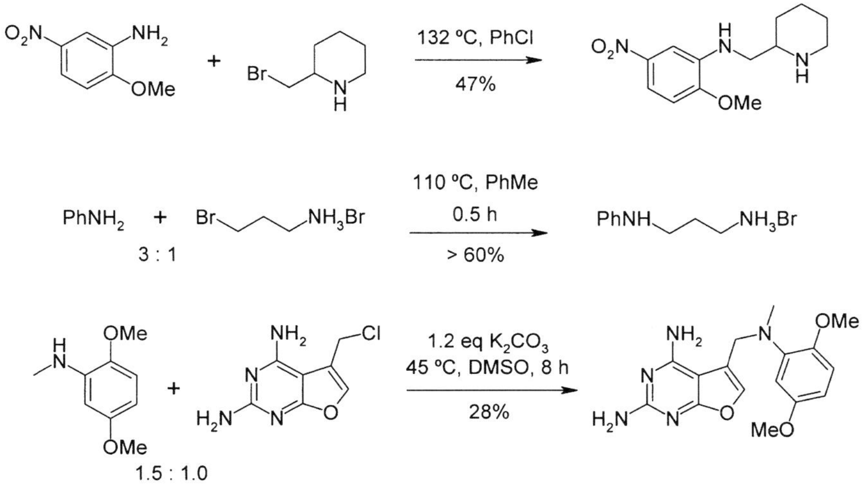

Scheme 6.11. Alkylation of anilines with haloalkylamines [35, 45, 46].

Treatment of aminophenols with alkylating agents can yield either *O*- or *N*-alkylated products, depending on the type of electrophile used and on the reaction conditions. If weak bases and hard electrophiles are used, either clean *O*-alkylation or mixtures of products can result (Scheme 6.20). Acid-catalyzed alkylation of aminophenols with epoxides usually yields *N*-alkylated products [47] (Scheme 6.12). Selective *N*-alkylation of aminophenols can also be achieved by using softer electrophiles or by conversion of the aminophenol into a dianion, followed by treatment with one equivalent of alkylating agent (Scheme 6.12). Comparison of the examples in Schemes 6.12 and 6.20 show that the chemoselectivity of these reactions is difficult to predict.

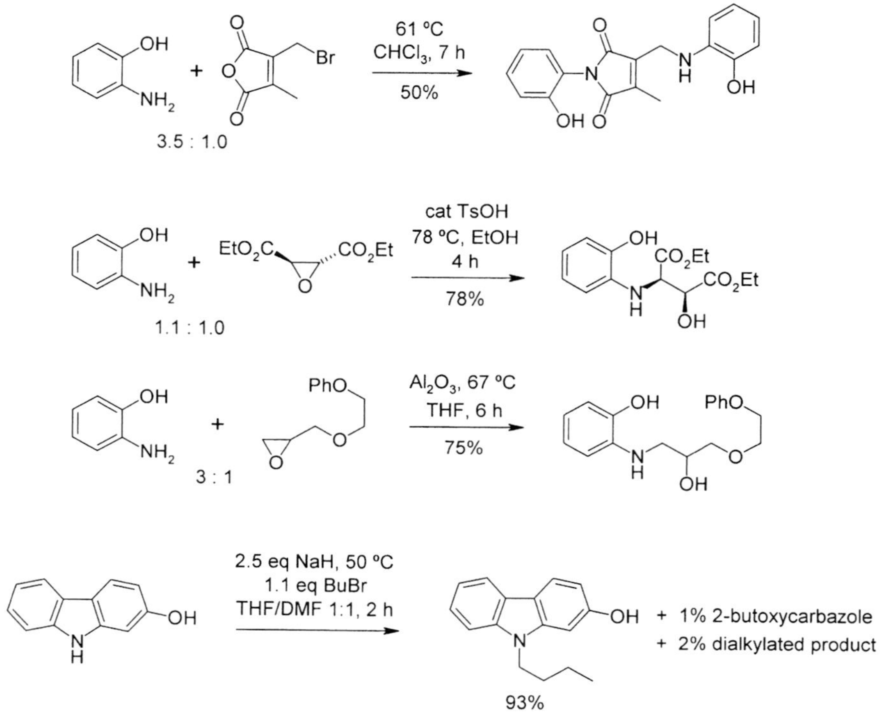

Scheme 6.12. *N*-Alkylation of aminophenols [48–51].

Polyaminoheteroarenes have also been selectively monoalkylated by metalation and treatment with an electrophile. As illustrated by the examples shown in Scheme 6.13, astonishing selectivity can sometimes be achieved.

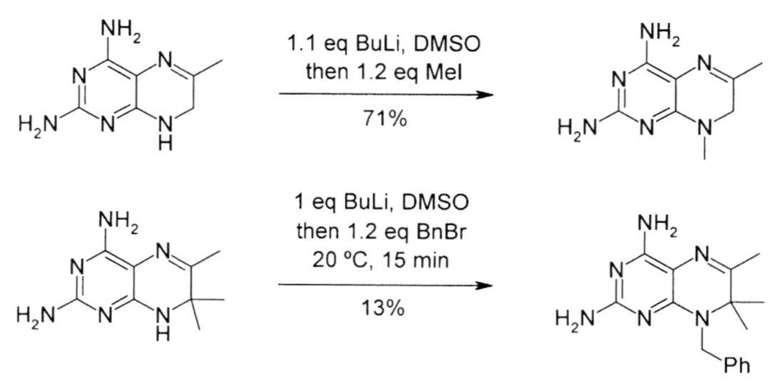

Scheme 6.13. Monoalkylation of diaminopteridines [52, 53].

6.4 Alkylation of Alcohols

In the absence of strong bases alcohols will only be alkylated by strong alkylating agents, for example trialkyloxonium salts and some triflates, or by carbocations via the SN1 mechanism. The scope of the Williamson ether synthesis, in which the alcohol is first converted stoichiometrically into an alcoholate which is then alkylated with an alkyl halide, is significantly broader. Most primary, secondary, and tertiary alcoholates can be readily alkylated, although for tertiary alcoholates this reaction proceeds only slowly (Scheme 6.14), and alkylating reagents with β hydrogen will

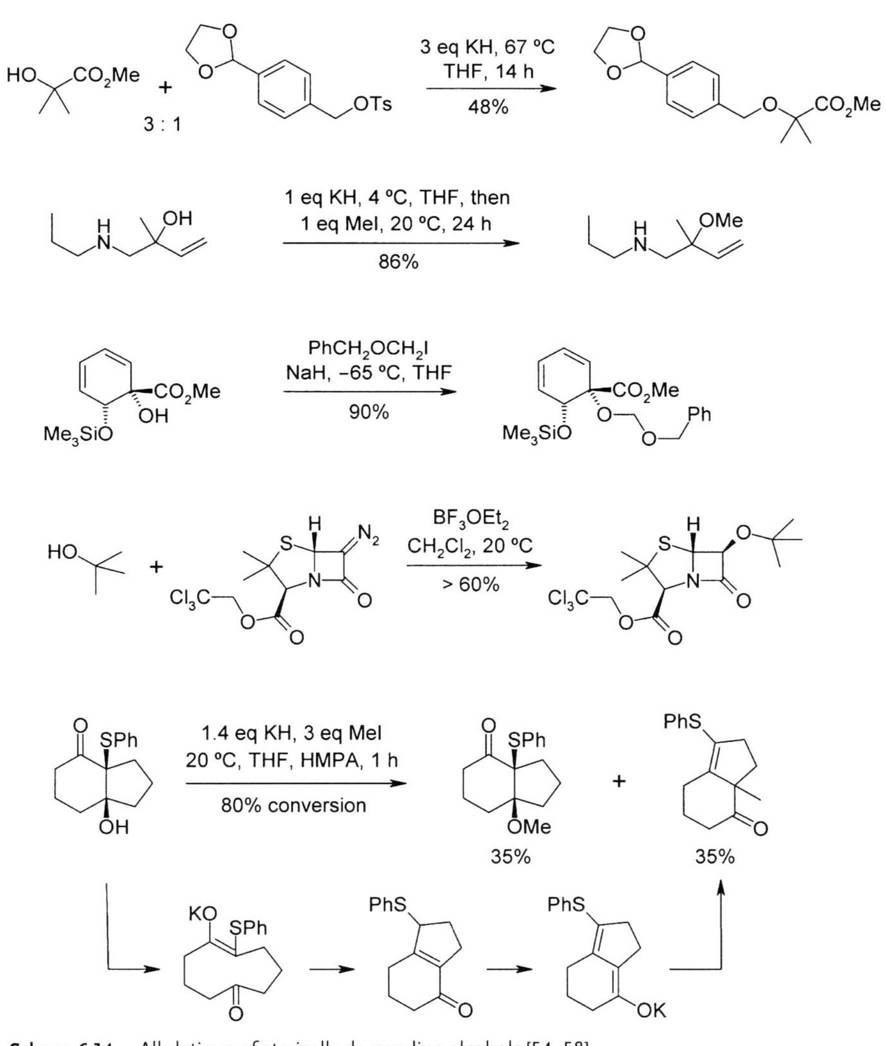

HO CO$_2$Me + OTs 3 : 1 → 3 eq KH, 67 °C, THF, 14 h, 48% → O CO$_2$Me

1 eq KH, 4 °C, THF, then 1 eq MeI, 20 °C, 24 h, 86%

PhCH$_2$OCH$_2$I, NaH, −65 °C, THF, 90%

BF$_3$OEt$_2$, CH$_2$Cl$_2$, 20 °C, > 60%

1.4 eq KH, 3 eq MeI, 20 °C, THF, HMPA, 1 h, 80% conversion → 35% + 35%

Scheme 6.14. Alkylations of sterically demanding alcohols [54–58].

usually undergo β-elimination when treated with tertiary alcoholates (Scheme 6.16). Alkali metal alcoholates are strong (but basic) nucleophiles which can be alkylated even with weak electrophiles. The best results are obtained with sodium or potassium alcoholates; lithium alcoholates, on the other hand, are less reactive, and are not always readily alkylated or acylated. The alkylation of alcoholates usually proceeds more rapidly than alkylation of non-metalated amines, and amino alcohols can be selectively *O*-alkylated without extensive *N*-alkylation (Scheme 6.14).

If the alcoholate or the alkylating reagent contains a carboxylic acid ester, acylation of the alcoholate can compete with alkylation. This potential side reaction does not cause trouble in the examples sketched in Scheme 6.14 (first and third reactions), because these esters are sterically hindered and devoid of α hydrogen (no ketene formation can occur) but, as illustrated in Scheme 6.15, less hindered esters can readily undergo transesterification with alcoholates.

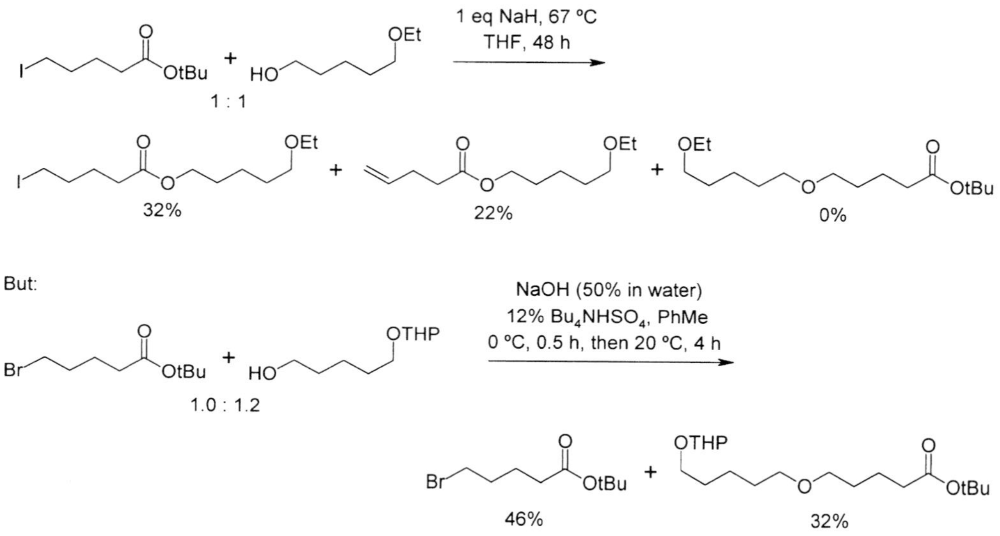

1 eq NaH, 67 °C
THF, 48 h

1 : 1

32% 22% 0%

But:

NaOH (50% in water)
12% Bu_4NHSO_4, PhMe
0 °C, 0.5 h, then 20 °C, 4 h

1.0 : 1.2

46% 32%

Scheme 6.15. Reactions of *tert*-butyl 5-halovalerates with alcoholates [59].

Secondary alkyl halides will mainly undergo elimination when treated with alcoholates, particularly if the alcoholate is sterically demanding and if the reaction is conducted at high temperatures [60]. Because of the high basicity of alcoholates, even primary alkyl halides can undergo dehydrohalogenation on treatment with alcoholates (Scheme 6.16).

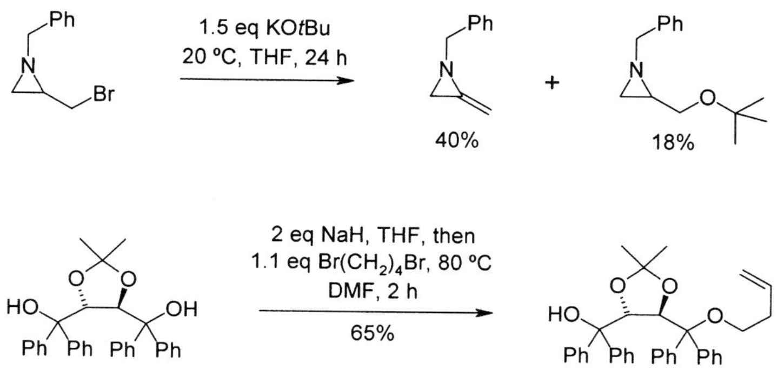

Scheme 6.16. The formation of alkenes during the Williamson ether synthesis [61, 62].

6.5
Alkylation of Phenols

The most common strategies for the alkylation of phenols are the Williamson ether synthesis and the Mitsunobu reaction. Phenolates are ambident nucleophiles which can yield products from either of *O*- or *C*-alkylation on treatment with an alkylating agent (Scheme 6.17). *C*-Alkylations will usually occur if the electrophile forms a bond with the oxygen atom reversibly (e.g. on treatment of phenols with aldehydes or benzylic electrophiles under acidic conditions [63], or with α,β-unsaturated ketones), or if the nucleophilicity of oxygen is reduced by a hard cation (e.g. Li^+ [64]) or an acidic, protic solvent (e.g. CF_3CH_2OH). Poorly cation-solvating solvents, for example THF, will also enhance the yield of *C*-alkylated product, because the phenolate salt will not dissociate and the cation will remain strongly bound to oxygen. Dipolar aprotic solvents, on the other hand, will enhance the accessibility to oxygen by solvating the cation, and will therefore promote the formation of aryl ethers [2, 65–67]. Treatment of phenols with alkenes at high temperatures under pressure mainly yields *C*-alkylated products [68, 69].

If the right reaction conditions are chosen, most phenols can be cleanly *O*-alkylated with a broad range of alkylating agents. Even sterically demanding phenols (Scheme 6.18) or highly acidic phenols, for example picric acid [73], react with alkyl halides to yield the expected ethers. It has been suggested that the regioselectivity of the last reaction in Scheme 6.18 is because the 2,6-diarylphenol is more acidic than the monoarylphenols [74].

Interestingly, it is possible to etherify hydroxybenzoic acids without the need to protect the carboxyl group (Scheme 6.19). The high charge delocalization of the carboxylate obviously leads to a sufficient decrease of nucleophilicity to enable clean ether formation under certain conditions. During the planning of such reactions it should, however, be kept in mind that carboxylates can be *O*-alkylated under conditions similar to those required for the *O*-alkylation of phenols (see Section 6.9).

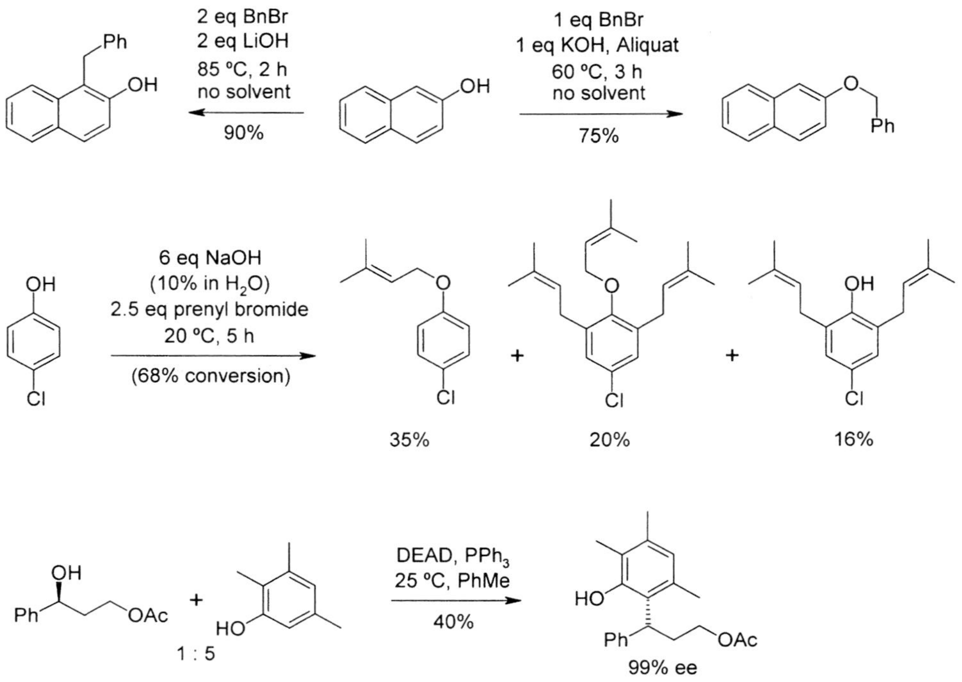

Ph
OH
2 eq BnBr
2 eq LiOH
85 °C, 2 h
no solvent
90%
OH
1 eq BnBr
1 eq KOH, Aliquat
60 °C, 3 h
no solvent
75%
O
Ph
OH
Cl
6 eq NaOH
(10% in H2O)
2.5 eq prenyl bromide
20 °C, 5 h
(68% conversion)
O
Cl
+
O
Cl
+
OH
Cl
35%
20%
16%
OH
Ph
OAc
+
HO
1 : 5
DEAD, PPh3
25 °C, PhMe
40%
HO
Ph
OAc
99% ee

Scheme 6.17. *O*- and *C*-alkylations of phenols [70–72].

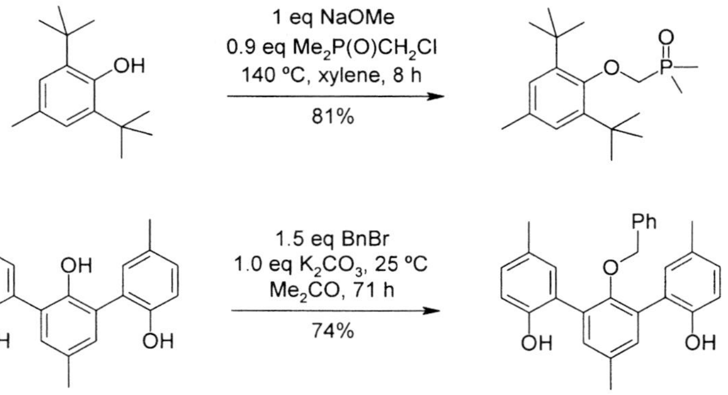

OH
1 eq NaOMe
0.9 eq Me2P(O)CH2Cl
140 °C, xylene, 8 h
81%
O
O
P
OH
OH
OH
1.5 eq BnBr
1.0 eq K2CO3, 25 °C
Me2CO, 71 h
74%
Ph
O
OH
OH

Scheme 6.18. Alkylation of sterically demanding phenols [74, 75].

Phenacyl halides tend, for instance, to alkylate carboxylates faster than anilines or phenols, as illustrated by the example shown in Scheme 6.19 [76].

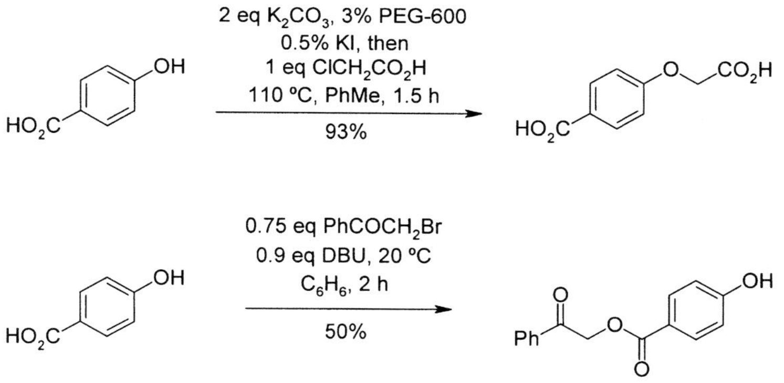

Scheme 6.19. Alkylation of 4-hydroxybenzoic acid [76, 77].

Aminophenols can be selectively *O*-alkylated under basic reaction conditions [78–81] (Scheme 6.20). If a large excess of alkylating agent or too little base is used, however, mixtures of *O*- and *N*-alkylated products can result [78].

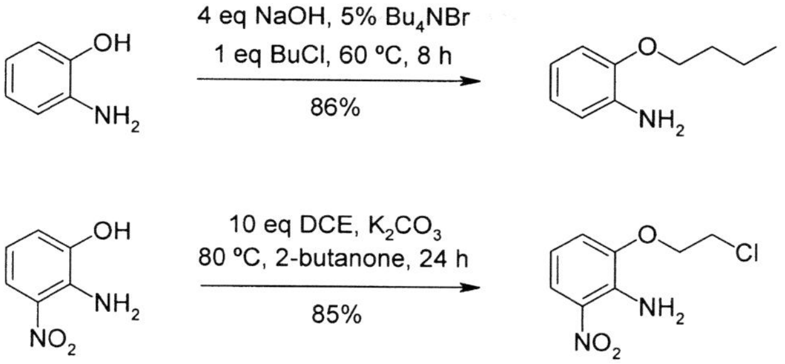

Scheme 6.20. *O*-Alkylation of aminophenols [79, 82].

6.6 Alkylation of Amides

Non-deprotonated amides are weak nucleophiles and are only alkylated by trialkyloxonium salts or dimethyl sulfate at oxygen or by some carbocations at nitrogen [16, 83]. Alkylation with primary or secondary alkyl halides under basic reaction conditions is usually rather difficult, because of the low nucleophilicity and high basicity of deprotonated amides. Non-cyclic amides are extremely difficult to *N*-alkylate, and few examples of such reactions (mainly methylations, benzylations, or allylations) have been reported (Scheme 6.21). 4-Halobutyramides, on the other hand, can often be cyclized to pyrrolidinones in high yield by treatment with bases (see Scheme 1.8) [84–86].

Secondary amines can be prepared by conversion of primary amines into sulfonamides, followed by *N*-alkylation and hydrolysis. Because sulfonamides are often difficult to hydrolyze, the *N*-alkylation of trifluoroacetamides has been investigated. Tri-

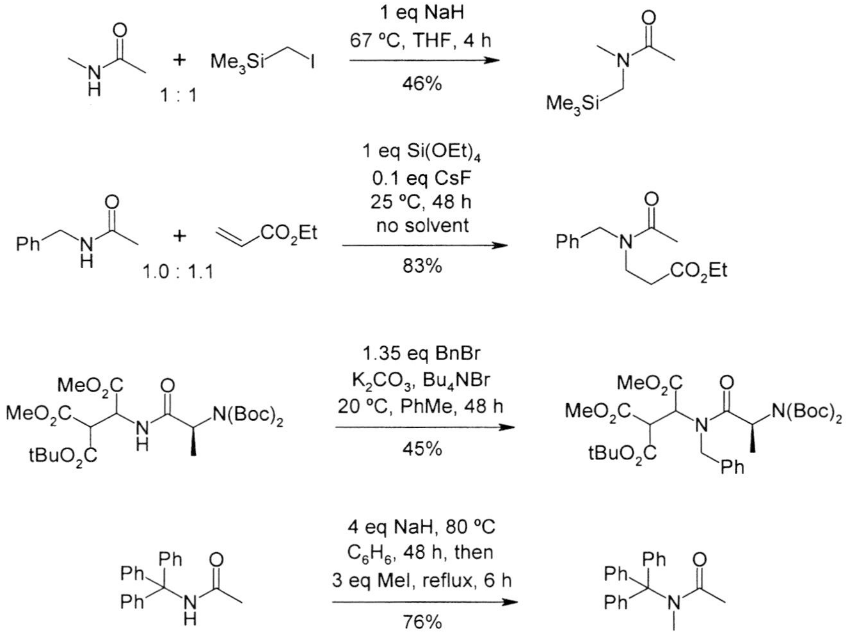

Scheme 6.21. Alkylation of acyclic amides [87–90].

fluoroacetamides are more acidic than, e.g., acetamides, and can sometimes be *N*-alkylated under mild reaction conditions and with high yields (Scheme 6.22). This reaction is a valuable tool for the preparation of secondary amines, because trifluoroacetamides are readily hydrolyzed and are therefore more suitable than most sulfonamides as transient protective and activating groups. As illustrated by the two last reactions in Scheme 6.22, some trifluoroacetamides are sufficiently acidic to undergo Pd(0) catalyzed *N*-allylation in the absence of stoichiometric amounts of base.

Lactams or related cyclic, conformationally fixed amides are more readily *N*-alkylated than acyclic amides [96]. As illustrated by the examples in Scheme 6.23, structurally elaborate alkylating agents can be used to alkylate lactams. During the work-up of such reactions it should be kept in mind that four- and six-membered lactams are readily hydrolyzed by aqueous base (Scheme 3.8), and most lactams are also readily hydrolyzed by aqueous acids. Prolonged treatment of lactams with alkali metal hydroxides or acids during the work-up should therefore be avoided.

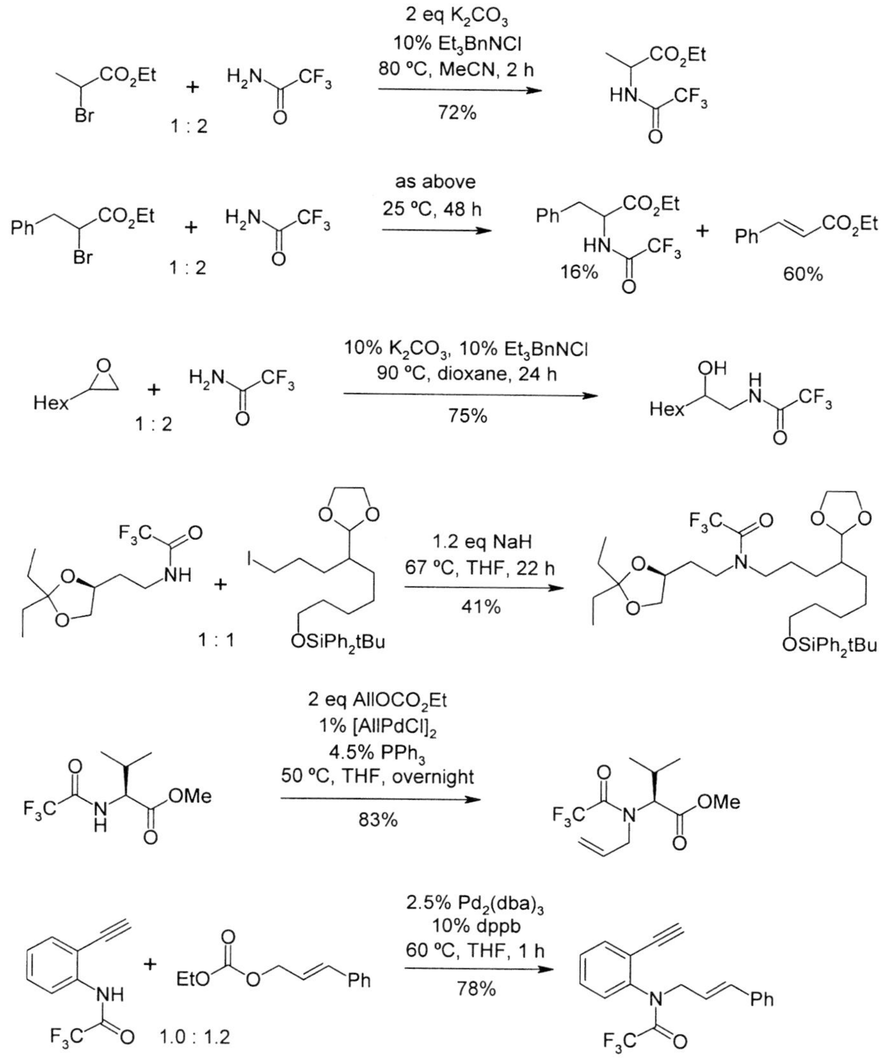

Scheme 6.22. N-Alkylation of trifluoroacetamides [91–95].

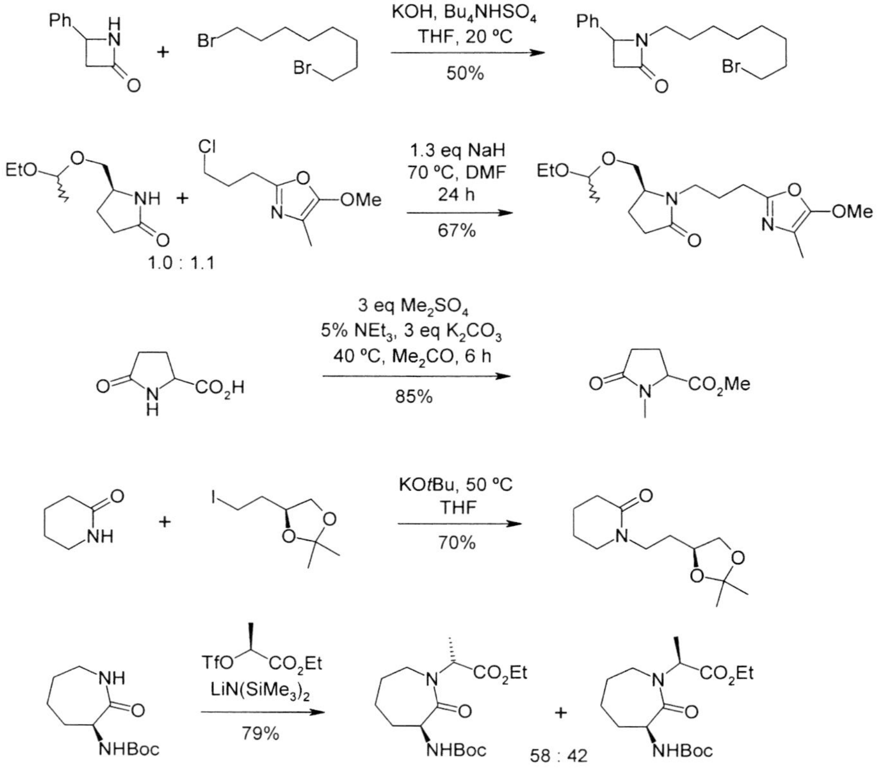

Scheme 6.23. *N*-Alkylation of lactams [97–101].

Cyclic or acyclic imides can be *N*-alkylated with soft alkylating agents (Scheme 6.24). Cyclic imides generally give better results than acyclic imides.

Imides are sufficiently acidic to enable *N*-alkylation via the Mitsunobu reaction. As with amides, only cyclic imides are readily *N*-alkylated whereas acyclic imides tend to yield mixtures of *N*- and *O*-alkylated products or *O*-alkylated products exclusively when treated with an alkylating agent (Scheme 6.25).

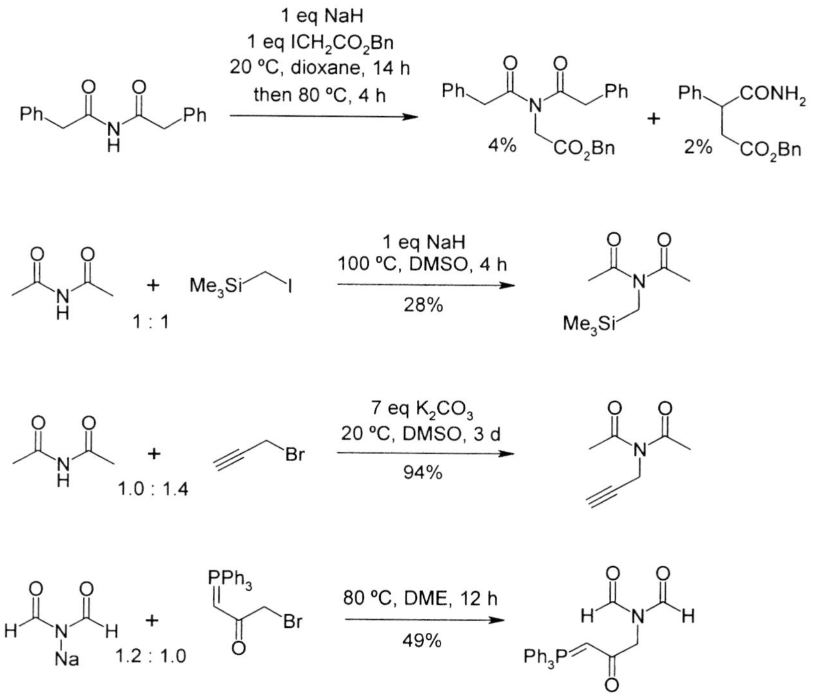

Scheme 6.24. Alkylation of imides [87, 102–104].

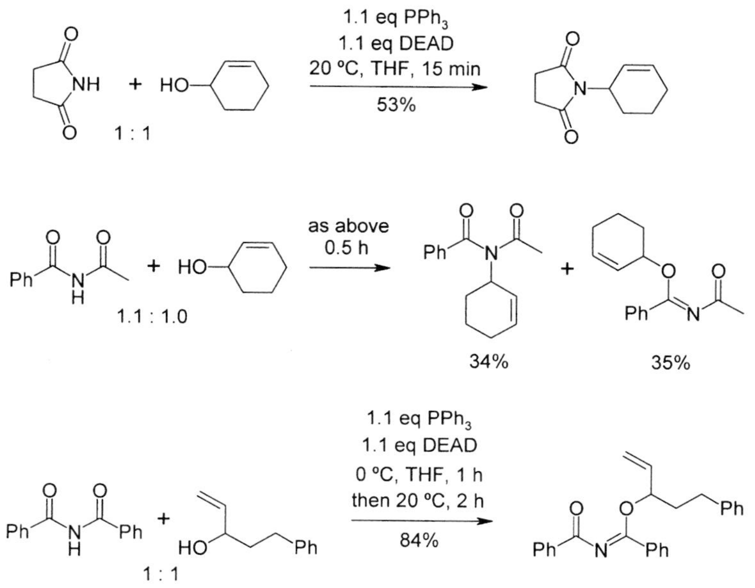

Scheme 6.25. Alkylation of imides by the Mitsunobu reaction [105, 106].

The alkylation of pyridones [107–109], quinolones [110, 111], and acridinones [112] can yield either *O*- or *N*-alkylated products (Scheme 1.7). Hard alkylating agents will generally have a higher tendency to alkylate oxygen whereas softer electrophiles will tend to alkylate nitrogen. The chemoselectivity (*O*- or *C*-alkylation) is, however, less easy to predict than for phenols, because the difference between the chemical hardness of oxygen and nitrogen is smaller than that for oxygen and carbon.

6.7
Alkylation of Carbamates and Ureas

The alkylation of carbamates and ureas is similarly difficult as the alkylation of amides. Non-deprotonated carbamates or ureas are weak nucleophiles which will react only with carbocations or other, similarly reactive alkylating agents. Metalated carbamates and ureas, on the other hand, are strong bases but poor nucleophiles, and most of the reported examples of their alkylation are limited to methylations, allylations, or benzylations, or to cyclic carbamates and ureas. Two examples of the alkylation of *N*-aryl carbamates are given in Scheme 6.26.

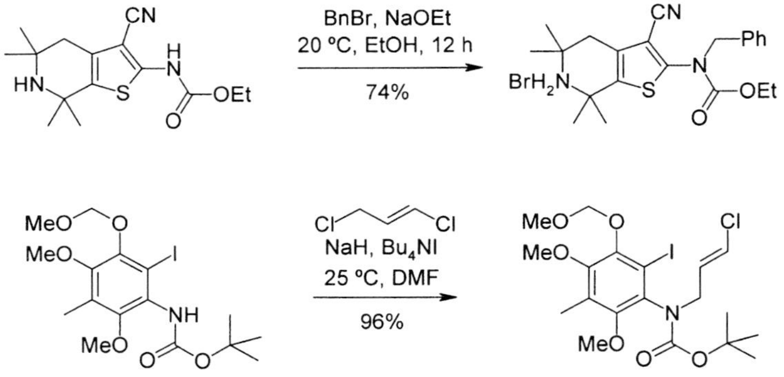

BnBr, NaOEt
20 °C, EtOH, 12 h
74%

Cl/CH=CH/Cl
NaH, Bu_4NI
25 °C, DMF
96%

Scheme 6.26. *N*-Alkylation of carbamates [29, 113].

The *N*-benzylation of *N*-phenylureas in liquid ammonia does not proceed smoothly and requires the formation of a dianion (Scheme 6.27). Only *N,N'*-diphenylurea can be benzylated under these conditions; *N*-phenylurea or *N*-acylureas, on the other hand, remain unchanged [114]. Attempts to alkylate ureas with alkyl halides in DMSO at room temperature using solid KOH as base yielded only small amounts of *N*-alkylated ureas [115].

Better results are obtained when the alkylation of ureas is performed in toluene in the presence of a phase-transfer catalyst (Scheme 6.28). This method enables the use of primary alkyl bromides as alkylating agents and both *N*-aryl or *N*-alkylureas as substrates.

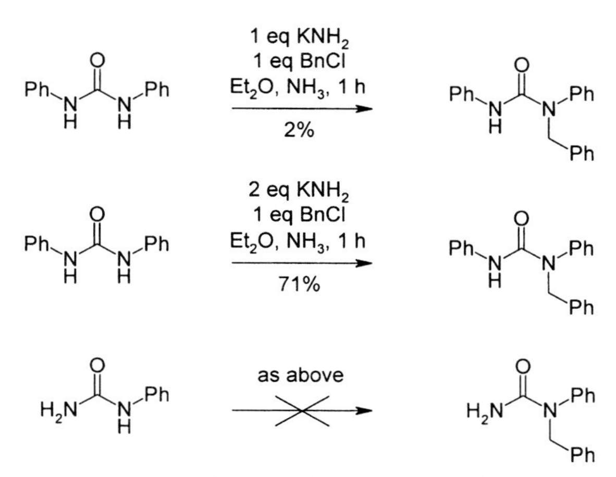

Scheme 6.27. Benzylation of phenylureas in liquid ammonia [114].

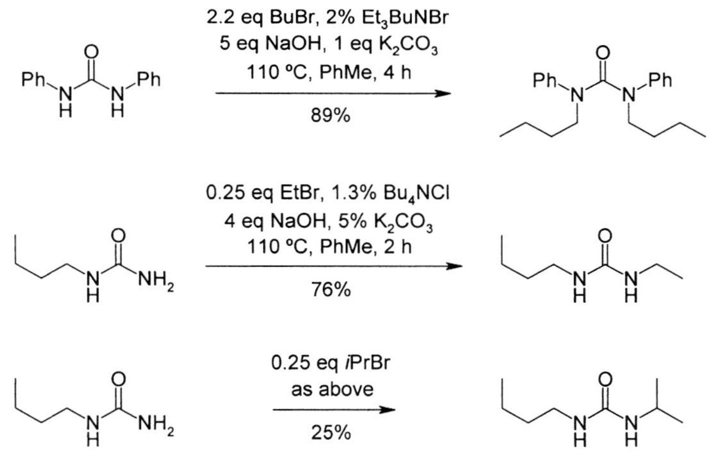

Scheme 6.28. *N*-Alkylation of ureas in the presence of phase-transfer catalysts [115, 116].

As alternative to alkylations under basic conditions, some ureas can also be *N*-alkylated by carbocations under acidic reaction conditions (Scheme 6.29). The scope of such reactions is, however, rather limited.

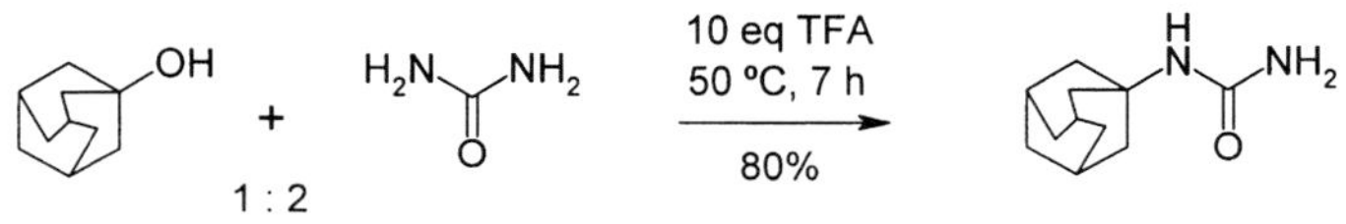

10 eq TFA
50 °C, 7 h
80%
1 : 2

Scheme 6.29. Alkylation of urea under acidic reaction conditions [83].

6.8
Alkylation of Amidines and Guanidines

Amidines and guanidines are slightly more basic than aliphatic amines, and sterically crowded amidines (e.g. DBU) or guanidines are often used to mediate dehydrohalogenations. Conditions can, however, sometimes be found which lead to the *N*-alkylation of these organic bases (Scheme 6.30). Identification of appropriate conditions for such reactions is mostly empirical, because small changes can have important but unforeseeable effects on the selectivity of a reaction (compare, e.g., the first and second reactions in Scheme 6.30). If the reactivity of a given substrate is too low, its nucleophilicity can be enhanced by deprotonation.

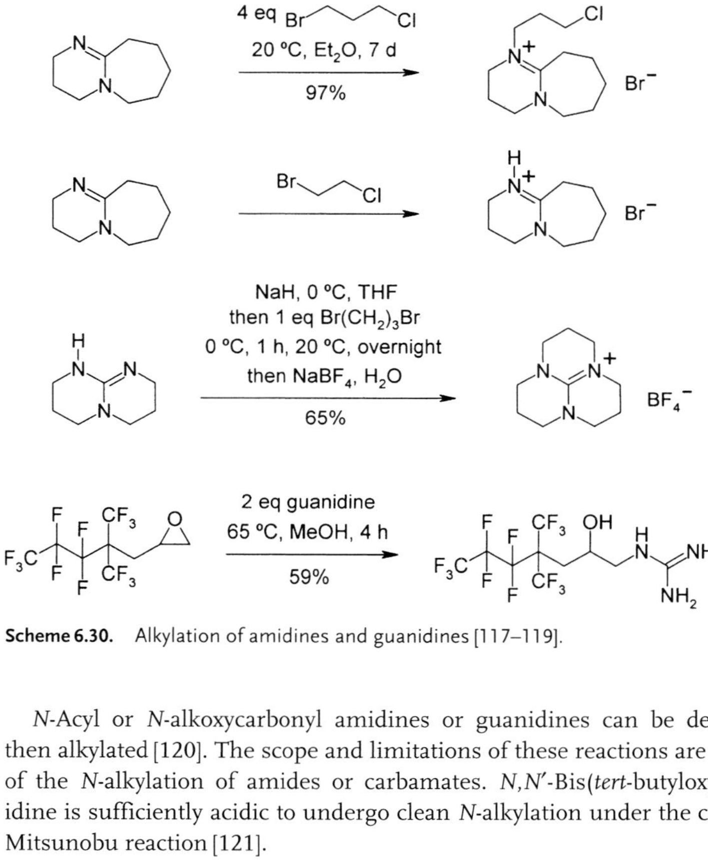

4 eq Br(CH$_2$)$_3$Cl, 20 °C, Et$_2$O, 7 d, 97%

NaH, 0 °C, THF, then 1 eq Br(CH$_2$)$_3$Br, 0 °C, 1 h, 20 °C, overnight, then NaBF$_4$, H$_2$O, 65%

2 eq guanidine, 65 °C, MeOH, 4 h, 59%

Scheme 6.30. Alkylation of amidines and guanidines [117–119].

N-Acyl or *N*-alkoxycarbonyl amidines or guanidines can be deprotonated and then alkylated [120]. The scope and limitations of these reactions are similar to those of the *N*-alkylation of amides or carbamates. *N,N'*-Bis(*tert*-butyloxycarbonyl)guanidine is sufficiently acidic to undergo clean *N*-alkylation under the conditions of the Mitsunobu reaction [121].

6.9 Alkylation of Carboxylates

The *O*-alkylation of carboxylates is a useful alternative to the acid-catalyzed esterification of carboxylic acids with alcohols. Carboxylates are weak, hard nucleophiles which are alkylated quickly by carbocations and by highly reactive, carbocation-like electrophiles (e.g. trityl or some benzhydryl halides). Suitable procedures include treatment of carboxylic acids with alcohols under the conditions of the Mitsunobu reaction [122], or with diazoalkanes. With soft electrophiles, such as alkyl iodides, alkylation of carboxylic acid salts proceeds more slowly, but in polar aprotic solvents, such as DMF, or with non-coordinating cations acceptable rates can still be achieved. Alkylating agents with a high tendency to *O*-alkylate carboxylates include α-halo ketones [42], dimethyl sulfate [100, 123], and benzyl halides (Scheme 6.31).

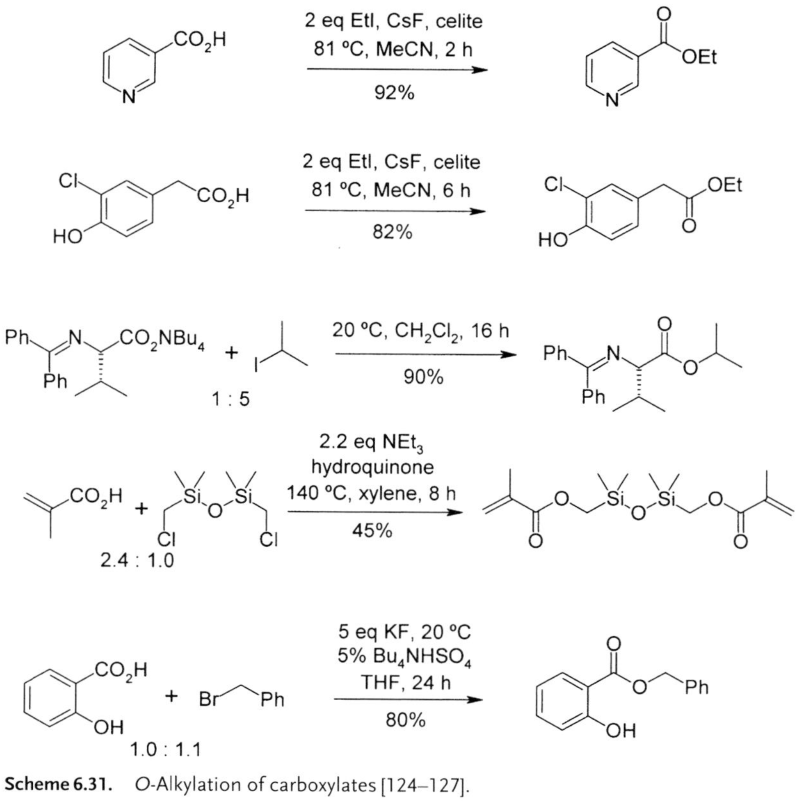

2 eq EtI, CsF, celite, 81 °C, MeCN, 2 h — 92%

2 eq EtI, CsF, celite, 81 °C, MeCN, 6 h — 82%

20 °C, CH_2Cl_2, 16 h — 90% (1 : 5)

2.2 eq NEt_3, hydroquinone, 140 °C, xylene, 8 h — 45% (2.4 : 1.0)

5 eq KF, 20 °C, 5% Bu_4NHSO_4, THF, 24 h — 80% (1.0 : 1.1)

Scheme 6.31. *O*-Alkylation of carboxylates [124–127].

Although sulfonamides (pK_a 9–11) are rather acidic and undergo deprotonation as quickly as carboxylic acids, selective *O*-alkylations of sulfonamide-containing carboxylic acids are possible under carefully controlled conditions (Scheme 6.32).

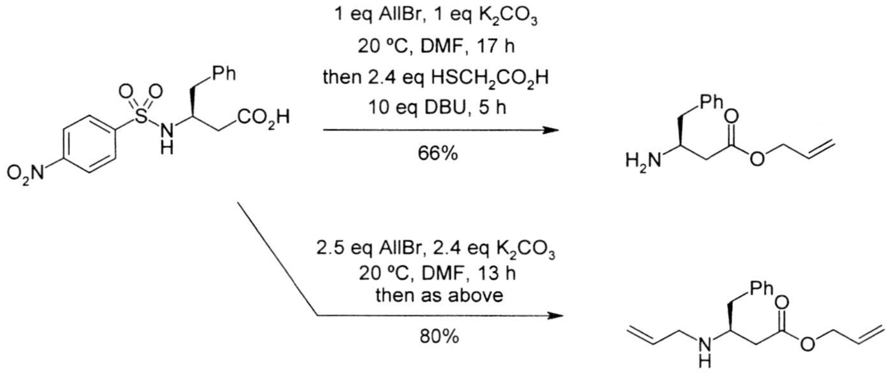

1 eq AllBr, 1 eq K_2CO_3
20 °C, DMF, 17 h
then 2.4 eq $HSCH_2CO_2H$
10 eq DBU, 5 h
66%

2.5 eq AllBr, 2.4 eq K_2CO_3
20 °C, DMF, 13 h
then as above
80%

Scheme 6.32. Alkylation of carboxylates in the presence of sulfonamides [128].

Carbocations such as those generated by halogen abstraction with Ag^+ or by protolysis of diazoalkanes preferentially alkylate negatively charged functional groups and can, therefore, be used to alkylate carboxylates with high chemoselectivity (Scheme 6.33). Under strongly acidic reaction conditions, however, no carboxylate but only the protonated carboxylic acid will be present in the reaction mixture, and other functional groups may be alkylated more rapidly by carbocations than the carboxyl group. Thus, treatment of acetamide with 4,4′-dimethoxybenzhydrol in acetic acid only yields the *N*-alkylated acetamide but no benzhydryl acetate (last reaction, Scheme 6.33; see also Scheme 6.3).

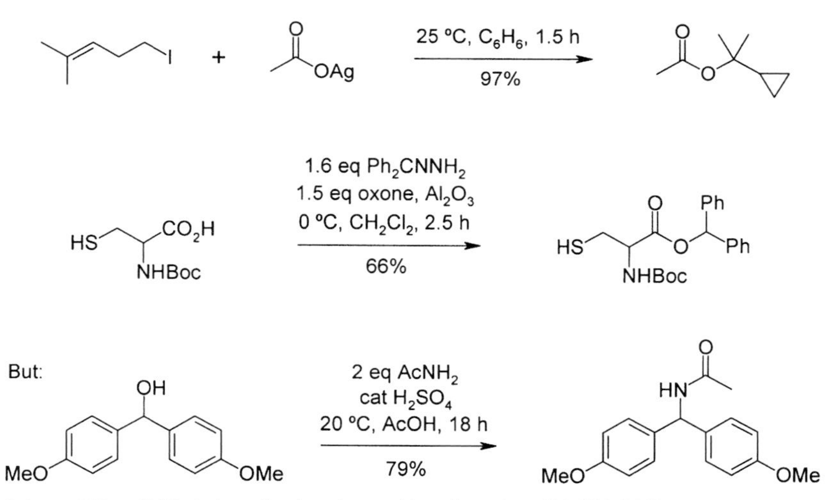

Scheme 6.33. *O*-Alkylation of carboxylates with carbocations [16, 129, 130].

Alkyl halides with a high tendency to form carbocations, for example trityl halides or α-halo ethers, can alkylate carboxylates under basic reaction conditions with high selectivity. Neutral functional groups, for example the amino group, are not alkylated as quickly as carboxylates by these reagents and do not always need to be protected. During carboxylate alkylation with simple alkyl halides amines of low nucleophilicity can, similarly, not compete efficiently with carboxylates, and might remain unchanged (last reaction, Scheme 6.34; see also Scheme 6.9).

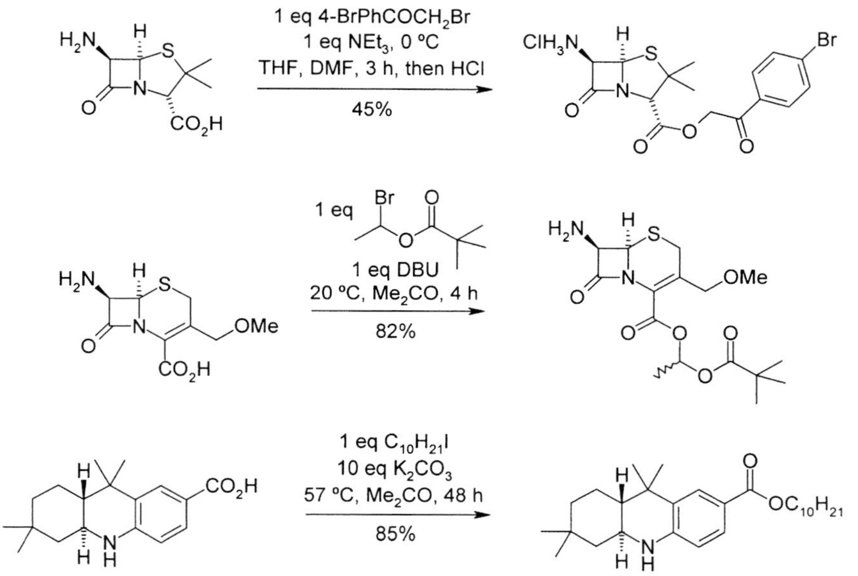

Scheme 6.34. *O*-Alkylation of unprotected amino acids [131–133].

One little-known alternative to diazoalkanes or to the Mitsunobu reaction is the alkylation of carboxylates or other negatively charged functional groups with *S*-propargyl xanthates (Scheme 6.35). On heating in the presence of an acid these propargyl xanthates cyclize, yielding cationic, *O*-alkylated dithiolanones which efficiently alkylate negatively charged, hard nucleophiles (including fluoride). Interestingly, even neopentyl esters can be prepared by means of this method (Scheme 6.35).

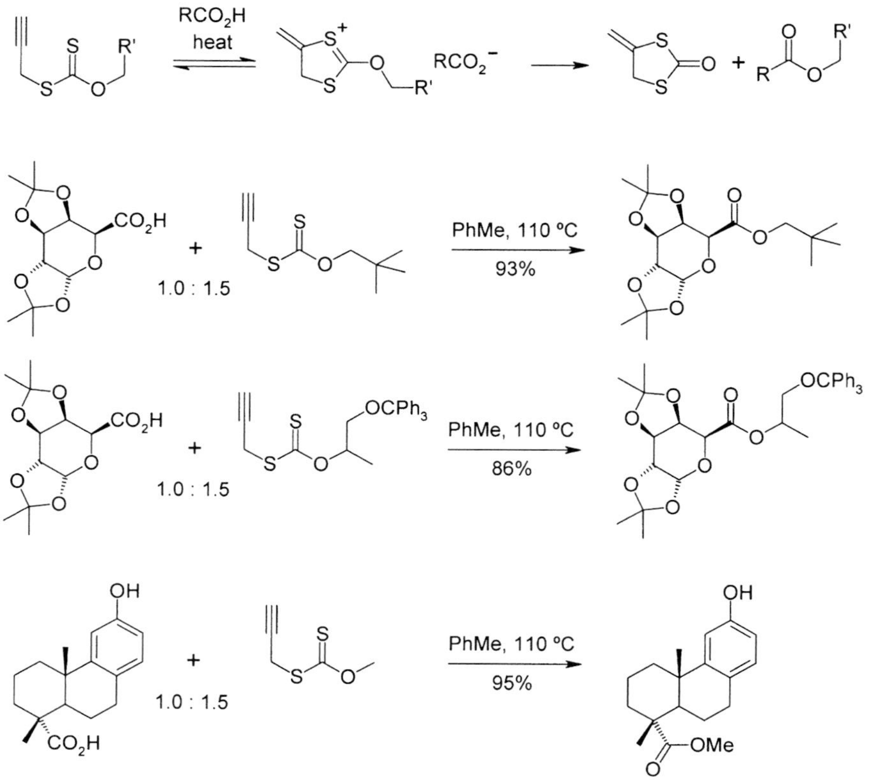

Scheme 6.35. *O*-Alkylation of carboxylic acids with *S*-propargyl xanthates [134].

References

1 Méndez, F.; Romero, M. d.; De Proft, F.; Geerlings, P. The basicity of *p*-substituted phenolates and the elimination–substitution ratio in *p*-nitrophenethyl bromide: a HSAB theoretical study. *J. Org. Chem.* **1998**, *63*, 5774–5778.

2 Le Noble, W. J. Conditions for the alkylation of ambident anions. *Synthesis* **1970**, 1–6.

3 Albanese, D.; Landini, D.; Penso, M. Hydrated TBAF as a powerful nucleophilic fluorinating agent. *J. Org. Chem.* **1998**, *63*, 9587–9589.

4 Sawaguchi, M.; Hara, S.; Nakamura, Y.; Ayuba, S.; Fukuhara, T.; Yoneda, N. Deiodinative fluorination of alkyl iodides with *p*-iodotoluene difluoride. *Tetrahedron* **2001**, *57*, 3315–3319.

5 Hoffmann, F. W. Aliphatic fluorides. I. ω,ω'-Difluoroalkanes. *J. Org. Chem.* **1949**, *14*, 105–110.

6 Kim, D. W.; Song, C. E.; Chi, D. Y. New method of fluorination using KF in ionic liquid: significantly enhanced reactivity of fluoride and improved selectivity. *J. Am. Chem. Soc.* **2002**, *124*, 10278–10279.

7 Kalaritis, P.; Regenye, R. W. Enantiomerically pure ethyl (*R*)- and (*S*)-2-fluorohexanoate by

enzyme-catalyzed kinetic resolution. *Org. Synth.* **1993**, *Coll. Vol. VIII*, 258–262.

8 Andres, P.; Marhold, A. A new synthesis of 5-trifluoromethyluracil. *J. Fluorine Chem.* **1996**, *77*, 93–95.

9 Kobayashi, Y.; Akashi, C. Studies on organic fluorine compounds. III. Conversion of alcohols to fluorides by Ph_3PF_2. *Chem. Pharm. Bull.* **1968**, *16*, 1009–1013.

10 Miller, W. T.; Fried, J. H.; Goldwhite, H. Substitution and addition reactions of the fluoroolefins. IV. Reactions of fluoride ion with fluoroolefins. *J. Am. Chem. Soc.* **1960**, *82*, 3091–3099.

11 Petrov, V. A.; Krespan, C. G.; Smart, B. E. Isomerization of halopolyfluoroalkanes by the action of aluminum chlorofluoride. *J. Fluorine Chem.* **1998**, *89*, 125–130.

12 Miller, W. T.; Frass, W.; Resnick, P. R. Cesium fluoride catalyzed rearrangement of perfluorodienes to perfluorodialkylacetylenes. *J. Am. Chem. Soc.* **1961**, *83*, 1767–1768.

13 Rodríguez, S.; Castillo, E.; Carda, M.; Marco, J. A. Synthesis of conjugated δ-lactams using ring-closing metathesis. *Tetrahedron* **2002**, *58*, 1185–1192.

14 Takeuchi, R.; Ue, N.; Tanabe, K.; Yamashita, K.; Shiga, N. Iridium complex-catalyzed allylic amination of allylic esters. *J. Am. Chem. Soc.* **2001**, *123*, 9525–9534.

15 Imada, Y.; Yuasa, M.; Nakamura, I.; Murahashi, S.-I. Copper(I)-catalyzed amination of propargyl esters. Selective synthesis of propargylamines, 1-alken-3-ylamines, and (*Z*)-allylamines. *J. Org. Chem.* **1994**, *59*, 2282–2284.

16 Henneuse, C.; Boxus, T.; Tesolin, L.; Pantano, G.; Marchand-Brynaert, J. One-step hydroxy substitution of 4,4′-dimethoxybenzhydrol with amides, lactams, carbamates, ureas, and anilines. *Synthesis* **1996**, 495–501.

17 Sommer, H. Z.; Lipp, H. I.; Jackson, L. L. Alkylation of amines. A general exhaustive alkylation method for the synthesis of quaternary ammonium compounds. *J. Org. Chem.* **1971**, *36*, 824–828.

18 Salvatore, R. N.; Schmidt, S. E.; Shin, S. I.; Nagle, A. S.; Worrell, J. H.; Jung, K. W. CsOH-promoted chemoselective mono-*N*-alkylation of diamines and polyamines. *Tetrahedron Lett.* **2000**, *41*, 9705–9708.

19 Salvatore, R. N.; Nagle, A. S.; Jung, K. W. Cesium effect: high chemoselectivity in direct *N*-alkylation of amines. *J. Org. Chem.* **2002**, *67*, 674–683.

20 Burguete, M. I.; Escuder, B.; Luis, S. V.; Miravet, J. F.; García-España, E. Selective monofunctionalization of polyaza[n]paracyclophanes. *Tetrahedron Lett.* **1994**, *35*, 9075–9078.

21 von Kiedrowski, G.; Zaragoza Dörwald, F. Synthesis of polyamine-carbodiimides. Potential activators for chemical ligations of oligodeoxynucleotide 3′-phosphates. *Liebigs Ann. Chem.* **1988**, 787–794.

22 Frydman, B.; Blokhin, A. V.; Brummel, S.; Wilding, G.; Maxuitenko, Y.; Sarkar, A.; Bhattacharya, S.; Church, D.; Reddy, V. K.; Kink, J. A.; Marton, L. J.; Valasinas, A.; Basu, H. S. Cyclopropane-containing polyamine analogues are efficient growth inhibitors of a human prostate tumor xenograft in nude mice. *J. Med. Chem.* **2003**, *46*, 4586–4600.

23 Amssoms, K.; Augustyns, K.; Yamani, A.; Zhang, M.; Haemers, A. An efficient synthesis of orthogonally protected spermidine. *Synth. Commun.* **2002**, *32*, 319–328.

24 Nagarajan, M.; Xiao, X.; Antony, S.; Kohlhagen, G.; Pommier, Y.; Cushman, M. Design, synthesis, and biological evaluation of indenoisoquinoline topoisomerase I inhibitors featuring polyamine side chains on the lactam nitrogen. *J. Med. Chem.* **2003**, *46*, 5712–5724.

25 DesMarteau, D. D.; Montanari, V. Easy preparation of bioactive peptides from the novel N^{α}-trifluoroethyl amino acids. *Chem. Lett.* **2000**, 1052–1053.

26 Henderson, A. P.; Riseborough, J.; Bleasdale, C.; Clegg, W.; Elsegood, M. R. J.; Golding, B. T. 4,4′-Dimethoxytrityl and 4,4′,4″-trimethoxytrityl as protecting groups for amino functions; selectivity for primary amino groups and application in ^{15}N-labelling. *J. Chem. Soc. Perkin Trans. 1* **1997**, 3407–3413.

27 Dugave, C.; Menez, A. Stereoconservative synthesis of orthogonally protected γ-functionalized amino acids using *N*-tritylserine derivatives. *J. Org. Chem.* **1996**, *61*, 6067–6070.

28 Gharbia, S. B.; Besbes, R.; Villiéras, J.; Amri, H. Selected reaction of (*Z*)-dimethyl α-(bromomethyl)fumarate with secondary amines. *Synth. Commun.* **1996**, *26*, 1685–1692.

29 Sensfuss, U.; Habicher, W. D. 2-Aminothiophenes from triacetonamine: a convenient way to novel sterically hindered piperidine derivatives. *Heteroatom Chem.* **1998**, *9*, 529–536.

30 Ballabio, M.; Barlocco, D.; Cignarella, G.; Colombo, D.; Toma, L. 2,2,6- And 2,3,5-trimethylpiperazines as monocyclic analogues of the μ-opioid agonist 3,8-diazabicyclo[3.2.1]octanes: synthesis, modeling, and activity. *Tetrahedron* **1997**, *53*, 1481–1490.

31 Srivastava, S. K.; Chauhan, P. M. S.; Bhaduri, A. P. A novel strategy for *N*-alkylation of primary amines. *Synth. Commun.* **1999**, *29*, 2085–2091.

32 Hayat, S.; Rahman, A.; Choudhary, M. I.; Khan, K. M.; Schumann, W.; Bayer, E. *N*-Alkylation of anilines, carboxamides, and several nitrogen heterocycles using CsF–celite/alkyl halides/MeCN combination. *Tetrahedron* **2001**, *57*, 9951–9957.

33 Curini, M.; Epifano, F.; Marcotullio, M. C.; Rosati, O. Zirconium sulfophenyl phosphonate as a heterogeneous catalyst in the preparation of β-amino alcohols from epoxides. *Eur. J. Org. Chem.* **2001**, 4149–4152.

34 Moreno-Mañas, M.; Morral, L.; Pleixats, R. Palladium(0)-catalyzed allylation of highly acidic and nonnucleophilic anilines. The origin of stereochemical scrambling when using allylic carbonates. *J. Org. Chem.* **1998**, *63*, 6160–6166.

35 Gangjee, A.; Guo, X.; Queener, S. F.; Cody, V.; Galitsky, N.; Luft, J. R.; Pangborn, W. Selective pneumocystis carinii dihydrofolate reductase inhibitors: design, synthesis, and biological evaluation of new 2,4-diamino-5-substituted-furo[2,3-*d*]pyrimidines. *J. Med. Chem.* **1998**, *41*, 1263–1271.

36 Chiellini, G.; Nguyen, N.-H.; Apriletti, J. W.; Baxter, J. D.; Scanlan, T. S. Synthesis and biological activity of novel thyroid hormone analogues: 5′-aryl substituted GC-1 derivatives. *Bioorg. Med. Chem.* **2002**, *10*, 333–346.

37 Endo, Y.; Yamaguchi, M.; Hirano, M.; Shudo, K. Role of the hydrophobic moiety of tumor promoters. Synthesis and activity of 9-alkylated benzolactams. *Chem. Pharm. Bull.* **1996**, *44*, 1138–1140.

38 DePue, J. S.; Collum, D. B. Structure and reactivity of lithium diphenylamide. Role of aggregates, mixed aggregates, monomers, and free ions on the rates and selectivities of *N*-alkylation and E2 elimination. *J. Am. Chem. Soc.* **1988**, *110*, 5524–5533.

39 Singh, S.; Nicholas, K. M. 2,4-Dinitrophenylhydroxylamine: an efficient and more general reagent for iron-catalyzed allylic amination. *Synth. Commun.* **2001**, *31*, 3087–3097.

40 Sharma, S.; Srivastava, V. K.; Kumar, A. Newer *N*-substituted anthranilic acid derivatives as potent anti-inflammatory agents. *Eur. J. Med. Chem.* **2002**, *37*, 689–697.

41 Gandhi, S. S.; Bell, K. L.; Gibson, M. S. Synthetic routes to 1,5-dihydro-5-oxo-4,1-benzoxazepines and to 5-oxooxazolo[3,2-*a*]quinolines. *Tetrahedron* **1995**, *51*, 13301–13308.

42 Hradil, P.; Hlavác, J.; Lemr, K. Preparation of 1,2-disubstituted 3-hydroxy-4(1*H*)-quinolines and the influence of substitution on the course of cyclization. *J. Heterocyclic Chem.* **1999**, *36*, 141–144.

43 Deng, S. L.; Liu, D. Z.; Huang, J. M.; Chen, R. Y. Studies on phosphoroheterocycle chemistry I: a facile synthesis of new tricyclic phosphoroheterocycles 1,3,2-oxazaphospholidino(or oxazaphosphorino)[3,2-*a*][1,3,2]benzodiazaphosphorines. *Synthesis* **2001**, 1631–1634.

44 Gardiner, J. M.; Loyns, C. R.; Schwalbe, C. H.; Barrett, G. C.; Lowe, P. R. Synthesis of 1-alkoxy-2-alkylbenzimidazoles from 2-nitroanilines via tandem *N*-alkylation-cyclization-*O*-alkylation. *Tetrahedron* **1995**, *51*, 4101–4110.

45 Bromidge, S. M.; Clarke, S. E.; King, F. D.; Lovell, P. J.; Newman, H.; Riley, G.; Routledge, C.; Serafinowska, H. T.; Smith, D. R.; Thomas, D. R. Bicyclic piperazinylbenzenesulfonamides are potent and selective 5-HT_6 receptor antagonists. *Bioorg. Med. Chem. Lett.* **2002**, *12*, 1357–1360.

46 Orelli, L. R.; García, M. B.; Niemevz, F.; Perillo, I. A. Selective monoformylation of 1,3-diaminopropane derivatives. *Synth. Commun.* **1999**, *29*, 1819–1833.

47 Sekar, G.; Singh, V. K. An efficient method for cleavage of epoxides with aromatic amines. *J. Org. Chem.* **1999**, *64*, 287–289.

48 Deshpande, A. M.; Natu, A. A.; Argade, N. P. Chemo- and regioselective nucleophilic reactions of (bromomethyl)methylmaleic anhydride: synthesis of α-quinoxalinyl- and α-benzothiazinylacrylic acids. *Heterocycles* **1999**, *51*, 2159–2162.

49 Woydowski, K.; Liebscher, J. Synthesis of optically active 1,4-benzoxazinones and 1,5-benzoxazepinones by regiocontrolled ring transformations of oxirane carboxylic acids and esters with aromatic *o*-hydroxyarylamines. *Tetrahedron* **1999**, *55*, 9205–9220.

50 Harrak, Y.; Pujol, M. D. Mild cleavage of aliphatic epoxides with substituted anilines on alumina. *Tetrahedron Lett.* **2002**, *43*, 819–822.

51 Albanese, D.; Landini, D.; Penso, M.; Spanò, G.; Trebicka, A. Chemoselective *N*-alkylation of 2-hydroxycarbazole as a model for the synthesis of *N*-substituted pyrrole derivatives containing acidic functions. *Tetrahedron* **1995**, *51*, 5681–5688.

52 Chaykovsky, M. Direct N^8-alkylation of 2,4-diamino-7,8-dihydropteridines. Preparation of 7,8-dihydro-8-methylmethotrexate. *J. Org. Chem.* **1975**, *40*, 145–146.

53 Al-Hassan, S. S.; Cameron, R.; Nicholson, S. H.; Robinson, D. H.; Suckling, C. J.; Wood, H. C. S. Specific inhibitors in vitamin biosynthesis. Part 9. Reactions of 7,7-dialkyl-7,8-dihydropteridines of use in the synthesis of potential inhibitors of tetrahydrofolate biosynthesis. *J. Chem. Soc. Perkin Trans. 1* **1985**, 2145–2150.

54 Barluenga, J.; del Pozo, C.; Olano, B. 2-Hydroxy-2,2-dimethylacetic acid ester derived heterobiaryl ethers containing 1,3,5-triazine substituents. *Synthesis* **1995**, 1529–1533.

55 Overman, L. E.; Kakimoto, M.; Okazaki, M. E.; Meier, G. P. Carbon–carbon bond formation under mild conditions via tandem cationic aza-Cope rearrangement–Mannich reactions. A convenient synthesis of polysubstituted pyrrolidines. *J. Am. Chem. Soc.* **1983**, *105*, 6622–6629.

56 Myers, A. G.; Siegel, D. R.; Buzard, D. J.; Charest, M. G. Synthesis of a broad array of highly functionalized, enantiomerically pure cyclohexanecarboxylic acid derivatives by microbial dihydroxylation of benzoic acid and subsequent oxidative and rearrangement reactions. *Org. Lett.* **2001**, *3*, 2923–2926.

57 Giddings, P. J.; John, D. I.; Thomas, E. J. Some Lewis acid catalyzed reactions of 2,2,2-trichloroethyl 6-diazopenicillanate. *Tetrahedron Lett.* **1976**, 995–998.

58 Caine, D.; McCloskey, C. J.; Atwood, J. L.; Bott, S. G.; Zhang, H.-M.; VanDerveer, D. Synthesis and base-induced methylation reactions of *cis*-7a-hydroxy-3a-(phenylsulfenyl)-3a,4,5,6,7,7a-hexahydro-4-indanone. *J. Org. Chem.* **1987**, *52*, 1280–1284.

59 Katoh, A.; Lu, T.; Devadas, B.; Adams, S. P.; Gordon, J. I.; Gokel, G. W. Nucleophile-dependent substitution reactions of 5-halovaleric acid esters: synthesis of 6,12-dioxamyristic acid. *J. Org. Chem.* **1991**, *56*, 731–735.

60 Bartsch, R. A.; Bunnett, J. F. Orientation of olefin-forming elimination in reactions of 2-substituted hexanes with KO*t*Bu–*t*BuOH and NaOMe–MeOH. *J. Am. Chem. Soc.* **1969**, *91*, 1376–1382.

61 De Kimpe, N.; De Smaele, D.; Sakonyi, Z. A new synthesis of 2-methyleneaziridines. *J. Org. Chem.* **1997**, *62*, 2448–2452.

62 Jha, S. C.; Joshi, N. N. Intramolecular dehydrohalogenation during base-mediated reaction of diols with dihaloalkanes. *J. Org. Chem.* **2002**, *67*, 3897–3899.

63 Buu-Hoi, N. P.; Demerseman, P. Zinc chloride-catalyzed benzylations of phenols and naphthols. *J. Org. Chem.* **1955**, *20*, 1129–1134.

64 Curtin, D. Y.; Crawford, R. J.; Wilhelm, M. Factors controlling position of alkylation of alkali metal salts of phenols, benzyl and allyl halides. *J. Am. Chem. Soc.* **1958**, *80*, 1391–1397.

65 Kornblum, N.; Seltzer, R.; Haberfield, P. Solvation as a factor in the alkylation of ambident anions: the importance of the dielectric factor. *J. Am. Chem. Soc.* **1963**, *85*, 1148–1154.

66 Kornblum, N.; Berrigan, P. J.; Le Noble, W. J. Solvation as a factor in the alkylation of ambident anions: the importance of the hydrogen bonding capacity of the solvent. *J. Am. Chem. Soc.* **1963**, *85*, 1141–1147.

67 Kornblum, N.; Berrigan, P. J.; Le Noble, W. J. Chemical effects arising from selective solvation: selective solvation as a factor in the alkylation of ambident anions. *J. Am. Chem. Soc.* **1960**, *82*, 1257–1258.

68 Kolka, A. J.; Napolitano, J. P.; Ecke, G. G. The *ortho*-alkylation of phenols. *J. Org. Chem.* **1956**, *21*, 712–713.

69 Goldsmith, E. A.; Schlatter, M. J.; Toland, W. G. Uncatalyzed thermal *ortho*-alkylation of phenols. *J. Org. Chem.* **1958**, *23*, 1871–1876.

70 Bram, G.; Loupy, A.; Sansoulet, J.; Vaziri-Zand, F. Highly selective benzylations of β-naphthoxide anion in heterogeneous media. *Tetrahedron Lett.* **1984**, *25*, 5035–5038.

71 Uto, Y.; Hirata, A.; Fujita, T.; Takubo, S.; Nagasawa, H.; Hori, H. First total synthesis of artepillin C established by *o,o'*-diprenylation of *p*-halophenols in water. *J. Org. Chem.* **2002**, *67*, 2355–2357.

72 Fukumoto, S.; Fukushi, S.; Terao, S.; Shiraishi, M. Direct and enantiospecific *ortho*-benzylation of phenols by the Mitsunobu reaction. *J. Chem. Soc. Perkin Trans. 1* **1996**, 1021–1026.

73 Bandgar, B. P.; Dhakne, V. D.; Nigal, J. N. Rapid synthesis of nitro substituted diaryl ethers under mild conditions. *Indian J. Chem.* **1999**, *38B*, 111–113.

74 Cram, D. J.; Dicker, I. B.; Lauer, M.; Knobler, C. B.; Trueblood, K. N. Host–guest complexation. 32. Spherands composed of cyclic urea and anisyl units. *J. Am. Chem. Soc.* **1984**, *106*, 7150–7167.

75 Varbanov, S.; Tosheva, T.; Borisov, G. New dimethyl(methyleneoxyaryl)phosphine oxides. *Phosphorus, Sulfur, and Silicon* **1991**, *63*, 397–402.

76 Banerjee, A.; Falvey, D. E. Protecting groups that can be removed through photochemical electron transfer: mechanistic and product studies on photosensitized release of carboxylates from phenacyl esters. *J. Org. Chem.* **1997**, *62*, 6245–6251.

77 Wang, X.; Li, Z.; Da, Y.; Chen, J.; Wei, T. Phase transfer catalyzed synthesis of 4-carboxylphenoxyacetic acid derivatives. *Synth. Commun.* **1999**, *29*, 4153–4161.

78 Loupy, A.; Sansoulet, J.; Díez-Barra, E.; Carrillo, J. R. Phase transfer catalysis without solvent. Alkylation of phenol and derivatives. *Synth. Commun.* **1991**, *21*, 1465–1471.

79 Carrillo, J. R.; Díez-Barra, E. A simple preparation of *O*-substituted *o*-aminophenols. *Synth. Commun.* **1994**, *24*, 945–950.

80 Bogdal, D.; Pielichowski, J.; Boron, A. New synthetic method of aromatic ethers under microwave irradiation in dry media. *Synth. Commun.* **1998**, *28*, 3029–3039.

81 Kato, S.; Morie, T.; Yoshida, N.; Matsumoto, J. Novel benzamides as selective and potent gastrokinetic agents. IV. Synthesis and structure–activity relationships of 2-substituted 4-amino-*N*-[(4-benzyl-2-morpholinyl)methyl]-5-chlorobenzamides. *Chem. Pharm. Bull.* **1992**, *40*, 1470–1475.

82 Mewshaw, R. E.; Nelson, J. A.; Shah, U. S.; Shi, X.; Mazandarani, H.; Coupet, J.; Marquis, K.; Brennan, J. A.; Andree, T. H. New generation dopaminergic agents. 7. Heterocyclic bioisosteres that exploit the 3-OH-phenoxyethylamine D_2 template. *Bioorg. Med. Chem. Lett.* **1999**, *9*, 2593–2598.

83 Shokova, E.; Mousoulou, T.; Luzikov, Y.; Kovalev, V. Adamantylation and adamantylalkylation of amides, nitriles and ureas in CF_3CO_2H. *Synthesis* **1997**, 1034–1040.

84 Ikuta, H.; Shirota, H.; Kobayashi, S.; Yamagichi, Y.; Yamada, K.; Yamatsu, I.; Katayama, K. Synthesis and antiinflammatory activities of 3-(3,5-di-*tert*-butyl-4-hydroxybenzylidene)pyrrolidin-2-ones. *J. Med. Chem.* **1987**, *30*, 1995–1998.

85 Sugiyama, S.; Morishita, K.; Chiba, M.; Ishii, K. Chemoselective debenzylation of the *N*-1-phenethyl group in 2-oxazolidinones by the anisole–$MeSO_3H$ system. *Heterocycles* **2002**, *57*, 637–648.

86 Reid, R. C.; Kelso, M. J.; Scanlon, M. J.; Fairlie, D. P. Conformationally constrained macrocycles that mimic tripeptide β-strands in water and aprotic solvents. *J. Am. Chem. Soc.* **2002**, *124*, 5673–5683.

87 Yoon, U. C.; Cho, S. J.; Lee, Y.-J.; Mancheno, M. J.; Mariano, P. S. Investigations of novel azomethine ylide-forming photoreactions of *N*-silylmethylimides. *J. Org. Chem.* **1995**, *60*, 2353–2360.

88 Ahn, K. H.; Lee, S. J. Conjugate addition of amides to α,β-unsaturated esters by the CsF–$Si(OEt)_4$ system. *Tetrahedron Lett.* **1994**, *35*, 1875–1878.

89 Robl, J. A.; Cimarusti, M. P.; Simpkins, L. M.; Weller, H. N.; Pan, Y. Y.; Malley, M.; DiMarco, J. D. Peptidomimetic synthesis: a novel, highly stereoselective route to substituted Freidinger lactams. *J. Am. Chem. Soc.* **1994**, *116*, 2348–2355.

90 Schreiber, K. C.; Fernandez, V. P. The $LiAlH_4$ reduction of some *N*-substituted succinimides. *J. Org. Chem.* **1961**, *26*, 1744–1747.

91 Landini, D.; Penso, M. *N*-Alkylation of CF_3CONH_2 with 2-bromo carboxylic esters under PTC conditions: a new procedure for the synthesis of α-amino acids. *J. Org. Chem.* **1991**, *56*, 420–423.

92 Albanese, D.; Landini, D.; Penso, M. Regioselective opening of epoxides to β-amino alcohols under solid–liquid PTC conditions. *Tetrahedron* **1997**, *53*, 4787–4790.

93 Börjesson, L.; Csöregh, I.; Welch, C. J. Synthesis and configurational analysis of 2,9-disubstituted 1-oxaquinolizidines. *J. Org. Chem.* **1995**, *60*, 2989–2999.

94 Zumpe, F. L.; Kazmaier, U. Application of the Pd-catalyzed *N*-allylation to the modification of amino acids and peptides. *Synthesis* **1999**, 1785–1791.

95 Cacchi, S.; Fabrizi, G.; Pace, P. Palladium-catalyzed cyclization of *o*-alkynyltrifluoroacetanilides with allyl esters. A regioselective synthe-

sis of 3-allylindoles. *J. Org. Chem.* **1998**, *63*, 1001–1011.

96 Kondo, K.; Seki, M.; Kuroda, T.; Yamanaka, T.; Iwasaki, T. 2-Substituted 2,3-dihydro-4*H*-1,3-benzoxazin-4-ones: novel auxiliaries for the stereoselective synthesis of 1-β-methylcarbapenems. *J. Org. Chem.* **1997**, *62*, 2877–2884.

97 Crombie, L.; Jones, R. C. F.; Haigh, D. Transamidation reactions of β-lactams: a synthesis of (±)-dihydroperiphylline. *Tetrahedron Lett.* **1986**, *27*, 5151–5154.

98 Jenhi, A.; Lavergne, J.-P.; Rolland, M.; Martinez, J.; Hasnaoui, A. 2,5-Imidazolidinedione formation from seven-membered cyclodipeptide rearrangement. *Synth. Commun.* **2001**, *31*, 1707–1714.

99 Jacobi, P. A.; Lee, K. Total synthesis of (±) and (–)-stemoamide. *J. Am. Chem. Soc.* **2000**, *122*, 4295–4303.

100 Cauliez, P.; Rigo, B.; Fasseur, D.; Couturier, D. Studies on pyrrolidones. Convenient synthesis of methyl, methyl *N*-methyl and methyl *N*-methoxymethylpyroglutamate. *J. Heterocyclic Chem.* **1991**, *28*, 1143–1146.

101 Ahn, K. H.; Lee, S. J. Synthesis of 1-oxaquinolizidines via reductive cyclization of hydroxylactams. *Tetrahedron Lett.* **1992**, *33*, 507–510.

102 Sheehan, J. C.; Corey, E. J. Synthesis and reactions of acyclic *N,N*-diacylglycines. *J. Am. Chem. Soc.* **1952**, *74*, 4555–4559.

103 Nilsson, B. M.; Ringdahl, B.; Hacksell, U. Derivatives of the muscarinic agent *N*-methyl-*N*-(1-methyl-4-pyrrolidino-2-butynyl)acetamide. *J. Med. Chem.* **1988**, *31*, 577–582.

104 Flitsch, W.; Hohenhorst, M. *N*-Geschützte 3-Hydroxypyrrole. *Liebigs Ann. Chem.* **1990**, 397–399.

105 Sammes, P. G.; Thetford, D. Stereocontrolled preparation of cyclohexane amino alcohols utilising a modified Mitsunobu reaction. *J. Chem. Soc. Perkin Trans. 1* **1989**, 655–661.

106 Overman, L. E.; Zipp, G. G. Allylic transposition of alcohol and amine functionality by thermal or Pd(II)-catalyzed rearrangements of allylic *N*-benzoylbenzimidates. *J. Org. Chem.* **1997**, *62*, 2288–2291.

107 Bergmann, R.; Eiermann, V.; Gericke, R. 4-Heterocyclyloxy-2*H*-1-benzopyran potassium channel activators. *J. Med. Chem.* **1990**, *33*, 2759–2767.

108 Comins, D. L.; Jianhua, G. *N*- vs *O*-Alkylation in the Mitsunobu reaction of 2-pyridone. *Tetrahedron Lett.* **1994**, *35*, 2819–2822.

109 Somekawa, K.; Okuhira, H.; Sendayama, M.; Suishu, T.; Shimo, T. Intramolecular [2 + 2]-photocycloadditions of 1-(ω-alkenyl)-2-pyridones possessing an ester group on the olefinic carbon chain. *J. Org. Chem.* **1992**, *57*, 5708–5712.

110 Hadjeri, M.; Mariotte, A.-M.; Boumendjel, A. Alkylation of 2-phenyl-4-quinolones: synthetic and structural studies. *Chem. Pharm. Bull.* **2001**, *49*, 1352–1355.

111 Fossa, P.; Mosti, L.; Menozzi, G.; Marzano, C.; Baccichetti, F.; Bordin, F. Novel angular furo and thienoquinolinones: synthesis and preliminary photobiological studies. *Bioorg. Med. Chem.* **2002**, *10*, 743–751.

112 Chen, J.-J.; Deady, L. W.; Mackay, M. F. Synthesis of some acridone alkaloids and related compounds. *Tetrahedron* **1997**, *53*, 12717–12728.

113 Boger, D. L.; Boyce, C. W.; Garbaccio, R. M.; Searcey, M. Synthesis of CC-1065 and duocarmycin analogs via intramolecular aryl radical cyclizatioon of a tethered vinyl chloride. *Tetrahedron Lett.* **1998**, *39*, 2227–2230.

114 Bryant, D. R.; Work, S. D.; Hauser, C. R. Formation and benzylation of the dianion of *sym*-diphenylurea by KNH_2 in liquid ammonia. Results with related compounds. *J. Org. Chem.* **1964**, *29*, 235–237.

115 Hackl, K. A.; Falk, H. The synthesis of *N*-substituted ureas I. The *N*-alkylation of ureas. *Monatsh. Chem.* **1992**, *123*, 599–606.

116 Kalkote, U. R.; Choudhary, A. R.; Ayyangar, N. R. Phase transfer catalyzed *N*-alkylation of *sym-N,N′*-diarylureas. *Organic Preparations and Procedures International* **1992**, *24*, 83–87.

117 Alder, R. W.; Sessions, R. B. Synthesis of medium-ring bicyclic diamines by the alkylation and cleavage of cyclic amidines. *Tetrahedron Lett.* **1982**, *23*, 1121–1124.

118 Alder, R. W.; Mowlam, R. W.; Vachon, D. J.; Weisman, G. R. New synthetic routes to macrocyclic triamines. *J. Chem. Soc. Chem. Commun.* **1992**, 507–508.

119 Plenkiewicz, H.; Dmowski, W. Synthetic utility of 3-(perfluoro-1,1-dimethylbutyl)-1-propene. Part V. Reactions of 3-(perfluoro-1,1-dimethylbutyl)-1,2-epoxypropane with inorganic anions, thiourea, and guanidine. *J. Fluorine Chem.* **1991**, *51*, 43–52.

120 Vaidyanathan, G.; Zalutsky, M. R. A new route to guanidines from bromoalkanes. *J. Org. Chem.* **1997**, *62*, 4867–4869.

121 Beumer, R.; Reiser, O. β-Aminocyclopropanecarboxylic acids with α-amino acid side chain functionality. *Tetrahedron* **2001**, *57*, 6497–6503.

122 Dodge, J. A.; Nissen, J. S.; Presnell, M. A general procedure for Mitsunobu inversion of sterically hindered alcohols: inversion of menthol. (1*S*,2*S*,5*R*)-5-Methyl-2-(1-methylethyl)cyclohexyl 4-nitrobenzoate. *Org. Synth.* **1998**, *Coll. Vol. IX*, 607–609.

123 Chakraborti, A. K.; Basak, A.; Grover, V. Chemoselective protection of carboxylic acids as methyl esters: a practical alternative to the diazomethane protocol. *J. Org. Chem.* **1999**, *64*, 8014–8017.

124 Lee, J. C.; Choi, Y. An improved method for the preparation of carboxylic esters using the CsF–celite/alkyl halide/MeCN combination. *Synth. Commun.* **1998**, *28*, 2021–2026.

125 O'Donnell, M. J.; Cook, G. K.; Rusterholz, D. B. Oxygen alkylation of Schiff base derivatives of amino acids. *Synthesis* **1991**, 989–993.

126 Merker, R. L.; Scott, M. J. The reaction of alkyl halides with carboxylic acids and phenols in the presence of tertiary amines. *J. Org. Chem.* **1961**, *26*, 5180–5182.

127 Ooi, T.; Sugimoto, H.; Doda, K.; Maruoka, K. Esterification of carboxylic acids catalyzed by in situ generated tetraalkylammonium fluorides. *Tetrahedron Lett.* **2001**, *42*, 9245–9248.

128 Farràs, J.; Ginesta, X.; Sutton, P. W.; Taltavull, J.; Egeler, F.; Romea, P.; Urpí, F.; Vilarrasa, J. β^3-Amino acids by nucleophilic ring-opening of *N*-nosyl aziridines. *Tetrahedron* **2001**, *57*, 7665–7674.

129 Previtera, L.; Monaco, P.; Mangoni, L. Cyclopropylcarbinyl compounds from homoallylic iodides. *Tetrahedron Lett.* **1984**, *25*, 1293–1294.

130 Curini, M.; Rosati, O.; Pisani, E. Preparation of diphenylmethyl esters by oxone oxidation of benzophenone hydrazone. *Tetrahedron Lett.* **1997**, *38*, 1239–1240.

131 Bamberg, P.; Ekström, B.; Sjöberg, B. Semisynthetic penicillins VII. The use of phenacyl 6-aminopenicillinates in penicillin synthesis. *Acta Chem. Scand.* **1967**, *21*, 2210–2214.

132 Defossa, E.; Fischer, G.; Gerlach, U.; Hörlein, R.; Isert, D.; Krass, N.; Lattrell, R.; Stache, U.; Wollmann, T. Synthesis of HR 916 K: an efficient route to the pure diastereomers of the 1-(pivaloyloxy)ethyl esters of cephalosporins. *Liebigs Ann. Chem.* **1996**, 1743–1749.

133 Schulte, J. L.; Laschat, S.; Kotila, S.; Hecht, J.; Fröhlich, R.; Wibbeling, B. Synthesis of η^6-(octahydroacridine)chromiumtricarbonyl complexes with non-polar tails via molecular sieves-catalyzed cyclization of *N*-arylimines and subsequent diastereoselective complexation. *Heterocycles* **1996**, *43*, 2713–2724.

134 Boivin, J.; Henriet, E.; Zard, S. Z. A highly efficient reaction for the synthesis of esters and for the inversion of secondary alcohols. *J. Am. Chem. Soc.* **1994**, *116*, 9739–9740.

7
The Acylation of Heteroatoms

One of the most common strategies for preparation of carboxylic acid derivatives is the acylation of nucleophiles of general formula RXH (X = NR, O, S). If the right reaction conditions are chosen and if the product is stable these reactions usually proceed swiftly and in high yield. Some carboxylic acids and nucleophiles, however, do not react smoothly or lead to unexpected results. The following sections cover a selection of representative examples of problematic starting materials.

7.1
Problematic Carboxylic Acids

7.1.1
Sterically Demanding Carboxylic Acids

Despite the huge structural diversity of known carboxylic acids, most of these are readily converted into esters or amides. Even sterically hindered acids, for example pivalic, triphenylacetic [1], or 2,6-disubstituted benzoic acids [1, 2], can be converted into suitable acylating reagents for alcohols or amines (Scheme 7.1). Esters of sterically demanding carboxylic acids can, alternatively, also be prepared by *O*-alkylation of the corresponding carboxylates [3, 4].

The high yields obtained in the examples shown in Scheme 7.1 are surprising, because of the strong steric shielding of the carbonyl group of the acylating agents. The difficult approach of nucleophiles to the carbonyl group renders esters or amides of such sterically shielded carboxylic acids highly resistant to hydrolysis, reduction, or nucleophilic attack by organometallic reagents [1]. The corresponding acyl chlorides or HOBt-esters, though, are obviously sufficiently reactive to undergo such an attack under mild reaction conditions.

Side Reactions in Organic Synthesis. Florencio Zaragoza Dörwald

ISBN: 3-527-31021-5

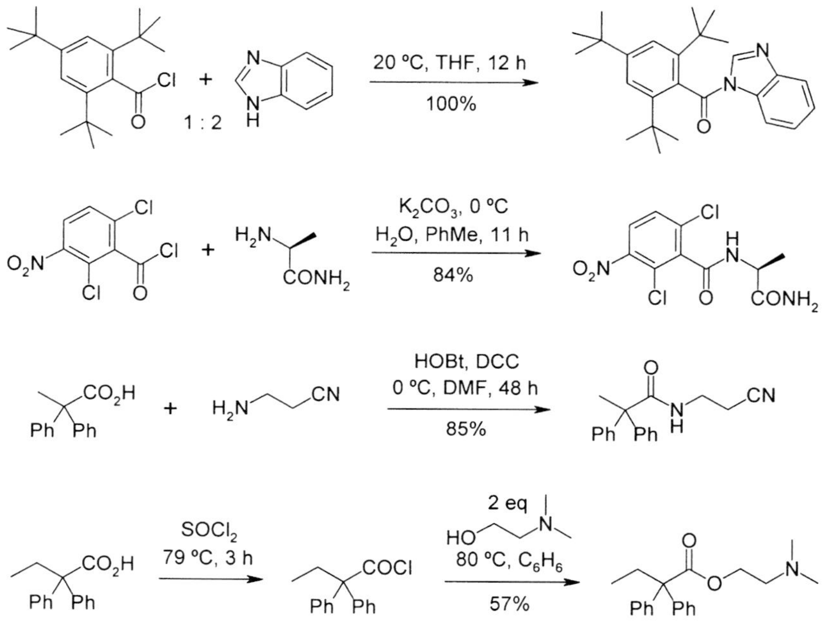

Scheme 7.1. Acylations with sterically hindered carboxylic acids [5–8].

7.1.2
Unprotected Amino and Hydroxy Carboxylic Acids

Carboxylic acids which contain a hydroxyl or amino group may oligomerize if an activation of the acyl group is attempted, and protection of the hydroxyl or amino group will usually be required. It is, however, sometimes possible to acylate a strong nucleophile with unprotected hydroxy or amino acids if these are converted into weak acylating agents only (Scheme 7.2). Unprotected hydroxy acids can usually be converted into amides in acceptable yields by in-situ activation with standard coupling reagents. Esterifications of hydroxy acids can also be achieved by treatment of the acid with excess alcohol under acidic reaction conditions [9, 10] (Scheme 7.2).

There have been few reports only of *N*-acylations with unprotected amino acids (Scheme 7.3). Such reactions will only yield substantial amounts of the desired product if the (unprotected) amino group in the acylating reagent is difficult to acylate. The examples in Scheme 7.3 show that reaction conditions for performing such transformations with success can sometimes be found; for large-scale or industrial preparations, in particular, such "protective-group-free" shortcuts should always be considered.

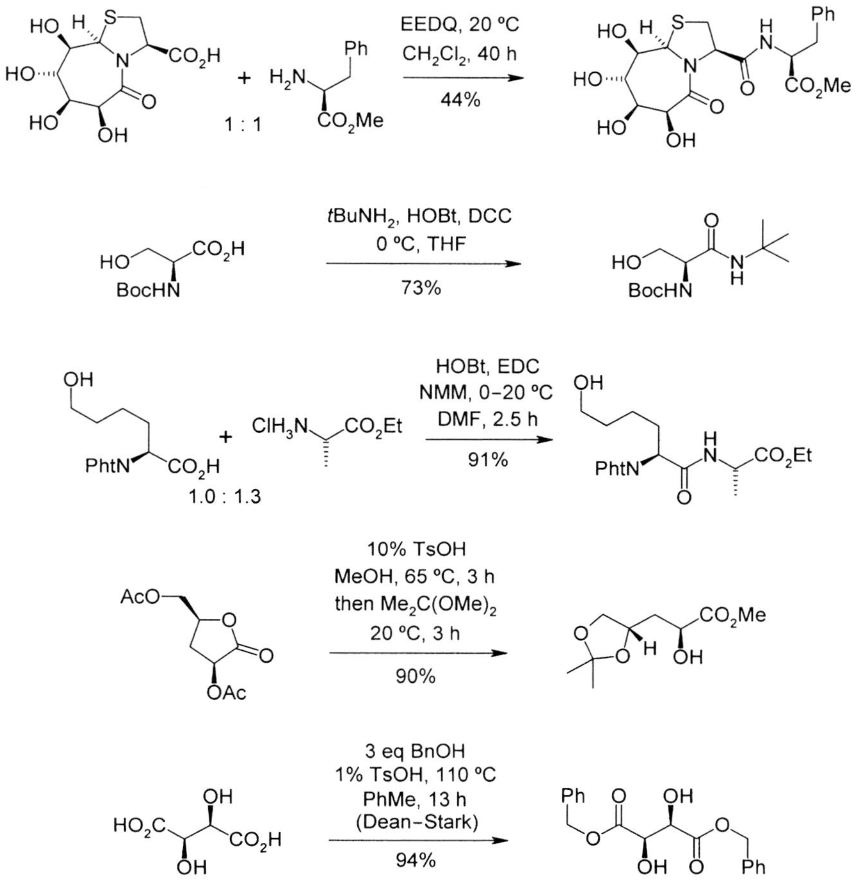

HO H S
HO N CO_2H
HO O
OH
+ H_2N Ph CO_2Me
1 : 1

EEDQ, 20 °C
CH_2Cl_2, 40 h
44%

HO H S H Ph
HO N N
HO O O CO_2Me
OH

HO CO_2H
BocHN

tBuNH$_2$, HOBt, DCC
0 °C, THF
73%

HO O
BocHN N H

OH
PhtN CO_2H
1.0 : 1.3
+ ClH_3N CO_2Et

HOBt, EDC
NMM, 0–20 °C
DMF, 2.5 h
91%

OH
PhtN H N CO_2Et
O

AcO O
O
OAc

10% TsOH
MeOH, 65 °C, 3 h
then $Me_2C(OMe)_2$
20 °C, 3 h
90%

O O H CO_2Me
OH

HO_2C OH CO_2H
OH

3 eq BnOH
1% TsOH, 110 °C
PhMe, 13 h
(Dean–Stark)
94%

Ph O OH
O O
OH O Ph

Scheme 7.2. Acylations with unprotected hydroxy acids [10–14].

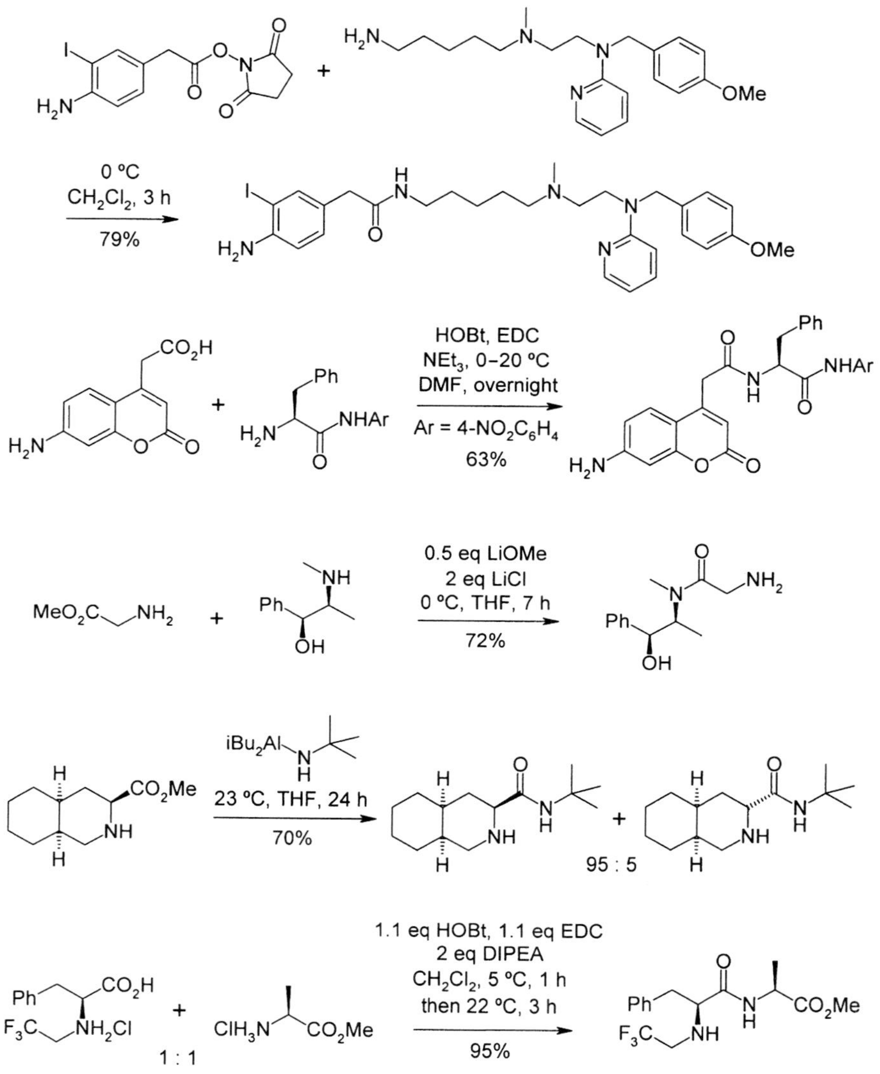

Scheme 7.3. Acylations with unprotected amino acids [15–19].

Another problem posed by unprotected amino acids is that most are insoluble in organic solvents. Their conversion into an acylating reagent will, therefore, either not proceed or will proceed more slowly than usual, and a careful optimization of the reaction conditions might be required. Successful examples of such reactions are shown in Scheme 7.4. Problems resulting from low solubility of unprotected amino acids can be avoided by silylation or by conversion into esters in a separate step and by using these esters as acylating agents.

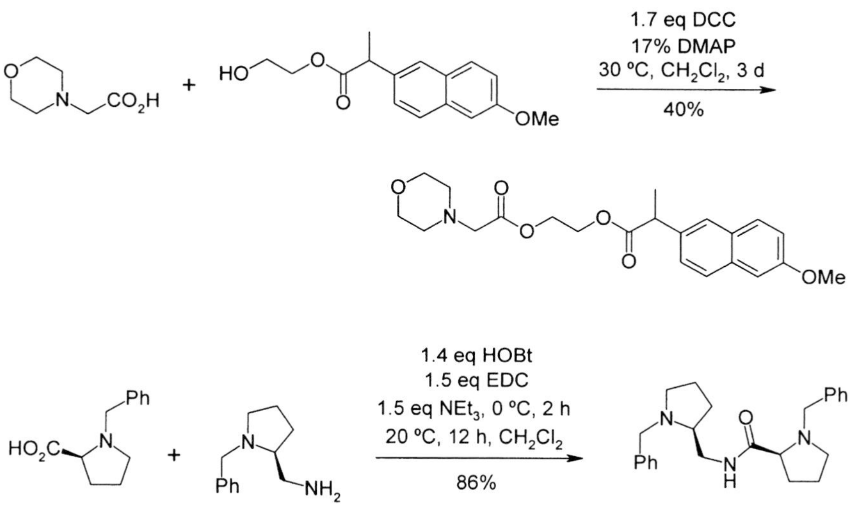
1.7 eq DCC
17% DMAP
30 °C, CH2Cl2, 3 d
40%
CO2H
HO
OMe
1.4 eq HOBt
1.5 eq EDC
1.5 eq NEt3, 0 °C, 2 h
20 °C, 12 h, CH2Cl2
86%
HO2C
Ph
NH2

Scheme 7.4. Acylations with dialkylamino acids [20, 21].

7.1.3
Carboxylic Acids with Additional Electrophilic Groups

Carboxylic acids with strongly electron-withdrawing groups, for example trifluoroacetic or 2,4,6-trinitrobenzoic acid [22], are readily converted into esters or amides. The products can, however, be unusually sensitive toward attack by nucleophiles and can readily undergo hydrolysis, transesterification, or transamidation. 2,4,6-Tris(trifluoromethyl)benzoic acid has been reported to undergo conversion into the acyl chloride or esters only with difficulty [23].

Acrylic acid esters can polymerize readily; this must be taken into account during their preparation. Thus, attempts to prepare pentafluorophenyl acrylate from acryloyl chloride in the presence of pyridine led to extensive polymerization of the product [24]. This polymerization can be prevented by using the less nucleophilic 2,6-dimethylpyridine as base and diethyl ether or pentane instead of THF as solvent (Scheme 7.5). Esterifications of acrylic acid under acidic conditions should be conducted in the presence of small amounts of hydroquinone as radical scavenger. Acrylic acid derivatives can also be prepared by acylation with a propionic acid with a leaving group at C-3 followed by β-elimination.

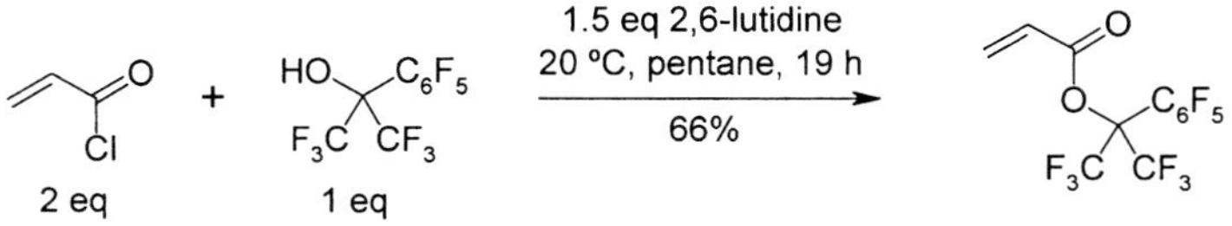
1.5 eq 2,6-lutidine
20 °C, pentane, 19 h
66%
Cl
2 eq
HO
C6F5
F3C CF3
1 eq

Scheme 7.5. Preparation of an acrylate [24].

Acrylic and related acid derivatives are good Michael acceptors, and will quickly react with amines or thiols if these are present in excess. Nevertheless, because acylations are usually much faster than Michael additions, acylation of these nucleophiles, even under basic conditions, can be performed with high yields. Similarly, acylating agents containing a functional group able to alkylate nucleophiles will usually act as acylating agents. Only with unreactive acylating functionalities will alkylation become the main reaction pathway (Scheme 7.6).

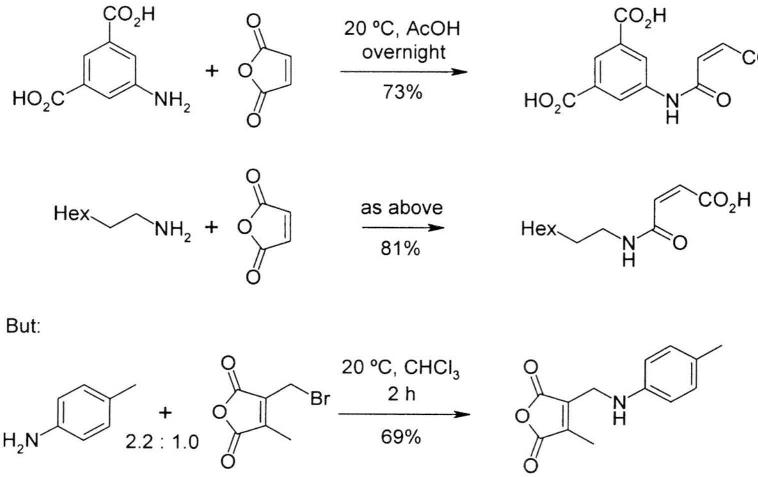

20 °C, AcOH overnight 73%

as above 81%

But:

2.2 : 1.0 20 °C, $CHCl_3$ 2 h 69%

Scheme 7.6. Reactivity of acylating agents containing an alkylating functional group [25, 26].

β-Keto acids and 2-alkyl- or 2-arylmalonic acids readily decarboxylate, and acylations with these proceed satisfactorily only at low temperatures. An alternative to direct acylation with β-keto acids is thermolysis of acyl Meldrum's acid (to generate an acylketene) in the presence of a nucleophile (Scheme 7.7). Esters of β-keto acids can also be conveniently prepared from other keto esters by transesterification [27, 28].

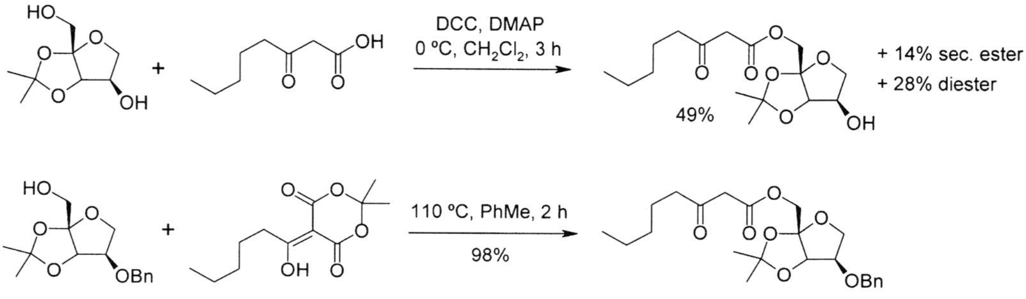

DCC, DMAP 0 °C, CH_2Cl_2, 3 h 49% + 14% sec. ester + 28% diester

110 °C, PhMe, 2 h 98%

Scheme 7.7. Esterification with β-keto acids [29].

7.2 Problematic Amines

7.2.1 Sterically or Electronically Deactivated Amines

Most aliphatic and aromatic amines can be acylated in high yields with a suitable carboxylic acid derivative, even in the presence of water or alcohols. Problems can arise, however, if the amine is sterically demanding or if its nucleophilicity is reduced by electron-withdrawing groups. In such instances highly reactive acylating agents, for example ketenes, acyl halides, or anhydrides, might be required and the acylation will have to be conducted under anhydrous conditions. Nucleophilic catalysts, for example DMAP, can, furthermore, be used to enhance the rate of difficult acylations. Some sterically hindered amines, for example *N*-alkyl-*N*-trityl amines, cannot even be acylated intermolecularly with ketenes, although examples of intramolecular acylations have been reported [30]. Examples of the acylation of sterically demanding amines are shown in Scheme 7.8.

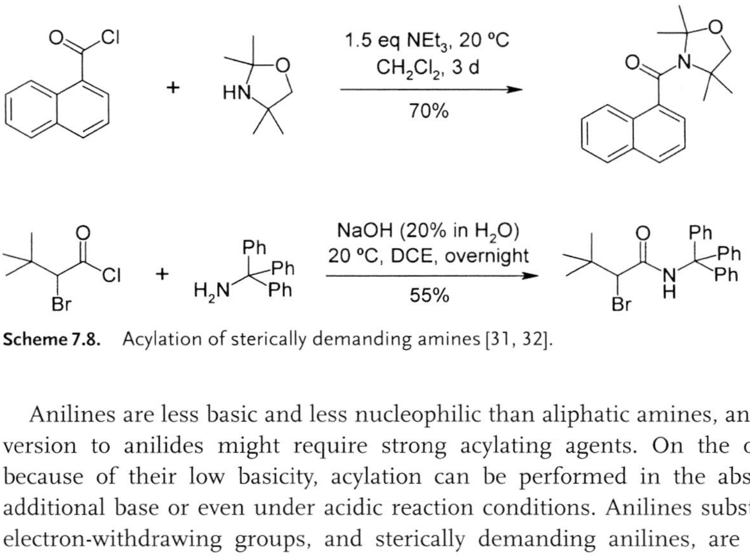

1.5 eq NEt_3, 20 °C, CH_2Cl_2, 3 d; 70%

NaOH (20% in H_2O), 20 °C, DCE, overnight; 55%

Scheme 7.8. Acylation of sterically demanding amines [31, 32].

Anilines are less basic and less nucleophilic than aliphatic amines, and their conversion to anilides might require strong acylating agents. On the other hand, because of their low basicity, acylation can be performed in the absence of an additional base or even under acidic reaction conditions. Anilines substituted with electron-withdrawing groups, and sterically demanding anilines, are particularly difficult to acylate (Scheme 7.9). If satisfactory conditions for acylating a given aniline cannot be found, its nucleophilicity (and basicity) can be enhanced by deprotonation with a strong base. Activated acids with α hydrogen (R_2CHCOX) will, however, be converted into ketenes by most metalated anilines.

Diarylamines will also usually require strong acylating agents to undergo acylation. If Lewis acids such as $ZnCl_2$ are used as catalysts the formation of an acridine can compete with *N*-acylation (Bernthsen acridine synthesis; last reaction, Scheme 7.10).

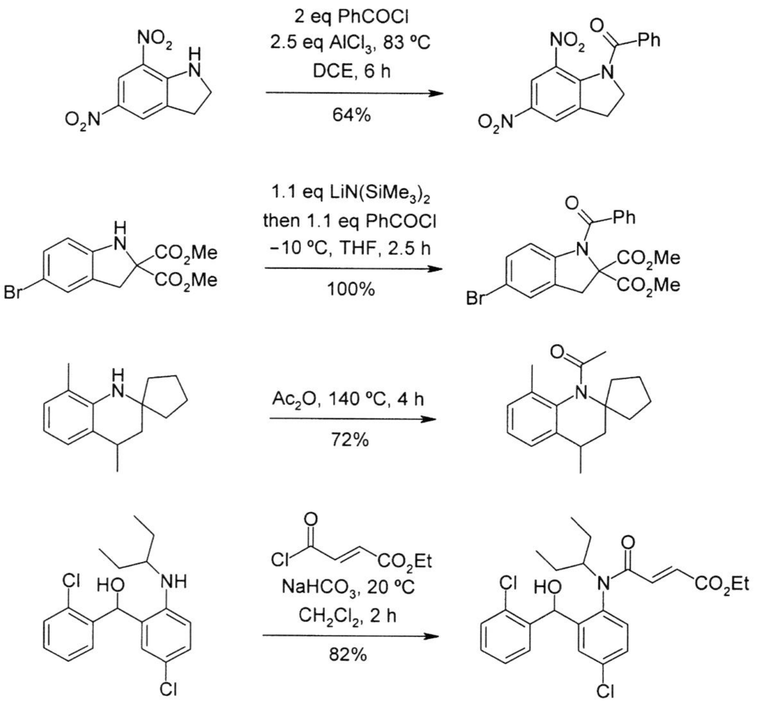

Scheme 7.9. Acylation of sterically or electronically deactivated anilines [33–36].

5 eq $PhCO_2H$, 4 eq $ZnCl_2$, microwaves (210 °C), 5 min 0 : 100 (37% yield; Ar = Ph)
5 eq $ArCO_2H$, 4 eq $ZnCl_2$, microwaves (210 °C), 10 min 100 : 1 (23% yield)
(Ar = 2,6-$(MeO)_2C_6H_3$)

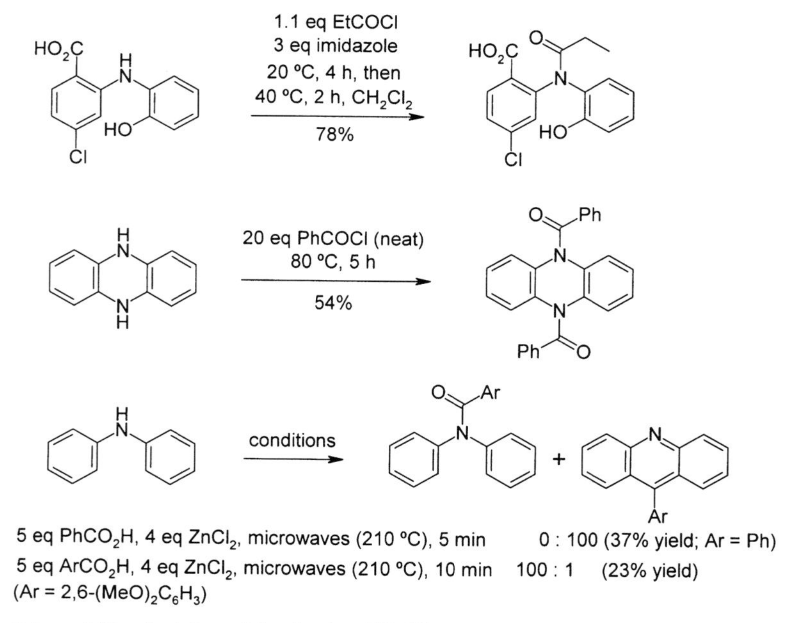

Scheme 7.10. Acylation of diarylamines [37–39].

Another class of problematic amines are α-amino nitriles, which are readily accessible from ketones, amines, and cyanide. Like α-amino acids, these amines are electronically deactivated and less basic and nucleophilic than purely aliphatic amines, and are therefore difficult to acylate. Some α-amino nitriles or the corresponding acylated derivatives can, furthermore, decompose into imines and cyanide if reaction temperatures are too high or if the bases used are too strong (Scheme 7.11).

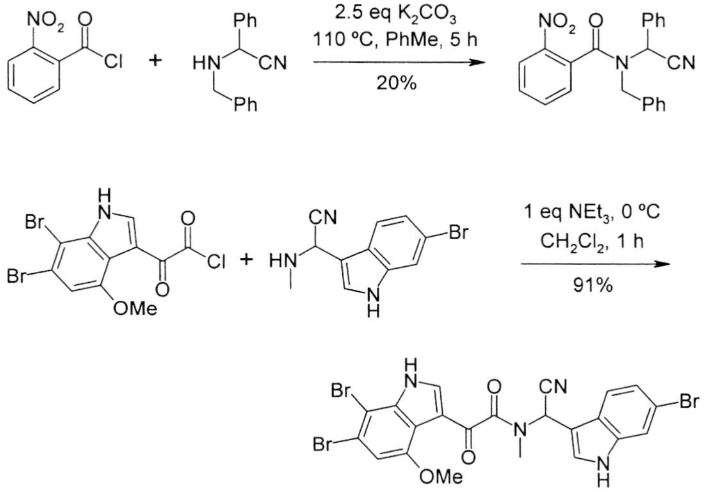

2.5 eq K_2CO_3, 110 °C, PhMe, 5 h; 20%

1 eq NEt_3, 0 °C, CH_2Cl_2, 1 h; 91%

Scheme 7.11. Acylation of α-amino nitriles [40, 41].

7.2.2
Amino Acids

As mentioned above, unprotected amino acids are often insoluble in aprotic solvents, which can render their *N*-acylation difficult. If Schotten–Baumann conditions are applicable, these will usually give the best results. The direct acylation of unprotected amino acids is, however, hampered by another problem: the formation of mixed anhydrides from the amino carboxylate and the added acylating agent. These anhydrides can react with the amine to yield unwanted oligomers of the amino acid, or lead to other unexpected reactions. In Scheme 7.12 two examples of such reactions have been sketched. These examples show that carboxylates react with acyl halides at similar rates as with amines, leading in these instances to excessive consumption of acyl chloride and to the formation of product mixtures if only one equivalent of acylating agent is used. For this reason more than two equivalents of acylating agent will be required for complete acylation of the amino group. Use of a large excess of acylating agent will, however, only be practicable if the *N*-acylamino acid can be readily separated from the hydrolyzed acylating agent (which is also an acid).

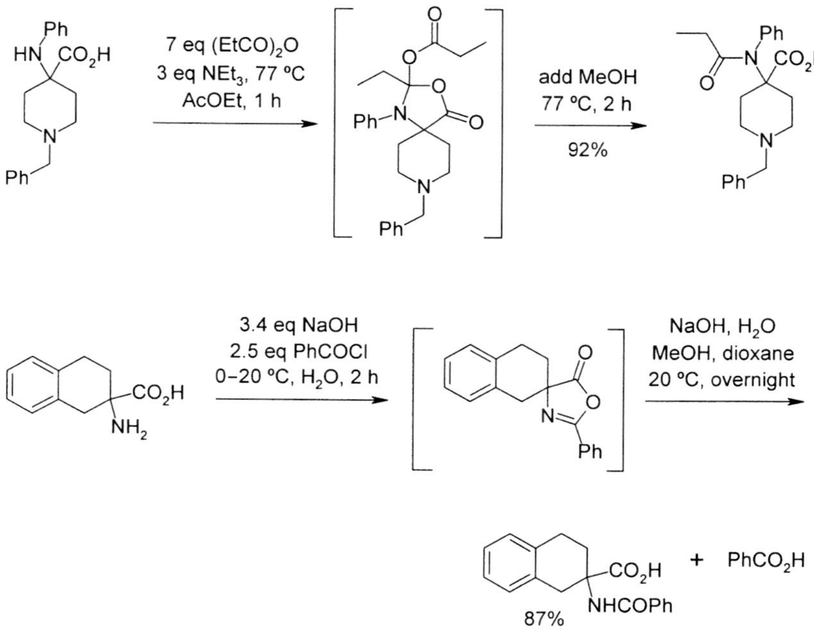

Scheme 7.12. Acylation of unprotected amino acids [42, 43].

7.2.3
Amines with Additional Nucleophilic Groups

Unsymmetric diamines can sometimes be acylated selectively at one amino group only, if one of the amino groups is sterically or electronically deactivated. If the difference in reactivity is only small, acylating agents of low reactivity will usually be required. Some polyamino heteroarenes can be monoacylated or monoalkylated with astonishingly high regioselectivity (Scheme 7.13), which is, however, difficult to predict.

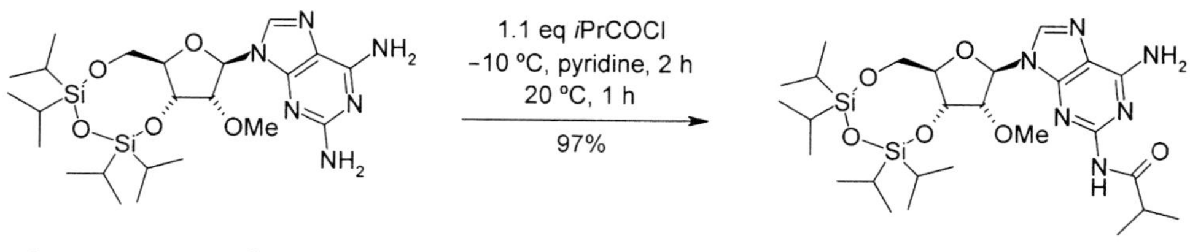

Scheme 7.13. Regioselective acylation of a 2,6-diaminopurine [44].

Amino alcohols or aminophenols can be acylated selectively either at the amino group or at the hydroxyl group. Under slightly basic or neutral reaction conditions most acylating reagents will acylate amino groups faster than hydroxyl groups, and amides will be the main products. Strongly basic conditions, however, under which the amino alcoholate is formed, or acidic reaction conditions will usually lead to the formation of amino esters (Scheme 7.17). Most amino esters with a primary or secondary amino group will be stable as ammonium salts only, and will rearrange to amides under neutral or basic conditions, especially if the amino group is in close proximity to the alkoxycarbonyl group, as is the case for 2-aminoethyl esters or 2-aminophenyl esters.

7.3 Problematic Alcohols

7.3.1 Sterically Deactivated and Base-labile Alcohols

Alcohols can be converted into esters under both acidic or basic reaction conditions, and numerous methods have been developed which will be suitable for most alcohols. Problems can, however, arise if the alcohol is sterically hindered. Strongly acidic alcohols or phenols can usually be readily converted into esters [45], but these will be reactive acylating reagents themselves, and may be hydrolyzed during aqueous work-up. (This can sometimes be avoided by performing the work-up with CH_2Cl_2/H_2O.)

Examples of the acylation of sterically demanding alcohols are given in Scheme 7.14. These reactions often proceed only slowly, and conditions suitable for most primary or secondary alcohols will fail with tertiary alcohols [46–48]. Trityl esters and similar compounds are difficult to prepare by acylation of the corresponding alcohols, but are readily accessible via *O*-alkylation of carboxylates.

If the hydroxyl group of an alcohol is β to an electron-withdrawing group, the resulting esters can undergo facile β-elimination. Such alcohols are difficult to acylate under basic reaction conditions but might undergo clean esterification in the presence of acids (Scheme 7.15).

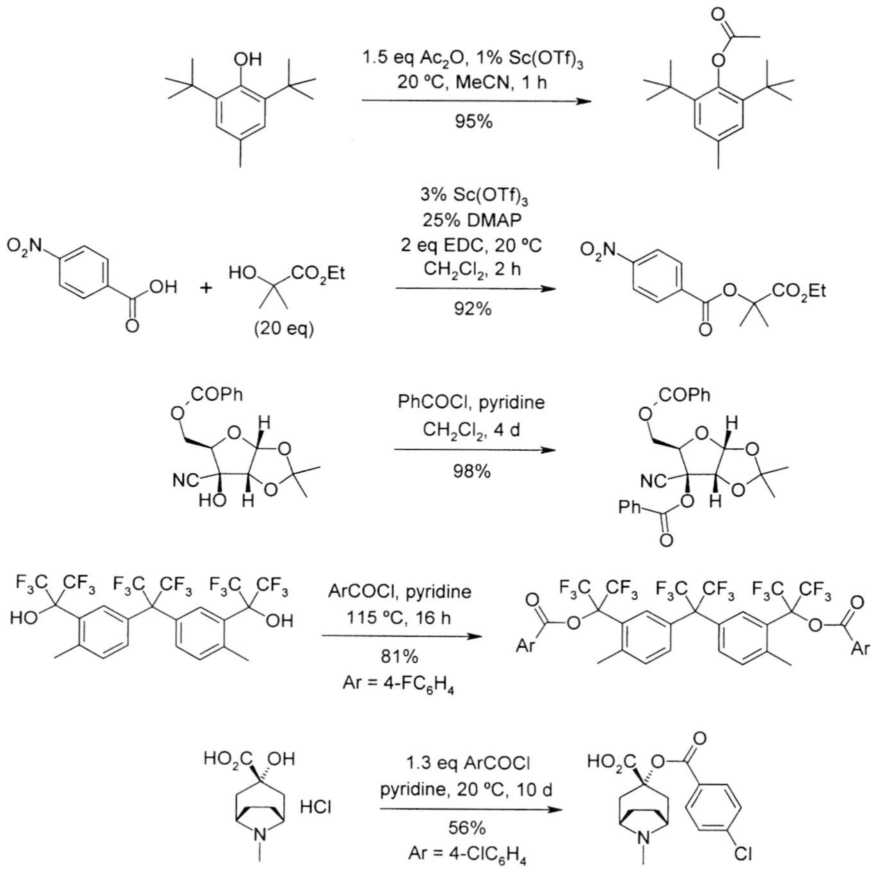

Scheme 7.14. Acylation of sterically demanding alcohols [49–53].

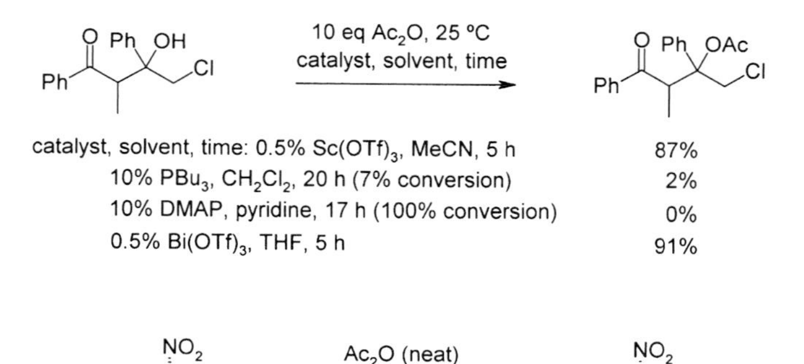

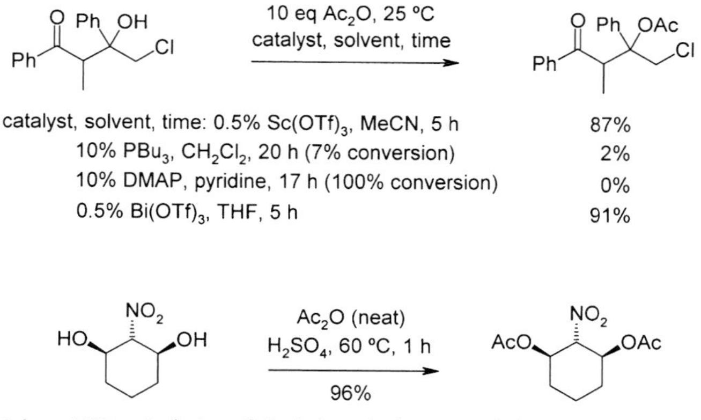

Scheme 7.15. Acylation of alcohols with electron-withdrawing groups in the β position [54, 55].

7.3.2 Alcohols with Additional Nucleophilic Groups

Unsymmetric polyols can be monoacylated selectively with acylating agents of moderate reactivity. For 1,2-diols the reactivity of the monoacylated product may be sufficiently lowered to prevent the formation of substantial amounts of diacylated product (first reaction, Scheme 7.16; see also Section 10.3). If the two hydroxyl groups are farther apart, however, selective monoacylations will require different nucleophilicity of both hydroxyl groups and an acylating reagent of low reactivity. If the hydroxyl groups are of different acidity, addition of a base can promote the esterification of the most acidic hydroxyl group (third example, Scheme 7.16). Polyols can also be partially protected against acylation by transient complexation with boron or tin derivatives [56].

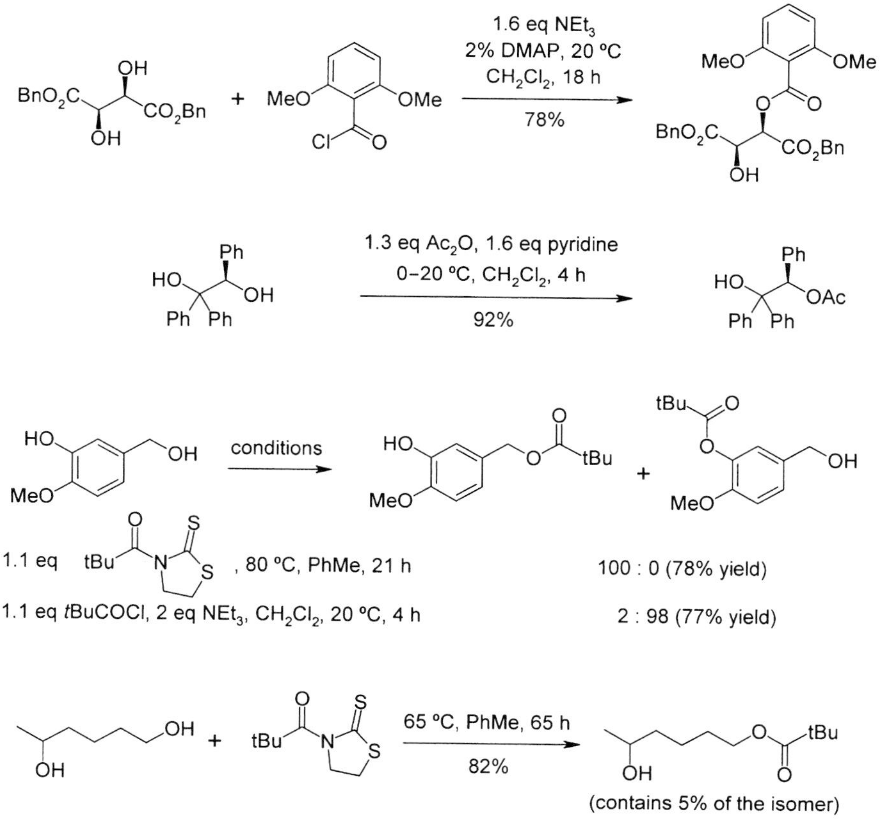

1.6 eq NEt_3, 2% DMAP, 20 °C, CH_2Cl_2, 18 h — 78%

1.3 eq Ac_2O, 1.6 eq pyridine, 0–20 °C, CH_2Cl_2, 4 h — 92%

conditions

1.1 eq tBu-CO-(2-thioxothiazolidin-3-yl), 80 °C, PhMe, 21 h — 100 : 0 (78% yield)

1.1 eq *t*BuCOCl, 2 eq NEt_3, CH_2Cl_2, 20 °C, 4 h — 2 : 98 (77% yield)

65 °C, PhMe, 65 h — 82% (contains 5% of the isomer)

Scheme 7.16. Selective monoacylation of diols [10, 57, 58].

Amino alcohols can be esterified, without simultaneous acylation of the amino group, by treatment with acids or acid anhydrides under acidic reaction conditions (Scheme 7.17) [59]. Reaction temperatures which are too high should be avoided, since these can lead to the acylation of ammonium salts to yield amides. Esters of 2-aminoethanols are stable only as ammonium salts, and conversion to the free amine will usually lead to migration of the acyl group (Section 3.3).

Aminophenols can be selectively esterified by deprotonation of the hydroxyl group followed by treatment with an acylating reagent of low reactivity (Scheme 7.17). Treatment with a strong acylating agent under acidic conditions will also result in clean esterification. In the presence of weak bases the regioselectivity of the acyla-

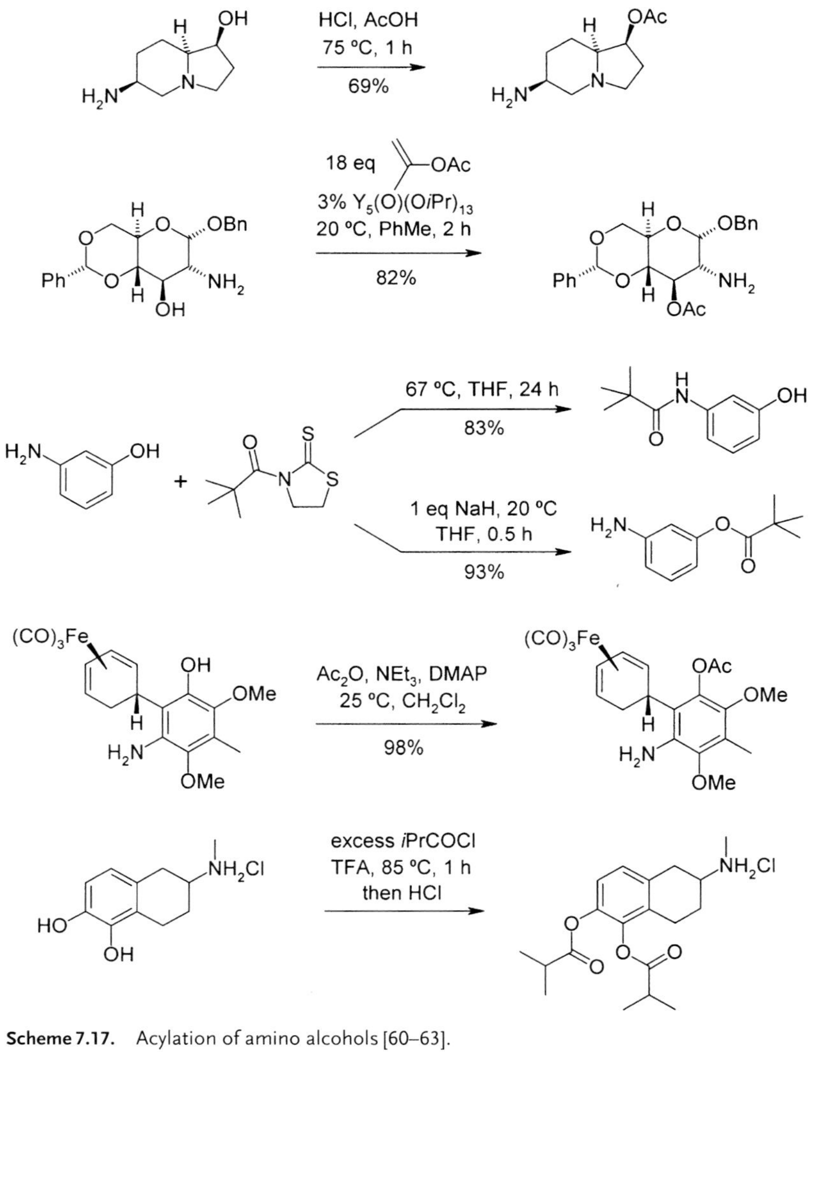

HCl, AcOH
75 °C, 1 h
69%

18 eq OAc
3% $Y_5(O)(OiPr)_{13}$
20 °C, PhMe, 2 h
82%

67 °C, THF, 24 h
83%

1 eq NaH, 20 °C
THF, 0.5 h
93%

Ac_2O, NEt_3, DMAP
25 °C, CH_2Cl_2
98%

excess *i*PrCOCl
TFA, 85 °C, 1 h
then HCl

Scheme 7.17. Acylation of amino alcohols [60–63].

tion of aminophenols seems to depend on the type of acylating reagent chosen and on the specific reaction conditions.

Mercaptoalcohols can also be acylated with high regioselectivity. Under neutral or acidic reaction conditions acylation of the hydroxyl group predominates [27, 64–66] whereas basic reaction conditions favor acylation of the mercapto group (Scheme 7.18).

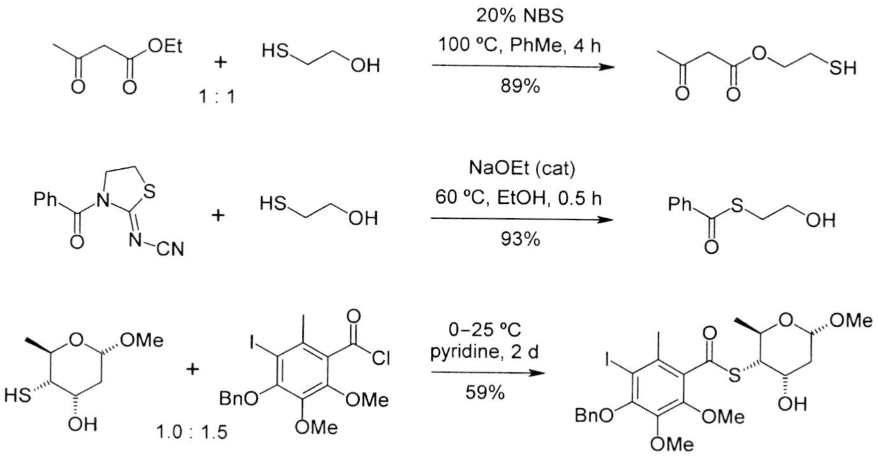

Scheme 7.18. Acylation of mercaptoalcohols [27, 67, 68].

References

1 Schlecker, R.; Seebach, D.; Lubosch, W. CH-Acidität in α-Stellung zum *N*-Atom in *N,N*-Dialkylamiden mit sterisch geschützter Carbonylgruppe zur nucleophilen Aminoalkylierung. *Helv. Chim. Acta* **1978**, *61*, 512–526.

2 Bengtsson, S.; Höberg, T. Pivaloyl-directed regioselective synthesis of 2,3,6-trioxygenated benzamides: phenolic metabolites of remoxipride. *J. Org. Chem.* **1993**, *58*, 3538–3542.

3 Ooi, T.; Sugimoto, H.; Doda, K.; Maruoka, K. Esterification of carboxylic acids catalyzed by in situ generated tetraalkylammonium fluorides. *Tetrahedron Lett.* **2001**, *42*, 9245–9248.

4 Shieh, W.-C.; Dell, S.; Repic, O. Nucleophilic catalysis with DBU for the esterification of carboxylic acids with dimethyl carbonate. *J. Org. Chem.* **2002**, *67*, 2188–2191.

5 Staab, H. A.; Lauer, D. Stabile rotationsisomere Carbonsäureamide. *Chem. Ber.* **1968**, *101*, 864–878.

6 Parker, K. A.; Coburn, C. A. Regioselectivity in intramolecular nucleophilic aromatic substitution. Synthesis of the potent anti HIV-1 8-halo TIBO analogues. *J. Org. Chem.* **1992**, *57*, 97–100.

7 Duncia, J. V.; Pierce, M. E.; Santella, J. B. Three synthetic routes to a sterically hindered tetrazole. A new one-step mild conversion of an amide into a tetrazole. *J. Org. Chem.* **1991**, *56*, 2395–2400.

8 Larsen, A. A.; Ruddy, A. W.; Elpern, B.; McMullin, M. Dialkylaminoalkyl esters of trisubstituted acetic acids. *J. Am. Chem. Soc.* **1949**, *71*, 532–533.

9 Drew, M. G. B.; Fengler-Veith, M.; Harwood, L. M.; Jahans, A. W. Highly selective chirally

templated isomünchnone cycloadditions of achiral aldehydes: synthesis of an enantiopure α,β-dihydroxy acid. *Tetrahedron Lett.* **1997**, *38*, 4521–4524.

10 Furuta, K.; Gao, Q.-Z.; Yamamoto, H. Chiral (acyloxy)borane complex catalyzed asymmetric Diels–Alder reaction: (1*R*)-1,3,4-trimethyl-3-cyclohexene-1-carboxaldehyde. *Org. Synth.* **1998**, *Coll. Vol. IX*, 722–727.

11 Geyer, A.; Moser, F. Polyol peptidomimetics. *Eur. J. Org. Chem.* **2000**, 1113–1120.

12 Robl, J. A.; Cimarusti, M. P.; Simpkins, L. M.; Weller, H. N.; Pan, Y. Y.; Malley, M.; DiMarco, J. D. Peptidomimetic synthesis: a novel, highly stereoselective route to substituted Freidinger lactams. *J. Am. Chem. Soc.* **1994**, *116*, 2348–2355.

13 Baldwin, J. E.; Farthing, C. N.; Russell, A. T.; Schofield, C. J.; Spivey, A. C. Use of (*S*)-*N-tert*-butoxycarbonylaziridine-2-carboxylate derivatives for α-amino acid synthesis. *Tetrahedron Lett.* **1996**, *37*, 3761–3764.

14 Sun, R. C.; Okabe, M. (2*S*,4*S*)-2,4,5-Trihydroxypentanoic acid 4,5-acetonide methyl ester. *Org. Synth.* **1998**, *Coll. Vol. IX*, 717–721.

15 Sorba, G.; Tertuik, W.; Ganellin, C. R. Synthesis and authentication of iodoazidophenpyramine, a photoaffinity reporter ligand previously used for histamine H_1-receptor labelling. *Arch. Pharm. (Weinheim)* **1995**, *328*, 677–680.

16 Myers, A. G.; Gleason, J. L.; Yoon, T.; Kung, D. W. Highly practical methodology for the synthesis of D- and L-α-amino acids, *N*-protected α-amino acids, and *N*-methyl-α-amino acids. *J. Am. Chem. Soc.* **1997**, *119*, 656–673.

17 Charitos, C.; Kokotos, G.; Tzougraki, C. Bifunctional coumarin derivatives in solution and solid phase synthesis of fluorogenic enzyme substrates. *J. Heterocycl. Chem.* **2001**, *38*, 153–158.

18 Houpis, I. N.; Molina, A.; Reamer, R. A.; Lynch, J. E.; Volante, R. P.; Reider, P. J. Towards the synthesis of HIV-protease inhibitors. Synthesis of optically pure 3-carboxydecahydroisoquinolines. *Tetrahedron Lett.* **1993**, *34*, 2593–2596.

19 DesMarteau, D. D.; Montanari, V. Easy preparation of bioactive peptides from the novel N^{α}-trifluoroethyl amino acids. *Chem. Lett.* **2000**, 1052–1053.

20 Rautio, J.; Nevalainen, T.; Taipale, H.; Vepsäläinen, J.; Gynther, J.; Laine, K.; Järvinen, T. Synthesis and in vitro evaluation of novel morpholinyl- and methylpiperazinylacyloxyalkyl prodrugs of 2-(6-methoxy-2-naphthyl)propionic acid (naproxen) for topical drug delivery. *J. Med. Chem.* **2000**, *43*, 1489–1494.

21 Ishihara, K.; Karumi, Y.; Kondo, S.; Yamamoto, H. Synthesis of C_3 symmetric, optically active triamidoamine and protetraazaphosphatrane. *J. Org. Chem.* **1998**, *63*, 5692–5695.

22 Zlotin, S. G.; Kislitsin, P. G.; Samet, A. V.; Serebryakov, E. A.; Konyushkin, L. D.; Semenov, V. V.; Buchanan, A. C.; Gakh, A. A. Synthetic utilization of polynitroaromatic compounds. 1. *S*-Derivatization of 1-substituted 2,4,6-trinitrobenzenes with thiols. *J. Org. Chem.* **2000**, *65*, 8430–8438.

23 Filler, R.; Gnandt, W. K.; Chen, W.; Lin, S. New aspects of the chemistry of 2,4,6-tris(trifluoromethyl)benzoic acid and related compounds. *J. Fluorine Chem.* **1991**, *52*, 99–105.

24 Blazejewski, J.-C.; Hofstraat, J. W.; Lequesne, C.; Wakselman, C.; Wiersum, U. E. Formation of monomeric halogenoaryl acrylates in the presence of hindered pyridine bases. *J. Fluorine Chem.* **1998**, *91*, 175–177.

25 Kalgutkar, A. S.; Crews, B. C.; Marnett, L. J. Design, synthesis, and biochemical evaluation of *N*-substituted maleimides as inhibitors of prostaglandin endoperoxide synthase. *J. Med. Chem.* **1996**, *39*, 1692–1703.

26 Deshpande, A. M.; Natu, A. A.; Argade, N. P. Chemo- and regioselective nucleophilic reactions of (bromomethyl)methylmaleic anhydride: synthesis of α-quinoxalinyl- and α-benzothiazinylacrylic acids. *Heterocycles* **1999**, *51*, 2159–2162.

27 Bandgar, B. P.; Uppalla, L. S.; Sadavarte, V. S. Chemoselective transesterification of β-keto esters under neutral conditions using NBS as a catalyst. *Synlett* **2001**, 1715–1718.

28 Seebach, D.; Hungerbühler, E.; Naef, R.; Schnurrenberger, P.; Weidmann, B.; Züger, M. Titanate-mediated transesterifications with functionalized substrates. *Synthesis* **1982**, 138–141.

29 Zeng, C.-m.; Midland, S. L.; Keen, N. T.; Sims, J. J. Concise syntheses of (–)- and (+)-syringolide 1 and (–)-Δ^7-syringolide 1. *J. Org. Chem.* **1997**, *62*, 4780–4784.

30 Fernández-Megía, E.; Iglesias-Pintos, J. M.; Sardina, F. J. Enantiomerically pure highly functionalized α-amino ketones from the reaction of chiral cyclic *N*-(9-phenylfluorenyl)

α-amido esters with organolithium reagents. *J. Org. Chem.* **1997**, *62*, 4770–4779.

31 Anstiss, M.; Clayden, J.; Grube, A.; Joussef, L. H. Lithiation and stereoselective transformations of 3-aroyl-2,2,4,4-tetramethyloxazolidines, a new class of acid-labile atropisomeric amides. *Synlett* **2002**, 290–294.

32 Lengyel, I.; Cesare, V.; Karram, H.; Taldone, T. About 1-triphenylmethyl-3-*tert*-butylaziridinone and some of its reactions. *J. Heterocycl. Chem.* **2001**, *38*, 997–1002.

33 Helgen, C.; Bochet, C. G. Preparation of secondary and tertiary amides under neutral conditions by photochemical acylation of amines. *Synlett* **2001**, 1968–1970.

34 Wright, S. W.; Dow, R. L.; McClure, L. D.; Hageman, D. L. A synthesis of functionalized indoline 2,2-biscarboxylates. *Tetrahedron Lett.* **1996**, *37*, 6965–6968.

35 Kouznetsov, V.; Zubkov, F.; Palma, A.; Restrepo, G. A simple synthesis of spiro-C_6-annulated hydrocyclopenta[g]indole derivatives. *Tetrahedron Lett.* **2002**, *43*, 4707–4709.

36 Miki, T.; Kori, M.; Mabuchi, H.; Banno, H.; Tozawa, R.; Nakamura, M.; Itokawa, S.; Sugiyama, Y.; Yukimasa, H. Novel 4,1-benzoxazepine derivatives with potent squalene synthase inhibitory activities. *Bioorg. Med. Chem.* **2002**, *10*, 401–414.

37 Chung, S. J.; Joo, K. C.; Kim, D. H. Convenient synthesis of 6-substituted 2-chloro-5,12-dihydro-5-oxobenzoxazolo[3,2-*a*]quinolines and *N*-acylated 3-chlorodibenz[*b*,*e*][1,4]oxazepin-11(5*H*)-ones. *J. Heterocycl. Chem.* **1997**, *34*, 485–488.

38 Mikulla, M.; Mülhaupt, R. Synthese von *N*-acylierten 5,10-Dihydrophenazin Verbindungen. *Chem. Ber.* **1994**, *127*, 1723–1728.

39 Koshima, H.; Kutsunai, K. Rapid synthesis of acridines using microwaves. *Heterocycles* **2002**, *57*, 1299–1302.

40 Broggini, G.; Garanti, L.; Molteni, G.; Zecchi, G. A new entry to [1,2,4]-triazolo[1,5-*a*][1,4]benzodiazepin-6-ones via intramolecular nitrilimine cycloaddition to the cyano group. *Tetrahedron* **1998**, *54*, 14859–14868.

41 Jiang, B.; Smallheer, J. M.; Amaral-Ly, C.; Wuonola, M. A. Total synthesis of (±)-dragmacidin: a cytotoxic bis(indole) alkaloid of marine origin. *J. Org. Chem.* **1994**, *59*, 6823–6827.

42 Coleman, M. J.; Goodyear, M. D.; Latham, D. W. S.; Whitehead, A. J. A convenient method of the *N*-acylation and esterification of hindered amino acids: synthesis of ultra short acting opioid agonist remifentanyl. *Synlett* **1999**, 1923–1924.

43 Obrecht, D.; Spiegler, C.; Schönholzer, P.; Müller, K.; Heimgartner, H.; Stierli, F. A new general approach to enantiomerically pure cyclic and open-chain (*R*)- and (*S*)- α,α-disubstituted α-amino acids. *Helv. Chim. Acta* **1992**, *75*, 1666–1696.

44 Beigelman, L.; Haeberli, P.; Sweedler, D.; Karpeisky, A. Improved synthetic approaches toward 2′-*O*-methyl adenosine and guanosine and their *N*-acyl derivatives. *Tetrahedron* **2000**, *56*, 1047–1056.

45 Wang, X.; Li, Z.; Da, Y.; Chen, J.; Wei, T. Phase transfer catalyzed synthesis of 4-carboxylphenoxyacetic acid derivatives. *Synth. Commun.* **1999**, *29*, 4153–4161.

46 Sano, T.; Ohashi, K.; Oriyama, T. Remarkably fast acylation of alcohols with PhCOCl promoted by TMEDA. *Synthesis* **1999**, 1141–1144.

47 Carrigan, M. D.; Freiberg, D. A.; Smith, R. C.; Zerth, H. M.; Mohan, R. S. A simple and practical method for large-scale acetylation of alcohols and diols using $Bi(OTf)_3$. *Synthesis* **2001**, 2091–2094.

48 Bandgar, B. P.; Kamble, V. T.; Sadavarte, V. S.; Uppalla, L. S. Selective sulfonylation of arenes and benzoylation of alcohols using $LiClO_4$ as a catalyst under neutral conditions. *Synlett* **2002**, 735–738.

49 Ishihara, K.; Kubota, M.; Kurihara, H.; Yamamoto, H. Scandium triflate as an extremely active Lewis acid catalyst in acylation of alcohols with acid anhydrides and mixed anhydrides. *J. Org. Chem.* **1996**, *61*, 4560–4567.

50 Zhao, H.; Pendri, A.; Greenwald, R. B. General procedure for acylation of tertiary alcohols: $Sc(OTf)_3$/DMAP reagent. *J. Org. Chem.* **1998**, *63*, 7559–7562.

51 Zhang, M.; Zhang, H.; Yang, Z.; Ma, L.; Min, J.; Zhang, L. Synthesis of 3-*C*-(methyl-β-D-xylofuranosid-3-yl)-5-phenyl-1,2,4-oxadiazole. *Carbohydr. Res.* **1999**, *318*, 157–161.

52 Reynolds, D. W.; Cassidy, P. E.; Johnson, C. G.; Cameron, M. L. Exploring the chemistry of the 2-arylhexafluoro-2-propanol group: synthesis and reactions of a new highly fluorinated monomer intermediate and its derivatives. *J. Org. Chem.* **1990**, *55*, 4448–4454.

53 Burgos, C.; Gálvez, E.; Izquierdo, M. L.; Arias, M. S.; López, P. Synthesis, structural, conformational, and biochemical study of some

3β-acyloxytropan-3α-carboxylic acid hydrochlorides. *J. Heterocycl. Chem.* **1992**, *29*, 1821–1827.

54 Orita, A.; Tanahashi, C.; Kakuda, A.; Otera, J. Highly powerful and practical acylation of alcohols with acid anhydrides catalyzed by $Bi(OTf)_3$. *J. Org. Chem.* **2001**, *66*, 8926–8934.

55 Eberle, M.; Missbach, M.; Seebach, D. Enantioselective saponification with pig liver esterase: (1*S*,2*S*,3*R*)-3-hydroxy-2-nitrocyclohexyl acetate. *Org. Synth.* **1993**, *Coll. Vol. VIII*, 332–338.

56 Bhaskar, V.; Duggan, P. J.; Humphrey, D. G.; Krippner, G. Y.; McCarl, V.; Offermann, D. A. Phenylboronic acid as a labile protective agent: the selective derivatisation of 1,2,3-triols. *J. Chem. Soc. Perkin Trans. 1* **2001**, 1098–1102.

57 Yamada, S.; Sugaki, T.; Matsuzaki, K. Twisted amides as selective acylating agents for hydroxyl groups under neutral conditions: models for activated peptides during enzymatic acyl transfer reactions. *J. Org. Chem.* **1996**, *61*, 5932–5938.

58 Braun, M.; Gräf, S.; Herzog, S. (*R*)-(+)-2-Hydroxy-1,2,2-triphenylethyl acetate. *Org. Synth.* **1998**, *Coll. Vol. IX*, 507–509.

59 Robinson, C. H.; Milewich, L.; Hofer, P. The oxidation of steroidal amines to nitro steroids. *J. Org. Chem.* **1966**, *31*, 524–528.

60 Lin, M.-H.; RajaBabu, T. V. Metal-catalyzed acyl transfer reactions of enol esters: role of $Y_5(OiPr)_{13}O$ and $(thd)_2Y(OiPr)$ as transesterification catalysts. *Org. Lett.* **2000**, *2*, 997–1000.

61 Dai, W.-M.; Cheung, Y. K.; Tang, K. W.; Choi, P. Y.; Chung, S. L. Highly chemoselective acylation of substituted aminophenols with 3-(trimethylacetyl)-1,3-thiazolidine-2-thione. *Tetrahedron* **1995**, *51*, 12263–12276.

62 Knölker, H.-J.; Fröhner, W. Transition metal complexes in organic synthesis, part 38. First total synthesis of carbazomycin G and H. *Tetrahedron Lett.* **1997**, *38*, 4051–4054.

63 Pighi, R.; Pivetti, F. Process for the manufacture of form I of nolomirole hydrochloride. Patent *EP 1384708A1*, **2004**.

64 Kumar, P.; Pandey, R. K.; Bodas, M. S.; Dongare, M. K. Yttria-zircona based Lewis acid: an efficient and chemoselective catalyst for acetylation reactions. *Synlett* **2001**, 206–209.

65 Li, T.-S.; Li, A.-X. Montmorillonite clay catalysis. Part 10. K-10 and KSF-catalyzed acylation of alcohols, phenols, thiols and amines: scope and limitation. *J. Chem. Soc. Perkin Trans. 1* **1998**, 1913–1917.

66 Iglesias, L. E.; Baldessari, A.; Gros, E. G. Lipase-catalyzed chemospecific *O*-acylation of 3-mercapto-1-propanol and 4-mercapto-1-butanol. *Bioorg. Med. Chem. Lett.* **1996**, *6*, 853–856.

67 Iwata, C.; Watanabe, M.; Okamoto, S.; Fujimoto, M.; Sakae, M.; Katsurada, M.; Imanishi, T. 3-Acyl-2-(*N*-cyanoimino)thiazolidines as an acylating agent. Preparation of amides, esters, and thioesters. *Heterocycles* **1988**, *27*, 323–326.

68 van Laak, K.; Rainer, H.; Scharf, H.-D. Synthesis of subunits of calichemicins γ_1^I. *Tetrahedron Lett.* **1990**, *31*, 4113–4116.

8
Palladium-catalyzed C–C Bond Formation

8.1
Introduction

The development in recent decades of numerous different Pd-catalyzed C–C bond-forming reactions has had a huge impact on synthetic organic chemistry [1–10]. Several classes of compounds, for example unsymmetrical biaryls or substituted polyenes, which were difficult if not impossible to synthesize in the 1950s, can now be readily prepared under mild conditions with high regio- and stereoselectivity via Pd-catalyzed cross-coupling reactions. The compatibility of Pd-catalysis with many functional groups and protic solvents, together with its broad scope, has secured a central role for this chemistry in organic synthesis.

The most important Pd-catalyzed C–C bond-forming reactions proceed according to the mechanisms outlined in Schemes 8.3 and 8.6. In these reactions a Pd(0) complex, which can be generated in situ by reduction of Pd(II), undergoes oxidative addition to an organic halide or sulfonate to yield an organopalladium(II) complex. This reaction is rapid with aryl or vinyl halides but proceeds more slowly with alkyl halides [9]. The reactivity decreases in the order $RI > ROTf \approx RBr \gg RCl \gg RF$ [11], and electron-poor aryl or vinyl halides react with Pd(0) complexes faster than electron-rich halides. The resulting organopalladium(II) complex can react with alkenes or a variety of organometallic compounds, forming the C–C bond and regenerating the catalytically active Pd(0) complex.

8.2
Chemical Properties of Organopalladium Compounds

The synthetically most important Pd-catalyzed C–C bond-forming reactions proceed via intermediate square planar organopalladium(II) complexes of the type $PdRL_3$, in which the ligands L are usually halide or phosphine ligands. Organopalladium(II) complexes are rather stable organometallic compounds which can be readily prepared and isolated [12–14]. Complexes such as $PdArL_2X$ with L representing phosphines or amines and X being a halogen do not usually react with water, and Pd-mediated C–C bond-forming reactions can be conducted in the presence of water

Side Reactions in Organic Synthesis. Florencio Zaragoza Dörwald

ISBN: 3-527-31021-5

or in water alone [15–20]. Water may even be beneficial for some types of Pd-catalyzed reaction [21]. Organopalladium complexes such as $PdArL_2X$ react with water only if L or X are very weak ligands (e.g. acetone [22]). These weak ligands can be displaced by water, and in the resulting complexes hydrolysis of the Pd–C bond can occur to yield a hydrocarbon.

The products of the thermolysis of $PdAr(Hal)(PAr_3)_2$ depend on the substitution pattern of the aryl groups, on the halide, on the solvent, and on the temperature chosen. Heating of $PdAr(Hal)(PAr_3)_2$, in the absence of other reagents, to temperatures above 50 °C leads to the reversible migration of aryl groups from palladium to phosphorus via intermediate formation of tetraarylphosphonium salts [11, 23, 24] (see Section 8.6). This migration proceeds more readily with electron-rich Pd-bound aryl groups [25]. Strong heating yields phosphonium salts, phosphines, biaryls Ar–Ar, reduced arenes ArH (the Ar group originating from both Pd-bound and phosphorus-bound aryl groups), and metallic palladium [11, 26], probably via homolytic Pd–C bond fission [27]. Heating in the presence of phosphines can lead to the reductive elimination of aryl halides [28].

Complexes such as $PdArL_2X$ are usually stable in air, but treatment with strong oxidants leads to cleavage of the Pd–C bond to yield phenols [29, 30]. Carbon monoxide cleanly converts organopalladium(II) complexes into acyl complexes (L_3PdCOR) [31, 32].

The metal-bound carbon atom in organopalladium(II) complexes can formally react either as an electrophile or as a nucleophile. Treatment of arylpalladium(II) complexes with alkyl halides, for example, yields products of homo- or cross-coupling, possibly via intermediate formation of hexacoordinated Pd(IV) complexes [31, 33] (Scheme 8.1). Treatment of the same type of complex with alkyl Grignard reagents or other carbon nucleophiles, on the other hand, also yields the corresponding alkyl arenes via nucleophilic displacement of a ligand followed by reductive elimination (Scheme 8.1).

Organopalladium complexes do usually not transfer a formal carbanion to aldehydes or ketones, but intramolecular Barbier-type reactions of this kind have been observed (Scheme 8.2).

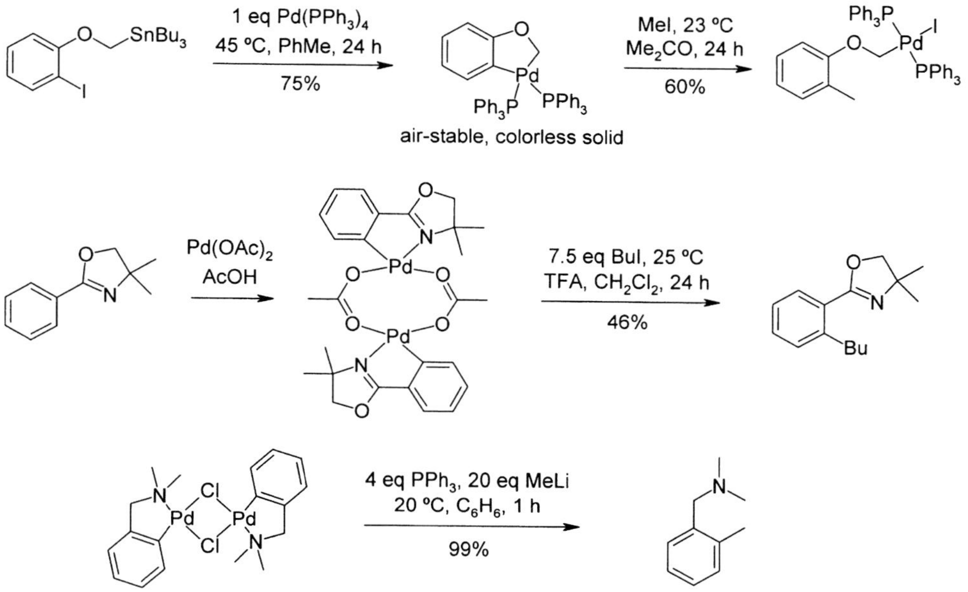

1 eq $Pd(PPh_3)_4$
45 °C, PhMe, 24 h
75%

air-stable, colorless solid

MeI, 23 °C
Me_2CO, 24 h
60%

$Pd(OAc)_2$
AcOH

7.5 eq BuI, 25 °C
TFA, CH_2Cl_2, 24 h
46%

4 eq PPh_3, 20 eq MeLi
20 °C, C_6H_6, 1 h
99%

Scheme 8.1. Reaction of arylpalladium(II) complexes with electrophilic and nucleophilic alkylating reagents [34–36].

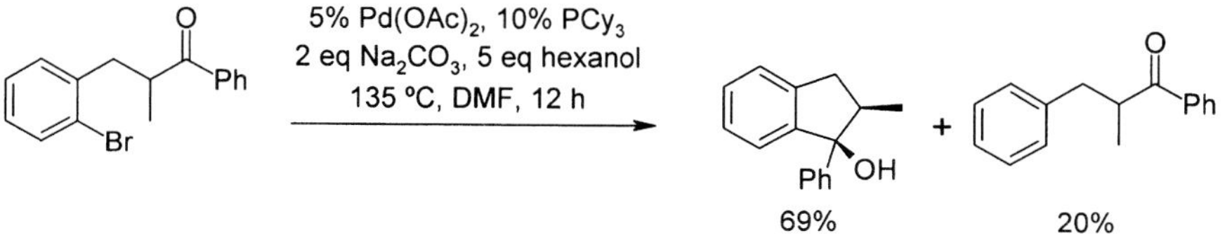

5% $Pd(OAc)_2$, 10% PCy_3
2 eq Na_2CO_3, 5 eq hexanol
135 °C, DMF, 12 h

69% 20%

Scheme 8.2. Palladium-mediated Barbier-type cyclization of a bromoaryl ketone [37].

8.3 Mechanisms of Pd-catalyzed C–C Bond Formation

8.3.1 Cross-coupling

For Pd-catalyzed cross-coupling reactions the organopalladium complex is generated from an organic electrophile RX and a Pd(0) complex in the presence of a carbon nucleophile. Not only organic halides but also sulfonium salts [38], iodonium salts [39], diazonium salts [40], or thiol esters (to yield acylpalladium complexes) [41] can be used as electrophiles. With allylic electrophiles (allyl halides, esters, or carbonates, or strained allylic ethers and related compounds) Pd-η^3-π-allyl complexes are formed; these react as soft, electrophilic allylating reagents.

Organopalladium complexes formed by oxidative insertion of Pd(0) into C–X bonds can undergo ligand exchange with other organometallic reagents R′M or with compounds with nucleophilic carbon (enolates, phenols, electron-rich heteroarenes) to yield diorganylpalladium(II) complexes (Scheme 8.3). These may undergo reductive elimination to yield the cross-coupled product R–R′ and the catalytically active Pd(0) complex (Scheme 8.3) [42].

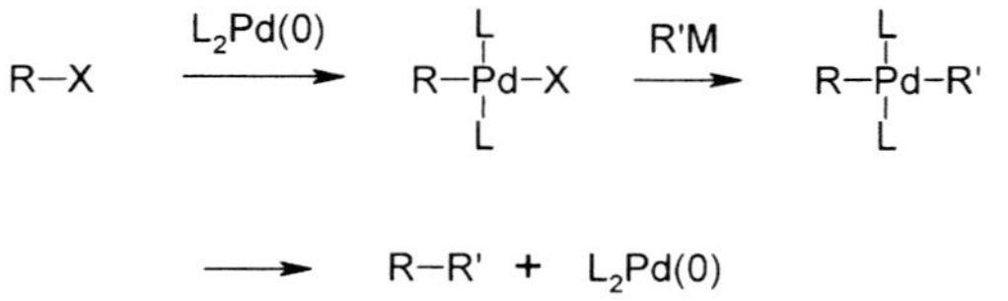

$$\mathrm{R{-}X} \xrightarrow{L_2Pd(0)} \mathrm{R{-}Pd(L)_2{-}X} \xrightarrow{R'M} \mathrm{R{-}Pd(L)_2{-}R'} \longrightarrow \mathrm{R{-}R'} + L_2Pd(0)$$

Scheme 8.3. Simplified mechanism of Pd-catalyzed cross-coupling reactions. R, R′ = aryl, vinyl, alkynyl, allyl, benzyl, alkyl, acyl; M = MgX, ZnX, $B(OR)_2$, BR_2, SnR_3.

The synthetically most exploited cross-coupling variants are those in which the organometallic components are organoboron (Suzuki reaction), organotin (Stille reaction), organozinc (Negishi reaction), or alkynylcopper derivatives (Sonogashira reaction). The advantages of these reagents, if compared with the more polar organolithium or Grignard reagents (which also undergo Pd-mediated cross-coupling reactions) are their low reactivity toward most functional groups, their stability, and the consequent simplicity of their handling. Other nucleophilic reagents which can displace the leaving group X from complexes $PdRL_2X$ include cyanide [43], enolates [3, 44–51], phenols [52–54], indoles [55], indolizines (Scheme 8.5) [56], pyrroles, furans, thiophenes, imidazoles, thiazoles, and related compounds [57–62]. With most of these reagents a new Pd–C bond is presumably formed on displacement of the ligand X.

Aryl-bound functional groups which are tolerated in Pd-mediated arylations include *ortho*-alkynyl [63], *ortho*-vinyl [64], *ortho*-nitro [65], and *ortho*-formyl groups [66]. Some examples of Pd-mediated cross-coupling reactions are depicted in Schemes 8.4 and 8.5, to illustrate the required conditions and scope of these reactions.

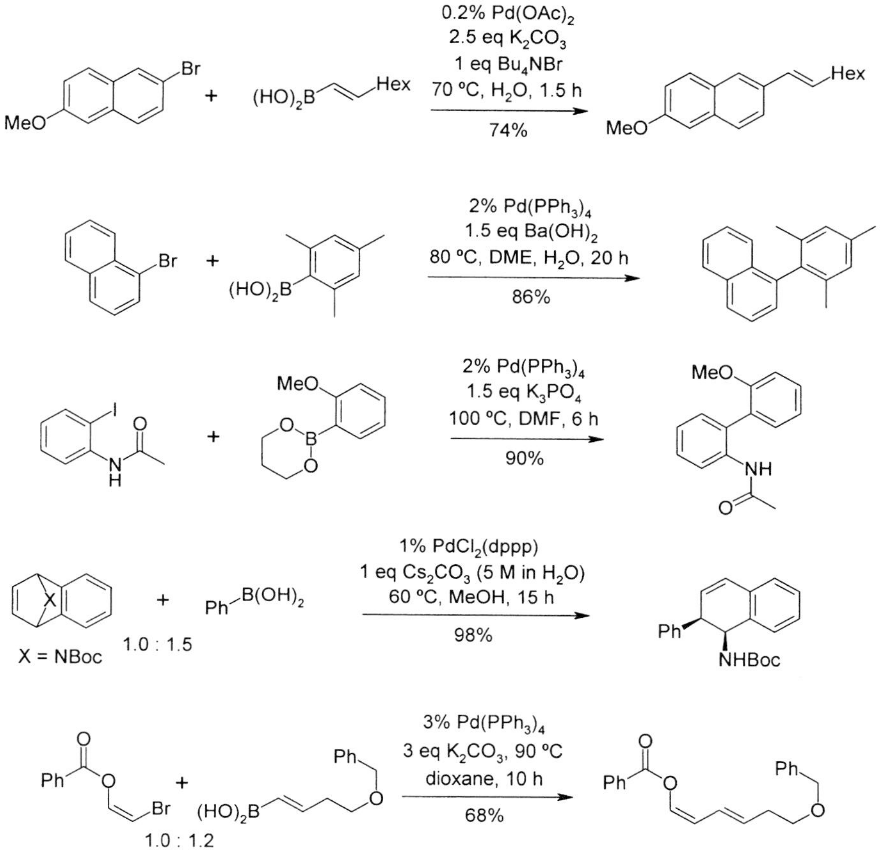

0.2% $Pd(OAc)_2$, 2.5 eq K_2CO_3, 1 eq Bu_4NBr, 70 °C, H_2O, 1.5 h; 74%

2% $Pd(PPh_3)_4$, 1.5 eq $Ba(OH)_2$, 80 °C, DME, H_2O, 20 h; 86%

2% $Pd(PPh_3)_4$, 1.5 eq K_3PO_4, 100 °C, DMF, 6 h; 90%

X = NBoc; 1.0 : 1.5; 1% $PdCl_2$(dppp), 1 eq Cs_2CO_3 (5 M in H_2O), 60 °C, MeOH, 15 h; 98%

1.0 : 1.2; 3% $Pd(PPh_3)_4$, 3 eq K_2CO_3, 90 °C, dioxane, 10 h; 68%

Scheme 8.4. Palladium-catalyzed cross-coupling reactions of boron derivatives [18, 66–68].

Organometallic reagents of low nucleophilicity, for example nitroaryl stannanes [69], undergo transmetalation only slowly; this can lead to low yields of cross-coupling products [69]. Catalytic amounts of CsF and CuI, together with a palladium source and a phosphine can bring about cross-coupling of such unreactive stannanes (first reaction, Scheme 8.5) [69].

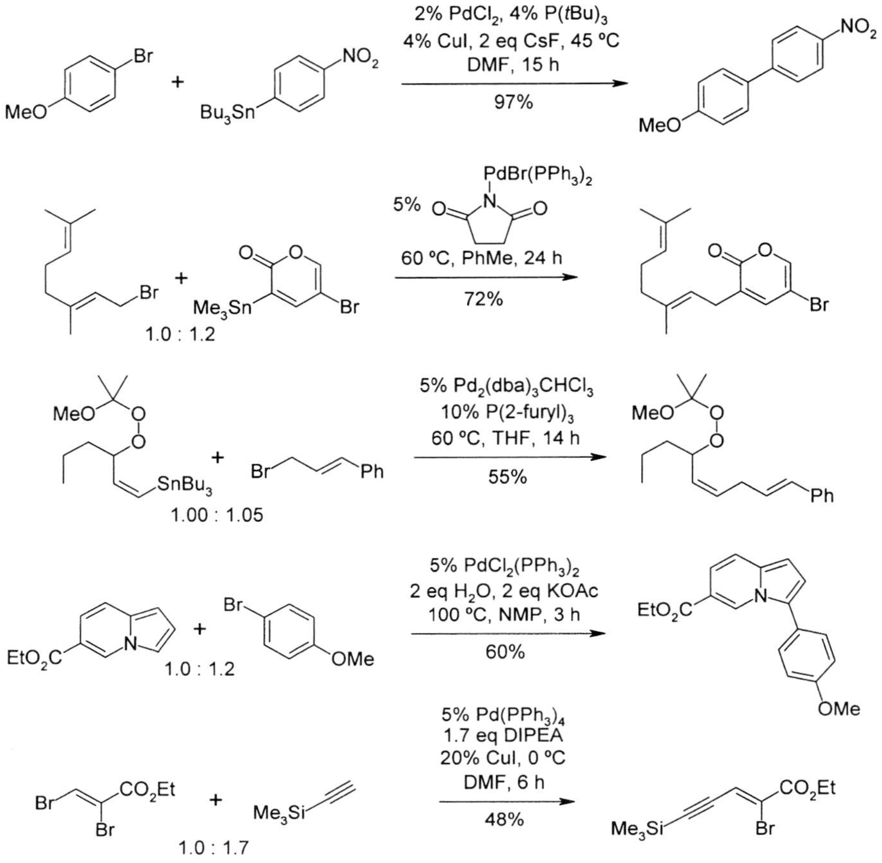

Scheme 8.5. Palladium-catalyzed cross-coupling reactions of stannanes and other carbon nucleophiles with aryl, allyl, and vinyl bromides [56, 69–72].

Alkylpalladium complexes generated by oxidative addition of Pd(0) to alkyl halides with a β hydrogen can undergo β-elimination to yield an alkene and a Pd-hydrido complex (as in the Heck reaction; Scheme 8.7). Nevertheless, this process is relatively slow compared with transmetalations and reductive eliminations, and simple alkyl halides or tosylates with β hydrogen can be cross-coupled with carbon nucleophiles under optimized conditions if the nucleophile is sufficiently reactive [9, 73–75] (Scheme 8.6).

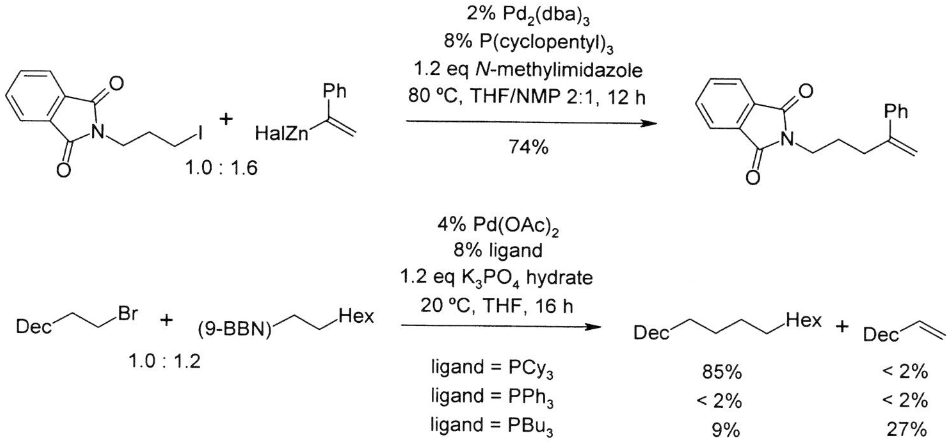

Scheme 8.6. Alkyl halides with β hydrogen in palladium-mediated cross-couplings [76, 77].

8.3.2 The Heck Reaction

If the initial organopalladium complex is formed in the presence of an alkene, transfer of the organyl group to the alkene can occur (Heck reaction, Scheme 8.7) [5, 78]. Electron-deficient and electron-rich alkenes and even some arenes [79] can be used. In the last step of this process a Pd-hydrido complex is formed by β-hydride elimination, leading to the formation of the C–C double bond. In the presence of bases the hydrido complex eliminates HX, regenerating the catalytically active Pd(0) complex (Scheme 8.7).

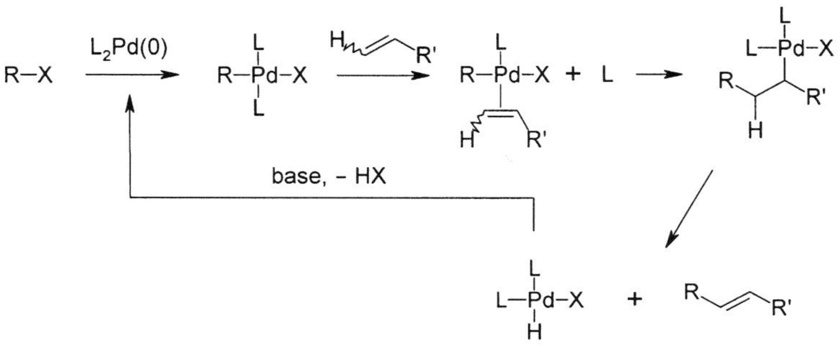

Scheme 8.7. Mechanism of the Heck reaction.

Representative examples of Heck reactions are depicted in Scheme 8.8. Terminal alkenes react faster than internal alkenes, and the formation of mixtures of products resulting from further arylation or vinylation of the initial product is therefore only observed when a large excess of halide and long reaction times are employed. Electron-poor alkenes usually react more rapidly than electron-rich alkenes, and the new C–C bond is usually formed at the most electron-deficient carbon atom.

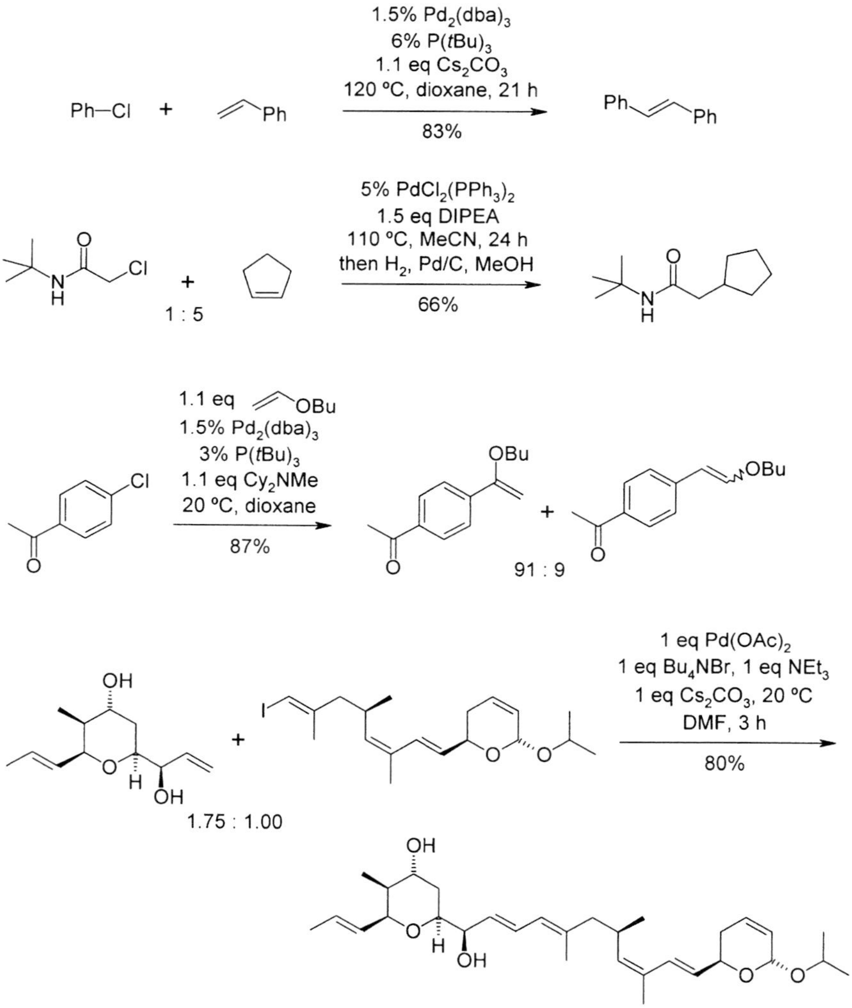

1.5% $Pd_2(dba)_3$
6% $P(tBu)_3$
1.1 eq Cs_2CO_3
120 °C, dioxane, 21 h
83%

Ph–Cl + Ph → Ph Ph

5% $PdCl_2(PPh_3)_2$
1.5 eq DIPEA
110 °C, MeCN, 24 h
then H_2, Pd/C, MeOH
66%

1 : 5

1.1 eq OBu
1.5% $Pd_2(dba)_3$
3% $P(tBu)_3$
1.1 eq Cy_2NMe
20 °C, dioxane
87%

OBu + OBu
91 : 9

1 eq $Pd(OAc)_2$
1 eq Bu_4NBr, 1 eq NEt_3
1 eq Cs_2CO_3, 20 °C
DMF, 3 h
80%

1.75 : 1.00

Scheme 8.8. Examples of Heck reactions [80–84].

Direct palladation of C–H bonds can be achieved by treatment of, for example, electron-rich arenes with Pd(II) salts (see also Section 8.11). After cross-coupling via reductive elimination the resulting Pd(0) must be reoxidized to Pd(II) if Pd-catalysis is the aim [85]. Reoxidation of Pd(0) with Cu or Ag salts (as in the Wacker process) is not always well suited for C–C bond-forming reactions [86], but other oxidants, for example peroxides, have been used with success (Scheme 8.9). The required presence of oxidants in the reaction mixture limits the scope of these reactions to oxidation-resistant starting materials.

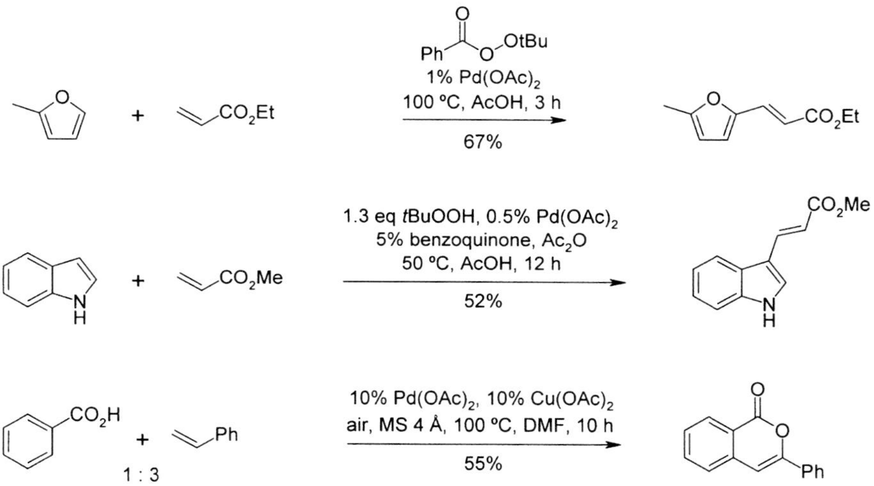

PhC(O)OOtBu, 1% $Pd(OAc)_2$, 100 °C, AcOH, 3 h — 67% (CO_2Et)

1.3 eq *t*BuOOH, 0.5% $Pd(OAc)_2$, 5% benzoquinone, Ac_2O, 50 °C, AcOH, 12 h — 52% (CO_2Me)

CO_2H + Ph (1 : 3); 10% $Pd(OAc)_2$, 10% $Cu(OAc)_2$, air, MS 4 Å, 100 °C, DMF, 10 h — 55%

Scheme 8.9. Formation of C–C bonds via intermediate palladation of C–H bonds [86–88].

The large number of highly diverse examples of high-yielding Pd-catalyzed organic reactions might give the non-specialist the impression that almost any conceivable transformation might work in the presence of a suitable Pd catalyst. This is, of course, not true, and even the most robust Pd-catalyzed processes have their limitations. Some of these will be discussed in the following sections. The most important unwanted processes which can compete with Pd(0)-catalyzed C–C bond formation include homocoupling or reduction of the halide and homocoupling, *C*-protonation, or oxidation of the organometallic reagent.

8.4 Homocoupling and Reduction of the Organyl Halide

As mentioned in Section 8.2, organopalladium(II) complexes can react with organyl halides to yield products of cross-coupling. The formation of large amounts of symmetric biaryls as a result of homocoupling of the aryl halide is often observed during the coupling of aryl halides with weak nucleophiles or unreactive alkenes. This reaction may proceed via transient formation of Pd(IV) complexes. Another possible mechanism of this homocoupling is reduction of the intermediate organopalladium(II) complex to a Pd(0) complex which then undergoes a second oxidative addition to the halide to yield a diorganylpalladium(II) complex (Scheme 8.10) [89]. Alternatively, the initially formed Pd(II) complex can undergo ligand exchange with itself to yield diorganylpalladium(II) complexes and complexes without organyl groups [90]. This process will compete most efficiently with Stille and Suzuki cross-couplings if the stannane or boronic acid is electron-deficient, sterically hindered, or otherwise unreactive and undergoes transmetalation only slowly. Benzylic halides and 2-halopyridines are particularly prone to undergo homocoupling (and reduction) [62, 91, 92].

Reduction of the halide to a hydrocarbon is another common side reaction of Pd-mediated cross-couplings. This hydrodehalogenation can proceed via reductive elimination from a hydrido complex $PdRL_2H$, which can be formed from the initial product of oxidative addition ($PdRL_2X$) by ligand exchange with an alkoxide, an amine, or an organometallic reagent containing β hydrogen, followed by β-hydride elimination (and formation of a carbonyl compound [93, 94], an imine [1], or an alkene [36], respectively). Hydrodehalogenation is, therefore, often observed in palladium-catalyzed reactions of aryl halides with alcoholates [95] or amines [96], but can be largely suppressed by use of sterically demanding phosphines, which accelerate reductive elimination of the arylamine from aryl(amido)palladium(II) complexes at the expense of β-hydride elimination [1, 97]. Alternative routes to hydrido complexes include the reduction of Pd(II) complexes to Pd(0) complexes, followed by protonation [98] (Scheme 8.10). It has been observed that arenes are formed during the reduction of a Pd(II) catalyst precursor to Pd(0) in the presence of aryl halides [97], but the precise mechanism of this process remains unknown.

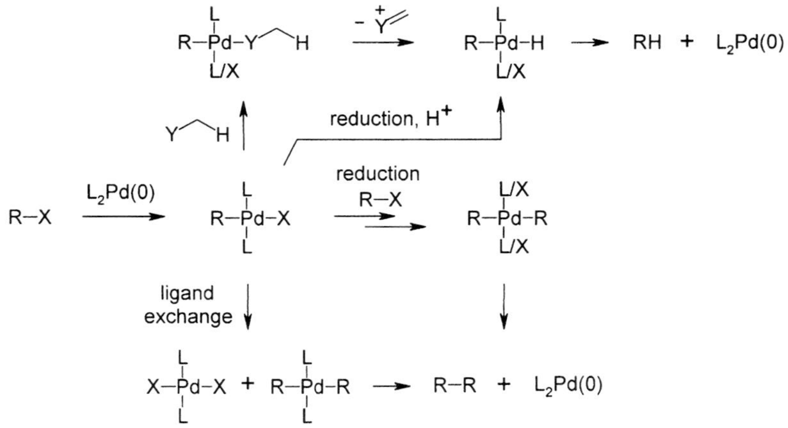

Scheme 8.10. Possible mechanisms of the Pd-catalyzed reduction and homocoupling of organic halides. X = halide, triflate; Y = R_2C^-, NHR, O^-.

Reducing agents which have been used to promote the Pd-mediated homocoupling of organyl halides include hydroquinone [99], zinc powder [15, 100, 101], magnesium [101], tetra(dimethylamino)ethene [102], tertiary aliphatic amines [103–105], and aliphatic alcohols [104, 106]. However, as mentioned above, homocoupling can also proceed in the absence of reducing agents [107]. As illustrated by the examples illustrated in Scheme 8.11, the reaction conditions for homocouplings do not differ

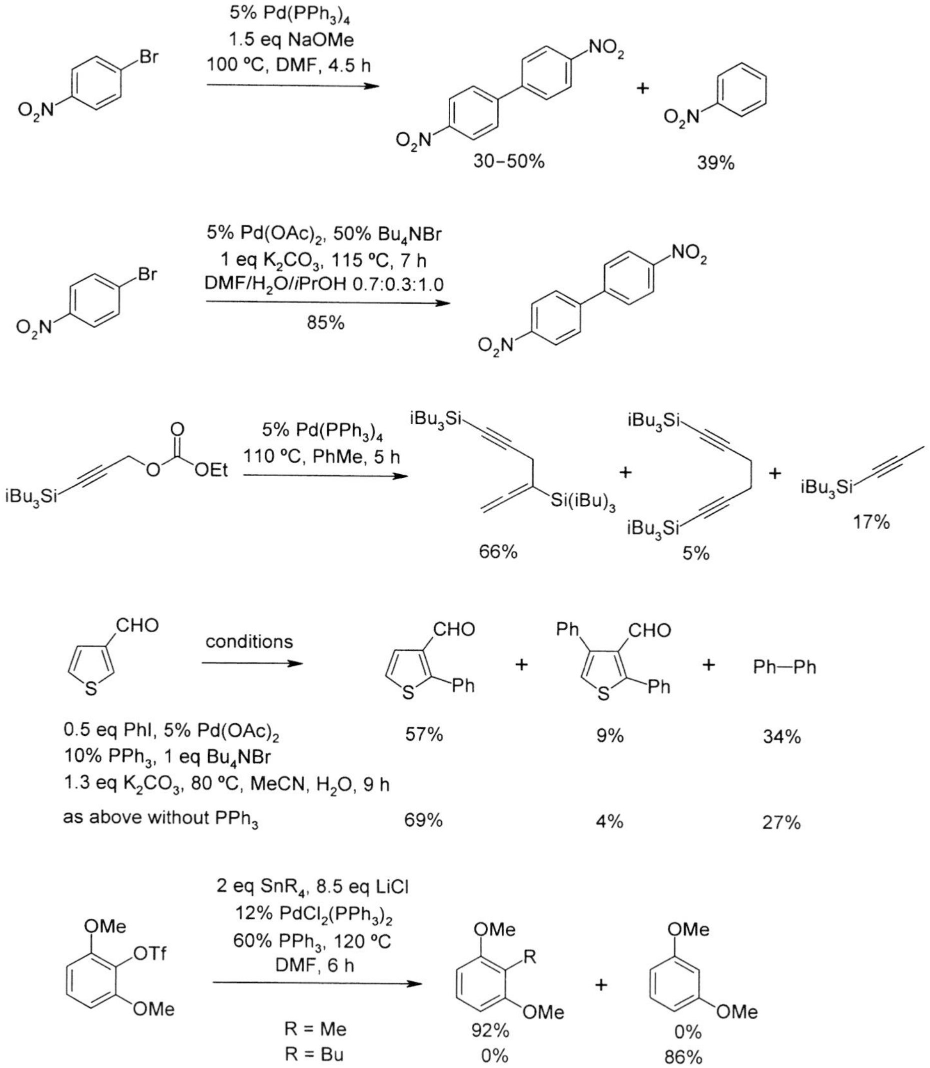

Scheme 8.11. Palladium-catalyzed homocoupling and reduction of organyl halides, carbonates, and triflates [57, 94, 104, 108, 109].

much from those used for Heck reactions or some cross-couplings, and homocoupling can therefore compete with these.

Reduction of the starting halide to a hydrocarbon should, at least in theory, be suppressible by strong ligands (e.g. chelating diphosphines), which will not be readily displaced by alcoholates or amines. Reduction should proceed swiftly if only weak, readily displaceable ligands are present in the reaction mixture [80], together with efficient reducing agents such as amines or alcoholates with α hydrogen.

The amount and type of base chosen can also have an impact on the extent of homocoupling and reduction of aryl halides in the presence of Pd(0). KO*t*Bu has been reported to increase the rate of homocoupling [46], whereas NaH can bring about quick reduction of some aryl halides [46, 47]. Higher reaction temperatures can lead to larger amounts of biaryls and to less hydrodehalogenation [15]. In the absence of strong reducing reagents or bases clean homocoupling of aryl halides without any reduction can occasionally be achieved [89] (Scheme 8.11). As illustrated by the examples depicted in Scheme 8.12, the amount of reduced product can also depend on the solvent. Reduction and homocoupling of the halide can, of course, also be suppressed by adding to the reaction mixture an efficient transmetalating reagent, a reactive alkene, or another reagent which can react quickly with the initially formed product of oxidative addition before it is converted to a hydrido com-

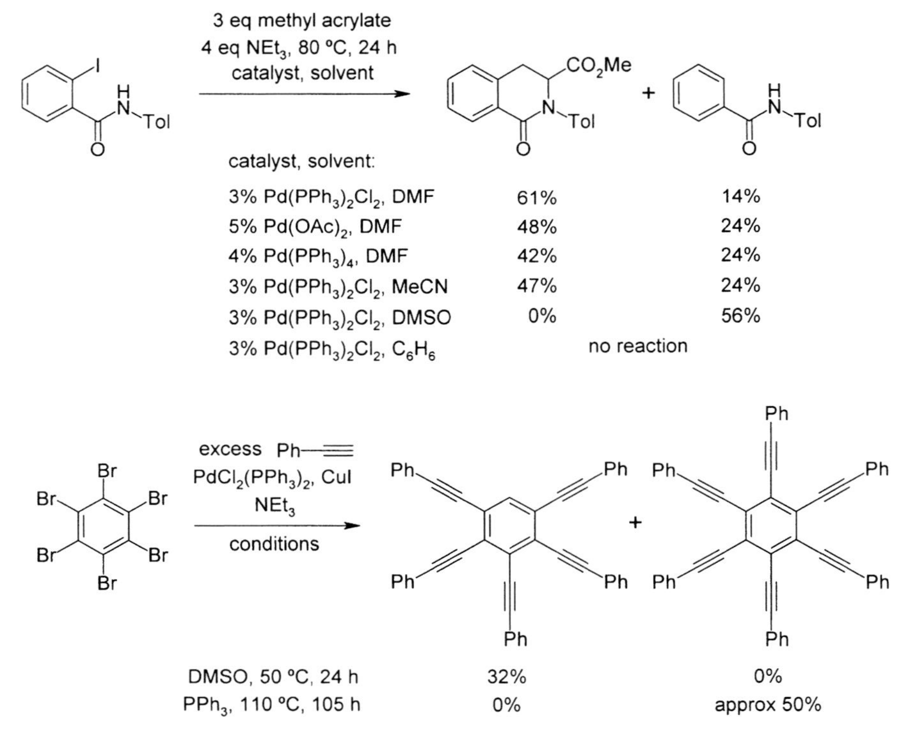

Scheme 8.12. Palladium-catalyzed reduction of organyl halides [112, 113].

plex. Thus, Heck or Sonogashira reactions are usually performed in the presence of amines, and even preformed complexes of palladium with aliphatic amines can be used as catalysts for Suzuki couplings, with no reduction of the halide being observed [110]. Similarly, carbon monoxide reacts much more rapidly with organopalladium complexes than do hydrogen or other reducing agents [111], and aryl or vinyl halides can be cleanly converted into aldehydes by treatment with mixtures of carbon monoxide and hydrogen without competing hydrodehalogenation [32].

8.5 Homocoupling and Oxidation of the Carbon Nucleophile

Most transition metal-catalyzed cross-coupling reactions also yield small quantities of the product of homocoupling of the nucleophilic reactant [16, 114, 115]. In particular terminal alkynes [116, 117] or metalated terminal alkynes [118] readily dimerize to the corresponding 1,3-butadiynes (Scheme 8.14).

Homocoupling of the carbon nucleophiles might proceed by the mechanism shown in Scheme 8.13. Stannanes or boronic acids, for instance, can sequentially displace either halides or organyl groups from Pd(II) complexes to yield new diorganylpalladium(II) complexes [67, 101, 116], which can undergo reductive elimination to yield a hydrocarbon and Pd(0). In the presence of an oxidant, a Pd(II) complex can be regenerated and homocoupling of the stannane or boronic acid will proceed with catalytic amounts of Pd [119, 120]. For arylboronic acids the type of base chosen can affect the extent of homocoupling: with NaOH and NEt_3 significantly more symmetric biaryl is sometimes formed than with CsF, K_3PO_4, or Na_2CO_3 as bases [16]. The reason for this behavior is not well understood.

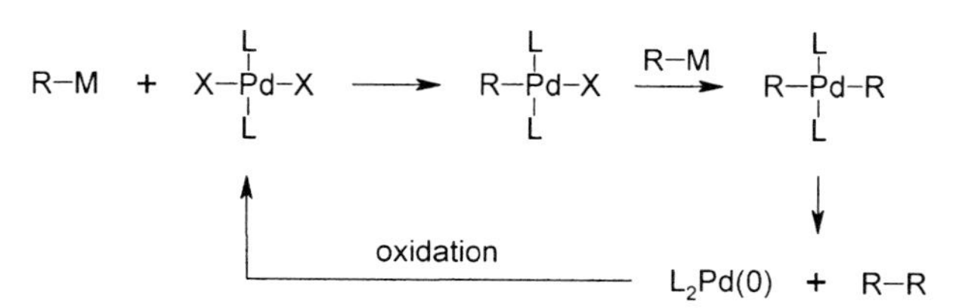

Scheme 8.13. Possible mechanism of the homocoupling of the organometallic component of Pd-mediated cross-couplings [119, 121].

Numerous examples of unwanted or deliberate Pd-mediated homocoupling of organometallic reagents have been reported (Scheme 8.14) [38, 122, 123]. Suitable oxidants include α-halo ketones [116, 124, 125] (which can, therefore, not be used as electrophilic component in cross-coupling reactions), oxygen [120, 124, 126], 1,2-diiodoethene [127], 2,3-dibromopropionic acid esters [119], and $CuCl_2$ [128].

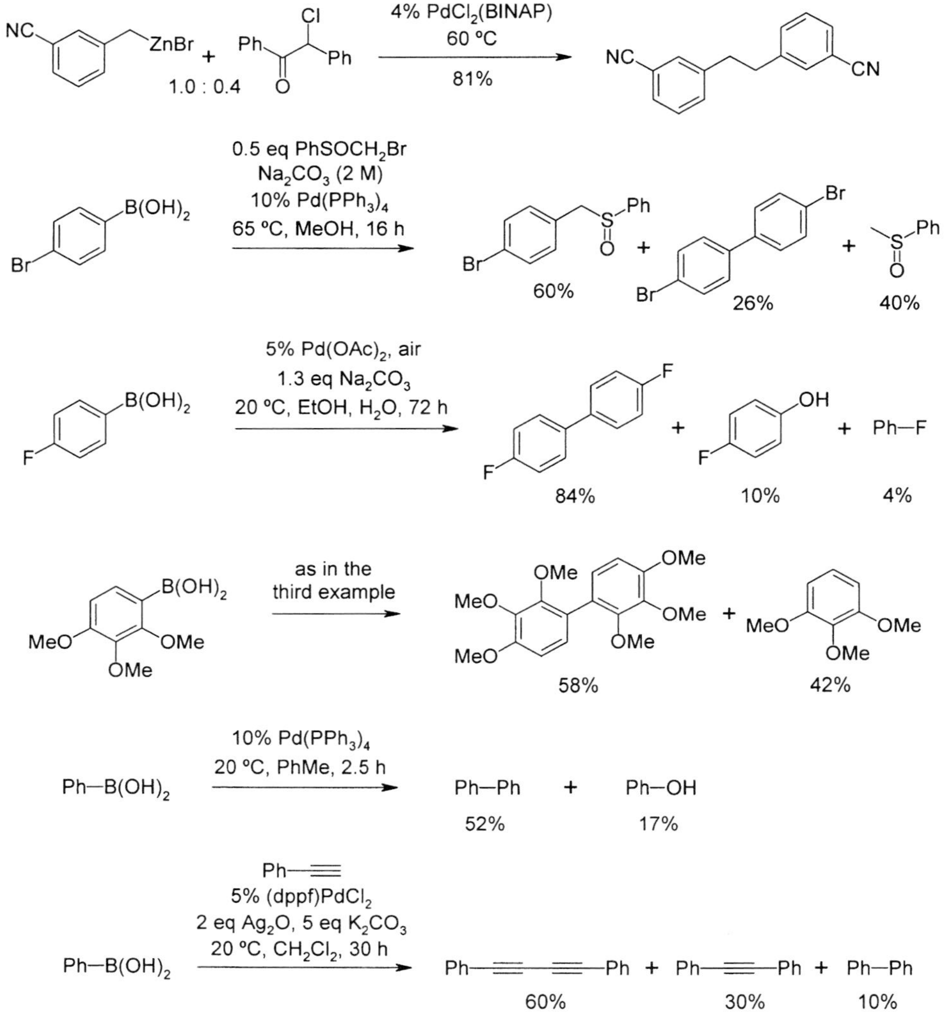

Scheme 8.14. Palladium-mediated homocoupling of carbon nucleophiles [117, 120, 121, 125, 129].

Occasionally hydroxylation of the carbon nucleophile is observed during Pd-catalyzed C–C bond formation (Scheme 8.15; third reaction in Scheme 8.14 [121]). These reactions may in some instances proceed by a mechanism analogous to the Wacker reaction [130], or to the hydroxylation of organometallic compounds or boranes by peroxides or air (Section 3.5).

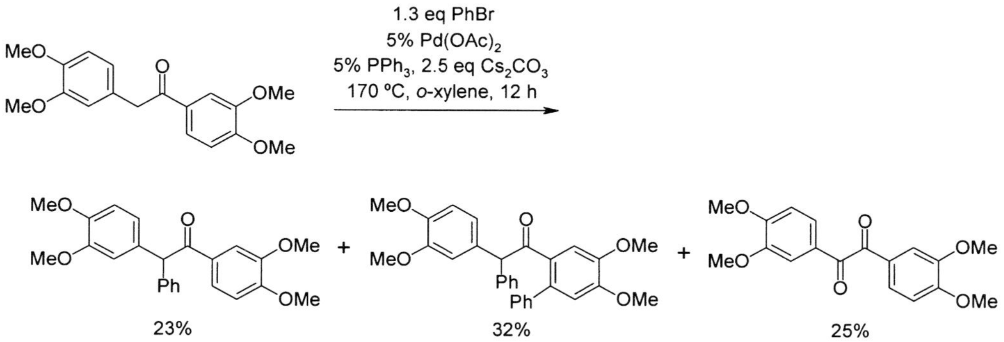

Scheme 8.15. Oxidation of a carbon nucleophile during Pd-catalyzed cross-coupling [48].

8.6 Transfer of Aryl Groups from the Phosphine Ligand

During arylations of carbon nucleophiles with aryl halides in the presence of palladium triarylphosphine complexes products containing the aryl group of the phosphine can result (Scheme 8.16). These reactions proceed via reversible arylation of the Pd-bound phosphine, which occurs at temperatures above 50 °C, particularly readily in the presence of iodide [11] (see Section 8.2). Electron-deficient aryl groups usually migrate less readily than electron-rich groups [23, 25].

Scheme 8.16. Aryl group exchange between phosphines and palladium [11, 131].

This unwanted side reaction can be suppressed by keeping the amount of catalyst as small as possible, by using sterically demanding trialkyl phosphine ligands, by conducting the reaction in solvents of low polarity, such as CH_2Cl_2 (instead of THF or DMF), or by using phosphine-free catalysts [23, 38, 62, 132–134].

8.7 *ipso*- vs *cine*-Substitution at Vinylboron and Vinyltin Derivatives

Cross-coupling reactions with vinylboronic acids can yield either the normal product (*ipso*-substitution of boron) or a regioisomer formed via a Heck-type reaction (*cine*-substitution; Scheme 8.17) [135]. Formation of the normal product (1-phenylhexene in Scheme 8.17) requires a base capable of binding to the boronic acid, thereby increasing the nucleophilicity of the boron-bound carbon atom (typically ROM, MOH, M_2CO_3, M_3PO_4, where M = alkali metal [136]). Products of *cine*-substitution result when tertiary amines are used as bases, i.e. under Heck-type reaction conditions.

Similarly, vinylstannanes can also yield products of *cine*-substitution (Scheme 8.17), specially if tin and an electron-withdrawing or aryl group are bound to the same carbon atom [40, 137–141]. It has been suggested that formation of these products proceeds via intermediate formation of a palladium carbene complex [138, 140] or via reversible β-hydride elimination [141], and can be avoided by addition of Cu(I) salts [142], which increase the rate of Stille coupling, or by protecting vinylic C–H groups by transient silylation [143].

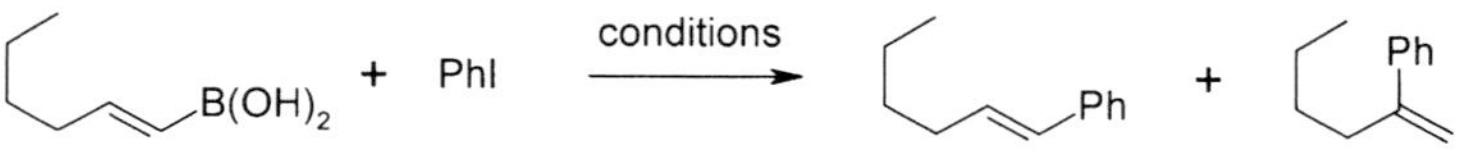

conditions	yield (ratio)
3% $Pd(PPh_3)_4$, 80 °C, C_6H_6, no base, 6 h	0%
3% $Pd(PPh_3)_4$, 80 °C, C_6H_6, 2 eq NaOH, 2 h	99% (100 : 0)
3% $Pd(PPh_3)_4$, 80 °C, DMF, 5 eq NEt_3, 20 h	54% (10 : 90)
3% Pd black, 80 °C, DMF, 5 eq NEt_3, 20 h	94% (4 : 96)
3% Pd black, 80 °C, DMF, 2 eq NaOH, 6 h	86% (56 : 44)

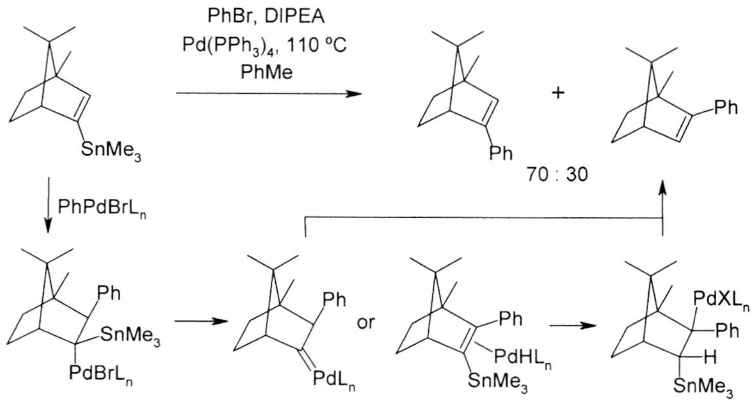

Scheme 8.17. Regioselectivity of cross-coupling reactions with vinylboronic acids and vinylstannanes [9, 144].

8.8
Allylic Arylation and Hydrogenation as Side Reactions of the Heck Reaction

Alkenes substituted with electron-withdrawing groups and with an allylic, acidified C–H group can react with organopalladium complexes under basic conditions either as carbon nucleophiles (cross-coupling) or as alkenes (Heck reaction). Highly substituted alkenes will undergo Heck reaction only slowly, so that their reaction as carbon nucleophiles might become the main reaction pathway (first two reactions, Scheme 8.18). Few examples of such reactions have been reported, however, and, as illustrated by the last two examples in Scheme 8.18, not all enones or enals react this way.

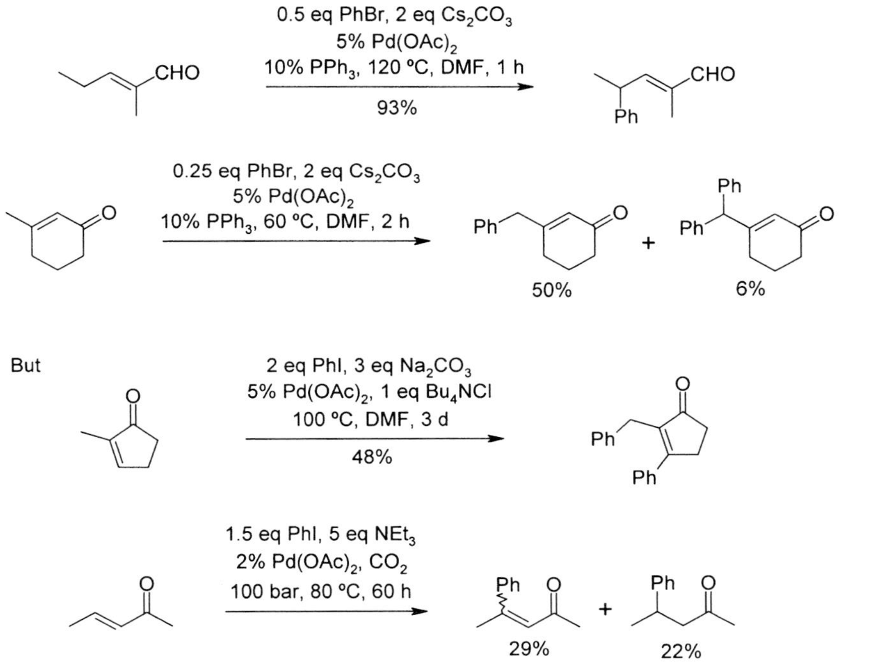

CHO
0.5 eq PhBr, 2 eq Cs_2CO_3
5% $Pd(OAc)_2$
10% PPh_3, 120 °C, DMF, 1 h
93%
Ph CHO

O
0.25 eq PhBr, 2 eq Cs_2CO_3
5% $Pd(OAc)_2$
10% PPh_3, 60 °C, DMF, 2 h
Ph O + Ph Ph O
50% 6%

But
O
2 eq PhI, 3 eq Na_2CO_3
5% $Pd(OAc)_2$, 1 eq Bu_4NCl
100 °C, DMF, 3 d
48%
Ph O Ph

O
1.5 eq PhI, 5 eq NEt_3
2% $Pd(OAc)_2$, CO_2
100 bar, 80 °C, 60 h
Ph O + Ph O
29% 22%

Scheme 8.18. Heck reaction vs allylic arylation [145–147].

Aliphatic amines can be readily oxidized by Pd(II) to imines or iminium salts and hydrido complexes. The latter can transfer hydrogen to alkenes, leading to the formation of alkanes as byproducts of the Heck reaction (last example, Scheme 8.18). Such reactions can be avoided by using alkali carbonates as base instead of aliphatic amines [148]. Treatment of stannanes or organoboron derivatives with electron-deficient alkenes under acidic reaction conditions can also lead to formal products of Michael addition instead of the products of a Heck-type reaction [149, 150] (Scheme 8.19).

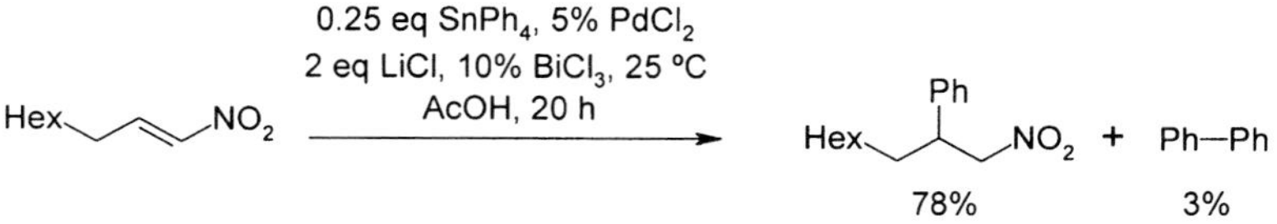

Hex–CH=CH–NO_2 → (0.25 eq $SnPh_4$, 5% $PdCl_2$, 2 eq LiCl, 10% $BiCl_3$, 25 °C, AcOH, 20 h) → Hex–CH(Ph)–NO_2 (78%) + Ph–Ph (3%)

Scheme 8.19. Arylation of electron-poor alkenes under acidic reaction conditions [150].

8.9 Protodemetalation of the Carbon Nucleophile

Boranes and, to a lesser extent, boronic acids can undergo slow hydrolysis (protodeboration) in the presence of protic solvents. This unwanted reaction can become predominant if a cross-coupling reaction only proceeds slowly (e.g. with electron-rich, sterically demanding, or unreactive halides; Scheme 8.20; see also Scheme 8.14) or if the boron derivative is particularly sensitive, for example 2-formylphenylboronic acid. In such instances the reaction should be performed under anhydrous conditions in an aprotic solvent with a boronic acid ester [151] or a stannane.

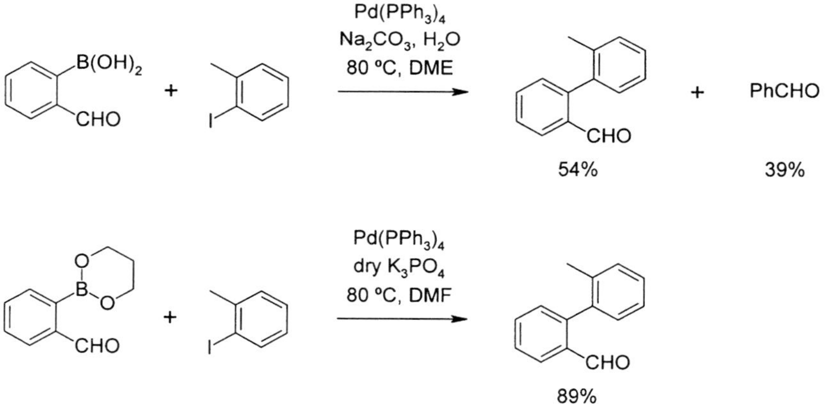

2-(CHO)C_6H_4B$(OH)_2$ + 2-iodotoluene → ($Pd(PPh_3)_4$, Na_2CO_3, H_2O, 80 °C, DME) → 2'-methylbiphenyl-2-carbaldehyde (54%) + PhCHO (39%)

2-(CHO)C_6H_4-(1,3,2-dioxaborinane) + 2-iodotoluene → ($Pd(PPh_3)_4$, dry K_3PO_4, 80 °C, DMF) → 2'-methylbiphenyl-2-carbaldehyde (89%)

Scheme 8.20. Protodeboration of boronic acids in the presence of water [66].

8.10 Sterically Hindered Substrates

Cross-couplings of 2,6-disubstituted aryl halides, stannanes, or boronic acids proceed less readily than with sterically less demanding starting materials and might require a careful optimization of the reaction conditions (Scheme 8.21) [109]. Reduction or hydrolysis to the corresponding arenes will usually compete effectively with these slow cross-couplings, which should therefore be conducted under anhydrous conditions and in the absence of reducing agents. It has been found that sterically demanding phosphines are particularly well suited to Suzuki or Stille couplings of sterically demanding starting materials. Thus, neither the coupling of mesitylboronic acid and iodomesitylene with $Pd(PPh_3)_4$ [66, 151] nor the coupling of mesitylmagnesium bromide and 2-chloro-*m*-xylene or bromomesitylene with 1,3-(2-mesityl)imidazol-2-ylidene Pd complexes [115] yield the desired 2,6,2′,6′-tetrasubstituted biaryls. Compounds of this type can, however, be prepared by using sterically demanding phosphines as ligands (Scheme 8.21). Heck reactions with sterically demanding aryl halides can, similarly, also be performed successfully with the aid of sterically demanding phosphines (last example, Scheme 8.21).

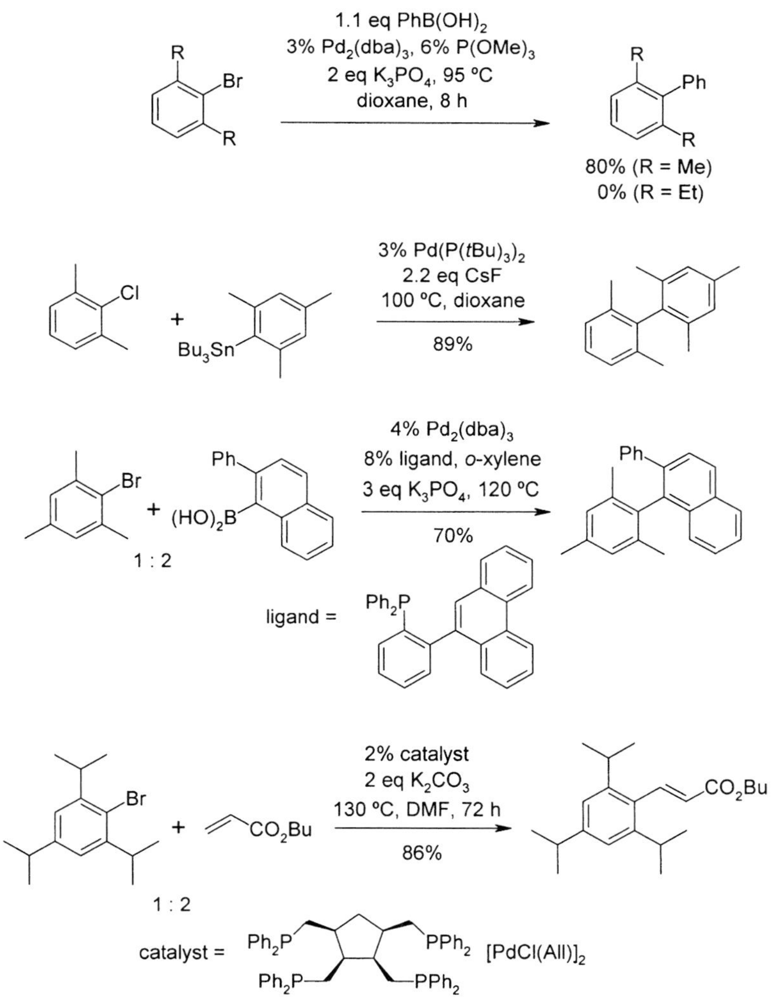

1.1 eq $PhB(OH)_2$
3% $Pd_2(dba)_3$, 6% $P(OMe)_3$
2 eq K_3PO_4, 95 °C
dioxane, 8 h

80% (R = Me)
0% (R = Et)

3% $Pd(P(tBu)_3)_2$
2.2 eq CsF
100 °C, dioxane
89%

4% $Pd_2(dba)_3$
8% ligand, *o*-xylene
3 eq K_3PO_4, 120 °C
70%
1 : 2
ligand =

2% catalyst
2 eq K_2CO_3
130 °C, DMF, 72 h
86%
1 : 2
catalyst = ... $[PdCl(All)]_2$

Scheme 8.21. Palladium-mediated C–C bond formation with sterically demanding substrates [152–155].

8.11 Cyclometalation

In transition metal complexes of suitable geometry the metal may undergo intramolecular oxidative insertion into C–H bonds. Intermediates of Pd-catalyzed C–C bond formation can also undergo such cyclometalations to yield "palladacycles". This can give rise to unexpected products or, if the palladacycles are too stable, the catalyst will be consumed and no further reaction will occur. At high temperatures reductive elimination from such complexes can occur to yield cyclic products.

In the first example sketched in Scheme 8.22 a palladacycle is formed by intramolecular insertion of Pd into a methyl C–H group after intermolecular oxidative addition to the C–I group. The resulting intermediate does not undergo reductive elimination to yield a benzocyclobutene, probably because it is less strained than the final intermediate of this process. First after cross-coupling with a second aryl iodide followed by a second cyclometalation the resulting palladacycle undergoes reductive elimination to yield a benzocyclobutene. Similarly, Heck reactions with norbornene can also give rise to cyclobutanes (second reaction, Scheme 8.22).

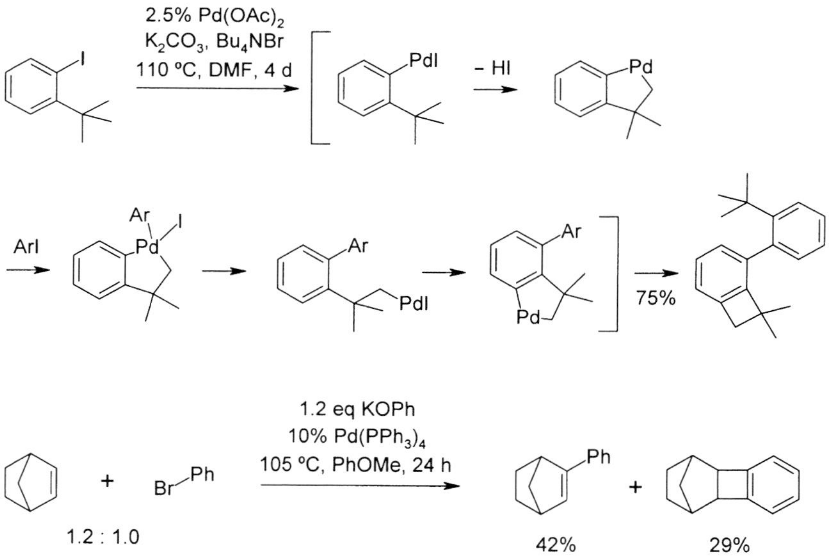

2.5% $Pd(OAc)_2$
K_2CO_3, Bu_4NBr
110 °C, DMF, 4 d
PdI
- HI
Pd
ArI
Ar
Pd
I
Ar
PdI
Ar
Pd
75%

1.2 eq KOPh
10% $Pd(PPh_3)_4$
105 °C, PhOMe, 24 h
+
Br Ph
1.2 : 1.0
Ph
+
42%
29%

Scheme 8.22. Cyclopalladation during Pd-mediated homocoupling of aryl iodides [156] and Heck reaction with norbornene [157]. Ar = 2-(*t*Bu)C_6H_4.

Other intermediates which can undergo similar cyclometalations include (2-methoxyphenyl)palladium [158, 159] and (1-naphthylmethyl)palladium complexes (Scheme 8.23) [160]. Cyclometalation is usually promoted by high temperatures, and if these are necessary the yield of cross-coupling reactions with substrates prone to undergo cyclopalladation (for example 2-bromoanisole) can drop dramatically [43].

Another example of transient formation of a palladacycle is the Pd-mediated *ortho*-alkylation and *ipso*-vinylation of aryl iodides depicted in Scheme 8.23. In this multicomponent reaction the ability of norbornene to undergo reversible arylation and palladacycle formation is exploited. This reaction also illustrates that aryl halides undergo oxidative addition to Pd faster than do alkyl halides, and that aryl–alkyl bond-formation by reductive elimination also proceeds faster than alkyl–alkyl bond-formation. The large excess of alkyl iodide used in these reactions prevents the formation of biaryls. Benzocyclobutenes can also be formed in this reaction, in particular when the alkyl group on the aryl iodide is sterically demanding or when a secondary alkyl iodide is used [161].

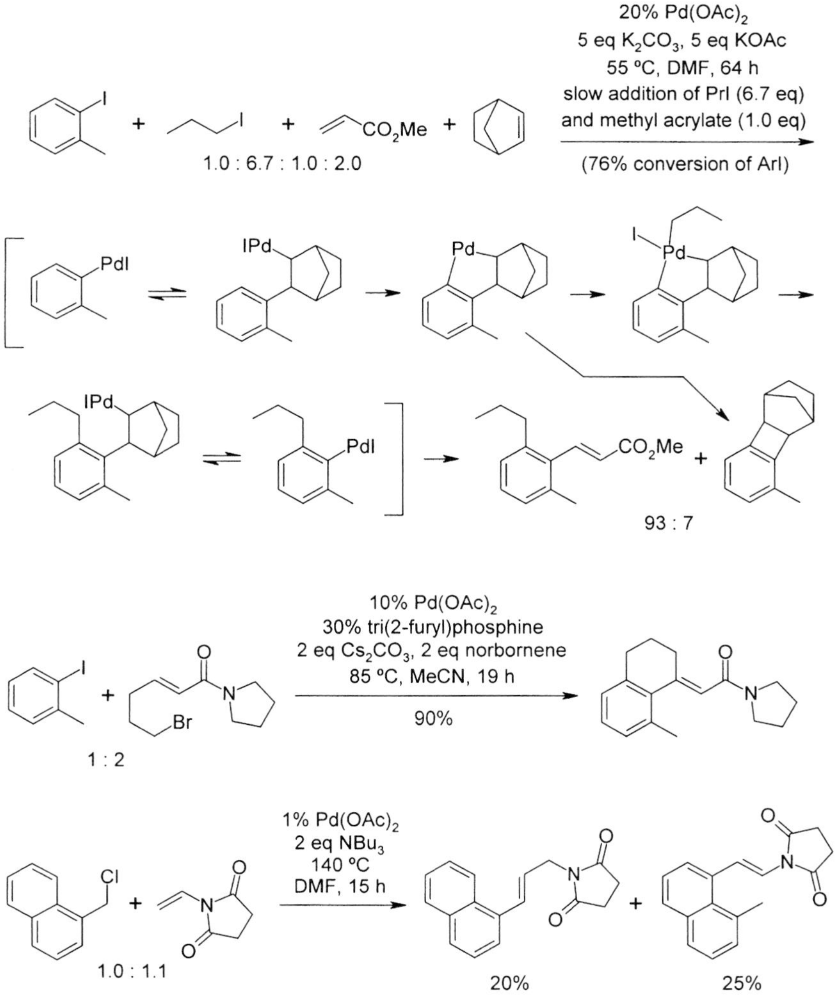

Scheme 8.23. Palladium-mediated C–C bond formation via cyclopalladation [160–162].

8.12 Chelate Formation

Organic compounds can react with Pd salts to form stable chelates via oxidative C–H insertion and simultaneous bond formation with a further coordination site in the molecule, for example an aliphatic amine, a hydrazone, a carboxylate, a phenolate, a phosphine, or a thioether [13, 14, 51, 163]. Such cyclic complexes can be significantly more stable than related, non-cyclic organopalladium complexes, and might no longer undergo cross-coupling or Heck reactions under the usual conditions [164]. For instance, although Heck reactions can usually be performed with internal alkenes [165], only terminal alkenes or allene react intermolecularly with isolated 2-(aminomethyl)arylpalladium complexes [166, 167]. 3-(Dialkylamino)propylpalladium complexes are even less reactive (Scheme 8.24) [168], and five-mem-

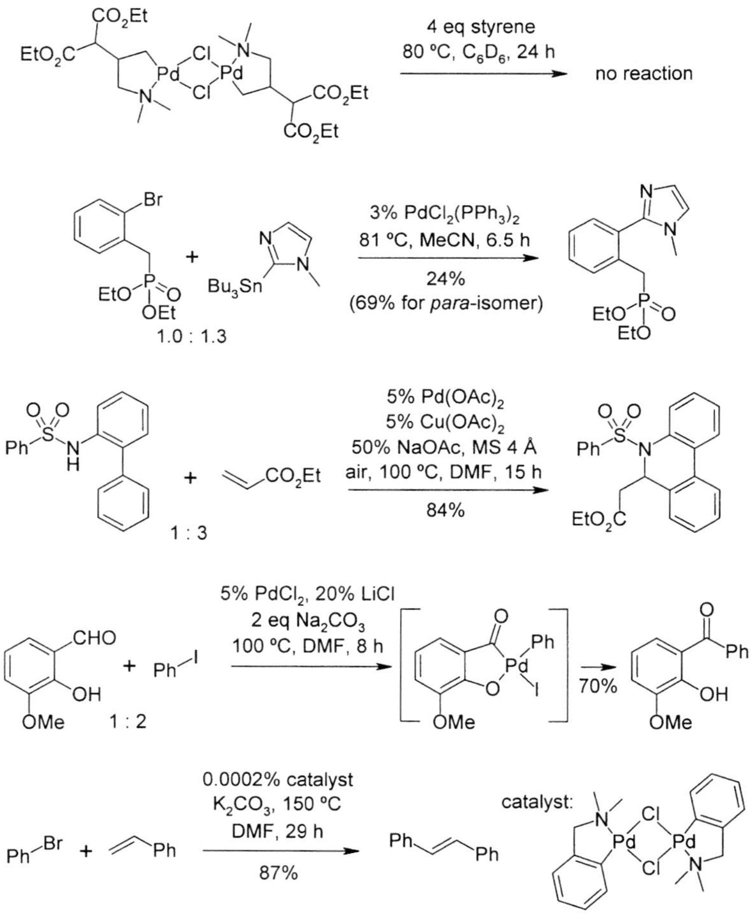

4 eq styrene, 80 °C, C_6D_6, 24 h → no reaction

3% $PdCl_2(PPh_3)_2$, 81 °C, MeCN, 6.5 h; 24% (69% for *para*-isomer); 1.0 : 1.3

5% $Pd(OAc)_2$, 5% $Cu(OAc)_2$, 50% NaOAc, MS 4 Å, air, 100 °C, DMF, 15 h; 84%; 1 : 3

5% $PdCl_2$, 20% LiCl, 2 eq Na_2CO_3, 100 °C, DMF, 8 h; 1 : 2; 70%

0.0002% catalyst, K_2CO_3, 150 °C, DMF, 29 h; 87%

Scheme 8.24. Reactions involving Pd-chelates [51, 88, 174–176].

bered Pd-chelates with sulfur or phosphorus ligands are unreactive towards alkenes [166, 169]. Accordingly, aryl halides with strongly chelating *ortho*-substituents will undergo transition metal-catalyzed C–C bond formation only sluggishly or not at all (Scheme 8.24). Some palladacycles or Pd-chelates are even sufficiently stable to be useful catalysts for cross-coupling or Heck reactions, and can be recovered unchanged because they do not undergo cross-coupling during these reactions (Scheme 8.24) [78, 170–172]. 2-(Hydroxymethyl)aryl halides or 2-(acylamino)aryl halides have, however, been used with success in Pd-mediated cross-coupling reactions [66, 173]. Similarly, in the third example in Scheme 8.24 the intermediate sulfonylamino chelate is sufficiently labile to undergo Heck reaction with ethyl acrylate.

References

1 Hartwig, J. F. Transition metal catalyzed synthesis of arylamines and aryl ethers from aryl halides and triflates: scope and mechanism. *Angew. Chem. Int. Ed.* **1998**, *37*, 2046–2067.
2 Suzuki, A. Recent advances in the cross-coupling reactions of organoboron derivatives with organic electrophiles. *J. Organomet. Chem.* **1999**, *576*, 147–168.
3 Prim, D.; Campagne, J.-M.; Joseph, D.; Andrioletti, B. Palladium-catalyzed reactions of aryl halides with soft, non-organometallic nucleophiles. *Tetrahedron* **2002**, *58*, 2041–2075.
4 Kotha, S.; Lahiri, K.; Kashinath, D. Recent applications of the Suzuki–Miyaura cross-coupling reaction in organic synthesis. *Tetrahedron* **2002**, *58*, 9633–9695.
5 Beletskaya, I. P.; Cheprakov, A. V. The Heck reaction as a sharpening stone of Pd catalysis. *Chem. Rev.* **2000**, *100*, 3009–3066.
6 Herrmann, W. A.; Öfele, K.; von Preysing, D.; Schneider, S. K. Phospha-palladacycles and *N*-heterocyclic carbene Pd complexes: efficient catalysts for CC-coupling reactions. *J. Organomet. Chem.* **2003**, *687*, 229–248.
7 Negishi, E.-I.; Anastasia, L. Palladium-catalyzed alkynylation. *Chem. Rev.* **2003**, *103*, 1979–2017.
8 Poli, G.; Giambastiani, G.; Heumann, A. Palladium in organic synthesis: fundamental transformations and domino processes. *Tetrahedron* **2000**, *56*, 5959–5989.
9 Miyaura, N.; Suzuki, A. Palladium-catalyzed cross-coupling reactions of organoboron compounds. *Chem. Rev.* **1995**, *95*, 2457–2483.
10 Hillier, A. C.; Grasa, G. A.; Viciu, M. S.; Lee, H. M.; Yang, C.; Nolan, S. P. Catalytic cross-coupling reactions mediated by Pd–nucleophilic carbene systems. *J. Organomet. Chem.* **2002**, *653*, 69–82.
11 Grushin, V. V. Thermal stability, decomposition paths, and Ph/Ph exchange reactions of $[(Ph_3P)_2Pd(Ph)X]$ (X = I, Br, Cl, F, and HF_2). *Organometallics* **2000**, *19*, 1888–1900.
12 de Graaf, W.; Boersma, J.; Smeets, W. J. J.; Spek, A. L.; van Koten, G. Dimethyl(TMEDA)-palladium(II) and dimethyl[1,2-bis(dimethylphosphino)ethane]palladium(II): synthesis, X-ray crystal structures, and thermolysis, oxidative addition, and ligand exchange reactions. *Organometallics* **1989**, *8*, 2907–2917.
13 Tamaru, Y.; Kagotani, M.; Yoshida, Z.-I. Palladation of sp^3-carbon atoms: preparation of *N*-palladiomethylthioamides. *Angew. Chem. Int. Ed.* **1981**, *20*, 980–981.
14 Holton, R. A.; Davis, R. G. Regiocontrolled aromatic palladation. *J. Am. Chem. Soc.* **1977**, *99*, 4175–4177.
15 Mukhopadhyay, S.; Rothenberg, G.; Gitis, D.; Sasson, Y. On the mechanism of Pd-catalyzed coupling of haloaryls to biaryls in water with zinc. *Org. Lett.* **2000**, *2*, 211–214.
16 Nishimura, M.; Ueda, M.; Miyaura, N. Palladium-catalyzed biaryl-coupling reaction of arylboronic acids in water using hydrophilic phosphine ligands. *Tetrahedron* **2002**, *58*, 5779–5787.
17 Genet, J. P.; Savignac, M. Recent developments of Pd(0)-catalyzed reactions in aqueous medium. *J. Organomet. Chem.* **1999**, *576*, 305–317.
18 Badone, D.; Baroni, M.; Cardamone, R.; Ielmini, A.; Guzzi, U. Highly efficient Pd-cata-

lyzed boronic acid coupling reactions in water: scope and limitations. *J. Org. Chem.* **1997**, *62*, 7170–7173.

19 Rai, R.; Aubrecht, K. B.; Collum, D. B. Palladium-catalyzed Stille couplings of aryl-, vinyl-, and alkyltrichlorostannanes in aqueous solution. *Tetrahedron Lett.* **1995**, *36*, 3111–3114.

20 Roshchin, A. I.; Bumagin, N. A.; Beletskaya, I. P. Palladium-catalyzed cross-coupling reaction of organostannoates with aryl halides in aqueous medium. *Tetrahedron Lett.* **1995**, *36*, 125–128.

21 Zhang, H.-C.; Daves, G. D. Water facilitation of Pd-mediated coupling reactions. *Organometallics* **1993**, *12*, 1499–1500.

22 Getty, A. D.; Goldberg, K. I. Reaction of a Pd(II) complex chelated by a tridentate PNC ligand with water to produce a $[(PN)Pd(\mu\text{-}OH)]_2^{2+}$ dimer: a rare observation of a well-defined hydrolysis of a Pd(II)–aryl compound. *Organometallics* **2001**, *20*, 2545–2551.

23 Goodson, F. E.; Wallow, T. I.; Novak, B. M. Mechanistic studies on the aryl–aryl interchange reaction of $ArPdL_2I$ (L= triarylphosphine) complexes. *J. Am. Chem. Soc.* **1997**, *119*, 12441–12453.

24 Kong, K.-C.; Cheng, C.-H. Facile aryl–aryl exchange between the Pd center and phosphine ligands in Pd(II) complexes. *J. Am. Chem. Soc.* **1991**, *113*, 6313–6315.

25 Herrmann, W. A.; Brossmer, C.; Priermeier, T.; Öfele, K. Oxidative Addition von Chloraromaten and Pd(0)-Komplexe: Synthese, Struktur und Stabilität von Arylpalladium(II)-chloriden der Phosphanreihe. *J. Organomet. Chem.* **1994**, *481*, 97–108.

26 Mann, G.; Hartwig, J. F. Palladium alkoxides: potential intermediacy in catalytic amination, reductive elimination of ethers, and catalytic etheration. Comments on alcohol elimination from Ir(III). *J. Am. Chem. Soc.* **1996**, *118*, 13109–13110.

27 Morvillo, A.; Turco, A. Thermal stability of organopalladium compounds: non-radical methyl elimination from $[PdXMe(PEt_3)_2]$. *J. Organomet. Chem.* **1984**, *276*, 431–438.

28 Roy, A. H.; Hartwig, J. F. Reductive elimination of aryl halides from Pd(II). *J. Am. Chem. Soc.* **2001**, *123*, 1232–1233.

29 Alsters, P. L.; Boersma, J.; van Koten, G. Direct alkoxylation of organopalladium compounds based on a new type of C–O coupling mediated by molybdenum peroxides. *Tetrahedron Lett.* **1991**, *32*, 675–678.

30 Pal, C. K.; Chattopadhyay, S.; Sinha, C.; Chakravorty, A. Synthesis and structure of a thioazobenzene palladacycle: oxygen insertion into the Pd–C bond by MCPBA. *J. Organomet. Chem.* **1992**, *439*, 91–99.

31 de Graaf, W.; Boersma, J.; van Koten, G. Cross-coupling vs homocoupling in the reactions of dimethyl(TMEDA)palladium with organic halides. *Organometallics* **1990**, *9*, 1479–1484.

32 Schoenberg, A.; Heck, R. F. Palladium-catalyzed formylation of aryl, heterocyclic, and vinylic halides. *J. Am. Chem. Soc.* **1974**, *96*, 7761–7764.

33 Tremont, S. J.; Rahman, H. U. *ortho*-Alkylation of acetanilides using alkyl halides and $Pd(OAc)_2$. *J. Am. Chem. Soc.* **1984**, *106*, 5759–5760.

34 Cárdenas, D. J.; Mateo, C.; Echavarren, A. M. Synthesis of oxa- and azapalladacycles from organostannanes. *Angew. Chem. Int. Ed.* **1994**, *33*, 2445–2447.

35 Clinet, J. C.; Balavoine, G. Palladium-mediated *ortho*-alkylation of 2-aryloxazolines. *J. Organomet. Chem.* **1991**, *405*, C29–C31.

36 Murahashi, S.-I.; Tamba, Y.; Yamamura, M.; Yoshimura, N. Reactions of cyclometalated Pd complexes with organolithium compounds or Grignard reagents. Selective *ortho* alkylation and arylation of benzaldehydes, azobenzenes, and tertiary benzylic amines. *J. Org. Chem.* **1978**, *43*, 4099–4106.

37 Quan, L. G.; Lamrani, M.; Yamamoto, Y. Intramolecular nucleophilic addition of aryl bromides to ketones catalyzed by Pd. *J. Am. Chem. Soc.* **2000**, *122*, 4827–4828.

38 Zhang, S.; Marshall, D.; Liebeskind, L. S. Efficient Pd-catalyzed heterobenzylic cross-coupling using sulfonium salts as substrates and $P(OPh)_3$ as a supporting ligand. *J. Org. Chem.* **1999**, *64*, 2796–2804.

39 Radhakrishnan, U.; Stang, P. J. Palladium-catalyzed arylation of enynes and electron-deficient alkynes using diaryliodonium salts. *Org. Lett.* **2001**, *3*, 859–860.

40 Kikukawa, K.; Umekawa, H.; Matsuda, T. Reaction of diazonium salts with transition metals XII. Palladium-catalyzed aryldestannylation of α-styrylstannanes by arenediazonium salts. *J. Organomet. Chem.* **1986**, *311*, C44–C46.

41 Savarin, C.; Srogl, J.; Liebeskind, L. S. Thiol ester–boronic acid cross-coupling. Catalysis using alkylative activation of the Pd thiolate intermediate. *Org. Lett.* **2000**, *2*, 3229–3231.

42 Amatore, C.; Jutand, A. Mechanistic and kinetic studies of Pd catalytic systems. *J. Organomet. Chem.* **1999**, *576*, 254–278.

43 Sundermeier, M.; Zapf, A.; Beller, M. A convenient procedure for the Pd-catalyzed cyanation of aryl halides. *Angew. Chem. Int. Ed.* **2003**, *42*, 1661–1664.

44 Lee, S.; Beare, N. A.; Hartwig, J. F. Palladium-catalyzed α-arylation of esters and protected amino acids. *J. Am. Chem. Soc.* **2001**, *123*, 8410–8411.

45 Culkin, D. A.; Hartwig, J. F. C–C Bond forming reductive elimination of ketones, esters, and amides from isolated arylpalladium(II) enolates. *J. Am. Chem. Soc.* **2001**, *123*, 5816–5817.

46 Kashin, A. N.; Mitin, A. V.; Beletskaya, I. P.; Wife, R. Palladium-catalyzed arylation of sulfonyl CH-acids. *Tetrahedron Lett.* **2002**, *43*, 2539–2542.

47 Beare, N. A.; Hartwig, J. F. Palladium-catalyzed arylation of malonates and cyanoesters using sterically hindered trialkyl- and ferrocenyldialkylphosphine ligands. *J. Org. Chem.* **2002**, *67*, 541–555.

48 Churruca, F.; SanMartin, R.; Carril, M.; Tellitu, I.; Domínguez, E. Towards a facile synthesis of triarylethanones: Pd-catalyzed arylation of ketone enolates under homogenous and heterogeneous conditions. *Tetrahedron* **2004**, *60*, 2393–2408.

49 Culkin, D. A.; Hartwig, J. F. Palladium-catalyzed α-arylation of carbonyl compounds and nitriles. *Acc. Chem. Res.* **2003**, *36*, 234–245.

50 Fox, J. M.; Huang, X.; Chieffi, A.; Buchwald, S. L. Highly active and selective catalysts for the formation of α-aryl ketones. *J. Am. Chem. Soc.* **2000**, *122*, 1360–1370.

51 Satoh, T.; Itaya, T.; Miura, M.; Nomura, M. Palladium-catalyzed coupling reaction of salicylaldehydes with aryl iodides via cleavage of the aldehyde C–H bond. *Chem. Lett.* **1996**, 823–824.

52 Satoh, T.; Kawamura, Y.; Miura, M.; Nomura, M. Palladium-catalyzed regioselective mono- and diarylation reactions of 2-phenylphenols and naphthols with aryl halides. *Angew. Chem. Int. Ed.* **1997**, *36*, 1740–1742.

53 Hennings, D. D.; Iwasa, S.; Rawal, V. H. Anion-accelerated Pd-catalyzed intramolecular coupling of phenols with aryl halides. *J. Org. Chem.* **1997**, *62*, 2–3.

54 Krotz, A.; Vollmüller, F.; Stark, G.; Beller, M. Salt-free C–C coupling reactions of arenes: Pd-catalyzed telomerization of phenols. *Chem. Commun.* **2001**, 195–196.

55 Miki, Y.; Shirokoshi, H.; Matsushita, K.-I. Intramolecular Pd-catalyzed cyclization of methyl 1-(2-bromobenzyl)indole-2-carboxylates: synthesis of pratosine and hippadine. *Tetrahedron Lett.* **1999**, *40*, 4347–4348.

56 Park, C.-H.; Ryabova, V.; Seregin, I. V.; Sromek, A. W.; Gevorgyan, V. Palladium-catalyzed arylation and heteroarylation of indolizines. *Org. Lett.* **2004**, *6*, 1159–1162.

57 Lavenot, L.; Gozzi, C.; Ilg, K.; Orlova, I.; Penalva, V.; Lemaire, M. Extension of the Heck reaction to the arylation of activated thiophenes. *J. Organomet. Chem.* **1998**, *567*, 49–55.

58 Aoyagi, Y.; Inoue, A.; Koizumi, I.; Hashimoto, R.; Tokunaga, K.; Gohma, K.; Komatsu, J.; Sekine, K.; Miyafuji, A.; Kunoh, J.; Honma, R.; Akita, Y.; Ohta, A. Palladium-catalyzed cross-coupling reactions of chloropyrazines with aromatic heterocycles. *Heterocycles* **1992**, *33*, 257–272.

59 Ohta, A.; Akita, Y.; Ohkuwa, T.; Chiba, M.; Fukunaga, R.; Miyafuji, A.; Nakata, T.; Tani, N.; Aoyagi, Y. Palladium-catalyzed arylation of furan, thiophene, benzo[*b*]furan and benzo[*b*]thiophene. *Heterocycles* **1990**, *31*, 1951–1958.

60 Pivsa-Art, S.; Satoh, T.; Kawamura, Y.; Miura, M.; Nomura, M. Palladium-catalyzed arylation of azole compounds with aryl halides in the presence of alkali metal carbonates and the use of CuI in the reaction. *Bull. Chem. Soc. Jpn.* **1998**, *71*, 467–473.

61 Akita, Y.; Itagaki, Y.; Takizawa, S.; Ohta, A. Cross-coupling reactions of chloropyrazines with 1-substituted indoles. *Chem. Pharm. Bull.* **1989**, *37*, 1477–1480.

62 Chabert, J. F. D.; Joucla, L.; David, E.; Lemaire, M. An efficient phosphine-free Pd coupling for the synthesis of new 2-arylbenzo[*b*]thiophenes. *Tetrahedron* **2004**, *60*, 3221–3230.

63 Chow, S.-Y.; Palmer, G. J.; Bowles, D. M.; Anthony, J. E. Perylene synthesis by the parallel cycloaromatization of adjacent enediynes. *Org. Lett.* **2000**, *2*, 961–963.

64 Müller, M.; Mauermann-Düll, H.; Wagner, M.; Enkelmann, V.; Müllen, K. A cycloaddition–cyclodehydrogenation route from stilbenoids to extended aromatic hydrocarbons. *Angew. Chem. Int. Ed.* **1995**, *34*, 1583–1586.

65 Lee, J.-H.; Kim, W.-S.; Lee, Y. Y.; Cho, C.-G. Stille couplings of 3-(trimethylstannyl)-5-bromo-2-pyrone for the syntheses of 3-aryl-5-bromo-2-pyrones and their ambident dienyl character. *Tetrahedron Lett.* **2002**, *43*, 5779–5782.

66 Watanabe, T.; Miyaura, N.; Suzuki, A. Synthesis of sterically hindered biaryls via the Pd-catalyzed cross-coupling reaction of arylboronic acids or their esters with haloarenes. *Synlett* **1992**, 207–210.

67 Lautens, M.; Dockendorff, C. Palladium(II) catalyst systems for the addition of boronic acids to bicyclic alkenes: new scope and reactivity. *Org. Lett.* **2003**, *5*, 3695–3698.

68 He, R.; Deng, M.-Z. A novel method for the construction of (*Z,E*)- or (*Z,Z*)-conjugated alkadienyl carboxylates. *Org. Lett.* **2002**, *4*, 2759–2762.

69 Mee, S. P. H.; Lee, V.; Baldwin, J. E. Stille coupling made easier. The synergic effect of Cu(I) salts and F^-. *Angew. Chem. Int. Ed.* **2004**, *43*, 1132–1136.

70 Crawforth, C. M.; Fairlamb, I. J. S.; Taylor, R. J. K. Efficient and selective Stille cross-coupling of benzylic and allylic bromides using bromobis(triphenylphosphine)(*N*-succinimide)Pd(II). *Tetrahedron Lett.* **2004**, *45*, 461–465.

71 Dussault, P. H.; Eary, C. T. Palladium-mediated C–C bond-forming reactions as a new method for the synthesis of peroxides and hydroperoxides. *J. Am. Chem. Soc.* **1998**, *120*, 7133–7134.

72 Myers, A. G.; Dragovich, P. S. Synthesis of functionalized enynes by Pd/Cu-catalyzed coupling reactions of acetylenes with (*Z*)-2,3-dibromopropenoic acid ethyl ester: (*Z*)-2-bromo-5-(trimethylsilyl)-2-penten-4-ynoic acid ethyl ester. *Org. Synth.* **1998**, *Coll. Vol. IX*, 117–120.

73 Cárdenas, D. J. Towards efficient and wide-scope metal-catalyzed alkyl–alkyl cross-coupling reactions. *Angew. Chem. Int. Ed.* **1999**, *38*, 3018–3020.

74 Kirchhoff, J. H.; Dai, C.; Fu, G. C. A method for Pd-catalyzed cross-couplings of simple alkyl chlorides: Suzuki reactions catalyzed by $[Pd_2(dba)_3]/PCy_3$. *Angew. Chem. Int. Ed.* **2002**, *41*, 1945–1947.

75 Potuzak, J. S.; Tan, D. S. Synthesis of *C*1-alkyl and acylglycals from glycals using a *B*-alkyl Suzuki–Miyaura cross coupling approach. *Tetrahedron Lett.* **2004**, *45*, 1797–1801.

76 Zhou, J.; Fu, G. C. Palladium-catalyzed Negishi cross-coupling reactions of unactivated alkyl iodides, bromides, chlorides, and tosylates. *J. Am. Chem. Soc.* **2003**, *125*, 12527–12530.

77 Netherton, M. R.; Dai, C.; Neuschütz, K.; Fu, G. C. Room-temperature alkyl–alkyl Suzuki cross-coupling of alkyl bromides that possess β hydrogens. *J. Am. Chem. Soc.* **2001**, *123*, 10099–10100.

78 Whitcombe, N. J.; Hii, K. K. (M.); Gibson, S. E. Advances in the Heck chemistry of aryl bromides and chlorides. *Tetrahedron* **2001**, *57*, 7449–7476.

79 Kametani, Y.; Satoh, T.; Miura, M.; Nomura, M. Regioselective arylation of benzanilides with aryl triflates of bromides under Pd catalysis. *Tetrahedron Lett.* **2000**, *41*, 2655–2658.

80 Casey, M.; Lawless, J.; Shirran, C. The Heck reaction: mechanistic insights and novel ligands. *Polyhedron* **2000**, *19*, 517–520.

81 Littke, A. F.; Fu, G. C. Heck reactions in the presence of P(*t*Bu)$_3$: expanded scope and milder conditions for the coupling of aryl chlorides. *J. Org. Chem.* **1999**, *64*, 10–11.

82 Glorius, F. Palladium-catalyzed Heck-type reaction of 2-chloro acetamides with olefins. *Tetrahedron Lett.* **2003**, *44*, 5751–5754.

83 Littke, A. F.; Fu, G. C. A versatile catalyst for Heck reactions of aryl chlorides and aryl bromides under mild conditions. *J. Am. Chem. Soc.* **2001**, *123*, 6989–7000.

84 Bhatt, U.; Christmann, M.; Quitschalle, M.; Claus, E.; Kalesse, M. The first total synthesis of (+)-ratjadone. *J. Org. Chem.* **2001**, *66*, 1885–1893.

85 Ritleng, V.; Sirlin, C.; Pfeffer, M. Ru-, Rh-, and Pd-catalyzed C–C bond formation involving C–H activation and addition on unsaturated substrates: reactions and mechanistic aspects. *Chem. Rev.* **2002**, *102*, 1731–1769.

86 Tsuji, J.; Nagashima, H. Palladium-catalyzed oxidative coupling of aromatic compounds with olefins using *tert*-butyl perbenzoate as hydrogen acceptor. *Tetrahedron* **1984**, *40*, 2699–2702.

87 Jia, C.; Lu, W.; Kitamura, T.; Fujiwara, Y. Highly efficient Pd-catalyzed coupling of arenes with olefins in the presence of *tert*-butyl hydroperoxide as oxidant. *Org. Lett.* **1999**, *1*, 2097–2100.

88 Miura, M.; Tsuda, T.; Satoh, T.; Pivsa-Art, S.; Nomura, M. Oxidative cross-coupling of *N*-(2′-phenylphenyl)benzenesulfonamides or benzoic or naphthoic acids with alkenes using a Pd–Cu catalyst system under air. *J. Org. Chem.* **1998**, *63*, 5211–5215.

89 Kuroboshi, M.; Waki, Y.; Tanaka, H. Tetrakis(dimethylamino)ethylene–Pd promoted reductive homocoupling of aryl halides. *Synlett* **2002**, 637–639.

90 Osakada, K.; Hamada, M.; Yamamoto, T. Intermolecular alkynyl ligand transfer in Pd(II) and platinum(II) complexes with $CCCO_2R$ and CCPh ligands. Relative stability of the alkynyl complexes and conproportionation of dialkynyl and diiodo complexes of these metals. *Organometallics* **2000**, *19*, 458–468.

91 Chen, H.; Deng, M.-Z. A novel Suzuki-type cross-coupling reaction of cyclopropylboronic esters with benzyl bromides. *J. Chem. Soc. Perkin Trans. 1* **2000**, 1609–1613.

92 Agrios, K. A.; Srebnik, M. Reduction of benzylic halides with Et_2Zn using $Pd(PPh_3)_4$ as catalyst. *J. Org. Chem.* **1993**, *58*, 6908–6910.

93 White, D. A.; Parshall, G. W. Reductive cleavage of aryl–Pd bonds. *Inorg. Chem.* **1970**, *9*, 2358–2361.

94 Zask, A.; Helquist, P. Palladium hydrides in organic synthesis. Reduction of aryl halides by NaOMe catalyzed by $Pd(PPh_3)_4$. *J. Org. Chem.* **1978**, *43*, 1619–1620.

95 Watanabe, M.; Nishiyama, M.; Koie, Y. Pd/$P(tBu)_3$-catalyzed synthesis of aryl *t*-butyl ethers and application to the first synthesis of 4-chlorobenzofuran. *Tetrahedron Lett.* **1999**, *40*, 8837–8840.

96 Desmarets, C.; Scheinder, R.; Fort, Y. Nickel-mediated amination chemistry. Part 1: Efficient aminations of (het)aryl 1,3-di and 1,3,5-trichlorides. *Tetrahedron Lett.* **2000**, *41*, 2875–2879.

97 Hartwig, J. F.; Richards, S.; Barañano, D.; Paul, F. Influences on the relative rates for C–N bond-forming reductive elimination and β-hydrogen elimination of amides. A case study on the origins of competing reduction in the Pd-catalyzed amination of aryl halides. *J. Am. Chem. Soc.* **1996**, *118*, 3626–3633.

98 Grushin, V. V. Hydrido complexes of Pd. *Chem. Rev.* **1996**, *96*, 2011–2033.

99 Hennings, D. D.; Iwama, T.; Rawal, V. H. Palladium-catalyzed (Ullmann-type) homocoupling of aryl halides: a convenient and general synthesis of symmetrical biaryls via inter- and intramolecular coupling reactions. *Org. Lett.* **1999**, *1*, 1205–1208.

100 Jutand, A.; Mosleh, A. Nickel- and Pd-catalyzed homocoupling of aryl triflates. Scope, limitation, and mechanistic aspects. *J. Org. Chem.* **1997**, *62*, 261–274.

101 van Asselt, R.; Elsevier, C. J. On the mechanism of formation of homocoupled products in the C–C cross-coupling reaction catalyzed by Pd complexes containing rigid bidentate nitrogen ligands: evidence for the exchange of organic groups between Pd and the transmetalating reagent. *Organometallics* **1994**, *13*, 1972–1980.

102 Kuroboshi, M.; Waki, Y.; Tanaka, H. Palladium-catalyzed tetrakis(dimethylamino)ethylene-promoted reductive coupling of aryl halides. *J. Org. Chem.* **2003**, *68*, 3938–3942.

103 Luo, F.-T.; Jeevanandam, A.; Basu, M. K. Efficient and high-turnover homocoupling reaction of aryl iodides by the use of palladacycle catalysts. A convenient way to prepare poly-*p*-phenylene. *Tetrahedron Lett.* **1998**, *39*, 7939–7942.

104 Penalva, V.; Hassan, J.; Lavenot, L.; Gozzi, C.; Lemaire, M. Direct homocoupling of aryl halides catalyzed by Pd. *Tetrahedron Lett.* **1998**, *39*, 2559–2560.

105 Clark, F. R. S.; Norman, R. O. C.; Thomas, C. B. Reactions of Pd(II) with organic compounds. Part III. Reactions of aromatic iodides in basic media. *J. Chem. Soc. Perkin Trans. 1* **1975**, 121–125.

106 Gitis, D.; Mukhopadhyay, S.; Rothenberg, G.; Sasson, Y. Solid/liquid Pd-catalyzed coupling of haloaryls using alcohols as reducing agents: kinetics and process optimization. *Org. Process Res. Dev.* **2003**, *7*, 109–114.

107 Damle, S. V.; Seomoon, D.; Lee, P. H. Palladium-catalyzed homocoupling reaction of 1-iodoalkynes: a simple and efficient synthesis of symmetrical 1,3-diynes. *J. Org. Chem.* **2003**, *68*, 7085–7087.

108 Ogoshi, S.; Nishiguchi, S.; Tsutsumi, K.; Kurosawa, H. Palladium-catalyzed reductive homocoupling reaction of 3-silylpropargyl carbon-

ates. New entry into allene-yne compounds. *J. Org. Chem.* **1995**, *60*, 4650–4652.

109 Saá, J. M.; Martorell, G.; García-Raso, A. Palladium-catalyzed cross-coupling reactions of highly hindered, electron-rich phenol triflates and organostannanes. *J. Org. Chem.* **1992**, *57*, 678–685.

110 Tao, B.; Boykin, D. W. *trans*-$Pd(OAc)_2(Cy_2NH)_2$-catalyzed Suzuki coupling reactions and its temperature-dependent activities toward aryl bromides. *Tetrahedron Lett.* **2003**, *44*, 7993–7996.

111 Schnyder, A.; Beller, M.; Mehltretter, G.; Nsenda, T.; Studer, M.; Indolese, A. F. Synthesis of primary aromatic amides by aminocarbonylation of aryl halides using formamide as an ammonia synthon. *J. Org. Chem.* **2001**, *66*, 4311–4315.

112 Mahanty, J. S.; De, M.; Kundu, N. G. Palladium-catalyzed heteroannulation of vinylic compounds: a highly convenient method for the synthesis of *N*-aryl-1,2,3,4-tetrahydro-1-oxoisoquinoline-3-carboxylic acids. *J. Chem. Soc. Perkin Trans. 1* **1997**, 2577–2579.

113 Nierle, J.; Barth, D.; Kuck, D. Pentakis(phenylethynyl)benzene and hexakis(phenylethynyl)-benzene: a revision concerning two far too similar prototype hydrocarbons. *Eur. J. Org. Chem.* **2004**, 867–872.

114 Saito, S.; Oh-tani, S.; Miyaura, N. Synthesis of biaryls via a Ni(0)-catalyzed cross-coupling reaction of chloroarenes with arylboronic acids. *J. Org. Chem.* **1997**, *62*, 8024–8030.

115 Huang, J.; Nolan, S. P. Efficient cross-coupling of aryl chlorides with aryl Grignard reagents (Kumada reaction) mediated by a Pd/imidazolium chloride system. *J. Am. Chem. Soc.* **1999**, *121*, 9889–9890.

116 Lei, A.; Srivastava, M.; Zhang, X. Transmetalation of Pd enolates and its application in the Pd-catalyzed homocoupling of alkynes: a room-temperature, highly efficient route to diynes. *J. Org. Chem.* **2002**, *67*, 1969–1971.

117 Zou, G.; Zhu, J.; Tang, J. Cross-coupling of arylboronic acids with terminal alkynes in air. *Tetrahedron Lett.* **2003**, *44*, 8709–8711.

118 Shirakawa, E.; Nakao, Y.; Murota, Y.; Hiyama, T. Palladium–iminophosphine-catalyzed homocoupling of alkynylstannanes and other organostannanes using allyl acetate or air as an oxidant. *J. Organomet. Chem.* **2003**, *670*, 132–136.

119 Yamaguchi, S.; Ohno, S.; Tamao, K. Palladium(II)-catalyzed oxidative homocoupling of aryl–metal compounds using acrylate dibromide derivatives as effective oxidants. *Synlett* **1997**, 1199–1201.

120 Smith, K. A.; Campi, E. M.; Jackson, W. R.; Marcuccio, S.; Naeslund, C. G. M.; Deacon, G. B. High yields of symmetrical biaryls from Pd catalysed homocoupling of arylboronic acids under mild conditions. *Synlett* **1997**, 131–132.

121 Moreno-Mañas, M.; Pérez, M.; Pleixats, R. Palladium-catalyzed Suzuki-type self-coupling of arylboronic acids. A mechanistic study. *J. Org. Chem.* **1996**, *61*, 2346–2351.

122 Koo, S.; Liebeskind, L. S. General synthetic entry to highly oxygenated, angularly fused polycyclic aromatic compounds. *J. Am. Chem. Soc.* **1995**, *117*, 3389–3404.

123 Barbarella, G.; Zambianchi, M. Polyhydroxyl oligothiophenes. I. Regioselective synthesis of 3,4′- and 3,3′-di(2-hydroxyethyl)-2,2′-bithiophene via Pd catalyzed coupling of thienylstannanes with thienyl bromides. *Tetrahedron* **1994**, *50*, 11249–11256.

124 Lei, A.; Zhang, X. A novel Pd-catalyzed homocoupling reaction initiated by transmetalation of Pd enolates. *Tetrahedron Lett.* **2002**, *43*, 2525–2528.

125 Lei, A.; Zhang, X. Palladium-catalyzed homocoupling reactions between two Csp^3–Csp^3 centers. *Org. Lett.* **2002**, *4*, 2285–2288.

126 Yoshida, H.; Yamaryo, Y.; Ohshita, J.; Kunai, A. Base-free oxidative homocoupling of arylboronic esters. *Tetrahedron Lett.* **2003**, *44*, 1541–1544.

127 Wright, M. E.; Porsch, M. J.; Buckley, C.; Cochran, B. B. A novel Pd-catalyzed homocoupling of alkynylstannanes: a new synthetic approach to extended linear-carbon polymers. *J. Am. Chem. Soc.* **1997**, *119*, 8393–8394.

128 Parrish, J. P.; Flanders, V. L.; Floyd, R. J.; Jung, K. W. Mild and efficient formation of symmetric biaryls via Pd(II) catalysis and Cu(II) oxidants. *Tetrahedron Lett.* **2001**, *42*, 7729–7731.

129 Rodríguez, N.; Cuenca, A.; Ramírez de Arellano, C.; Medio-Simón, M.; Asensio, G. Unprecedented Pd-catalyzed cross-coupling reaction of α-bromo sulfoxides with boronic acids. *Org. Lett.* **2003**, *5*, 1705–1708.

130 March, J. *Advanced Organic Chemistry*, John Wiley and Sons: New York, 1992.

131 O'Keefe, D. F.; Dannock, M. C.; Marcuccio, S. M. Palladium catalyzed coupling of halobenzenes with arylboronic acids: role of the PPh_3 ligand. *Tetrahedron Lett.* **1992**, *33*, 6679–6680.

132 LeBlond, C. R.; Andrews, A. T.; Sun, Y.; Sowa, J. R. Activation of aryl chlorides for Suzuki cross-coupling by ligandless, heterogeneous Pd. *Org. Lett.* **2001**, *3*, 1555–1557.

133 Zapf, A.; Beller, M. Palladium/phosphite catalyst system for efficient cross coupling of aryl bromides and chlorides with $PhB(OH)_2$. *Chem. Eur. J.* **2000**, *6*, 1830–1833.

134 Gürtler, C.; Buchwald, S. L. A phosphane-free catalyst system for the Heck arylation of disubstituted alkenes: application to the synthesis of trisubstituted olefins. *Chem. Eur. J.* **1999**, *5*, 3107–3122.

135 Peyroux, E.; Berthiol, F.; Doucet, H.; Santelli, M. Suzuki cross-coupling reactions between alkenylboronic acids and aryl bromides catalysed by a tetraphosphane-Pd catalyst. *Eur. J. Org. Chem.* **2004**, 1075–1082.

136 Miyaura, N.; Yamada, K.; Suginome, H.; Suzuki, A. Novel and convenient method for the stereo- and regiospecific synthesis of conjugated alkadienes and alkenynes via the Pd-catalyzed cross-coupling reaction of 1-alkenylboranes with bromoalkenes and bromoalkynes. *J. Am. Chem. Soc.* **1985**, *107*, 972–980.

137 Busacca, C. A.; Swestock, J.; Johnson, R. E.; Bailey, T. R.; Musza, L.; Rodger, C. A. The anomalous Stille reactions of methyl α-(tributylstannyl)acrylate: evidence for a Pd carbene intermediate. *J. Org. Chem.* **1994**, *59*, 7553–7556.

138 Fillion, E.; Taylor, N. J. *Cine*-substitution in the Stille coupling: evidence for the carbenoid reactivity of *sp*^{3}-*gem*-organodimetallic iodopalladio-trialkylstannylalkane intermediates. *J. Am. Chem. Soc.* **2003**, *125*, 12700–12701.

139 Chen, S.-H. On the Stille vinylation reactions with α-styryltrimethyltin. *Tetrahedron Lett.* **1997**, *38*, 4741–4744.

140 Quayle, P.; Wang, J.; Xu, J.; Urch, C. J. On the *cine* substitution of 1,1-bis(tributylstannyl)ethenes in an intramolecular Stille reaction. *Tetrahedron Lett.* **1998**, *39*, 489–492.

141 Flohr, A. Stille coupling vs *cine* substitution. Electronic effects also influence coupling sterically hindered stannanes. *Tetrahedron Lett.* **1998**, *39*, 5177–5180.

142 Levin, J. I. Palladium-catalyzed coupling of an α-stannyl acrylate to aryl iodides and triflates. A one-step synthesis of aryl propenoic esters. *Tetrahedron Lett.* **1993**, *34*, 6211–6214.

143 Belema, M.; Nguyen, V. N.; Zusi, F. C. Synthesis of 1,1-diarylethylenes from an α-stannyl β-silylstyrene. *Tetrahedron Lett.* **2004**, *45*, 1693–1697.

144 Stork, G.; Isaacs, R. C. A. *Cine* substitution in vinylstannane cross-coupling reactions. *J. Am. Chem. Soc.* **1990**, *112*, 7399–7400.

145 Terao, Y.; Satoh, T.; Miura, M.; Nomura, M. Regioselective arylation on the γ-position of α,β-unsaturated carbonyl compounds with aryl bromides by Pd catalysis. *Tetrahedron Lett.* **1998**, *39*, 6203–6206.

146 Larock, R. C.; Yum, E. K.; Yang, H. Palladium-catalyzed intermolecular arylation of functionally-substituted cycloalkenes by aryl iodides. *Tetrahedron* **1994**, *50*, 305–321.

147 Cacchi, S.; Fabrizi, G.; Gasparrini, F.; Pace, P.; Villani, C. The utilization of supercritical CO_2 in the Pd-catalyzed hydroarylation of β-substituted-α,β-enones. *Synlett* **2000**, 650–652.

148 Amorese, A.; Arcadi, A.; Bernocchi, E.; Cacchi, S.; Cerrini, S.; Fedeli, W.; Ortar, G. Conjugate addition vs vinylic substitution in Pd-catalyzed reaction of aryl halides with β-substituted-α,β-enones and -enals. *Tetrahedron* **1989**, *45*, 813–828.

149 Cho, C. S.; Motofusa, S.-I.; Ohe, K.; Uemura, S.; Shim, S. C. A new catalytic activity of $SbCl_3$ in Pd(0)-catalyzed conjugate addition of aromatics to α,β-unsaturated ketones and aldehydes with $NaBPh_4$ and arylboronic acids. *J. Org. Chem.* **1995**, *60*, 883–888.

150 Ohe, T.; Uemura, S. A novel catalytic activity of Bi(III) salts in Pd(II)-catalyzed atom economical Michael-type hydroarylation of nitroalkenes with aryltin compounds. *Tetrahedron Lett.* **2002**, *43*, 1269–1271.

151 Chaumeil, H.; Signorella, S.; Le Drian, C. Suzuki cross-coupling reaction of sterically hindered aryl boronates with 3-iodo-4-methoxybenzoic acid methyl ester. *Tetrahedron* **2000**, *56*, 9655–9662.

152 Griffiths, C.; Leadbeater, N. E. Palladium and Ni-catalyzed Suzuki cross-coupling of sterically hindered aryl bromides with $PhB(OH)_2$. *Tetrahedron Lett.* **2000**, *41*, 2487–2490.

153 Littke, A. F.; Schwarz, L.; Fu, G. C. Pd/$P(tBu)_3$: A mild and general catalyst for Stille

reactions of aryl chlorides and aryl bromides. *J. Am. Chem. Soc.* **2002**, *124*, 6343–6348.

154 Yin, J.; Rainka, M. P.; Zhang, X.-X.; Buchwald, S. L. A highly active Suzuki catalyst for the synthesis of sterically hindered biaryls: novel ligand coordination. *J. Am. Chem. Soc.* **2002**, *124*, 1162–1163.

155 Feuerstein, M.; Doucet, H.; Santelli, M. Palladium/tetraphosphine catalyzed Heck reaction with *ortho*-substituted aryl bromides. *Synlett* **2001**, 1980–1982.

156 Dyker, G. Palladium-catalyzed C–H activation of *tert*-butyl groups: a simple synthesis of 1,2-dihydrocyclobutabenzene derivatives. *Angew. Chem. Int. Ed.* **1994**, *33*, 103–105.

157 Catellani, M.; Chiusoli, G. P.; Ricotti, S. A new Pd-catalyzed synthesis of 1,2,3,4,4a,8b-hexahydro-1,4-methanobiphenylenes and 2-phenylbicyclo[2.2.1]hept-2-enes. *J. Organomet. Chem.* **1985**, *296*, C11–C15.

158 Dyker, G. Palladium-catalyzed C–H activation of methoxy groups: a facile synthesis of substituted 6*H*-dibenzo[*b,d*]pyrans. *Angew. Chem. Int. Ed.* **1992**, *31*, 1023–1025.

159 Dyker, G. Palladium-catalyzed C–H activation at methoxy groups for cross-coupling reactions: a new approach to substituted benzo[*b*]-furans. *J. Org. Chem.* **1993**, *58*, 6426–6428.

160 Wang, L.; Pan, Y.; Jiang, X.; Hu, H. Palladium-catalyzed reaction of 1-(chloromethyl)naphthalene with olefins. *Tetrahedron Lett.* **2000**, *41*, 725–727.

161 Catellani, M.; Cugini, F. A catalytic process based on sequential *ortho*-alkylation and vinylation of *ortho*-alkylaryl iodides via palladacycles. *Tetrahedron* **1999**, *55*, 6595–6602.

162 Lautens, M.; Paquin, J.-F.; Piguel, S.; Dahlmann, M. Palladium-catalyzed sequential alkylation–alkenylation reactions and their application to the synthesis of fused aromatic rings. *J. Org. Chem.* **2001**, *66*, 8127–8134.

163 Cárdenas, D. J.; Echavarren, A. M. Selectivity in the aliphatic palladation of ketone hydrazones. An example of Pd-promoted intramolecular addition of a *N,N*-dimethylhydrazone to an alkene. *Organometallics* **1995**, *14*, 4427–4430.

164 Pfeffer, M. Selected applications to organic synthesis of intramolecular C–H activation reactions by transition metals. *Pure Appl. Chem.* **1992**, *64*, 335–342.

165 Kondolff, I.; Doucet, H.; Santelli, M. Tetraphosphine/Pd-catalyzed Heck reactions of aryl halides with disubstituted alkenes. *Tetrahedron Lett.* **2003**, *44*, 8487–8491.

166 Ryabov, A. D. Cyclopalladated complexes in organic synthesis. *Synthesis* **1985**, 233–252.

167 Grigg, R.; MacLachlan, W. S.; MacPherson, D. T.; Sridharan, V.; Suganthan, S.; Thornton-Pett, M.; Zhang, J. Pictet–Spengler/Pd catalyzed allenylation and carbonylation processes. *Tetrahedron* **2000**, *56*, 6585–6594.

168 McCrindle, R.; McAlees, A. J. Some reactions of di-μ-chlorobis[3-(dimethylamino)-1-formyl-2,2-dimethylpropyl-*C,N*]dipalladium(II). *J. Organomet. Chem.* **1983**, *244*, 97–106.

169 Gruber, A. S.; Zim, D.; Ebeling, G.; Monteiro, A. L.; Dupont, J. Sulfur-containing palladacycles as catalyst precursors for the Heck reaction. *Org. Lett.* **2000**, *2*, 1287–1290.

170 Dupont, J.; Pfeffer, M.; Spencer, J. Palladacycles; an old organometallic family revisited: new, simple, and efficient catalyst precursors for homogeneous catalysis. *Eur. J. Inorg. Chem.* **2001**, 1917–1927.

171 Bergbreiter, D. E.; Osburn, P. L.; Wilson, A.; Sink, E. M. Palladium-catalyzed C–C coupling under thermomorphic conditions. *J. Am. Chem. Soc.* **2000**, *122*, 9058–9064.

172 Ohff, M.; Ohff, A.; Milstein, D. Highly active Pd(II) cyclometalated imine catalysts for the Heck reaction. *Chem. Commun.* **1999**, 357–358.

173 Leadbeater, N. E.; Tominack, B. J. Rapid, easy Cu-free Sonogashira couplings using aryl iodides and activated aryl bromides. *Tetrahedron Lett.* **2003**, *44*, 8653–8656.

174 Kennedy, G.; Perboni, A. D. The preparation of heterobiaryl phosphonates via the Stille coupling reaction. *Tetrahedron Lett.* **1996**, *37*, 7611–7614.

175 Weinberg, E. L.; Hunter, B. K.; Baird, M. C. Reactions of a chelated alkylpalladium compound. *J. Organomet. Chem.* **1982**, *240*, 95–102.

176 Iyer, S.; Ramesh, C. Aryl–Pd covalently bonded palladacycles, novel amino and oxime catalysts for the Heck reaction. *Tetrahedron Lett.* **2000**, *41*, 8981–8984.

9
Cyclizations

9.1
Introduction

Cyclic compounds are mainly prepared from non-cyclic starting materials either by cycloaddition reactions or by cyclization. Whereas in cycloadditions the ring size of the product is determined by the reaction mechanism of a given cycloaddition, cyclizations enable the use of most bond-forming reactions and, in principle, enable access to any ring size. Certain rings, however, are more difficult to prepare than others, and because oligomerization competes with all cyclizations, slow cyclizations will have a detrimental effect on the yield of cyclic products. In this section some of the critical structural considerations for successful cyclizations will be presented and examples of difficult cyclizations will be discussed.

9.2
Baldwin's Cyclization Rules

In the 1970s J.E. Baldwin proposed a set of qualitative generalizations for the probability of formation of cyclic compounds, based on the favored trajectories for the approach of one reactant to another [1–5]. Cyclizations were organized according to the ring size formed, whether the breaking bond is located in (endo) or outside (exo) the newly formed ring, and the hybridization of the (mostly uncharged) carbon atom with the highest *s* character involved in bond formation (tet, trig, or dig for sp^3-, sp^2-, or *sp*-hybridized carbon, respectively). Baldwin's rules state that 3 to 6-endo-tet, 3 to 5-endo-trig, and 3 and 4-exo-dig processes are unfavorable, whereas all remaining cyclizations are allowed (Scheme 9.1).

These rules do not apply strictly, but provide useful guidelines for synthesis design. The rules are generally not applicable to electrocyclic reactions or to substrates containing non-second-period elements (e.g. P or S), because their longer bond lengths imply different geometric constraints.

An example of a forbidden 6-endo-tet process would be the intramolecular alkylation of a carbanion via a six-membered transition state (Scheme 9.2). Because such an alkylation would proceed via the SN2 mechanism and would require a linear

Side Reactions in Organic Synthesis. Florencio Zaragoza Dörwald

ISBN: 3-527-31021-5

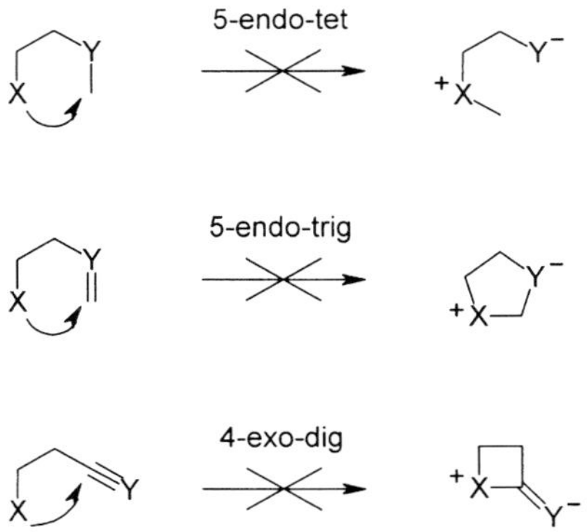

5-endo-tet

5-endo-trig

4-exo-dig

Scheme 9.1. Unfavorable processes according to Baldwin's rules.

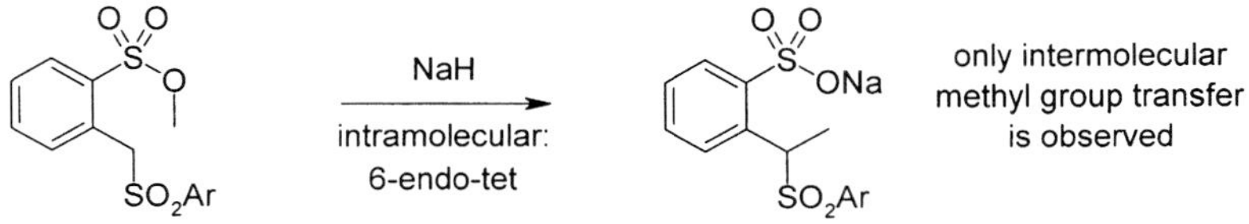

NaH

intramolecular: 6-endo-tet

only intermolecular methyl group transfer is observed

Scheme 9.2. Methylation of a metalated sulfone [6].

arrangement of the incoming nucleophile (the carbanion), the central carbon, and the leaving group, it cannot occur intramolecularly in rings with fewer than approximately nine atoms. The reaction sketched in Scheme 9.2 was shown by Eschenmoser to proceed exclusively intermolecularly.

Further unfavorable processes are 3 to 5-endo-trig cyclizations (Scheme 9.3). Although Michael additions of amines or alcohols to acrylates are usually faster than reaction of these nucleophiles with esters, 2-(2-amino/hydroxy)ethyl)acrylates do not undergo intramolecular Michael additions to yield pyrrolidines or furans, but yield lactams or lactones instead. The reason for this is that the nucleophilic functional groups cannot approach the π* orbital of the alkene sufficiently without undue generation of strain [7]. Interestingly, the reverse reactions (e.g. the base-induced scission of 3-methoxycarbonylfuran) do not proceed either. Similarly, the hydroxy ketone shown in Scheme 9.3 (third example) does not cyclize on treatment with bases, and the expected furanone, which can be prepared by acid-catalyzed cyclization of the hydroxy ketone, does not undergo ring opening on treatment with bases, because the energy barrier between these two compounds is too high. That the acid-catalyzed cyclization proceeds smoothly might be because of the reduced double bond character and increased flexibility of the alkene in the protonated hydroxy ketone [3, 8].

Another type of cyclization which follows Baldwin's rules is the intramolecular alkylation of enolates. Cyclopentanones cannot usually be prepared this way because the electrophilic carbon atom cannot approach the nucleophilic carbon atom sufficiently (Scheme 9.4). If the C–C double bond of the enolate remains exocyclic,

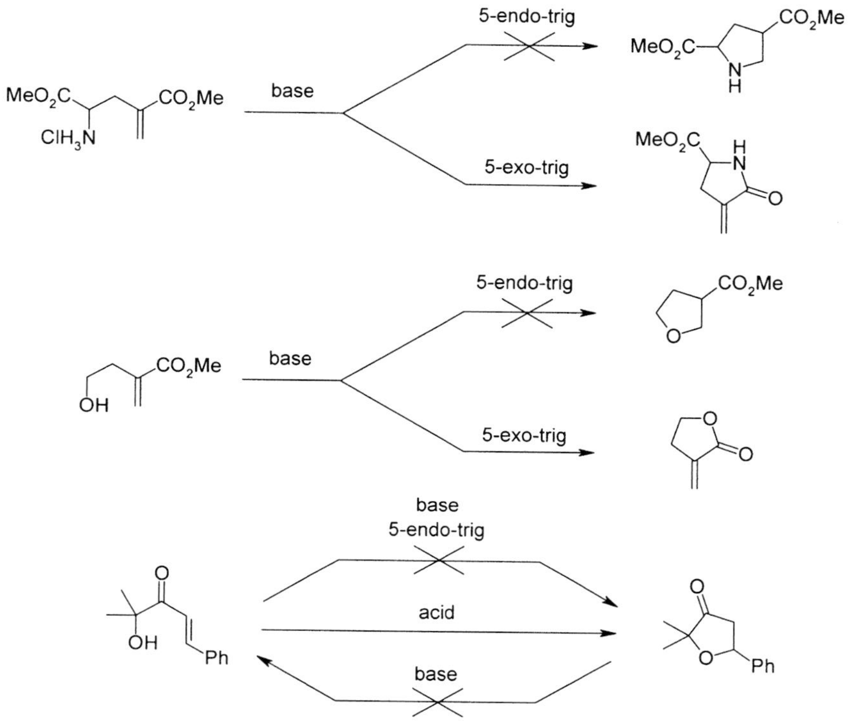

Scheme 9.3. Examples of allowed and forbidden cyclizations according to Baldwin's rules [2, 5].

however, cyclization proceeds smoothly to yield a cyclic enol ether. Because 6-endo-trig cyclization is usually favored, cyclohexanones can be formed by intramolecular *C*-alkylation of enolates.

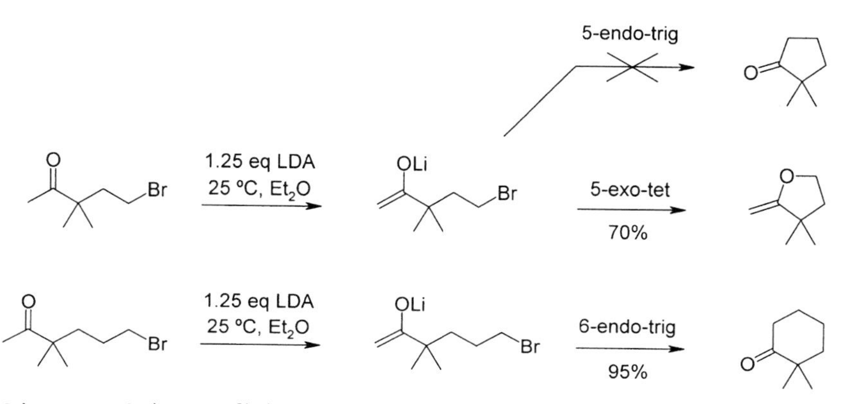

Scheme 9.4. Cyclization of haloalkyl ketones [9].

Examples of "forbidden" 5-endo-trig cyclizations which proceed anyway are shown in Scheme 9.5. Thiols do undergo 5-endo-trig cyclizations, because C–S bonds (1.82 Å) are substantially longer than C–N (1.47 Å) or C–O bonds (1.43 Å). Moreover, the intramolecular addition of carbon nucleophiles to imines to yield pyrrolidines [10, 11] or the easy conversion of aldehydes or ketones into five-membered cyclic aminals also involve an unfavorable 5-endo-trig cyclization [12], and suggest that Baldwin's rules are not applicable to cyclizations involving imines. The C–Si bond (1.88 Å) is also sufficiently long to enable 5-endo-trig cyclizations to proceed smoothly (last reaction, Scheme 9.5).

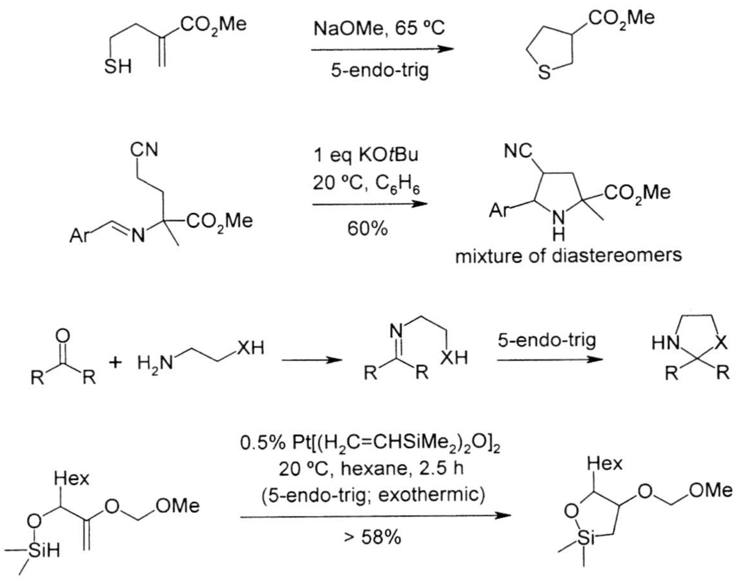

Scheme 9.5. Examples of forbidden cyclizations according to Baldwin's rules [2, 5, 10, 13]. Ar = 4-MeOC$_6$H$_4$; X = O, S, NH.

Whereas 5-endo-trig processes are unfavorable, 6-endo-trig cyclizations are allowed. For this reason the reaction conditions used for several well-known preparations of six-membered rings cannot be used to prepare the corresponding five-membered rings. For example, treatment of phenethylamines with aldehydes and acid yields tetrahydroisochinolines (Pictet–Spengler synthesis), but treatment of benzylamines under these conditions does not yield dihydroisoindoles (Scheme 9.6).

It might, at first glance, come as a surprise that, despite the ban on 5-endo-trig cyclizations, 5-endo-dig cyclizations are allowed. A possible reason for this observation is that intramolecular nucleophilic or electrophilic attack at an alkyne can proceed with the substrate in a flat conformation, because two orthogonal π^* (or π) orbitals are available. Similarly, hydroxymethyl allenes cyclize on treatment with a base (Scheme 9.7), because the terminal π^* orbital of the allene and the hydroxymethyl group can readily achieve a synperiplanar orientation. Intramolecular

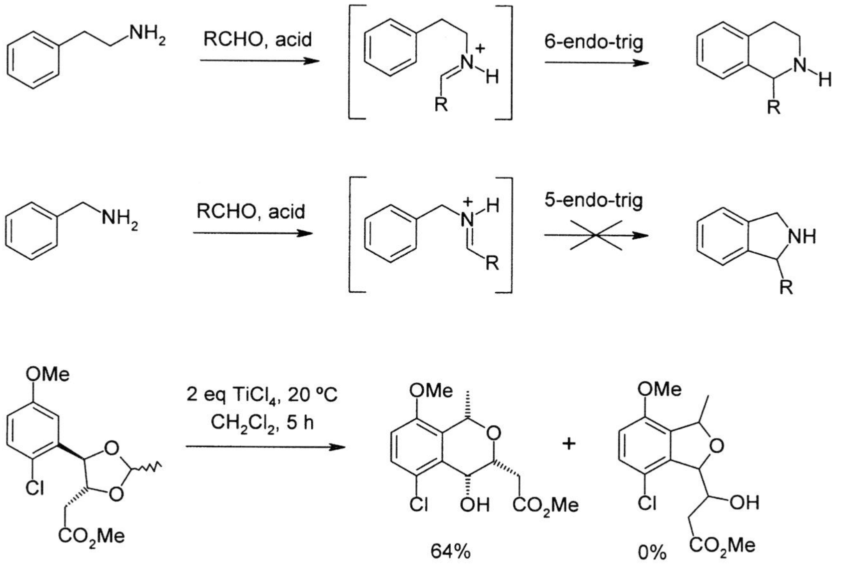

Scheme 9.6. Formation of heterocycles by 6-endo-trig cyclization [14].

nucleophilic attack at the C–C double bond in a homoallyic alcohol or amine, on the other hand, requires an out-of-plane approach of the nucleophile to the alkene, with a significant build-up of strain and thus a high barrier of activation. The cyclization of 3-butyn-1-ols or alkynylamines can, alternatively, also be catalyzed by a variety of transition metals [15, 16].

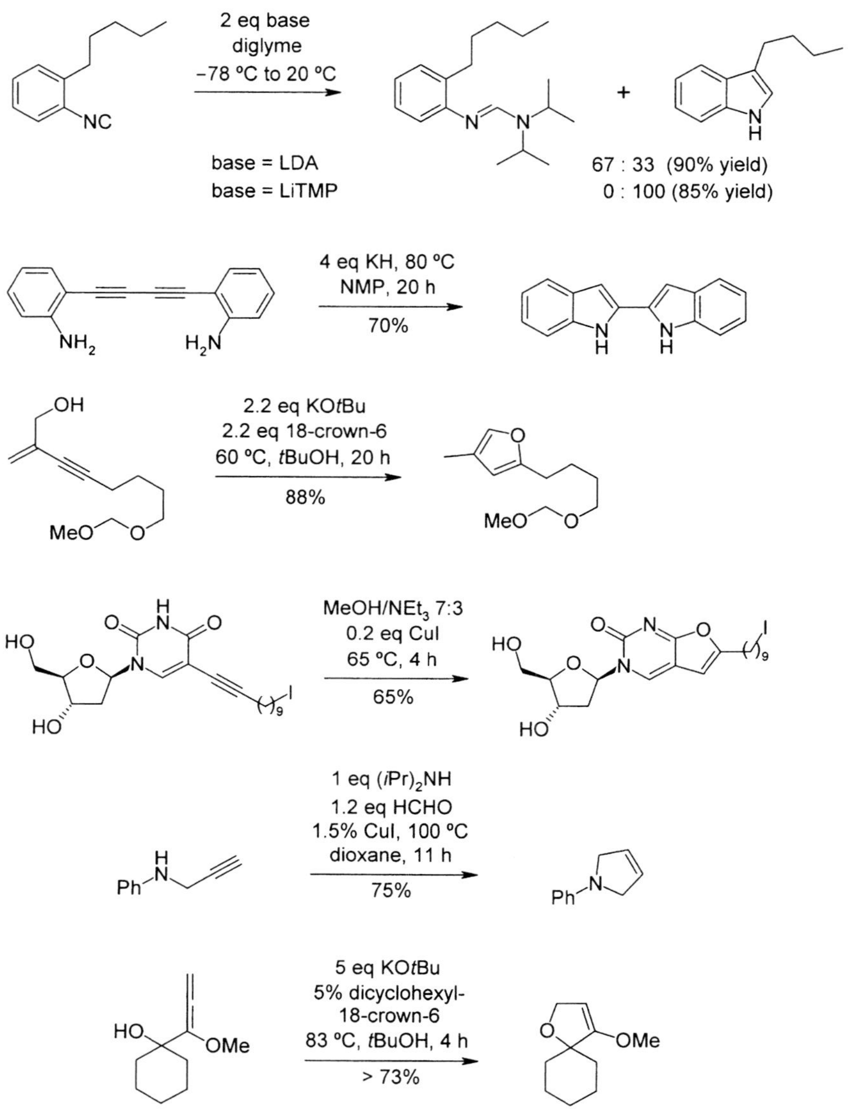

2 eq base
diglyme
−78 °C to 20 °C

NC

base = LDA
base = LiTMP

N N

+

N
H

67 : 33 (90% yield)
0 : 100 (85% yield)

NH_2 H_2N

4 eq KH, 80 °C
NMP, 20 h
70%

N N
H H

OH

MeO O

2.2 eq KO*t*Bu
2.2 eq 18-crown-6
60 °C, *t*BuOH, 20 h
88%

O

MeO O

HO O N O N O
H
HO
I
( )9

MeOH/NEt_3 7:3
0.2 eq CuI
65 °C, 4 h
65%

HO O N O N O
HO
( )9 I

Ph N H

1 eq (iPr)$_2$NH
1.2 eq HCHO
1.5% CuI, 100 °C
dioxane, 11 h
75%

Ph N

HO OMe

5 eq KO*t*Bu
5% dicyclohexyl-
18-crown-6
83 °C, *t*BuOH, 4 h
> 73%

O OMe

Scheme 9.7. Examples of 5-endo-dig cyclizations [17–21] and 5-endo-trig cyclizations of allenes [22].

9.3
Structural Features of the Chain

Cyclizations will only proceed if the substrate has or can readily attain a conformation in which the two reacting groups are in close proximity. Hence, substrates of low flexibility must already have the correct arrangement of functional groups to enable their reaction. Such preorganization can significantly enhance the rate of cyclizations, and even promote reactions which would not take place intermolecularly (Section 3.3). On the other hand, if the conformation required for cyclization is energetically unfavorable, the cyclization will not proceed, even if the two functional groups would readily react intermolecularly. Not only alkenes, but also oximes, *N*-monosubstituted amides (RCONHR′), or esters, for example, are flat and have a high barrier of rotation for the interconversion of *Z* and *E* conformers [23]. Because the *Z* conformation is preferred by amides and esters [24] (Scheme 9.8), cyclizations of amides or esters to yield lactams or lactones, respectively, do not always proceed smoothly. Catalysis by acids may in some cases facilitate the required change of conformation of amides [25, 26].

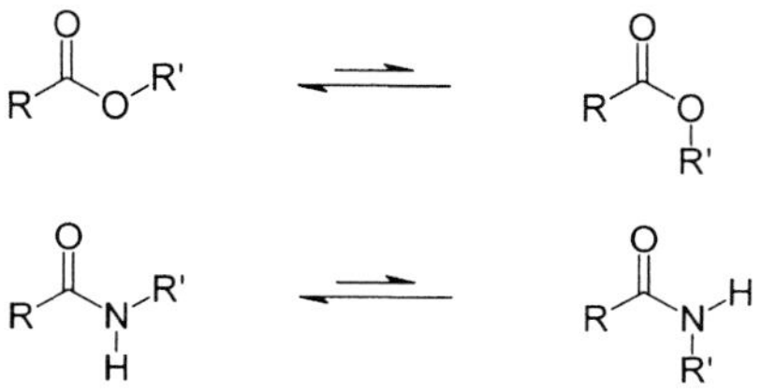

Scheme 9.8. Preferred conformations of esters and amides.

Some examples of photolytic intramolecular [2 + 2] cycloadditions which lead to the formation of lactones and lactams are sketched in Scheme 9.9. The allyl ester and *N*-monosubstituted amide do not undergo cyclization, probably because of the high energy of the required *E* conformers. Cyclization of the homoallyl ester does,

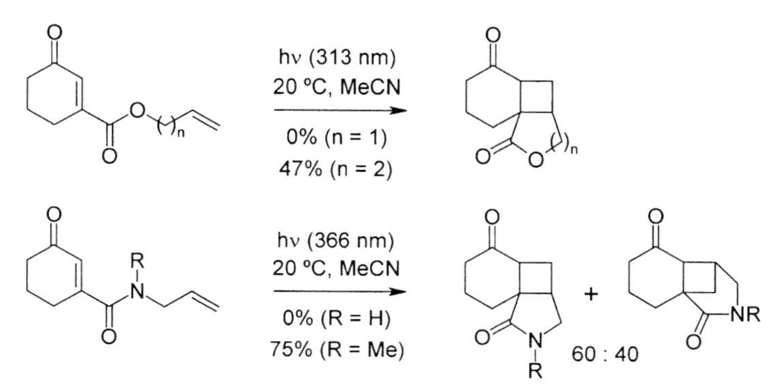

Scheme 9.9. Formation of lactones and lactams by intramolecular [2 + 2] cycloaddition [27].

however, proceed well because of the higher flexibility of this substrate. For the *N*-allyl-*N*-methylamide both possible conformations have similar energy and will be equally populated; cyclization also proceeds smoothly in this case.

Similarly, only *N*-substituted lactams are formed readily by intramolecular Diels–Alder or ene reactions (Scheme 9.10). Lactams without a substituent on the nitrogen can either not be prepared, or require higher reaction temperatures or longer reaction times than the preparation of similar, *N*-substituted lactams.

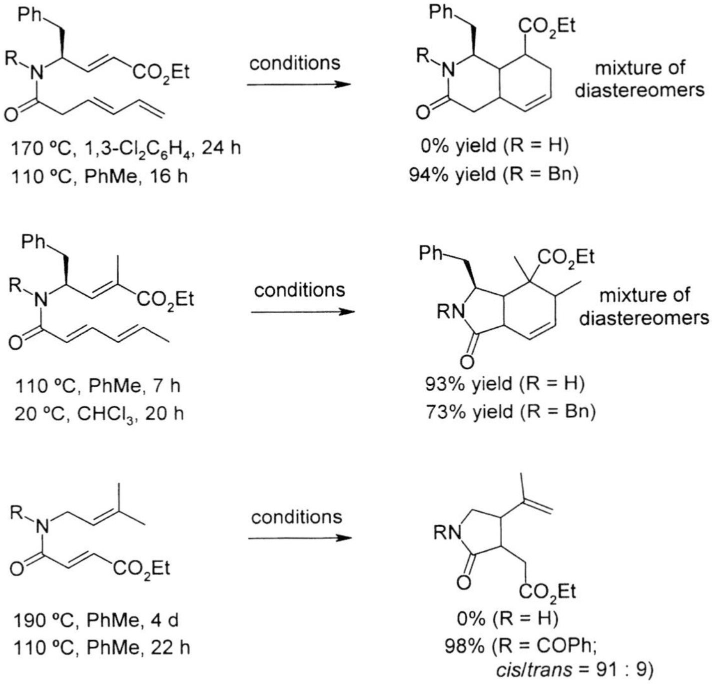

Ph R N CO_2Et O conditions Ph CO_2Et R N O mixture of diastereomers

170 °C, 1,3-$Cl_2C_6H_4$, 24 h — 0% yield (R = H)
110 °C, PhMe, 16 h — 94% yield (R = Bn)

Ph R N CO_2Et O conditions Ph CO_2Et RN O mixture of diastereomers

110 °C, PhMe, 7 h — 93% yield (R = H)
20 °C, $CHCl_3$, 20 h — 73% yield (R = Bn)

R N O CO_2Et conditions RN O CO_2Et

190 °C, PhMe, 4 d — 0% (R = H)
110 °C, PhMe, 22 h — 98% (R = COPh; *cis/trans* = 91 : 9)

Scheme 9.10. Formation of lactams by intramolecular Diels–Alder and ene reaction [25, 28]

Cyclizations proceeding via radicals are, apparently, better suited to the preparation of *N*-unsubstituted lactams (Scheme 9.11). Nevertheless, in this type of cyclization also the formation of *N*-substituted lactams proceeds more rapidly and more cleanly.

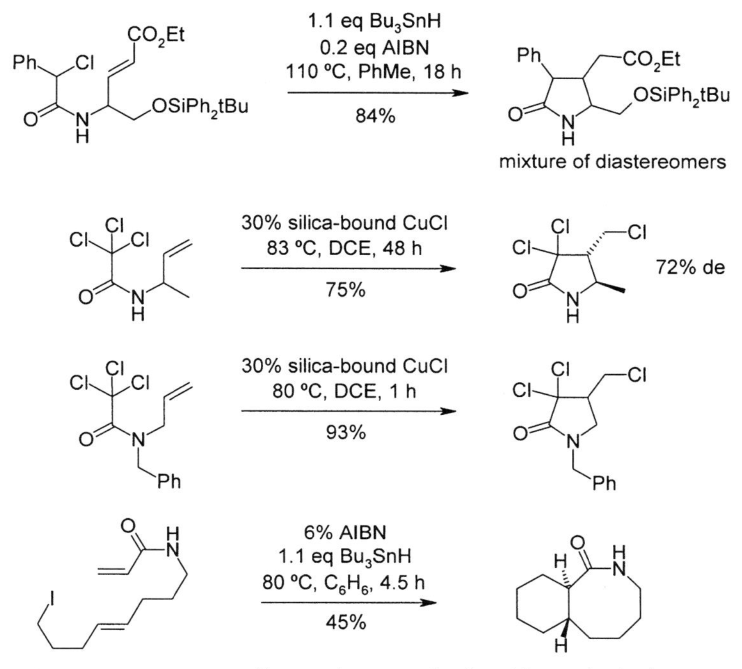

Scheme 9.11. Formation of lactams by intramolecular addition of radicals to C–C double bonds [29–31].

Alkylene chains with geminal dialkyl substitution have been shown to cyclize more readily than the corresponding unsubstituted compounds (Scheme 9.12). This effect can be used to achieve cyclization which might not occur without geminal disubstitution (see, e.g., Scheme 9.21) [32]. The magnitude of this effect is, however, highly dependent on the type of reaction and product, as illustrated by the examples in Scheme 9.12.

Two possible reasons for this effect have been suggested – angle compression at the substituted carbon and "the reactive rotamer effect" [34], i.e. facilitation, by the two alkyl groups, of the generation of an energy-rich conformation suitable for promoting the cyclization. The two last examples of the intramolecular Diels–Alder reaction in Scheme 9.12 indicate that angle compression does not have a decisive effect on reaction rate enhancement in this system, because in these two instances (1,1-disubstituted cyclopropane and cyclobutane) angle enlargement rather than angle compression will occur [34].

Cyclizations proceeding by intramolecular SN2 reactions are usually irreversible, and will, therefore, not necessarily yield the thermodynamically most stable product but that which is formed most quickly. Scheme 9.13 depicts a cyclization in which the outcome depends on the configuration of the starting alkene. Interestingly, the *cis* isomer gives mainly rise to a strained, eight-membered ether even though a path to the less strained tetrahydropyran is, in principle, accessible.

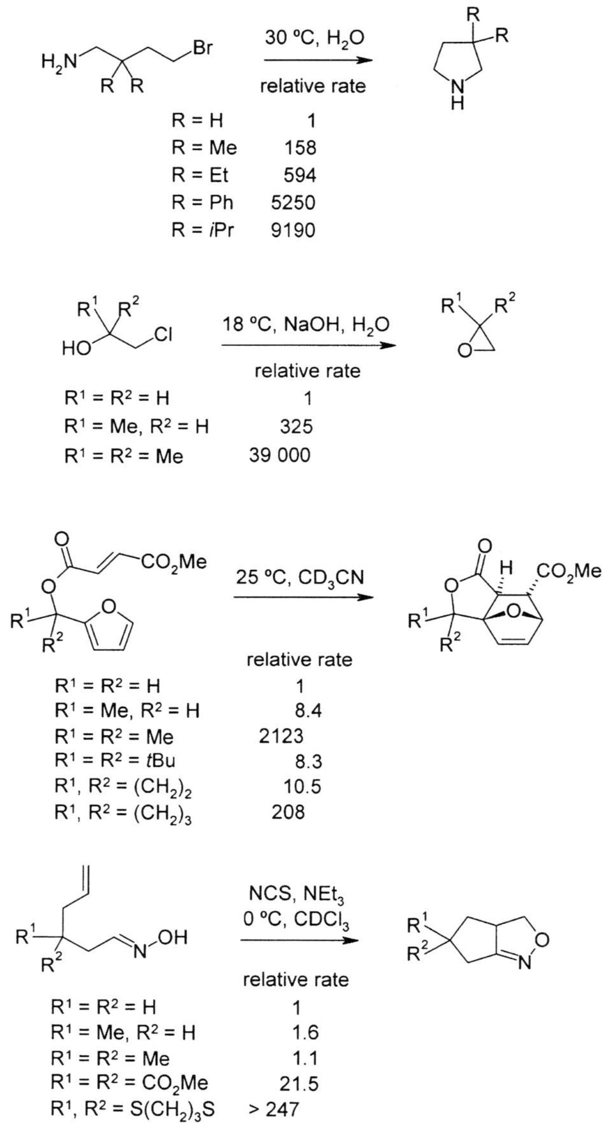

30 °C, H_2O

relative rate

R	relative rate
R = H	1
R = Me	158
R = Et	594
R = Ph	5250
R = *i*Pr	9190

18 °C, NaOH, H_2O

	relative rate
$R^1 = R^2 = H$	1
$R^1 = Me, R^2 = H$	325
$R^1 = R^2 = Me$	39 000

25 °C, CD_3CN

	relative rate
$R^1 = R^2 = H$	1
$R^1 = Me, R^2 = H$	8.4
$R^1 = R^2 = Me$	2123
$R^1 = R^2 = t\text{Bu}$	8.3
$R^1, R^2 = (CH_2)_2$	10.5
$R^1, R^2 = (CH_2)_3$	208

NCS, NEt_3, 0 °C, $CDCl_3$

	relative rate
$R^1 = R^2 = H$	1
$R^1 = Me, R^2 = H$	1.6
$R^1 = R^2 = Me$	1.1
$R^1 = R^2 = CO_2Me$	21.5
$R^1, R^2 = S(CH_2)_3S$	> 247

Scheme 9.12. Promotion of cyclizations by geminal disubstitution in the chain [33–36].

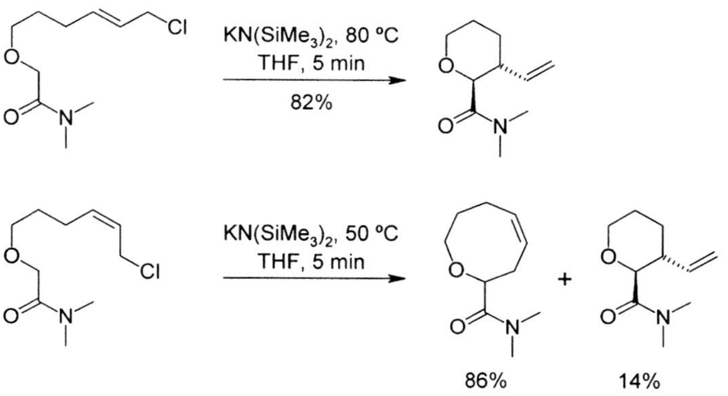

KN(SiMe3)2, 80 °C, THF, 5 min, 82%

KN(SiMe3)2, 50 °C, THF, 5 min; 86% + 14%

Scheme 9.13. Configuration-dependent cyclization of a chloroalkenyloxy acetamide [37].

9.4 Ring Size

Comparison of the heats of combustion of cycloalkanes (Table 9.1) shows that cyclopropane, cyclobutane, and cyclononane yield more energy per methylene group than the other cycloalkanes. This can be attributed to strain resulting from bond-angle distortion (Baeyer strain), eclipsed conformations (Pitzer strain), and transannular, repulsive van der Waals interactions. Common (five- and six-membered) rings and large (more than twelve-membered) rings have little or no strain. This

Table 9.1. Heats of combustion (ΔH_c, kcal mol^{-1}) in the gas phase of *n*-membered cycloalkanes per methylene group, and enthalpies ($\Delta H_{cy}°$, kcal mol^{-1}) and entropies ($\Delta S_{cy}°$, cal mol^{-1} K^{-1}) for the cyclization $H(CH_2)_nH \rightarrow$ cyclo-$(CH_2)_n + H_2$ [42, 43].

n	$-\Delta H_c$	$-\Delta H_c + \Delta H_c\ (C_6H_{12})$	$\Delta H_{cy}°$	$-\Delta S_{cy}°$
3	166.3	8.9	37.6	7.7
4	163.9	6.5	37.1	10.9
5	158.7	1.3	16.7	13.3
6	157.4	0.0	7.9	21.2
7	158.3	0.9	16.6	19.8
8	158.6	1.2	20.1	19.0
9	158.8	1.4	17.8	
10	158.6	1.2	22.8	
11	158.4	1.0	21.9	
12	157.8	0.4	14.3	
13	157.7	0.3	15.6	
14	157.4	0.0	22.3	
15	157.5	0.1	12.3	
16	157.5	0.1	12.9	
17			7.1	

does, however, not mean that small rings or seven- to eleven-membered rings will always be more difficult to prepare than the other ring sizes. Cyclopropanes, for instance, are quite readily formed by cyclization because the reacting carbon atoms are already in close proximity and few bonds are required to assume a suitable conformation [38]. The entropy change during cyclopropane formation is, therefore, only small compared with those for formation of larger rings (Table 9.1). Very large rings, on the other hand, are difficult to prepare, despite their lack of strain, because the two functional groups which have to react with each other are far apart and unlikely to come into close proximity [39]. Macrocyclizations, therefore, often require high dilution or a template which can bring the two ends of the precursor together [40, 41].

The relative stability of cycloalkylmethyl carbocations is illustrated by the rearrangements illustrated in Scheme 9.14. Primary cycloalkylmethyl cations tend to undergo ring enlargement to yield secondary cycloalkyl cations if the resulting, larger cycloalkane is not significantly more strained than the starting cycloalkane. Thus, the cyclobutylmethyl cation rearranges cleanly to the cyclopentyl cation. Ring enlargement of the cyclopropylmethyl cation to a cyclobutyl cation or ring scission to the homoallyl cation are, however, not particularly favorable processes, because of the strong carbocation-stabilizing effect of the cyclopropyl group. The solvolysis of

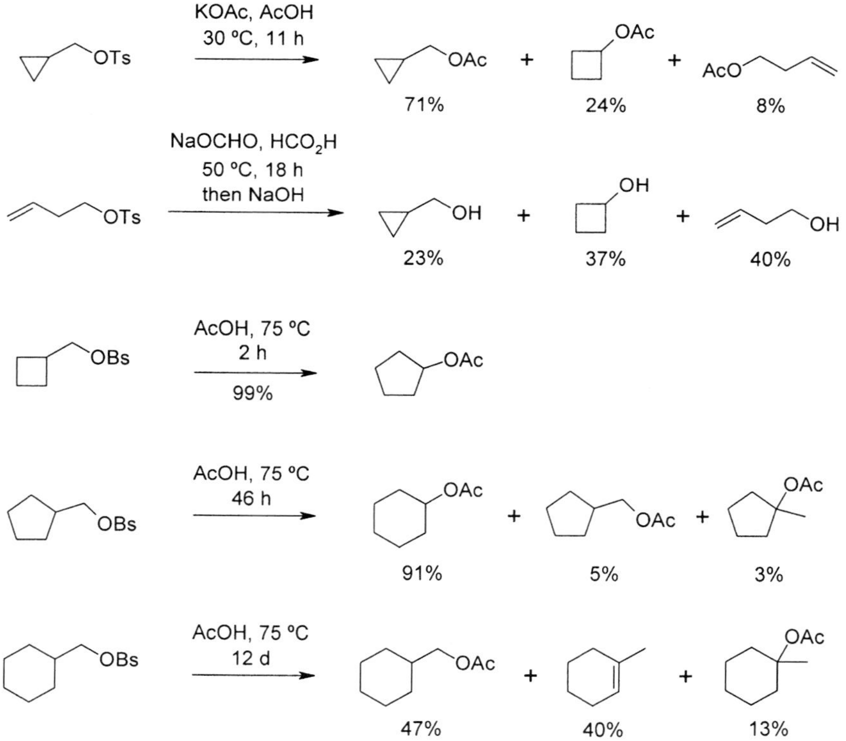

Scheme 9.14. Rearrangement of cycloalkylmethyl carbocations [45–47].

cyclopropylmethyl derivatives usually gives mixtures of cyclopropylmethyl, cyclobutyl, and homoallyl derivatives [44], and some homoallyl cations can even cyclize to yield cyclopropane derivatives (second reaction, Scheme 9.14). That the cyclohexylmethyl cation does not undergo ring expansion to a cycloheptyl cation results from the stability of cyclohexane derivatives – hydride migration to yield a tertiary 1-methyl-1-cyclohexyl cation is the preferred reaction pathway in this instance (last example, Scheme 9.14).

9.4.1 Formation of Cyclopropanes

The high strain energy of cyclopropane markedly affects its chemistry. As shown in Scheme 9.14, the cyclopropylmethyl carbocation is a relatively stable intermediate, which does not readily undergo ring expansion or ring scission. The cyclopropylmethyl radical [48] or cyclopropylmethyllithium [49] or -magnesium derivatives, on the other hand, tend to undergo rapid ring scission to the corresponding homoallylic intermediates [50], because the cyclopropyl group does not effectively stabilize either radicals or carbanions [51]. This behavior contrasts with the irreversible formation of cyclopentylmethyllithium from hexenyllithium (Scheme 9.15) or of other five- or six-membered rings from the corresponding alkenyllithium derivatives [52, 53].

The cyclopropylmethyl Grignard reagent can only be trapped with electrophiles before rearrangement if the Grignard reagent is generated and treated with the electrophile at –75 °C [54].

The facile ring scission of cyclopropylmethyl radicals or carbanions does not mean that cyclopropanes are kinetically unstable or synthetically inaccessible. Despite their high strain, cyclopropanes are not difficult to prepare by cyclization, and are sometimes even formed when reaction pathways to larger, less strained rings are available. The reason for this is the small entropy change which accompanies cyclopropane formation (Table 9.1) and the close proximity of the reacting centers [38]. This proximity of both reacting carbon atoms accelerates cyclizations to cyclopropanes significantly, and cyclization rates to cyclopropanes can be up to 1,000 times higher than the formation of cyclopentanes from homologous substrates [38] (Table 9.2).

Although the parent cyclopropylmethyl radical or lithium intermediates readily rearrange to homoallylic species, cyclopropanes can be prepared via similar intermediates if the corresponding cyclopropylmethyl derivatives are stabilized by suitable functional groups or trapped by an irreversible reaction (Scheme 9.16). Thus, 1-metallo-3-butenes cyclize readily to cyclopropanes if the double bond is substituted with an electron-withdrawing group [60], with an allylic leaving group [61], or is otherwise reactive towards nucleophiles [62]. Homoallyl radicals can cyclize irreversibly to cyclopropanes if the C–C double bond is substituted in the allylic position with a group with a weak bond to carbon, for example a phenylthio group [63]. Cyclopropanes can also be obtained in high yields by intramolecular addition of carbocations to alkenes (last reaction, Scheme 9.16).

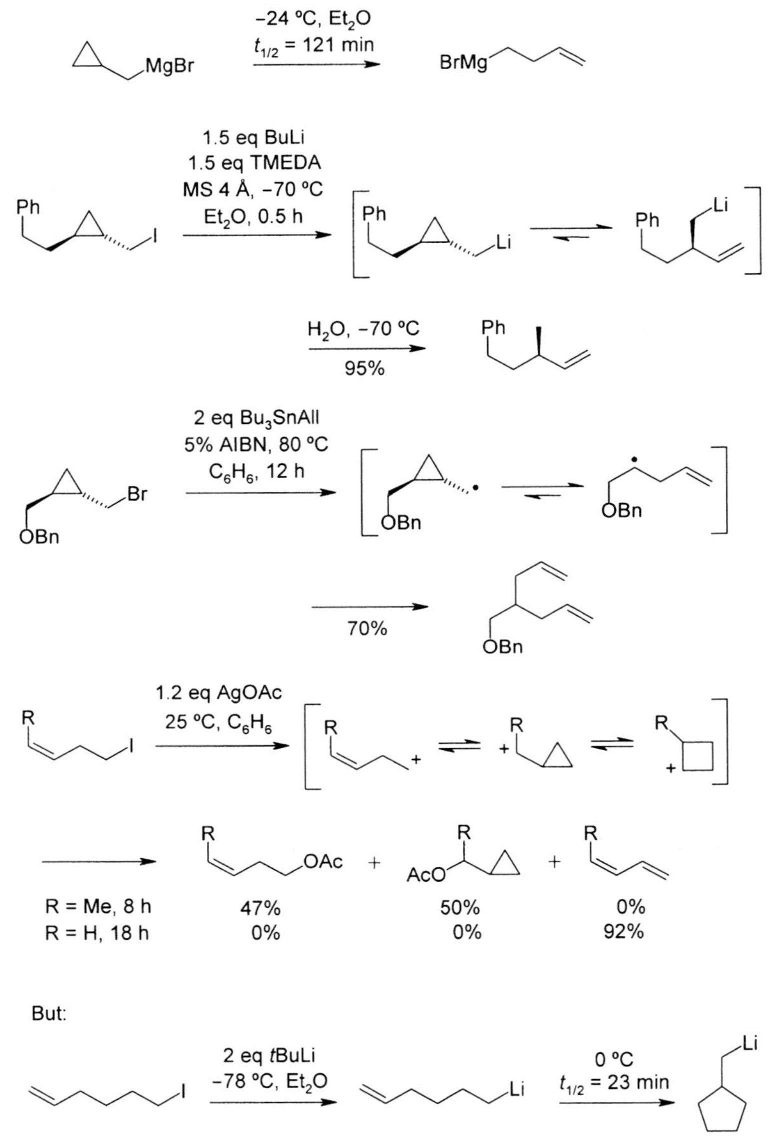

Scheme 9.15. Reactivity of cyclopropylmethyl carbanions, radicals, and carbocations [55–59].

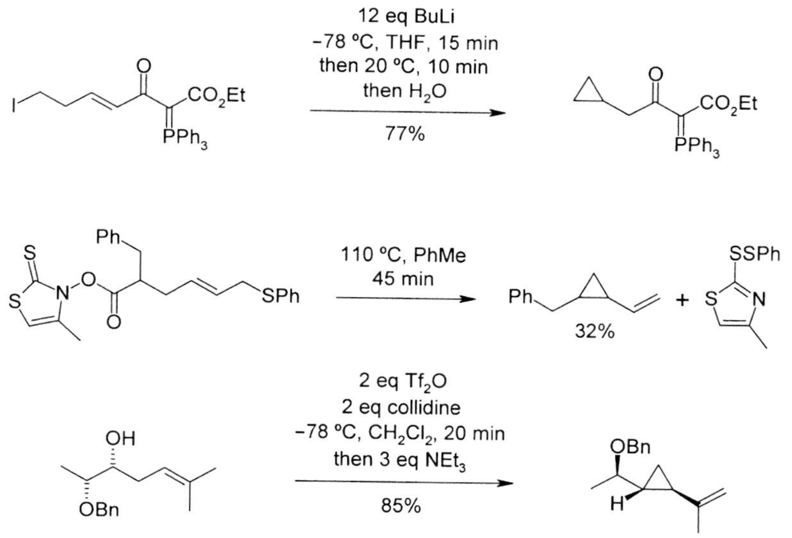

12 eq BuLi
−78 °C, THF, 15 min
then 20 °C, 10 min
then H_2O
77%

110 °C, PhMe
45 min
32%

2 eq Tf_2O
2 eq collidine
−78 °C, CH_2Cl_2, 20 min
then 3 eq NEt_3
85%

Scheme 9.16. Preparation of cyclopropanes by intramolecular addition of radicals, carbanions, or carbocations to C–C double bonds [63–65].

Examples of the preparation of cyclopropanes by intramolecular nucleophilic substitution are illustrated in Scheme 9.17. The first example is a synthesis of [1.1.1]propellane, which yields the product in acceptable yields, despite the high strain and poor stability of this compound [66]. The second and third examples illustrate the remarkable ease with which 3-halopropyl ketones cyclize to yield cyclopropanes instead of cyclic, five-membered enol ethers or ketones. Similarly, carbamates of 2-haloethylglycine esters do not undergo intramolecular *N*- or *O*-alkylation on treatment with bases, but yield cyclopropanes instead [67, 68]. Some nucleophiles can undergo Michael addition to 3-halomethyl acrylates faster than direct SN2 reaction, to yield cyclopropanes by cyclization of the intermediate enolates (fourth example, Scheme 9.17) [69].

Because of the readiness with which cyclopropanes are formed from 3-halopropyl ketones, cyclization of the latter to dihydrofurans is difficult, and few examples of such cyclizations have been reported (Scheme 9.18). Acylsilanes, on the other hand, are more nucleophilic at oxygen than ketones, and readily undergo intramolecular *O*-alkylations [73–75].

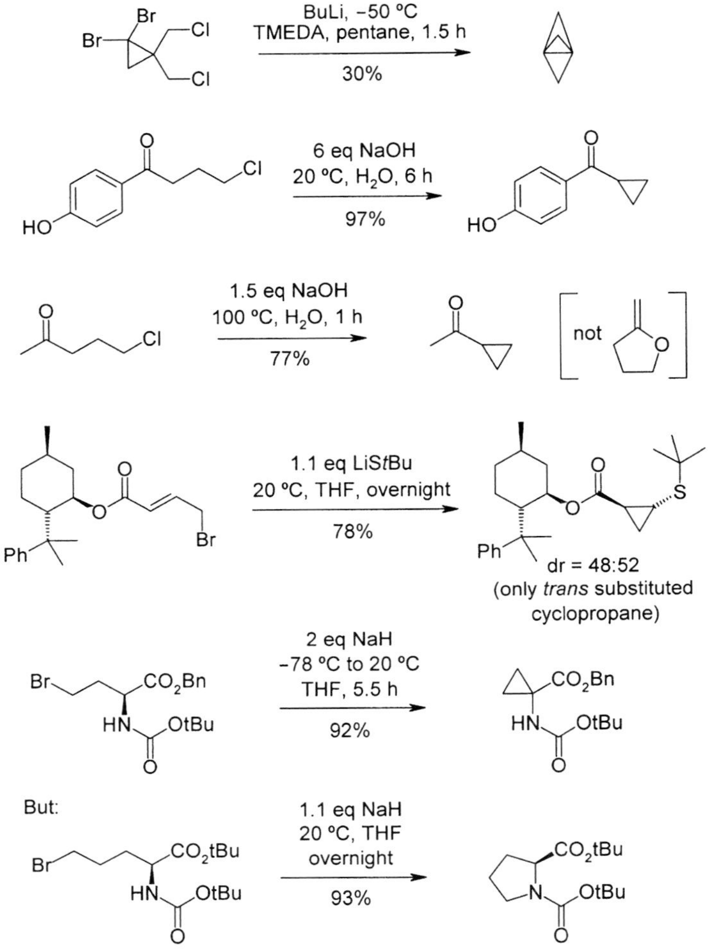

Scheme 9.17. Preparation of cyclopropanes by intramolecular SN2 reactions [66–68, 70–72].

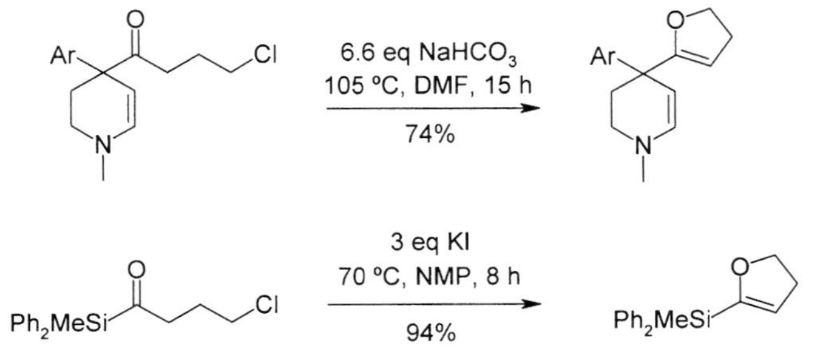

Scheme 9.18. Formation of dihydrofurans from 3-chloropropyl ketones and acylsilanes [74, 76]. Ar = 2,3-dimethoxyphenyl.

9.4.2 Formation of Cyclobutanes

Cyclobutanes are usually more difficult than cyclopropanes to prepare by cyclization. Although their ring strain is as high as that of cyclopropanes, more bonds must assume a suitable conformation for cyclobutane formation, resulting in a higher entropic barrier (Table 9.1).

Ring scission of cyclobutylmethyl radicals or carbanions is significantly slower than that of the corresponding cyclopropylmethyl intermediates. Thus, rearrangement of the cyclopropylmethyl radical to the homoallyl radical is extremely rapid ($k = 1.3 \times 10^8\ s^{-1}$ at 25 °C), and has often been used as a radical clock [48, 50, 77]. Ring scission of cyclobutylmethyl radicals, on the other hand, is approximately five orders of magnitude slower [78], and these radicals can therefore be trapped intermolecularly before ring opening has occurred (Scheme 9.19). Similarly, although cyclopropylmethyl Grignard reagents or organolithium compounds often rearrange even at low temperatures, the corresponding cyclobutyl derivatives are substantially more stable (Scheme 9.19).

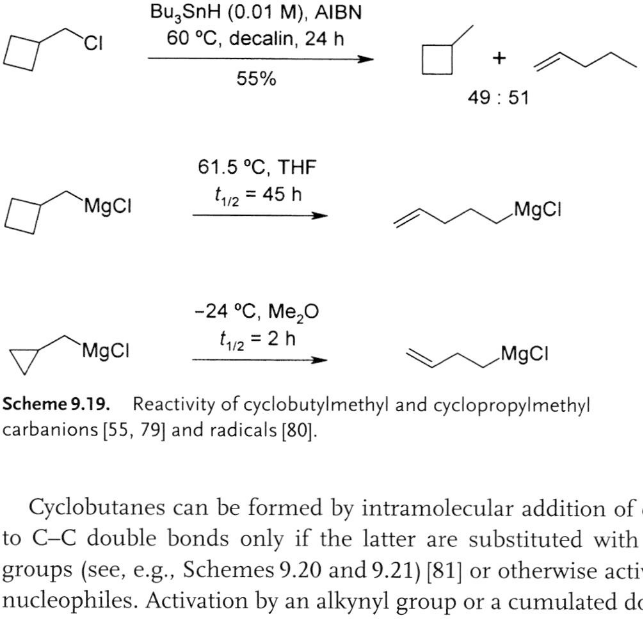

Scheme 9.19. Reactivity of cyclobutylmethyl and cyclopropylmethyl carbanions [55, 79] and radicals [80].

Cyclobutanes can be formed by intramolecular addition of carbanions or radicals to C–C double bonds only if the latter are substituted with electron-withdrawing groups (see, e.g., Schemes 9.20 and 9.21) [81] or otherwise activated toward attack by nucleophiles. Activation by an alkynyl group or a cumulated double bond can be sufficient to promote cyclobutane formation (Scheme 9.20). Unactivated alkenes, however, do not usually undergo cyclization to cyclobutanes via intramolecular addition of carbanions or radicals.

The fact that cyclobutanes are not readily formed is further illustrated by the reactions shown in Scheme 9.21. The radicals formed from the sketched (haloalkyl)acrylates or related (haloalkyl)acrylonitriles [85] can, in principle, cyclize to yield cyclobutane derivatives, but instead these intermediates react intermolecularly [86] or are

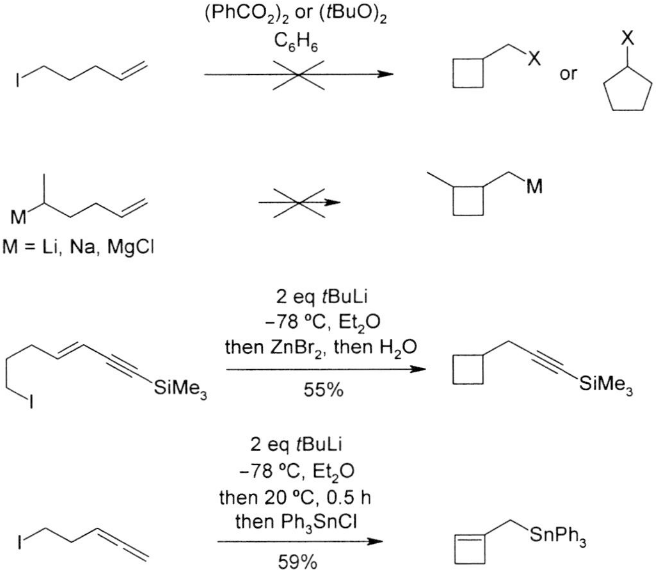

Scheme 9.20. Formation of cyclobutanes by intramolecular addition of organometallic reagents to C–C double bonds [60, 82–84].

reduced before cyclization. Cyclobutane formation can sometimes be promoted by introduction of substituents at the chain [32, 85].

Intramolecular nucleophilic displacements are sometimes better suited to difficult cyclizations than additions to C–C multiple bonds, because nucleophilic substitutions are usually irreversible. Some metalated 4-halobutyl imines cyclize to yield cyclobutanes rather than six-membered cyclic enamines (Scheme 9.22). If alkoxides are used as bases, however, exclusive *N*-alkylation is observed. No examples could be found of the cyclization of 4-halobutyl ketones to cyclobutyl ketones, but 5-halopen-

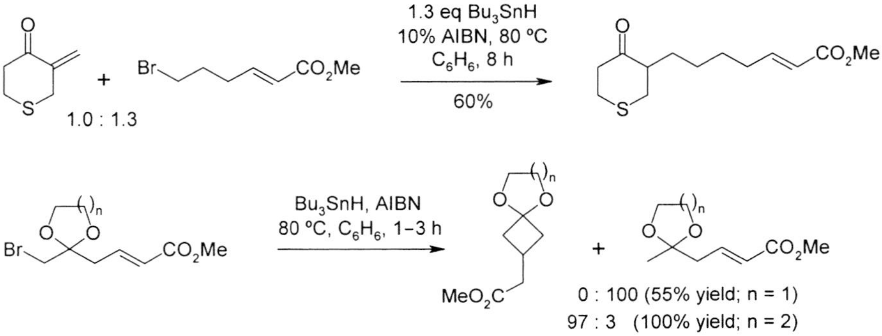

Scheme 9.21. Reactions of 3-(3-bromopropyl)acrylates [32, 87].

tanoic esters and amides, 5,6-epoxynitriles [88, 89], and 2-(2,3-epoxy-1-propyl)malonates [90] do undergo intramolecular *C*-alkylation to yield cyclobutane derivatives (Scheme 9.22).

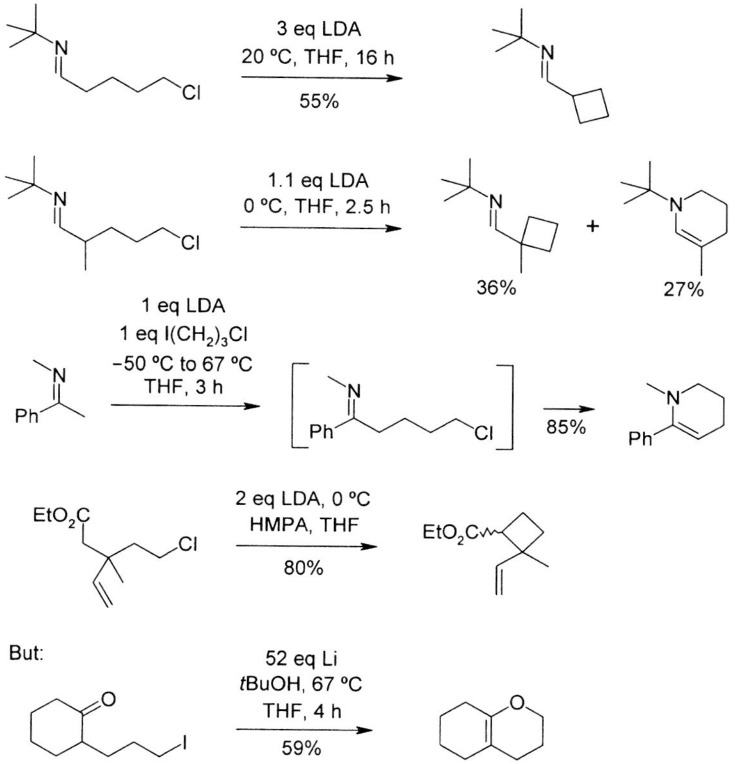

3 eq LDA
20 °C, THF, 16 h
55%

1.1 eq LDA
0 °C, THF, 2.5 h
36% + 27%

1 eq LDA
1 eq $I(CH_2)_3Cl$
-50 °C to 67 °C
THF, 3 h
85%

2 eq LDA, 0 °C
HMPA, THF
80%

But:

52 eq Li
*t*BuOH, 67 °C
THF, 4 h
59%

Scheme 9.22. Formation of cyclobutanes from enolates and metalated imines [91–94].

9.5 Heterocycles

The relative rates of formation of saturated heterocycles by intramolecular alkylation of the heteroatom do not correlate exactly with those of formation of carbocycles. For heterocycles, five-membered rings are usually formed faster than three-membered rings, whereas for carbocycles the opposite is observed (i.e. cyclopropanes are formed faster than cyclopentanes) (Table 9.2). For heterocycles and for carbocycles the formation of five-membered rings is faster than the formation of six-membered rings, and four-membered rings are usually formed more slowly than six-membered rings. The formation of five- and six-membered hetero- or carbocycles is usually highly favored over competing intermolecular reactions.

The relative rates given in Table 9.2 should be used as a rough guideline only, because small structural variations or different reaction conditions can alter the cyclization rates significantly.

Table 9.2. Relative rates of the reaction $HNu(CH_2)_nX \rightarrow$ cyclo-$Nu(CH_2)_n + HX$ as a function of product ring size.

Ring size	HNu/X NH_2/Br[a]	OH/Cl[b]	PhS/Cl[c]	CO_2H/Br[d]	E_2CH/Cl[e]
3	0.29	215	5.0	8.46×10^{-4}	125,000
4	3.98×10^{-3}	0.05	0.026	0.92	0.19
5	72.5	285	1.0	100	1,250
6	1.0	1.0		1.0	1.0
7	4.78×10^{-3}			3.73×10^{-3}	
8				3.85×10^{-5}	
9				4.23×10^{-5}	
10				1.31×10^{-4}	

a) H_2O, 25 °C [95, 96]
b) H_2O, NaOH, 30 °C [97]
c) rate of hydrolysis; H_2O, dioxane, 100 °C [98]
d) DMSO, 50 °C [36]
e) KO*t*Bu, *t*BuOH, 25 °C; E = CO_2Et [99]

References

1 Baldwin, J. E. Rules for ring closure. *J. Chem. Soc. Chem. Commun.* **1976**, 734–736.

2 Baldwin, J. E.; Cutting, J.; Dupont, W.; Kruse, L.; Silberman, L.; Thomas, R. C. 5-Endo-trigonal reactions: a disfavoured ring closure. *J. Chem. Soc. Chem. Commun.* **1976**, 736–738.

3 Johnson, C. D. Stereoelectronic effects in the formation of 5- and 6-membered rings: the role of Baldwin's rules. *Acc. Chem. Res.* **1993**, *26*, 476–482.

4 Baldwin, J. E.; Lusch, M. J. Rules for ring closure: application to intramolecular aldol condensations in polyketonic substrates. *Tetrahedron* **1982**, *38*, 2939–2947.

5 Baldwin, J. E.; Thomas, R. C.; Kruse, L. I.; Silberman, L. Rules for ring closure: ring formation by conjugate addition of oxygen nucleophiles. *J. Org. Chem.* **1977**, *42*, 3846–3852.

6 Tenud, L.; Farooq, S.; Seibl, J.; Eschenmoser, A. Endocyclische SN-Reaktionen am gesättigten Kohlenstoff? *Helv. Chim. Acta* **1970**, *53*, 2059–2069.

7 Menger, F. M. Directionality of organic reactions in solution. *Tetrahedron* **1983**, *39*, 1013–1040.

8 Ellis, G. W. L.; Johnson, C. D.; Rogers, D. N. Stereoelectronic control of intramolecular Michael additions. *J. Am. Chem. Soc.* **1983**, *105*, 5090–5095.

9 Baldwin, J. E.; Kruse, L. I. Rules for ring closure. Stereoelectronic control in the endocyclic alkylation of ketone enolates. *J. Chem. Soc. Chem. Commun.* **1977**, 233–235.

10 Grigg, R.; Kemp, J.; Malone, J.; Tangthongkum, A. 5-Endo-trig cyclization and 1,3-anionic cycloaddition in arylimine derivatives of α-amino acid esters. *J. Chem. Soc. Chem. Commun.* **1980**, 648–650.

11 Hodges, J. C.; Wang, W.; Riley, F. Synthesis of a spirocyclic indoline lactone. *J. Org. Chem.* **2004**, *69*, 2504–2508.

12 Astudillo, M. E. A.; Chokotho, N. C. J.; Jarvis, T. C.; Johnson, C. D.; Lewis, C. C.; McDonnell, P. D. Hydroxy Schiff base–oxazolidine tautomerism: apparent breakdown of Baldwin's rules. *Tetrahedron* **1985**, *41*, 5919–5928.

13 Tamao, K.; Nakagawa, Y.; Ito, Y. Regio- and stereoselective intramolecular hydrosilylation of α-hydroxy enol ethers: 2,3-*syn*-2-methoxymethoxy-1,3-nonanediol. *Org. Synth.* **1998**, *Coll. Vol. IX*, 539–547.

14 Giles, R. G. F.; Rickards, R. W.; Senanayake, B. S. Synthesis of isochroman-3-ylacetates and isochromane-γ-lactones through rearrangement of aryldioxolanylacetates. *J. Chem. Soc. Perkin Trans. 1* **1998**, 3949–3956.

15 McDonald, F. E.; Gleason, M. M. Asymmetric synthesis of nucleosides via molybdenum-catalyzed alkynol cycloisomerization coupled with stereoselective glycosylations of deoxyfuranose glycals and 3-amidofuranose glycals. *J. Am. Chem. Soc.* **1996**, *118*, 6648–6659.

16 Cacchi, S.; Fabrizi, G.; Pace, P. Palladium-catalyzed cyclization of *o*-alkynyltrifluoroacetanilides with allyl esters. A regioselective synthesis of 3-allylindoles. *J. Org. Chem.* **1998**, *63*, 1001–1011.

17 Ito, Y.; Kobayashi, K.; Saegusa, T. An efficient synthesis of indole. *J. Am. Chem. Soc.* **1977**, *99*, 3532–3534.

18 Brancale, A.; McGuigan, C.; Andrei, G.; Snoeck, R.; De Clercq, E.; Balzarini, J. Bicyclic nucleoside inhibitors of varicella-zoster virus: the effect of a terminal halogen substitution in the side-chain. *Bioorg. Med. Chem. Lett.* **2000**, *10*, 1215–1217.

19 Marshall, J. A.; DuBay, W. J. Base-catalyzed isomerization of alkynyloxiranes. A general synthesis of furans. *J. Am. Chem. Soc.* **1992**, *114*, 1450–1456.

20 Koradin, C.; Dohle, W.; Rodriguez, A. L.; Schmid, B.; Knochel, P. Synthesis of polyfunctional indoles and related heterocycles mediated by cesium and potassium bases. *Tetrahedron* **2003**, *59*, 1571–1587.

21 Jayaprakash, K.; Venkatachalam, C. S.; Balasubramanian, K. K. A convenient one-pot synthesis of *N*-aryl-3-pyrrolines. *Tetrahedron Lett.* **1999**, *40*, 6493–6496.

22 Gange, D.; Magnus, P. An unusual new allene cyclization reaction. Synthesis of dihydrofuran-3(2*H*)-ones. *J. Am. Chem. Soc.* **1978**, *100*, 7746–7747.

23 Staab, H. A.; Lauer, D. Stabile rotationsisomere Carbonsäureamide. *Chem. Ber.* **1968**, *101*, 864–878.

24 Pawar, D. M.; Khalil, A. A.; Hooks, D. R.; Collins, K.; Elliott, T.; Stafford, J.; Smith, L.; Noe, E. A. *E* and *Z* conformations of esters, thiol esters, and amides. *J. Am. Chem. Soc.* **1998**, *120*, 2108–2112.

25 Xia, Q.; Ganem, B. Asymmetric total synthesis of (–)-α-kainic acid using an enantioselective, metal-promoted ene cyclization. *Org. Lett.* **2001**, *3*, 485–487.

26 Magnus, P.; Lacour, J.; Evans, P. A.; Rigollier, P.; Tobler, H. Applications of the β-azidonation reaction to organic synthesis. α,β-Enones, conjugate addition, and γ-lactam annulation. *J. Am. Chem. Soc.* **1998**, *120*, 12486–12499.

27 Le Blanc, S.; Pete, J.-P.; Piva, O. Intramolecular [2 + 2] photocycloaddition of oxoesters and oxoamides. *Tetrahedron Lett.* **1993**, *34*, 635–638.

28 Murray, W. V.; Sun, S.; Turchi, I. J.; Brown, F. K.; Gauthier, A. D. Diels–Alder reactions of amino acid-derived trienes. *J. Org. Chem.* **1999**, *64*, 5930–5940.

29 Bryans, J. S.; Large, J. M.; Parsons, A. F. Efficient tin hydride-mediated radical cyclization of secondary amides leading to substituted pyrrolidinones. Part 2. Application to the synthesis of aromatic kainic acid analogues. *J. Chem. Soc. Perkin Trans. 1* **1999**, 2905–2910.

30 Clark, A. J.; Filik, R. P.; Haddleton, D. M.; Radigue, A.; Sanders, C. J.; Thomas, G. H.; Smith, M. E. Solid-supported catalysts for atom-transfer radical cyclization of 2-haloacetamides. *J. Org. Chem.* **1999**, *64*, 8954–8957.

31 Blake, A. J.; Hollingworth, G. J.; Pattenden, G. Radical mediated macrocyclization-transannulation cascades leading to ring-fused lactones and lactams. *Synlett* **1996**, 643–644.

32 Jung, M. E.; Marquez, R. *gem*-Substituent effects in small ring formation: novel ketal ring size effect. *Tetrahedron Lett.* **1997**, *38*, 6521–6524.

33 Brown, R. F.; van Gulick, N. M. The geminal alkyl effect on the rates of ring closure of bromobutylamines. *J. Org. Chem.* **1956**, *21*, 1046–1049.

34 Jung, M. E.; Gervay, J. *gem*-Dialkyl effect in the intramolecular Diels–Alder reaction of 2-furfuryl methyl fumarates: the reactive rotamer effect, enthalpic basis for acceleration, and evi-

dence for a polar transition state. *J. Am. Chem. Soc.* **1991**, *113*, 224–232.
35 Jung, M. E.; Vu, B. T. Substituent effects on intramolecular dipolar cycloadditions: the *gem*-dicarboalkoxy effect. *Tetrahedron Lett.* **1996**, *37*, 451–454.
36 Gilchrist, T. L. *Heterocyclenchemie*; Verlag Chemie: Weinheim, 1995.
37 Kim, H.; Choi, W. J.; Jung, J.; Kim, S.; Kim, D. Construction of eight-membered ether rings by olefin geometry-dependent internal alkylation: first asymmetric total synthesis of (+)-3-(*E*)- and (+)-3-(*Z*)-pinnatifidenyne. *J. Am. Chem. Soc.* **2003**, *125*, 10238–10240.
38 Gronert, S.; Azizian, K.; Friedman, M. A. Cyclizations of 3-chlorocarbanions to cyclopropanes: 'strain-free' transition states for forming highly strained rings. *J. Am. Chem. Soc.* **1998**, *120*, 3220–3226.
39 Lightstone, F. C.; Bruice, T. C. Enthalpy and entropy of ring closure reactions. *Bioorg. Chem.* **1998**, *26*, 193–199.
40 Roxburgh, C. J. The synthesis of large-ring compounds. *Tetrahedron* **1995**, *51*, 9767–9822.
41 Drandarov, K.; Hesse, M. Lithium and proton templated ω-polyazamacrolactamization, new general routes to macrocyclic polyamines. *Tetrahedron Lett.* **2002**, *43*, 7213–7216.
42 March, J. *Advanced Organic Chemistry*; John Wiley and Sons: New York, 1992.
43 Winnik, M. A. Cyclization and the conformation of hydrocarbon chains. *Chem. Rev.* **1981**, *81*, 491–524.
44 Kevill, D. N.; Abduljaber, M. H. Correlation of the rates of solvolysis of cyclopropylcarbinyl and cyclobutyl bromides using the extended Grünwald–Winstein equation. *J. Org. Chem.* **2000**, *65*, 2548–2554.
45 Roberts, D. D. Solvent effects. The solvolysis rates of cyclopropylcarbinyl, 1-methylcyclopropylcarbinyl, and 1-phenylcyclopropylcarbinyl arenesulfonate derivatives. *J. Org. Chem.* **1964**, *29*, 294–297.
46 Servis, K. L.; Roberts, D. D. Small-ring compounds XLI. The formolysis of allylcarbinyl tosylate. *J. Am. Chem. Soc.* **1964**, *86*, 3773–3777.
47 Roberts, D. D.; Wu, C.-H. A solvolytic investigation of cyclobutylcarbinyl and related *p*-bromobenzenesulfonates. *J. Org. Chem.* **1974**, *39*, 1570–1575.
48 Cooksy, A. L.; King, H. F.; Richardson, W. H. Molecular orbital calculations of ring opening of the isoelectronic cyclopropylcarbinyl radical, cyclopropoxy radical, cyclopropylaminium radical cation series of radical clocks. *J. Org. Chem.* **2003**, *68*, 9441–9452.
49 Mudryk, B.; Cohen, T. Generation, some synthetic uses, and 1,2-vinyl rearrangements of secondary and tertiary homoallyllithiums, including ring contractions and a ring expansion. Remarkable acceleration of the rearrangement by an oxyanionic group. *J. Am. Chem. Soc.* **1993**, *115*, 3855–3865.
50 Smith, D. M.; Nicolaides, A.; Golding, B. T.; Radom, L. Ring opening of the cyclopropylcarbinyl radical and its *N*- and *O*-substituted analogues: a theoretical examination of very fast unimolecular reactions. *J. Am. Chem. Soc.* **1998**, *120*, 10223–10233.
51 Perkins, M. J.; Peynircioglu, N. B. The rates of deprotonation of some cyclopropyltoluene derivatives in search for the effect of the conformation of cyclopropyl groups on the stabilization of the carbanionic centers. *Tetrahedron* **1985**, *41*, 225–227.
52 Coldham, I.; Vennall, G. P. Chiral organolithium species: determination of the rate of cyclization and extent of racemization. *Chem. Commun.* **2000**, 1569–1570.
53 Bailey, W. F.; Khanolkar, A. D.; Gavaskar, K.; Ovaska, T. V.; Rossi, K.; Thiel, Y.; Wiberg, K. B. Stereoselectivity of cyclization of substituted 5-hexen-1-yllithiums: regiospecific and highly stereoselective insertion of an unactivated alkene into a C–Li bond. *J. Am. Chem. Soc.* **1991**, *113*, 5720–5727.
54 Kündig, E. P.; Perret, C. Low temperature Grignard reactions with pure Mg slurries. Trapping of cyclopropylmethyl and benzocyclobutenylmethyl Grignard reagents with CO_2. *Helv. Chim. Acta* **1981**, *64*, 2606–2613.
55 Patel, D. J.; Hamilton, C. L.; Roberts, J. D. Small ring compounds. XLIV. Interconversion of cyclopropylcarbinyl and allylcarbinyl Grignard reagents. *J. Am. Chem. Soc.* **1965**, *87*, 5144–5148.
56 Gurjar, M. K.; Ravindranadh, S. V.; Karmakar, S. Mild and efficient methodology for installation of *gem*-diallyl functionality on carbohydrate synthons. *Chem. Commun.* **2001**, 241–242.
57 Charette, A. B.; Naud, J. Regioselective opening of substituted (cyclopropylmethyl)lithiums derived from cyclopropylmethyl iodides. *Tetrahedron Lett.* **1998**, *39*, 7259–7262.

58 Previtera, L.; Monaco, P.; Mangoni, L. Cyclopropylcarbinyl compounds from homoallylic iodides. *Tetrahedron Lett.* **1984**, *25*, 1293–1294.
59 Bailey, W. F.; Patricia, J. J.; DelGobbo, V. C.; Jarret, R. M.; Okarma, P. J. Cyclization of 5-hexenyllithium to (cyclopentylmethyl)lithium. *J. Org. Chem.* **1985**, *50*, 1999–2000.
60 Lorthiois, E.; Marek, I.; Normant, J. F. Intramolecular carbolithiation of silylated enynes. *Tetrahedron Lett.* **1996**, *37*, 6693–6694.
61 Bailey, W. F.; Tao, Y. Efficient preparation of vinylcyclopropane by SN′ cyclization of the organolithium derived from (*E*)-5-iodo-1-methoxy-2-pentene. *Tetrahedron Lett.* **1997**, *38*, 6157–6158.
62 Richey, H. G.; Kossa, W. C. Intramolecular cyclization of an allenic Grignard reagent. *Tetrahedron Lett.* **1969**, 2313–2314.
63 Cekovic, Z.; Saicic, R. Radical cyclization reactions. Cyclopropane ring formation by 3-exo-cyclization of 5-phenylthio-3-pentenyl radicals. *Tetrahedron Lett.* **1990**, *31*, 6085–6088.
64 Cooke, M. P.; Widener, R. K. Lithium–halogen exchange-initiated cyclization reactions. 3. Intramolecular conjugate addition reactions of unsaturated acylphosphoranes. *J. Org. Chem.* **1987**, *52*, 1381–1396.
65 Nagasawa, T.; Handa, Y.; Onoguchi, Y.; Suzuki, K. Stereoselective synthesis of cyclopropanes via homoallylic participation. *Bull. Chem. Soc. Jpn.* **1996**, *69*, 31–39.
66 Kaszynski, P.; Friedli, A. C.; Michl, J. Toward a molecular-size 'tinkertoy' construction set. Preparation of terminally functionalized [n]staffanes from [1.1.1]propellane. *J. Am. Chem. Soc.* **1992**, *114*, 601–620.
67 Yamaguchi, J.-I.; Ueki, M. Facile new method for the preparation of optically active protected proline. *Chem. Lett.* **1996**, 621–622.
68 Kienzler, T.; Strazewski, P.; Tamm, C. A new synthesis of coprine and *O*-ethylcoprine. *Helv. Chim. Acta* **1992**, *75*, 1078–1084.
69 Toru, T.; Nakamura, S.; Takemoto, H.; Ueno, Y. Stereoselective conjugate addition of an α-sulfinyl carbanion to α,β-unsaturated esters: asymmetric synthesis of cycloalkanecarboxylates. *Synlett* **1997**, 449–450.
70 Chang, S.-J.; Fernando, D. P.; King, S. Preparation of cyclopropyl *p*-hydroxyphenyl ketone and its precursor 3-chloropropyl *p*-hydroxyphenyl ketone. *Org. Process Res. Dev.* **2001**, *5*, 141–143.
71 Amputch, M. A.; Matamoros, R.; Little, R. D. Asymmetric induction in the Michael initiated ring closure reaction. *Tetrahedron* **1994**, *50*, 5591–5614.
72 Cannon, G. W.; Ellis, R. C.; Leal, J. R. Methyl cyclopropyl ketone. *Org. Synth.* **1963**, *Coll. Vol. IV*, 597–600.
73 Tsai, M.-Y.; Cherng, C.-D.; Nieh, H.-C.; Sieh, J.-A. The formation of oxygen-containing heterocycles via intramolecular cyclizations of halo-substituted acylsilanes and unsaturated acylsilanes. *Tetrahedron* **1999**, *55*, 14587–14598.
74 Tsai, Y.-M.; Nieh, H.-C.; Cherng, C.-D. Cyclizations of functionalized acylsilanes to form 2-silyldihydropyrans and 2-silyldihydrofurans. *J. Org. Chem.* **1992**, *57*, 7010–7012.
75 Yates, K.; Agolini, F. The importance of inductive effects in α-silyl and α-germyl ketones. *Can. J. Chem.* **1966**, *44*, 2229–2231.
76 Shenvi, A. B.; Ciganek, E. Convenient synthesis of 3-methyl-2,3,4,4aα,5,6,7,7aα-octahydro-1*H*-benzofuro[3,2-*e*]isoquinoline. *J. Org. Chem.* **1984**, *49*, 2942–2947.
77 Griller, D.; Ingold, K. U. Free-radical clocks. *Acc. Chem. Res.* **1980**, *13*, 317–323.
78 Stevenson, J. P.; Jackson, W. F.; Tanko, J. M. Cyclopropylcarbinyl-type ring openings. Reconciling the chemistry of neutral radicals and radical anions. *J. Am. Chem. Soc.* **2002**, *124*, 4271–4281.
79 Hill, E. A.; Davidson, J. A. Kinetics of the ring opening of cyclobutylmethylorganomagnesium compounds. *J. Am. Chem. Soc.* **1964**, *86*, 4663–4669.
80 Beckwith, A. L. J.; Moad, G. The kinetics and mechanism of ring opening of radicals containing the cyclobutylcarbinyl system. *J. Chem. Soc. Perkin Trans. 2* **1980**, 1083–1092.
81 Baker, J. M.; Dolbier, W. R. Density functional theory calculations of the effect of fluorine substitution on the cyclobutylcarbinyl to 4-pentenyl radical rearrangement. *J. Org. Chem.* **2001**, *66*, 2662–2666.
82 Hill, E. A.; Richey, H. G.; Rees, T. C. Intramolecular cleavages and double bond additions of polar organometallic compounds. *J. Org. Chem.* **1963**, *28*, 2161–2162.
83 Kaplan, L. Reactions of the cyclobutylcarbinyl radical. *J. Org. Chem.* **1968**, *33*, 2531–2532.
84 Crandall, J. K.; Ayers, T. A. Cyclizations of 3,4-pentadien-1-yllithium reagents. *J. Org. Chem.* **1992**, *57*, 2993–2995.

85 Park, S.-U.; Varick, T. R.; Newcomb, M. Acceleration of the 4-exo radical cyclization to a synthetically useful rate. Cyclization of the 2,2-dimethyl-5-cyano-4-pentenyl radical. *Tetrahedron Lett.* **1990**, *31*, 2975–2978.

86 Ono, N.; Yoshida, Y.; Tani, K.; Okamoto, S.; Sato, F. A highly efficient approach to prostaglandins via radical addition of α side chains to methylenecyclopentanones. Total synthesis of natural PGE_1, limaprost and new prostaglandin derivatives. *Tetrahedron Lett.* **1993**, *34*, 6427–6430.

87 Ward, D. E.; Gai, Y.; Lai, Y. A general method for the synthesis of 3-substituted tetrahydro- and 2,3-dihydro-4*H*-thiopyran-4-ones. *Synlett* **1996**, 261–262.

88 Stork, G.; Cohen, J. F. Ring size in epoxynitrile cyclization. A general synthesis of functionally substituted cyclobutanes. Application to (±)-grandisol. *J. Am. Chem. Soc.* **1974**, *96*, 5270–5272.

89 Lallemand, J. Y.; Onanga, M. Ring size in δ-epoxynitrile cyclization. *Tetrahedron Lett.* **1975**, 585–588.

90 Zucco, M.; Le Bideau, F.; Malacria, M. Palladium-catalyzed intramolecular cyclization of vinyloxirane. Regioselective formation of cyclobutanol derivatives. *Tetrahedron Lett.* **1995**, *36*, 2487–2490.

91 Stevens, C. V.; De Kimpe, N.; Katritzky, A. R. α,α-Cyclobisalkylation of aldehydes via ω-haloaldimines. *Tetrahedron Lett.* **1994**, *35*, 3763–3766.

92 Evans, D. A. A new endocyclic enamine synthesis. *J. Am. Chem. Soc.* **1970**, *92*, 7593–7595.

93 Crandall, J. K.; Magaha, H. S. Magnesium-induced cyclizations of 2-(3-iodopropyl)cycloalkanones. A cyclopentane annelation method. *J. Org. Chem.* **1982**, *47*, 5368–5371.

94 Babler, J. H. Base-promoted cyclization of a δ-chloro ester: application to the total synthesis of (±)-grandisol. *Tetrahedron Lett.* **1975**, 2045–2048.

95 DeTar, D. F.; Brooks, W. Cyclization and polymerization of ω-(bromoalkyl)dimethylamines. *J. Org. Chem.* **1978**, *43*, 2245–2248.

96 DeTar, D. F.; Luthra, N. P. Quantitative evaluation of steric effects in S_N2 ring closure reactions. *J. Am. Chem. Soc.* **1980**, *102*, 4505–4512.

97 Richardson, W. H.; Golino, C. M.; Wachs, R. H.; Yelvington, M. B. Neighboring oxide ion and fragmentation reactions of 1,3-chlorohydrins. *J. Org. Chem.* **1971**, *36*, 943–948.

98 Böhme, H.; Sell, K. Die Hydrolyse halogenierter Äther und Thioäther in Dioxan-Wasser-Gemischen. *Chem. Ber.* **1948**, *81*, 123–130.

99 Knipe, A. C.; Stirling, C. J. M. Intramolecular reactions. Part VI. Rates of ring formation in reactions of ω-halogenoalkylmalonic esters with bases. *J. Chem. Soc. (B)* **1968**, 67–71.

10
Monofunctionalization of Symmetric Difunctional Substrates

10.1
Introduction

Many symmetric, difunctional organic compounds, for example diols, dihalides, diamines, or dicarboxylic acids, are commercially available, and the organic chemist is often tempted to base a synthetic strategy on the "desymmetrization" (i.e. monofunctionalization) of these, often attractively cheap, starting materials. Unfortunately, such monofunctionalization can quickly turn into a nightmare. Because unoptimized procedures will often yield product mixtures containing only small amounts of the desired product, its successful isolation will require perseverance and sometimes a good portion of luck. Therefore, before embarking in such ventures, alternative synthetic routes based on an unsymmetric starting material should always be seriously considered.

Difunctional compounds can be roughly divided into two groups – those in which chemical modification of one group significantly changes the reactivity of the other group and those in which no such change will occur. Monofunctionalization of the former can be easy and high-yielding whereas the second group will usually yield mixtures of starting material and the mono- and difunctionalized products when monofunctionalization is attempted. For instance, cyclic anhydrides such as succinic or glutaric anhydride are readily converted into monoesters or monoamides, because after acylation the remaining carboxyl group is no longer activated. Preparation of monoesters from other activated diacids (e.g. isophthalic acid dichloride) will, on the other hand, be more difficult, because the functional groups have little effect on each other. Similarly, Friedel–Crafts acylation of benzene leads to a strong decrease of electron density on the arene, which prevents further acylation. In Friedel–Crafts alkylations, however, no such deactivation occurs, and polyalkylated products will be obtained unless a large excess of arene is used.

Theoretical product ratios of monofunctionalized and difunctionalized product, as a function of the excess of difunctional reagent, for completely independent reactivity of both functional groups, are given in Table 10.1. These ratios show that purer monofunctionalized products will result when a large excess of difunctional reagent is used. The strategy of using a large excess of difunctional reagent will, however, be practicable only if this excess can be readily separated from the desired product.

Side Reactions in Organic Synthesis. Florencio Zaragoza Dörwald

ISBN: 3-527-31021-5

Table 10.1. Theoretical product ratios of the reaction of a difunctional reagent A with a monofunctional reagent B ($A + B \rightarrow A + AB + AB_2$).

Starting material	Products			
A/B	A	AB	AB_2	AB/AB_2
1:2	0%	0%	100%	0
1:1.5	6%	38%	56%	0.7
1:1.33	11%	44%	44%	1
1:1	25%	50%	25%	2
1.5:1	44%	44%	11%	4
2:1	56%	38%	6%	6
3:1	69%	28%	3%	10
4:1	76.5%	22%	1.5%	14
5:1	81%	18%	1%	18

A further requisite for achieving the theoretical yield of monofunctional product is that both starting reagents are thoroughly mixed *before* extensive conversion to the products has occurred. If the reaction is rapid, it might proceed to completion as soon as the two reagents come into contact, and large amounts of difunctionalized product will result because the initially formed monofunctionalized product cannot diffuse away from the added reagent.

10.2 Monofunctionalization of Dicarboxylic Acids

Monoesters of symmetric dicarboxylic acids can be prepared either by monoesterification of a diacid [1] or by monosaponification of a diester. Dicarboxylic acids which can form five- or six-membered cyclic anhydrides are readily transformed into monoesters via these intermediates, but for diacids which cannot be converted into such cyclic anhydrides monosaponification of diesters seems to be more reliable than selective monoesterification. Monoesters or monoamides of succinic, maleic, glutaric, or related diacids can be rather unstable, because of the close proximity of a carboxyl group (see Section 3.3).

Diesters of α,ω-alkanedicarboxylic acids can be monosaponified by addition of one equivalent of hydroxide. Best results are obtained if conditions can be found under which the desired product precipitates from the reaction mixture. This is, for instance, observed for all the examples shown in Scheme 10.1. The products of monosaponification of malonic esters are significantly less electrophilic than the diesters, which leads to a highly selective reaction. This is, however, not so for the undecanoic diacid diester. With this compound precipitation of the monoester salt from the reaction mixture is presumably the main reason for the high yield obtained.

Symmetric diesters with enantiotopic alkoxycarbonyl groups can sometimes be monosaponified enantioselectively by use of esterases (Scheme 10.2) [5]. Enzymes

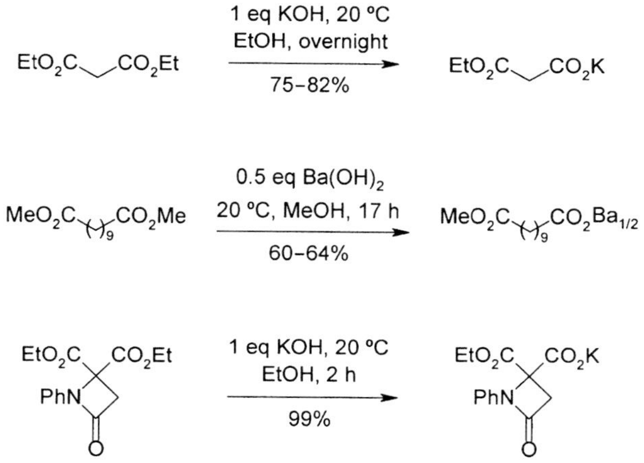

$EtO_2C\frown CO_2Et$ —(1 eq KOH, 20 °C, EtOH, overnight; 75–82%)→ $EtO_2C\frown CO_2K$

$MeO_2C(CH_2)_9CO_2Me$ —(0.5 eq $Ba(OH)_2$, 20 °C, MeOH, 17 h; 60–64%)→ $MeO_2C(CH_2)_9CO_2Ba_{1/2}$

EtO_2C, CO_2Et, PhN, O —(1 eq KOH, 20 °C, EtOH, 2 h; 99%)→ EtO_2C, CO_2K, PhN, O

Scheme 10.1. Monosaponification of diesters [2–4].

are highly selective and can often discriminate between the diester and the monoester, even if the two functionalities are far apart from each other. This elegant method requires the diester to be soluble in water to some extent, because most enzymes require water as solvent to unfold their catalytic activity. Alternatively, addition of detergents might help to attain useful reaction rates [6]. A related strategy is the enzymatic monoesterification of symmetric dicarboxylic acids [7]. The main problem of these enzymatic reactions is that their outcome with a new substrate is difficult to predict and must always be established empirically.

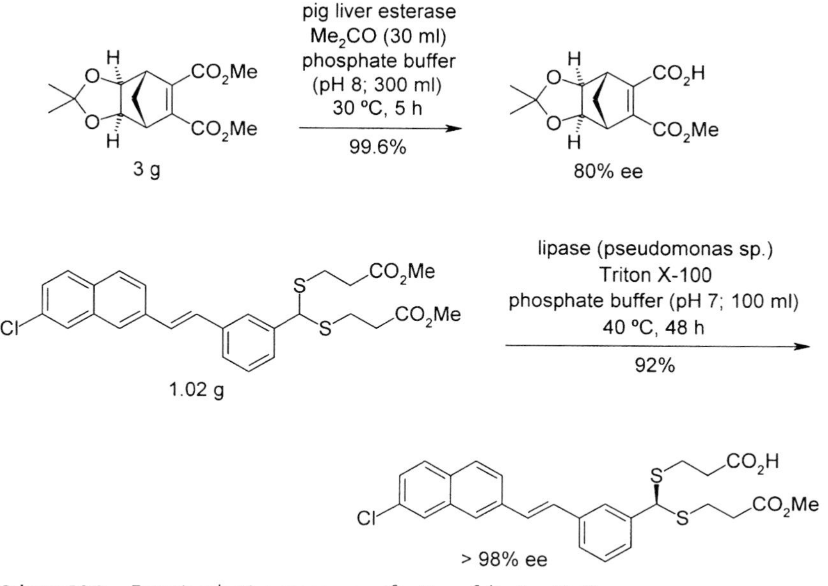

Scheme 10.2. Enantioselective monosaponification of diesters [6, 8].

Dinitriles and diamides also can sometimes be partially hydrolyzed by use of enzymes. As with other enzymatic reactions, small structural modifications can substantially modify the suitability of a given substrate. Scheme 10.3 gives illustrative examples of the highly substrate-dependent hydrolysis of dinitriles and diamides by use of a microorganism.

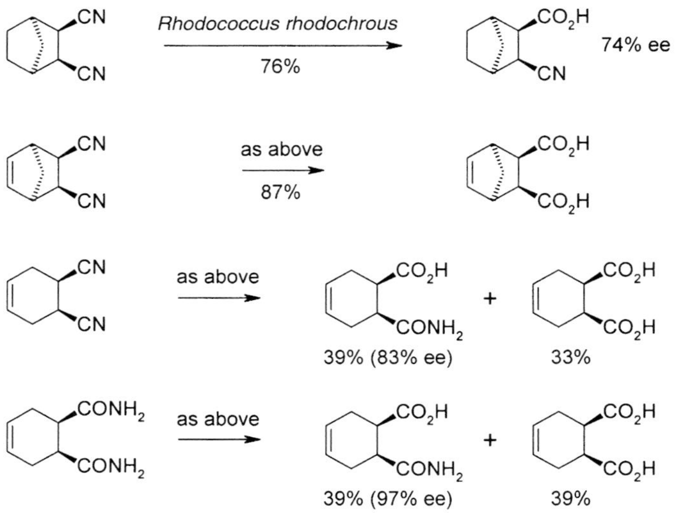

Rhodococcus rhodochrous 76% — 74% ee

as above 87%

as above — 39% (83% ee) + 33%

as above — 39% (97% ee) + 39%

Scheme 10.3. Microorganism-mediated hydrolysis of dinitriles and diamides [9].

10.3 Monofunctionalization of Diols

The monoacylation of symmetrical diols has been extensively investigated, and several useful methods have been developed. 1,2-Diols can be monoacylated readily, presumably because the product is acylated more slowly than the starting diol (Scheme 10.4). A variety of catalysts, for example lanthanide salts [10, 11], zeolites [12, 13], and tin derivatives [14, 15] have been used to further improve the selectivity of this reaction. Orthoesters have also been successfully used as acylating agents for the monoacylation of 1,2-diols [16].

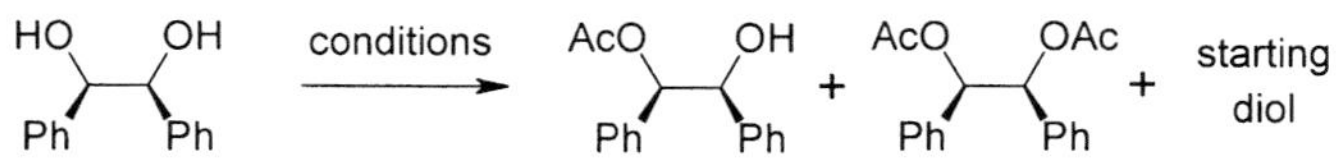

conditions	monoacetate : diacetate : starting diol
10 eq Ac_2O, THF, 44 h	14 : 2 : 84
10 eq Ac_2O, 10% $CeCl_3$, THF, 23 h	90 : 5 : 5
10 eq Ac_2O, 10% $YbCl_3$, THF, 4 h	95 : 5 : 0

Scheme 10.4. Acetylation of *meso*-hydrobenzoin [11].

If the monoacylation of a diol with enantiotopic hydroxyl groups is performed in the presence of a chiral tertiary amine, a chiral diamine [17], or a chiral phosphine [18], the acylation can proceed with high enantioselectivity, as illustrated by the example shown in Scheme 10.5. These reactions require stoichiometric amounts of a tertiary amine (DIPEA or NEt_3), and the chiral tertiary amine must, therefore, be significantly more nucleophilic than DIPEA or NEt_3, in order to react faster with the acylating agent to generate a chiral *N*-acyl ammonium salt. This salt is believed to be the intermediate which transfers the acyl group to the alcohol. Pyrrolidines, quinuclidines, or analogs of DMAP are usually much more nucleophilic than noncyclic tertiary amines, and are therefore effective nucleophilic catalysts [19].

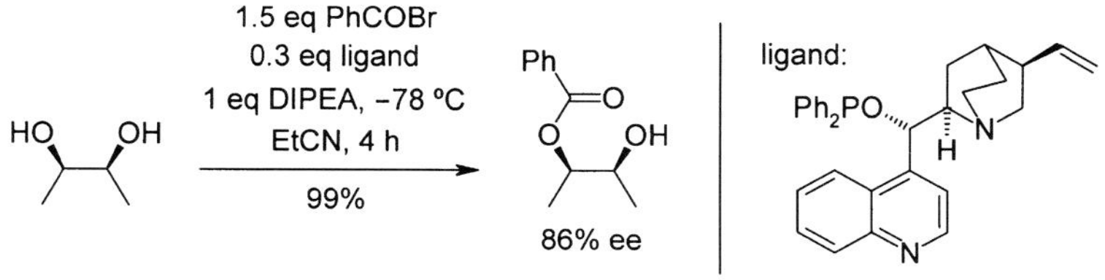

1.5 eq PhCOBr
0.3 eq ligand
1 eq DIPEA, −78 °C
EtCN, 4 h
99%

86% ee

ligand: Ph_2PO

Scheme 10.5. Enantioselective monobenzoylation of a *meso*-1,2-diol [20].

Diols with more than two carbon atoms between the two hydroxyl groups may require more carefully controlled reaction conditions if a monoacylation is desired (Scheme 10.6). Suitable catalysts which promote the formation of monoacylated products include silica gel [21], zeolites [12], and tin(IV) derivatives [14, 22].

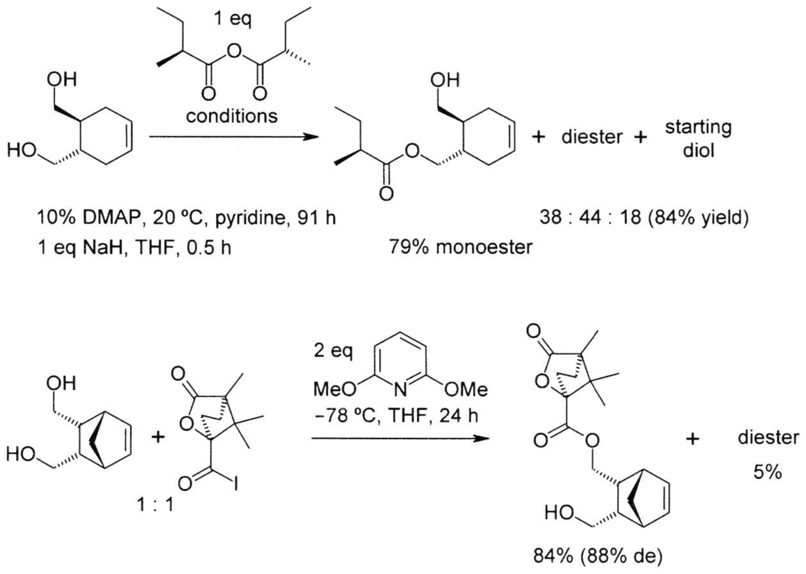

1 eq
conditions
+ diester + starting diol

10% DMAP, 20 °C, pyridine, 91 h — 38 : 44 : 18 (84% yield)
1 eq NaH, THF, 0.5 h — 79% monoester

1 : 1
2 eq MeO N OMe
−78 °C, THF, 24 h
+ diester 5%
84% (88% de)

Scheme 10.6. Monoacylation of 1,4-diols [23, 24].

An alternative strategy for preparation of monoacylated 1,2- and 1,3-diols is oxidative cleavage of cyclic acetals prepared from a diol and an aliphatic or aromatic aldehyde (Scheme 10.7). For this purpose the required acetal does not need to be isolated, but can be generated in situ [25]. Acetals prepared from strongly electrophilic aldehydes, for example nitrobenzaldehydes, will, however, usually be difficult to oxidize (and to hydrolyze).

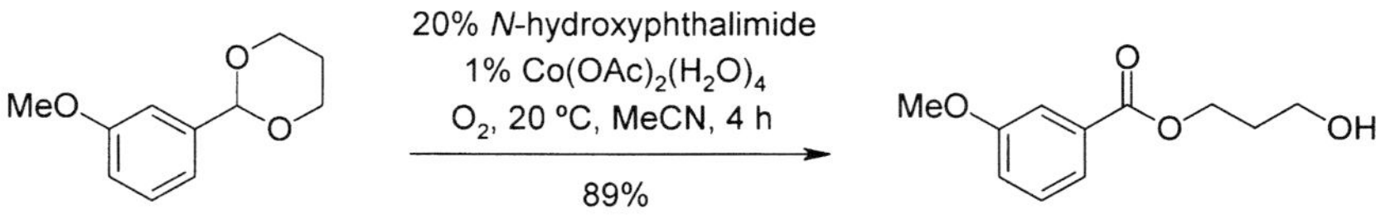

20% *N*-hydroxyphthalimide
1% $Co(OAc)_2(H_2O)_4$
O_2, 20 °C, MeCN, 4 h
89%

Scheme 10.7. Oxidative cleavage of acetals to yield monoacylated diols [26].

Diols can also be selectively monoacylated by treatment with an acylating agent of low reactivity (vinyl, 2,2,2-trifluoroethyl, phenyl, or ethyl esters) in the presence of catalytic amounts of an esterase (Scheme 10.8) [27]. Surprisingly, these enzymatic

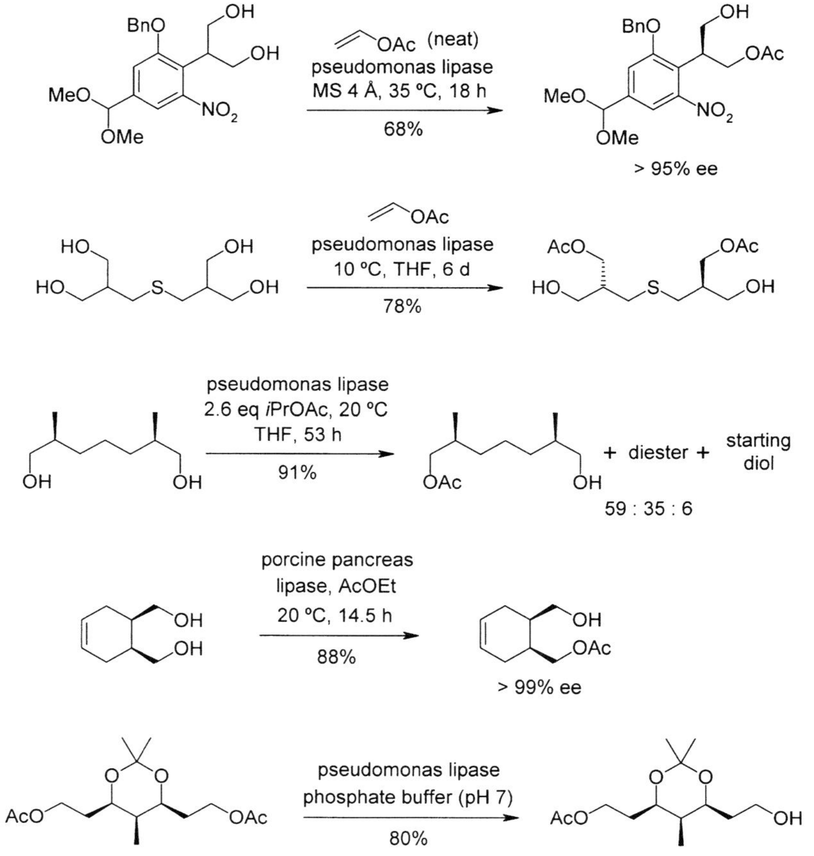

OAc (neat)
pseudomonas lipase
MS 4 Å, 35 °C, 18 h
68%
> 95% ee

OAc
pseudomonas lipase
10 °C, THF, 6 d
78%

pseudomonas lipase
2.6 eq *i*PrOAc, 20 °C
THF, 53 h
91%
+ diester + starting diol
59 : 35 : 6

porcine pancreas lipase, AcOEt
20 °C, 14.5 h
88%
> 99% ee

pseudomonas lipase
phosphate buffer (pH 7)
80%

Scheme 10.8. Enzymatic acetylation of diols and deacetylation of diacetates [30–34].

acylations can be conducted in pure organic solvents, but reaction times can be quite long. The absence of water drives these reactions in a direction opposite to that of normal esterase activity. The monoacylation will usually fail if the two hydroxyl groups are too far apart (e.g. >10 carbon atoms [28]). If the two hydroxyl groups are enantiotopic, enantiomerically enriched products can result [29]. Esterases can also be used to partially saponify diesters of symmetrical diols (last reaction, Scheme 10.8).

The reaction of arenesulfonyl chlorides with alcohols to yield sulfonates is relatively slow (if compared, e.g., with the formation of mesylates or sulfonamides), and treatment of diols with tosyl chloride can readily yield the statistically expected amount of monotosylated product [35, 36]. The selectivity of such reactions can sometimes be enhanced by additives such as AgO [37] or Bu_2SnO [38]. Deprotonation of the diol can be used to increase its nucleophilicity and thereby reduce the reaction time (Scheme 10.9). This strategy can, however, lead to problems, because the products are sensitive toward strong bases, and may cyclize, oligomerize, or undergo elimination.

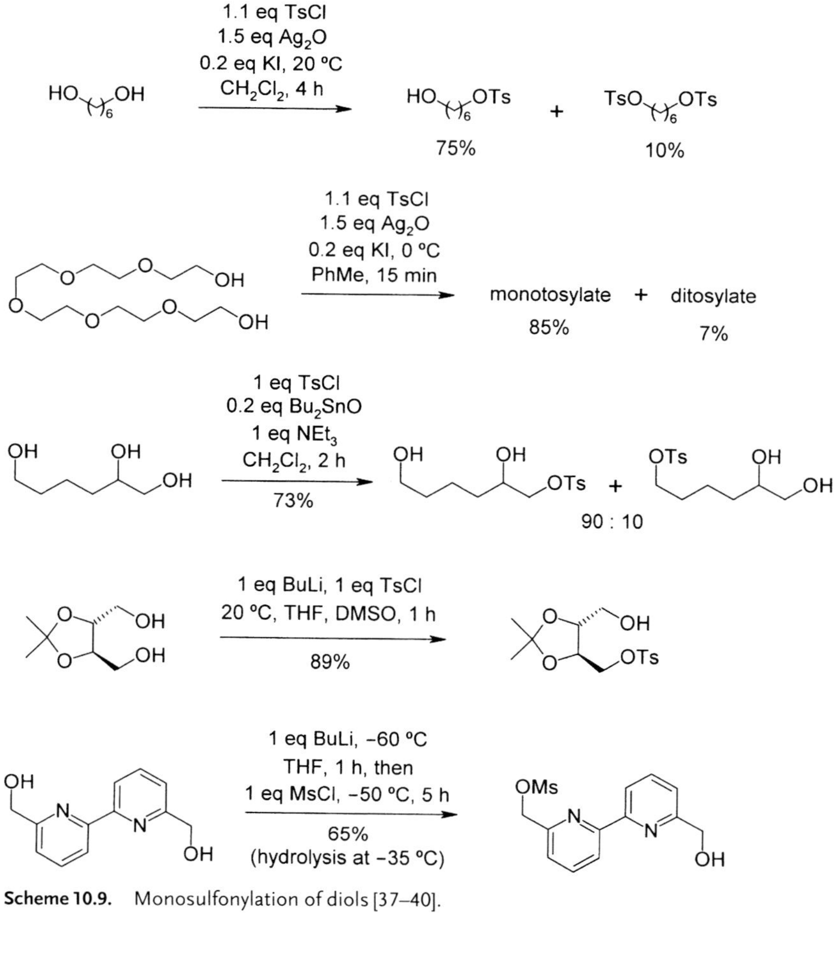

1.1 eq TsCl, 1.5 eq Ag_2O, 0.2 eq KI, 20 °C, CH_2Cl_2, 4 h — 75% + 10%

1.1 eq TsCl, 1.5 eq Ag_2O, 0.2 eq KI, 0 °C, PhMe, 15 min — monotosylate 85% + ditosylate 7%

1 eq TsCl, 0.2 eq Bu_2SnO, 1 eq NEt_3, CH_2Cl_2, 2 h, 73% — 90 : 10

1 eq BuLi, 1 eq TsCl, 20 °C, THF, DMSO, 1 h — 89%

1 eq BuLi, −60 °C, THF, 1 h, then 1 eq MsCl, −50 °C, 5 h — 65% (hydrolysis at −35 °C)

Scheme 10.9. Monosulfonylation of diols [37–40].

The reaction of diols with alkylating agents can also be performed such that approximately the theoretical amount of monoalkylated product is obtained. Some substrates can even be monoalkylated [36, 41, 42] or monoarylated [43] with substantially higher yields than statistics would predict what must be due to the slowness of the second alkylation. This behavior, however, is not usually observed, and other strategies must usually be used to achieve clean monoalkylations. Use of excess diol will often furnish pure monoalkylated diol, but a suitable means of removal of excess diol will be required. In the examples sketched in Scheme 10.10 the products were purified by column chromatography, recrystallization, or distillation. As with acylations, the formation of monoalkylated diols can also be promoted by Ag_2O [44] or by transient protection with boron [45] or tin [46] derivatives.

As mentioned above, the statistically predicted amount of monofunctionalized product can only be obtained if both reactants are thoroughly mixed. If the difunctional reagent is poorly soluble under the reaction conditions chosen, its concentration might be lower than that of the monofunctional reagent. This can lead to large

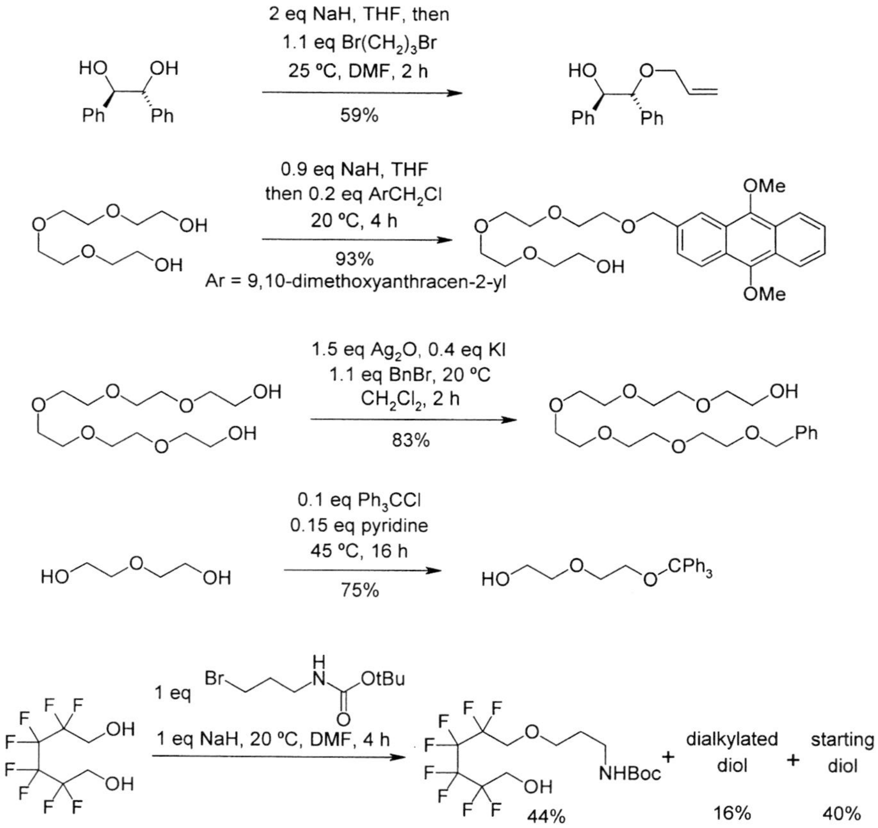

2 eq NaH, THF, then 1.1 eq $Br(CH_2)_3Br$ 25 °C, DMF, 2 h — 59%

0.9 eq NaH, THF then 0.2 eq $ArCH_2Cl$ 20 °C, 4 h — 93%
Ar = 9,10-dimethoxyanthracen-2-yl

1.5 eq Ag_2O, 0.4 eq KI 1.1 eq BnBr, 20 °C CH_2Cl_2, 2 h — 83%

0.1 eq Ph_3CCl 0.15 eq pyridine 45 °C, 16 h — 75%

1 eq Br(CH2)3NHC(O)OtBu; 1 eq NaH, 20 °C, DMF, 4 h — 44% + dialkylated diol 16% + starting diol 40%

Scheme 10.10. Monoalkylation of diols [36, 42, 47–50].

amounts of difunctionalized product, even if a large excess of difunctional reagent has been used. This can happen readily during the monofunctionalization of polyhydroxybenzenes, because the salts of these compounds are often poorly soluble in organic solvents. Thus, the monoalkylation of hydroquinone is complicated by the low solubility of the dianion in organic solvents, which leads to heterogeneous reaction mixtures and large amounts of dialkylated products, even if a large excess of hydroquinone is used. This problem can be solved for certain electrophiles by using an aqueous solvent in which the dianion of hydroquinone is soluble (Scheme 10.11) [50]. Alternatively, partially alkylated polyhydroxybenzenes can also be prepared by partial dealkylation of polyalkoxybenzenes [51].

An elegant strategy for the preparation of monoalkylated 1,2- and 1,3-diols is the nucleophilic ring-opening of cyclic acetals by hydride or by other nucleophiles (Scheme 10.12) [46, 56]. Not all acetals undergo these reactions cleanly, and careful

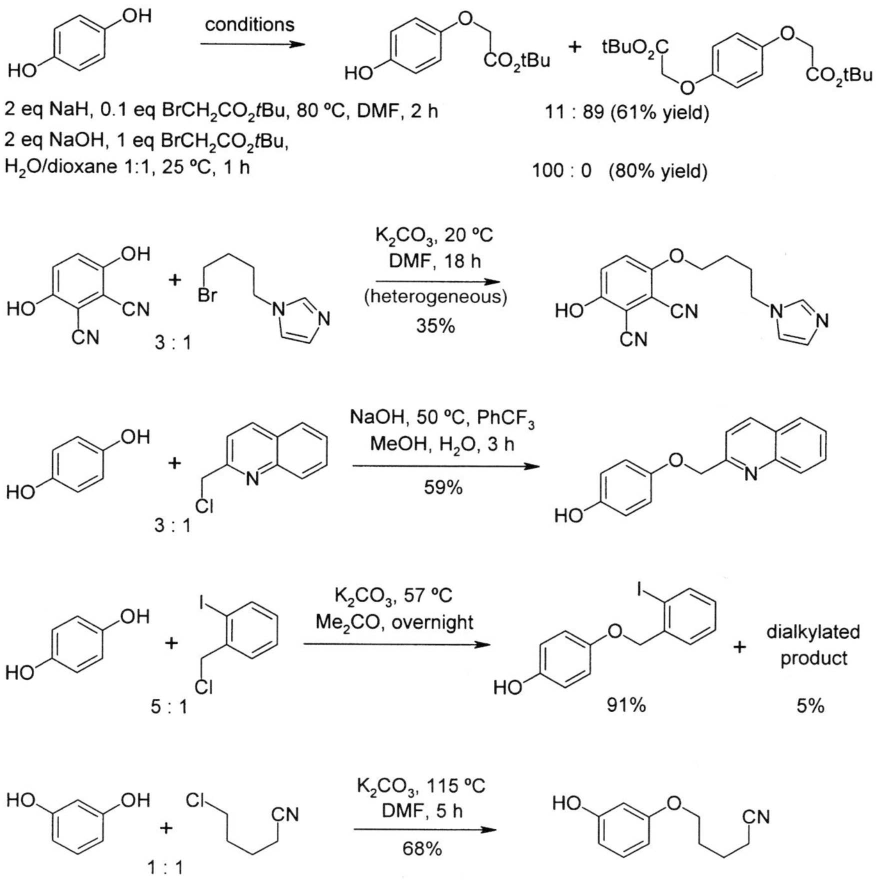

2 eq NaH, 0.1 eq $BrCH_2CO_2tBu$, 80 °C, DMF, 2 h — 11 : 89 (61% yield)
2 eq NaOH, 1 eq $BrCH_2CO_2tBu$, H_2O/dioxane 1:1, 25 °C, 1 h — 100 : 0 (80% yield)

K_2CO_3, 20 °C, DMF, 18 h (heterogeneous) 35%; 3 : 1

NaOH, 50 °C, $PhCF_3$, MeOH, H_2O, 3 h, 59%; 3 : 1

K_2CO_3, 57 °C, Me_2CO, overnight; 5 : 1; 91% + dialkylated product 5%

K_2CO_3, 115 °C, DMF, 5 h, 68%; 1 : 1

Scheme 10.11. Partial alkylation of polyhydroxybenzenes [50, 52–55].

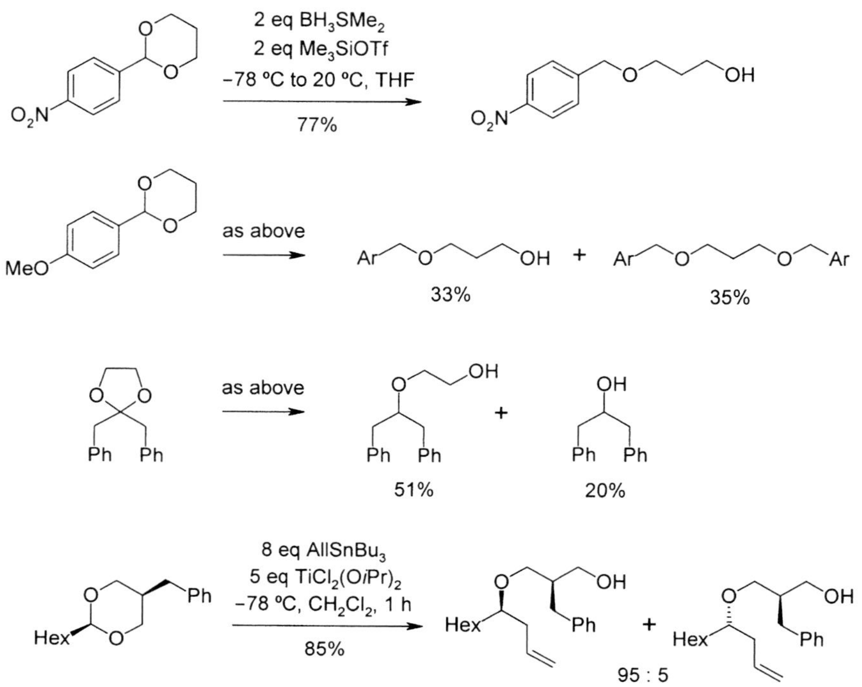

2 eq BH_3SMe_2
2 eq Me_3SiOTf
−78 °C to 20 °C, THF
77%

as above
33%
35%

as above
51%
20%

8 eq $AllSnBu_3$
5 eq $TiCl_2(OiPr)_2$
−78 °C, CH_2Cl_2, 1 h
85%
95 : 5

Scheme 10.12. Nucleophilic ring-opening of acetals [57, 58]. Ar = 4-$MeOC_6H_4$.

optimization might be required for some substrates. Six-membered acetals are usually more reactive than five-membered acetals [57].

Diols can be protected as monotetrahydropyranyl ethers, and good yields have even been reported for substrates in which the two hydroxyl groups are far away from each other [59] (Scheme 10.13). The reasons for the higher reaction rate of the diol compared with the monoethers is unclear.

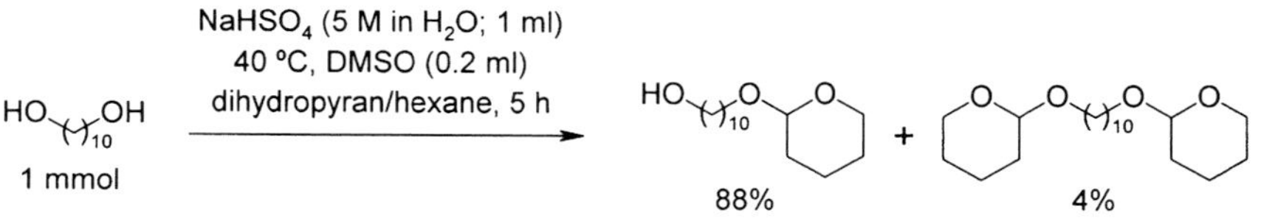

$NaHSO_4$ (5 M in H_2O; 1 ml)
40 °C, DMSO (0.2 ml)
dihydropyran/hexane, 5 h
1 mmol
88%
4%

Scheme 10.13. Monotetrahydropyranylation of diols [60].

10.4 Monofunctionalization of Diamines

If diamines are treated with strong acylating or sulfonylating reagents, high yields of bis-derivatized products are usually obtained, even if a large excess of diamine is used [61] (Scheme 10.14). Because the derivatization reaction is very fast, it goes to completion after each added droplet, before the monoacylated product can diffuse away from the added portion of acylating agent. The statistically expected yields of monoacylated diamines can, however, be obtained if less reactive acylating agents are used (e.g. benzyl esters in the presence of a lipase [62], *N*-hydroxysuccinimidyl esters [61, 63], methyl esters [64], ethyl esters [65], or amides [66]) or if the reagents are mixed at low temperatures, to prevent extensive reaction before complete mixing of the reacting partners has occurred. The reactivity of amines can be reduced by complexation with Lewis acids [67].

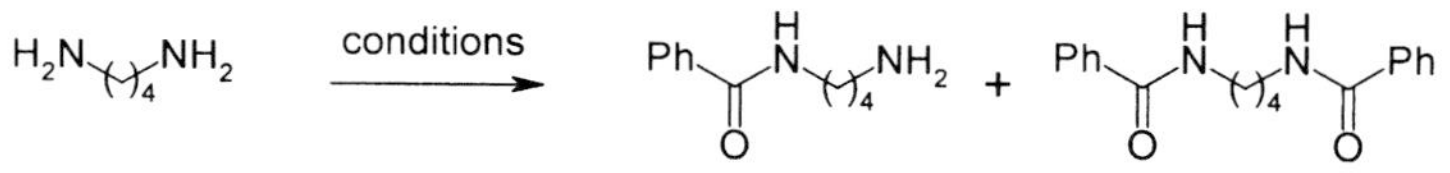

conditions	mono : di
0.95 eq PhCOCl, THF	< 5 : 95 (95% yield)
0.2 eq PhCOCl, −78 °C, CH_2Cl_2, overnight	35 : 65
0.2 eq $(PhCO)_2O$, −78 °C, CH_2Cl_2, overnight	84 : 16
1 eq 9-BBN, THF, then 0.95 eq PhCOCl	90 : 10 (88% yield)

Scheme 10.14. Mono- and diacylation of putrescine [61, 67].

A further strategy for the monoacylation of symmetric diamines consists in deprotonating the diamine twice and treating the resulting dianion with one equivalent of acylating reagent (Scheme 10.15). In 1,2- and 1,3-diaminoalkanes the dianion will be significantly more reactive than the product, and high yields of monoacylated diamines can be obtained. This strategy is particularly suitable for monoacylation of piperazine and related diamines, but will not be suitable for diamines with a large distance between the two amino groups.

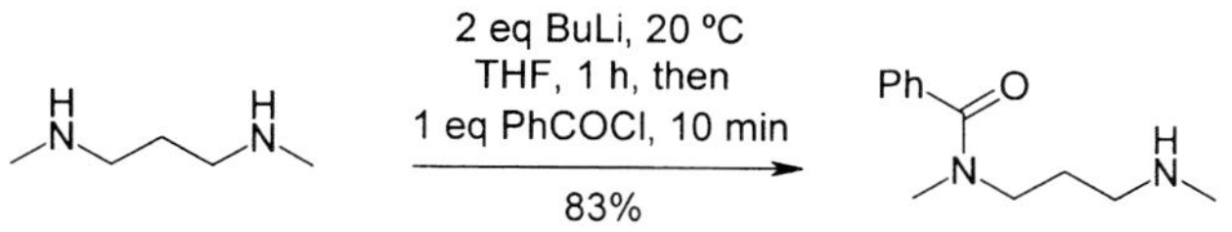

2 eq BuLi, 20 °C, THF, 1 h, then 1 eq PhCOCl, 10 min; 83%

Scheme 10.15. Twofold deprotonation and monoacylation of diamines [68].

As for acylations with reactive acylating agents, the reaction of diamines with sulfonyl chlorides usually leads to large amounts of bis-sulfonylated diamine. Notable exceptions are shown in Scheme 10.16. The last example in Scheme 10.16, in which the electrophile was added slowly with a syringe pump, indicates that monosulfonylations can be successfully performed if the reagents are mixed efficiently. Lower reaction temperatures or less reactive sulfonylating reagents should further facilitate the formation of monosulfonylated diamines.

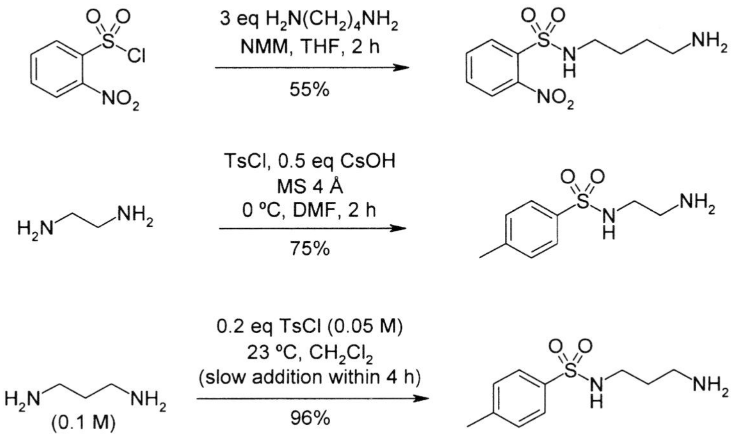

3 eq $H_2N(CH_2)_4NH_2$
NMM, THF, 2 h
55%

TsCl, 0.5 eq CsOH
MS 4 Å
0 °C, DMF, 2 h
75%

0.2 eq TsCl (0.05 M)
23 °C, CH_2Cl_2
(slow addition within 4 h)
96%
(0.1 M)

Scheme 10.16. Monosulfonylation of diamines [69–71].

Alkylation of amines usually proceeds more slowly than acylations, and alkylating reagents can usually be thoroughly mixed with an amine before reaction is complete. Diamines or polyamines can therefore be transformed into mixtures of the statistically expected amounts of alkylated derivatives (Scheme 10.17). The use of CsOH as base has been found to be particularly conducive to the formation of monoalkylated products [72]. Some polyamines have also been monoalkylated selectively after complexation with Zn(II) [73].

The selectivity of acylations or alkylations of polyamines can also be modulated by protonation [78] or by conversion to cyclic aminals [71, 79, 80]. The latter strategy has, for instance, been used successfully for the preparation of monofunctionalized 1,4,7-triazacyclononanes (Scheme 10.18). In this reaction the first alkylation leads to the formation of an amidinium salt, which is more difficult to alkylate than a tertiary amine. Thus, the use of only one equivalent of alkylating agent leads to a clean monoalkylation.

Monoarylation of diamines has been achieved by treatment of excess diamine with an arylating reagent (Scheme 10.19) [82–84]. If the arylation is slow the statistically expected amounts of mono- and diarylated product will usually result. An effective means removing excess diamine will, however, be required the purification of the products.

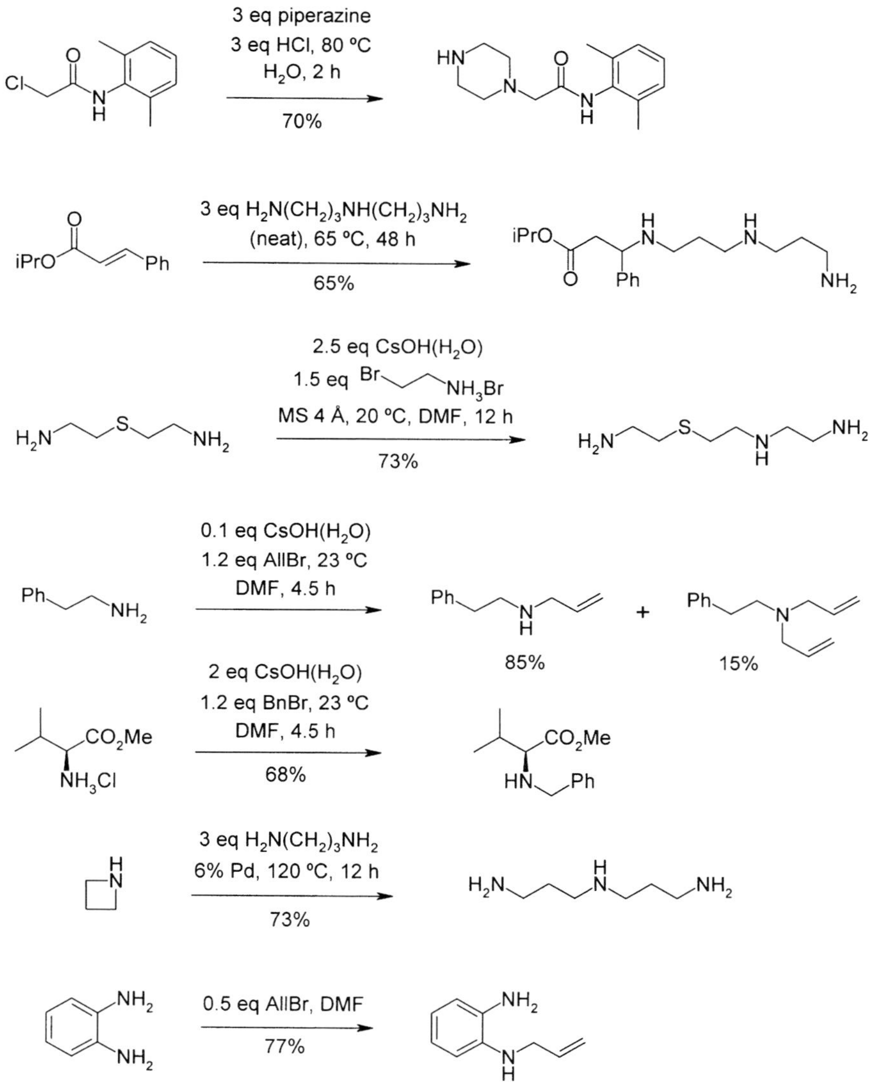

3 eq piperazine
3 eq HCl, 80 °C
H_2O, 2 h
70%

3 eq $H_2N(CH_2)_3NH(CH_2)_3NH_2$
(neat), 65 °C, 48 h
65%

2.5 eq CsOH(H_2O)
1.5 eq Br–$(CH_2)_2$–NH_3Br
MS 4 Å, 20 °C, DMF, 12 h
73%

0.1 eq CsOH(H_2O)
1.2 eq AllBr, 23 °C
DMF, 4.5 h
85% + 15%

2 eq CsOH(H_2O)
1.2 eq BnBr, 23 °C
DMF, 4.5 h
68%

3 eq $H_2N(CH_2)_3NH_2$
6% Pd, 120 °C, 12 h
73%

0.5 eq AllBr, DMF
77%

Scheme 10.17. Monoalkylation of amines [70, 72, 74–77].

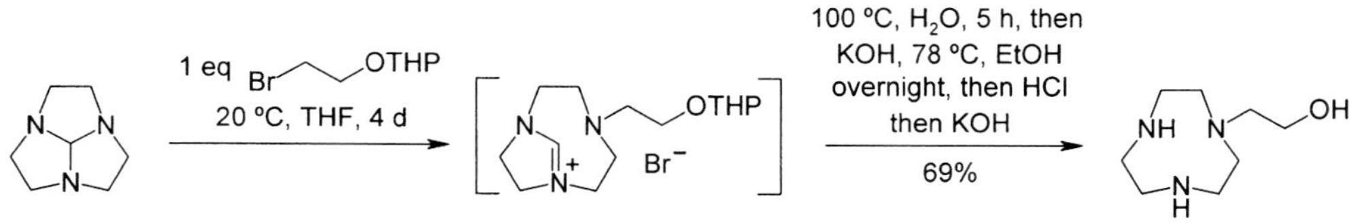

Scheme 10.18. Monoalkylation of a tricyclic orthoamide derived from 1,4,7-triazacyclononane [81].

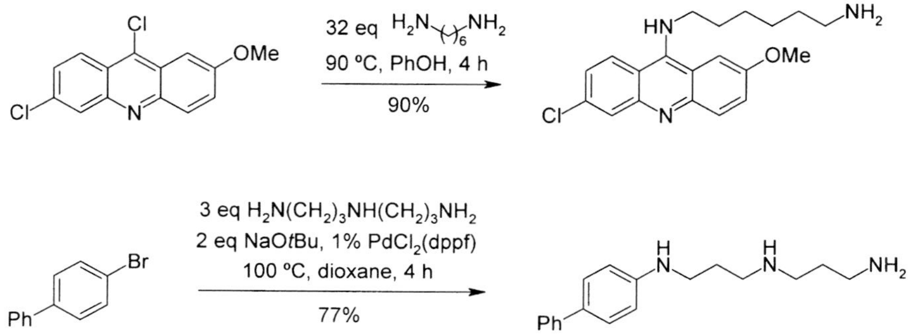

Scheme 10.19. Monoarylation of polyamines [85, 86].

10.5 Monoalkylation of C,H-Acidic Compounds

The alkylation of carbanions can sometimes lead to the formation of di- and trialkylated products (Scheme 10.20). Such polyalkylations occur more readily if alkylation is slow compared with the rate of proton exchange between the carbanion and the product. Because the acidity of organic compounds usually decreases on alkylation, the nucleophilicity of the monoalkylated carbanion will often be substantially higher than that of the starting carbanion, and will therefore be able to compete efficiently with the latter, despite its low concentration. Polyalkylations such as those shown in Scheme 10.20 can be suppressed by increasing the reactivity of the carbanion, for instance by choosing a different counter-ion (K^+ or R_4N^+ instead of Li^+ [87]) or by adding HMPA or similar cosolvents to the reaction mixture, which enhances the rate of SN2 reactions. Alternatively, protective groups can be used to avoid multiple alkylation of carbanions [88].

The multiple alkylation of carbanions with electron-deficient alkenes (Michael addition) only yields the expected products if the carbanion is less basic than the initial product of Michael addition. If the attacking carbanion and the carbanion resulting from Michael addition have similar basicity, oligomerization of the Michael acceptor can occur instead of multiple alkylations of the same carbon atom (Scheme 10.21).

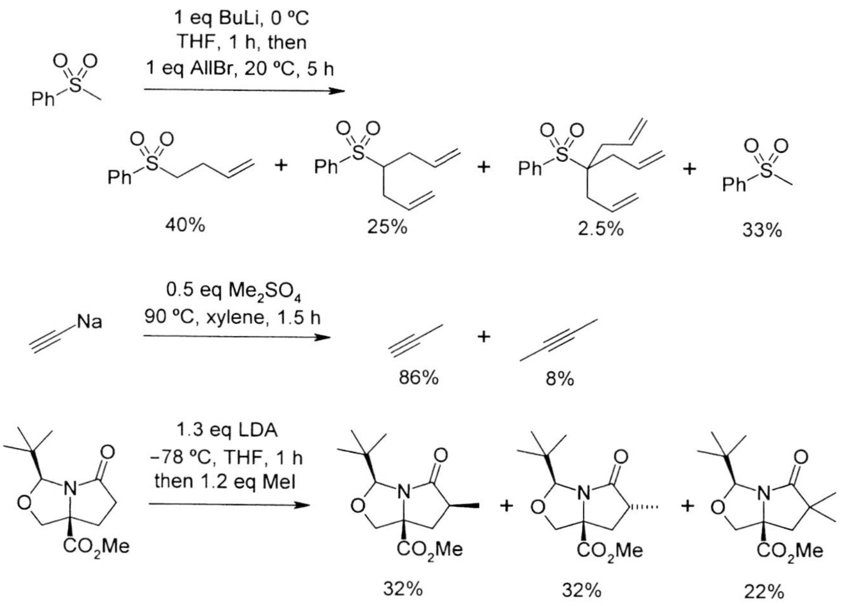

1 eq BuLi, 0 °C
THF, 1 h, then
1 eq AllBr, 20 °C, 5 h

40% + 25% + 2.5% + 33%

0.5 eq Me_2SO_4
90 °C, xylene, 1.5 h

86% + 8%

1.3 eq LDA
−78 °C, THF, 1 h
then 1.2 eq MeI

32% + 32% + 22%

Scheme 10.20. Polyalkylation of carbanions [87–89].

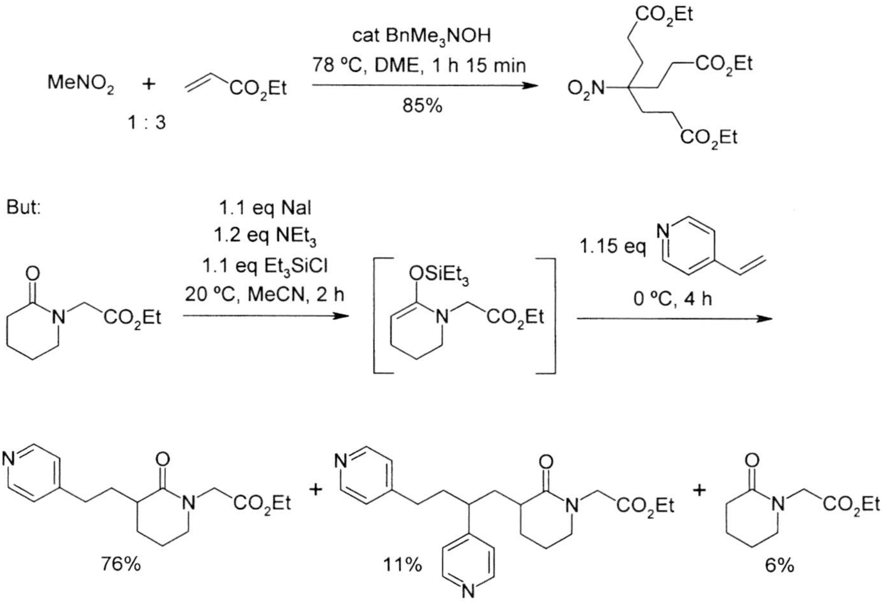

cat $BnMe_3NOH$
78 °C, DME, 1 h 15 min
85%

$MeNO_2$ + CO_2Et (1 : 3)

But:

1.1 eq NaI
1.2 eq NEt_3
1.1 eq Et_3SiCl
20 °C, MeCN, 2 h

1.15 eq
0 °C, 4 h

76% + 11% + 6%

Scheme 10.21. Dependence of the outcome of Michael additions on the basicity of the intermediates [90, 91].

10.6
Monoderivatization of Dihalides

α,ω-Dichloro- and dibromoalkanes are cheap, readily available reagents. Their low boiling points and low heats of evaporation enable their facile separation from product mixtures, and thus their use as reactants in excess or even as solvents. Numerous successful monoderivatizations of α,ω-dihaloalkanes have been reported.

As mentioned in Section 4.3.4, 1,2-dihaloethanes undergo nucleophilic substitutions less readily than *n*-alkyl halides. Nevertheless, under optimum conditions high-yielding monosubstitutions can be performed with these reagents (Scheme 10.22) [92].

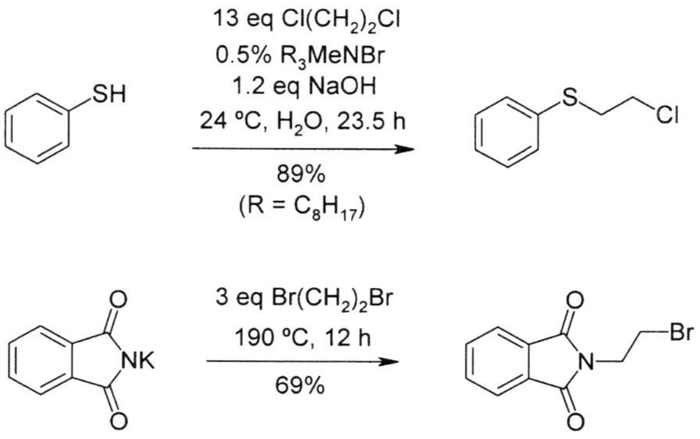

13 eq $Cl(CH_2)_2Cl$
0.5% R_3MeNBr
1.2 eq NaOH
24 °C, H_2O, 23.5 h
89%
(R = C_8H_{17})

3 eq $Br(CH_2)_2Br$
190 °C, 12 h
69%

Scheme 10.22. Nucleophilic substitutions at symmetric 1,2-dihaloethanes [93, 94].

Longer α,ω-dihaloalkanes undergo nucleophilic substitutions readily, and a slight excess of dihalide can lead to acceptable yields of haloalkylated nucleophile (Scheme 10.23). Yields are sometimes higher than those predicted theoretically. Further examples are given in Schemes 4.39 and 6.23.

Halogen–metal exchange at polyhalogenated arenes significantly reduces the ability of the remaining halogen atoms to be displaced by the metal. Hence, the conversion of symmetric 1,2- and 1,3-dibromoarenes and heteroarenes into mono-Grignard reagents proceeds without problems [98–100]. Dihaloalkanes are, however, more difficult to convert into haloalkyl Grignard reagents – vicinal dihalides are usually converted into alkenes on treatment with a metalating reagent (Section 5.4.3) whereas longer dihaloalkanes can undergo cyclization upon metalation (Section 5.4.4).

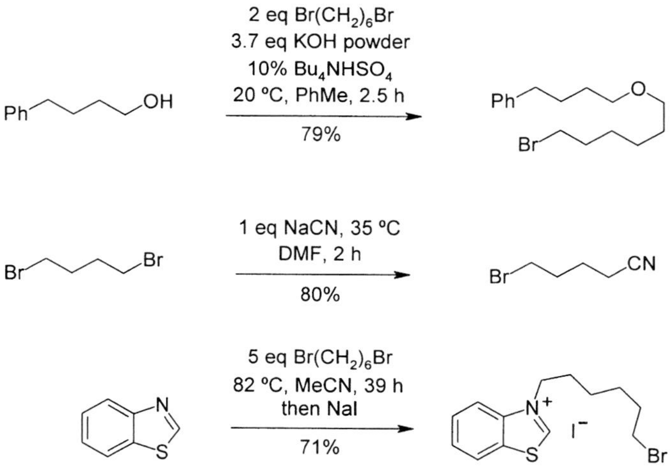

2 eq $Br(CH_2)_6Br$
3.7 eq KOH powder
10% Bu_4NHSO_4
20 °C, PhMe, 2.5 h
79%

1 eq NaCN, 35 °C
DMF, 2 h
80%

5 eq $Br(CH_2)_6Br$
82 °C, MeCN, 39 h
then NaI
71%

Scheme 10.23. Monoderivatization of α,ω-dibromoalkanes [95–97].

References

1 Diaz, H.; Kelly, J. W. The synthesis of dibenzofuran based diacids and amino acids designed to nucleate parallel and antiparallel β-sheet formation. *Tetrahedron Lett.* **1991**, *32*, 5725–5728.

2 Sheehan, J. C.; Bose, A. K. A new synthesis of β-lactams. *J. Am. Chem. Soc.* **1950**, *72*, 5158–5161.

3 Durham, L. J.; McLeod, D. J.; Cason, J. Methyl hydrogen hendecanedioate. *Org. Synth.* **1963**, *Coll. Vol. IV*, 635–638.

4 Strube, R. E. Ethyl *tert*-butyl malonate. *Org. Synth.* **1963**, *Coll. Vol. IV*, 417–419.

5 Kuhn, T.; Tamm, C.; Riesen, A.; Zehnder, M. Stereoselective hydrolysis of the dimethyl 4,5-epoxy-1,2-*cis*-cyclohexanedicarboxylates with pig liver esterase. *Tetrahedron Lett.* **1989**, *30*, 693–696.

6 Hughes, D. L.; Bergan, J. J.; Amato, J. S.; Reider, P. J.; Grabowski, E. J. J. Synthesis of chiral dithioacetals: a chemoenzymic synthesis of a novel LTD_4 antagonist. *J. Org. Chem.* **1989**, *54*, 1787–1788.

7 Fabre, J.; Betbeder, D.; Paul, F.; Monsan, P.; Perie, J. Versatile enzymatic diacid ester synthesis of butyl α-D-glucopyranoside. *Tetrahedron* **1993**, *49*, 10877–10882.

8 Arita, M.; Kunitomo, A.; Ito, Y.; Sawai, H.; Ohno, M. Enantioselective synthesis of the carbocyclic nucleosides (–)-aristeromycin and (–)-neplanocin A by a chemicoenzymatic approach. *J. Am. Chem. Soc.* **1983**, *105*, 4049–4055.

9 Matoishi, K.; Sano, A.; Imai, N.; Yamazaki, T.; Yokoyama, M.; Sugai, T.; Ohta, H. *Rhodococcus rhodochrous* IFO 15564-mediated hydrolysis of alicyclic nitriles and amides: stereoselectivity and use for kinetic resolution and asymmetrization. *Tetrahedron Asymmetry* **1998**, *9*, 1097–1102.

10 Clarke, P. A. Selective monoacylation of *meso*- and C_2-symmetric 1,3- and 1,4-diols. *Tetrahedron Lett.* **2002**, *43*, 4761–4763.

11 Clarke, P. A.; Kayaleh, N. E.; Smith, M. A.; Baker, J. R.; Bird, S. J.; Chan, C. A one-step procedure for the monoacylation of symmetrical 1,2-diols. *J. Org. Chem.* **2002**, *67*, 5226–5231.

12 Srinivas, K. V. N. S.; Mahender, I.; Das, B. Selective monoacetylation of symmetrical diols and selective monodeacetylation of symmetrical diacetates using HY-zeolite as reusable heterogeneous catalyst. *Synlett* **2003**, 2419–2421.

13 Ding, Y.; Wu, R.; Lin, Q. The chemoselective preparation of the substituted phenyl benzoates using RE-Y zeolite as catalyst. *Synth. Commun.* **2002**, *32*, 2149–2153.

14 Iwasaki, F.; Maki, T.; Onomura, O.; Nakashima, W.; Matsumura, Y. Chemo- and stereoselective monobenzoylation of 1,2-diols catalyzed by organotin compounds. *J. Org. Chem.* **2000**, *65*, 996–1002.

15 Roelens, S. Organotin-mediated monoacylation of diols with reversed chemoselectivity. Mechanism and selectivity. *J. Org. Chem.* **1996**, *61*, 5257–5263.

16 Oikawa, M.; Wada, A.; Okazaki, F.; Kusumoto, S. Acidic, selective monoacylation of *vic*-diols. *J. Org. Chem.* **1996**, *61*, 4469–4471.

17 Oriyama, T.; Imai, K.; Sano, T.; Hosoya, T. Highly efficient catalytic asymmetric acylation of *meso*-1,2-diols with PhCOCl in the presence of a chiral diamine combined with NEt_3. *Tetrahedron Lett.* **1998**, *39*, 3529–3532.

18 Vedejs, E.; Daugulis, O.; Tuttle, N. Desymmetrization of *meso*-hydrobenzoin using chiral, nucleophilic phosphine catalysts. *J. Org. Chem.* **2004**, *69*, 1389–1392.

19 France, S.; Guerin, D. J.; Miller, S. J.; Lectka, T. Nucleophilic chiral amines as catalysts in asymmetric synthesis. *Chem. Rev.* **2003**, *103*, 2985–3012.

20 Mizuta, S.; Sadamori, M.; Fujimoto, T.; Yamamoto, I. Asymmetric desymmetrization of *meso*-1,2-diols by phosphinite derivatives of cinchona alkaloids. *Angew. Chem. Int. Ed.* **2003**, *42*, 3383–3385.

21 Ogawa, H.; Amano, M.; Chihara, T. Facile and highly selective monoacylation of symmetric diols adsorbed on silica gel with MeCOCl. *Chem. Commun.* **1998**, 495–496.

22 Green, J.; Woodward, S. Selective monoacylation of 2,2′-binaphthol as a route to new monothiobinaphthol-based Cu catalysts for conjugate addition. *Synlett* **1995**, 155–156.

23 Heathcock, C. H.; Davis, B. R.; Hadley, C. R. Synthesis and biological evaluation of a monocyclic, fully functional analogue of compactin. *J. Med. Chem.* **1989**, *32*, 197–202.

24 Ishihara, K.; Kubota, M.; Yamamoto, H. First application of hydrogen bonding interactions to the design of asymmetric acylation of *meso*-diols with optically active acyl halides. *Synlett* **1994**, 611–614.

25 Sharghi, H.; Sarvari, M. H. Highly selective methodology for the direct conversion of aromatic aldehydes to glycol monoesters. *J. Org. Chem.* **2003**, *68*, 4096–4099.

26 Karimi, B.; Rajabi, J. Efficient aerobic oxidation of acetals to esters catalyzed by *N*-hydroxy phthalimide and Co(II) under mild conditions. *Synthesis* **2003**, 2373–2377.

27 Baldessari, A.; Mangone, C. P.; Gros, E. G. Lipase-catalyzed acylation and deacylation reactions of pyridoxine, a member of the vitamin-B_6 group. *Helv. Chim. Acta* **1998**, *81*, 2407–2413.

28 Theil, F.; Sonnenschein, H.; Kreher, T. Lipase-catalyzed resolution of 3,3′-bi-indolizines: the first preparative access to enantiomerically pure samples. *Tetrahedron Asymmetry* **1996**, *7*, 3365–3370.

29 Andreu, C.; Villarroya, J.-P.; García-Gastaldi, A.; Medio-Simón, M.; Server-Carrió, J.; Varea, T. Enzymatic esterification of bicyclic *meso*-diols derived from 2,5-bis(hydroxymethyl)-furan. An enantioselective Diels–Alder reaction equivalent. *Tetrahedron Asymmetry* **1998**, *9*, 3105–3114.

30 Fellows, I. M.; Kaelin, D. E.; Martin, S. F. Application of ring-closing metathesis to the formal total synthesis of (+)-FR900482. *J. Am. Chem. Soc.* **2000**, *122*, 10781–10787.

31 Schmittberger, T.; Uguen, D. A formal synthesis of brassinolide. *Tetrahedron Lett.* **1997**, *38*, 2837–2840.

32 Chênevert, R.; Desjardins, M. Enzymatic desymmetrization of *meso*-2,6-dimethyl-1,7-heptanediol. Enantioselective formal synthesis of the vitamin E side chain and the insect pheromone tribolure. *J. Org. Chem.* **1996**, *61*, 1219–1222.

33 Danieli, B.; Lesma, G.; Mauro, M.; Palmisano, G.; Passarella, D. First enantioselective synthesis of (–)-akagerine by a chemoenzymatic approach. *J. Org. Chem.* **1995**, *60*, 2506–2513.

34 Bonini, C.; Chiummiento, L.; Funicello, M.; Marconi, L.; Righi, G. Chiral polypropionate subunit by a chemoenzymatic approach. *Tetrahedron Asymmetry* **1998**, *9*, 2559–2561.

35 Hoye, T. R.; Hanson, P. R.; Kovelesky, A. C.; Ocain, T. D.; Zhuang, Z. Synthesis of (+)-(15,16,19,20,23,24)-hexepi-uvaricin: a bis(tetrahydrofuranyl) annonaceous acetogenin analogue. *J. Am. Chem. Soc.* **1991**, *113*, 9369–9371.

36 Loiseau, F. A.; Hii, K. K.; Hill, A. M. Multigram synthesis of well-defined extended

bifunctional polyethylene glycol chains. *J. Org. Chem.* **2004**, *69*, 639–647.
37 Bouzide, A.; Sauvé, G. Silver(I) oxide mediated highly selective monotosylation of symmetrical diols. Application to the synthesis of polysubstituted cyclic ethers. *Org. Lett.* **2002**, *4*, 2329–2332.
38 Martinelli, M. J.; Vaidyanathan, R.; Pawlak, J. M.; Nayyar, N. K.; Dhokte, U. P.; Doecke, C. W.; Zollars, L. M. H.; Moher, E. D.; Khau, V. V.; Kosmrlj, B. Catalytic regioselective sulfonylation of α-chelatable alcohols: scope and mechanistic insight. *J. Am. Chem. Soc.* **2002**, *124*, 3578–3585.
39 Kotsuki, H.; Kadota, I.; Ochi, M. A novel C–C bond-forming reaction of triflates with Cu(I)-catalyzed Grignard reagents. A new concise and enantiospecific synthesis of (+)-*exo*-brevicomin, (5*R*,6*S*)-(–)-6-acetoxy-5-hexadecanolide, and L-factor. *J. Org. Chem.* **1990**, *55*, 4417–4422.
40 Schubert, U. S.; Kersten, J. L.; Pemp, A. E.; Eisenbach, C. D.; Newkome, G. R. A new generation of 6,6′-disubstituted 2,2′-bipyridines: towards novel oligo(bipyridine) building blocks for potential applications in materials science and supramolecular chemistry. *Eur. J. Org. Chem.* **1998**, 2573–2581.
41 Lecourt, T.; Mallet, J.-M.; Sinaÿ, P. Efficient synthesis of doubly connected primary face-to-face cyclodextrin homo-dimers. *Eur. J. Org. Chem.* **2003**, 4553–4560.
42 Jha, S. C.; Joshi, N. N. Intramolecular dehydrohalogenation during base-mediated reaction of diols with dihaloalkanes. *J. Org. Chem.* **2002**, *67*, 3897–3899.
43 Fedorov, A. Y.; Finet, J.-P. Synthesis and reactivity of pentavalent biphenyl-2,2′-ylenebismuth derivatives. *J. Chem. Soc. Perkin Trans. 1* **2000**, 3775–3778.
44 Bouzide, A.; Sauvé, G. Highly selective Ag_2O-mediated monoprotection of symmetrical diols. *Tetrahedron Lett.* **1997**, *38*, 5945–5948.
45 Bhaskar, V.; Duggan, P. J.; Humphrey, D. G.; Krippner, G. Y.; McCarl, V.; Offermann, D. A. Phenylboronic acid as a labile protective agent: the selective derivatisation of 1,2,3-triols. *J. Chem. Soc. Perkin Trans. 1* **2001**, 1098–1102.
46 Martínez-Bernhardt, R.; Castro, P. P.; Godjoian, G.; Gutiérrez, C. G. Substituted diether diols by ring-opening of carbocyclic and stannylene acetals. *Tetrahedron* **1998**, *54*, 8919–8932.
47 Jiang, H.; Xu, H.; Ye, J. Synthesis and cation-mediated electron transfer in intramolecular fluorescence quenching of donor-acceptor podands: observation of Marcus inverted region in forward electron transfer reactions. *J. Chem. Soc. Perkin Trans. 2* **2000**, 925–930.
48 Keegstra, E. M. D.; Zwikker, J. W.; Roest, M. R.; Jenneskens, L. W. A highly selective synthesis of monodisperse oligo(ethylene glycols). *J. Org. Chem.* **1992**, *57*, 6678–6680.
49 Mougenot, P.; Mertens, P.; Nguyen, M.; Touillaux, R.; Marchand-Brynaert, J. α,ω-Difunctional perfluorinated spacer arms for polymeric material derivatization. *J. Org. Chem.* **1996**, *61*, 408–412.
50 Newman, M. S.; Cella, J. A. Studies on the monoalkylation of hydroquinone. *J. Org. Chem.* **1974**, *39*, 214–215.
51 Mirrington, R. N.; Feutrill, G. I. Orcinol monomethyl ether. *Org. Synth.* **1988**, *Coll. Vol. VI*, 859–861.
52 Miller, S. P. F.; French, S. A.; Kaneski, C. R. Novel lysosomotropic enzyme substrate. *J. Org. Chem.* **1991**, *56*, 30–34.
53 Bridge, A. W.; Jones, R. H.; Kabir, H.; Kee, A. A.; Lythgoe, D. J.; Nakach, M.; Pemberton, C.; Wrightman, J. A. The process development of RG 12525 (2-{[tetrazol-5-ylmethylphenyl)-methoxy]phenoxymethyl}quinoline). *Org. Process Res. Dev.* **2001**, *5*, 9–15.
54 Imbos, R.; Minnaard, A. J.; Feringa, B. L. A highly enantioselective intramolecular Heck reaction with a monodentate ligand. *J. Am. Chem. Soc.* **2002**, *124*, 184–185.
55 Mazzei, M.; Dondero, R.; Ledda, B.; Demontis, F.; Vargiu, L. Disoxaril-related 3-(diethylamino)-5-phenylisoxazoles. *Il Farmaco* **2000**, *55*, 119–124.
56 Gardiner, J. M.; Mather, P.; Morjan, R.; Pritchard, R. G.; Warren, J. E.; Cooper, M. L.; Ferwanah, A. E.-R. S.; Abu-Tiem, O. S. Synthesis and structure of 2-aryl-5,5-disubstituted-1,3-dioxanes and conversion into chiral (1,1,1-trishydroxymethyl)methane derivatives. *Tetrahedron Lett.* **2002**, *43*, 2091–2094.
57 Bartels, B.; Hunter, R. A selectivity study of activated ketal reduction with $Me_2S–BH_3$. *J. Org. Chem.* **1993**, *58*, 6756–6765.
58 Denmark, S. E.; Almstead, N. G. On the stereoselective opening of achiral dioxane acetals. *J. Org. Chem.* **1991**, *56*, 6458–6467.
59 Nishiguchi, T.; Fujisaki, S.; Kuroda, M.; Kajisaki, K.; Saitoh, M. Selective monotetrahydro-

pyranylation of symmetrical diols catalyzed by ion-exchange resins. *J. Org. Chem.* **1998**, *63*, 8183–8187.

60 Nishiguchi, T.; Hayakawa, S.; Hirasaka, Y.; Saitoh, M. Selective monotetrahydropyranylation of 1,n-diols catalyzed by aqueous acids. *Tetrahedron Lett.* **2000**, *41*, 9843–9846.

61 Jacobson, A. R.; Makris, A. N.; Sayre, L. M. Monoacylation of symmetrical diamines. *J. Org. Chem.* **1987**, *52*, 2592–2594.

62 Adamczyk, M.; Grote, J. Regioselective pseudomonas cepacia lipase mediated amidations of benzyl esters with diamines. *Tetrahedron Asymmetry* **1997**, *8*, 2099–2100.

63 Sadownik, A.; Deng, G.; Janout, V.; Regen, S. L.; Bernard, E. M.; Kikuchi, K.; Armstrong, D. Rapid construction of a squalamine mimic. *J. Am. Chem. Soc.* **1995**, *117*, 6138–6139.

64 Barnett, M.; Secondo, P.; Collier, H. Synthesis and characterization of new 1,1′-diester, diketone, and dinitrile derivatives of 2,2′-biimidazole. *J. Heterocyclic Chem.* **1996**, *33*, 1363–1365.

65 Blagbrough, I. S.; Geall, A. J. Practical synthesis of unsymmetrical polyamine amides. *Tetrahedron Lett.* **1998**, *39*, 439–442.

66 Zaragoza Dörwald, F.; von Kiedrowski, G. A simple and economic synthesis of monoacylated alkanediamines by thermal transamidation. *Synthesis* **1988**, 917–918.

67 Zhang, Z.; Yin, Z.; Meanwell, N. A.; Kadow, J. F.; Wang, T. Selective monoacylation of symmetrical diamines via prior complexation with boron. *Org. Lett.* **2003**, *5*, 3399–3402.

68 Wang, T.; Zhang, Z.; Meanwell, N. A. Benzoylation of dianions: preparation of monobenzoylated derivatives of symmetrical secondary diamines. *J. Org. Chem.* **1999**, *64*, 7661–7662.

69 Amssoms, K.; Augustyns, K.; Yamani, A.; Zhang, M.; Haemers, A. An efficient synthesis of orthogonally protected spermidine. *Synth. Commun.* **2002**, *32*, 319–328.

70 Salvatore, R. N.; Schmidt, S. E.; Shin, S. I.; Nagle, A. S.; Worrell, J. H.; Jung, K. W. CsOH-promoted chemoselective mono-*N*-alkylation of diamines and polyamines. *Tetrahedron Lett.* **2000**, *41*, 9705–9708.

71 Jentgens, C.; Bienz, S.; Hesse, M. Selective monoderivatization of propane-1,3-diamine with acid chlorides: ‘hexahydropyrimidine method’ vs statistic methods. *Helv. Chim. Acta* **1997**, *80*, 1133–1143.

72 Salvatore, R. N.; Nagle, A. S.; Jung, K. W. Cesium effect: high chemoselectivity in direct *N*-alkylation of amines. *J. Org. Chem.* **2002**, *67*, 674–683.

73 Burguete, M. I.; Escuder, B.; Luis, S. V.; Miravet, J. F.; García-España, E. Selective monofunctionalization of polyaza[n]paracyclophanes. *Tetrahedron Lett.* **1994**, *35*, 9075–9078.

74 Guillaume, M.; Cuypers, J.; Vervest, I.; De Smaele, D.; Leurs, S. Synthesis of T2288: from bench synthesis to pilot production. *Org. Process Res. Dev.* **2003**, *7*, 939–941.

75 Drandarov, K.; Hesse, M. Lithium and proton templated ω-polyazamacrolactamization, new general routes to macrocyclic polyamines. *Tetrahedron Lett.* **2002**, *43*, 7213–7216.

76 Murahashi, S.-I.; Yoshimura, N.; Tsumiyama, T.; Kojima, T. Catalytic alkyl group exchange reaction of primary and secondary amines. *J. Am. Chem. Soc.* **1983**, *105*, 5002–5011.

77 Lai, G.; Anderson, W. K. A concise synthesis of a benzimidazole analogue of mycophenolic acid using a BF_3–Et_2O catalyzed amino-Claisen rearrangement. *Tetrahedron Lett.* **1993**, *34*, 6849–6852.

78 Fensterbank, H.; Berthault, P.; Larpent, C. A tunable one-step *N,N′*-disubstitution of 1,4,8,11-tetraazacyclotetradecane with acrylamide. *Eur. J. Org. Chem.* **2003**, 3985–3990.

79 Tripier, R.; Lagrange, J.-M.; Espinosa, E.; Denat, F.; Guilard, R. From new tricyclic bisaminal derivatives to *trans-N,N′*-disubstituted cyclams. *Chem. Commun.* **2001**, 2728–2729.

80 Alder, R. W.; Mowlam, R. W.; Vachon, D. J.; Weisman, G. R. New synthetic routes to macrocyclic triamines. *J. Chem. Soc. Chem. Commun.* **1992**, 507–508.

81 McGowan, P. C.; Podesta, T. J.; Thornton-Pett, M. *N*-Monofunctionalized 1,4,7-triazacyclononane macrocycles as building blocks in inorganic crystal engineering. *Inorg. Chem.* **2001**, *40*, 1445–1453.

82 De, D.; Byers, L. D.; Krogstad, D. J. Antimalarials: synthesis of 4-aminoquinolines that circumvent drug resistance in malaria parasites. *J. Heterocyclic Chem.* **1997**, *34*, 315–320.

83 Beletskaya, I. P.; Bessmertnykh, A. G.; Guilard, R. Halo-substituted aminobenzenes prepared by Pd-catalyzed amination. *Synlett* **1999**, 1459–1461.

84 Sharghi, H.; Tamaddon, F. The synthesis and nucleophilic substitutions of haloxanthones. *J. Heterocyclic Chem.* **2001**, *38*, 617–622.

85 Chibale, K.; Haupt, H.; Kendrick, H.; Yardley, V.; Saravanamuthu, A.; Fairlamb, A. H.; Croft, S. L. Antiprotozoal and cytotoxicity evaluation of sulfonamide and urea analogues of quinacrine. *Bioorg. Med. Chem. Lett.* **2001**, *11*, 2655–2657.

86 Beletskaya, I. P.; Bessmertnykh, A. G.; Guillard, R. Palladium-catalyzed synthesis of aryl-substituted polyamine compounds from aryl halides. *Tetrahedron Lett.* **1997**, *38*, 2287–2290.

87 Pine, S. H.; Shen, G.; Bautista, J.; Sutton, C.; Yamada, W.; Apodaca, L. Monoalkylation vs dialkylation of a sulfone-stabilized carbanion. *J. Org. Chem.* **1990**, *55*, 2234–2237.

88 Rutledge, T. F. Sodium acetylide. II. Reactions of sodium acetylide in organic diluents. Preparation of monoalkyl acetylenes. *J. Org. Chem.* **1959**, *24*, 840–842.

89 Zhang, J.; Flippen-Anderson, J. L.; Kozikowski, A. P. A tandem Michael addition ring-closure route to the metabotropic receptor ligand α-(hydroxymethyl)glutamic acid and its γ-alkylated derivatives. *J. Org. Chem.* **2001**, *66*, 7555–7559.

90 Weis, C. D.; Newkome, G. R. Facile elimination of nitrous acid from quaternary nitroalkanes. *J. Org. Chem.* **1990**, *55*, 5801–5802.

91 Chung, J. Y. L.; Hughes, D. L.; Zhao, D.; Song, Z.; Mathre, D. J.; Ho, G.-J.; McNamara, J. M.; Douglas, A. W.; Reamer, R. A.; Tsay, F.-R.; Varsolona, R.; McCauley, J.; Grabowski, E. J. J.; Reider, P. J. A highly efficient synthesis of fibrinogen receptor antagonist L-734,217 via a novel chemoselective silyl-mediated conjugate addition of δ-lactams to 4-vinylpyridine. *J. Org. Chem.* **1996**, *61*, 215–222.

92 Settambolo, R.; Mariani, M.; Caiazzo, A. Synthesis of 1,2- and 1,3-divinylpyrrole. *J. Org. Chem.* **1998**, *63*, 10022–10026.

93 Brace, N. O. An economical and convenient synthesis of phenyl vinyl sulfone from benzenethiol and DCE. *J. Org. Chem.* **1993**, *58*, 4506–4508.

94 Salzberg, P. L.; Supniewski, J. V. β-Bromoethylphthalimide. *Org. Synth.* **1941**, *Coll. Vol. I*, 119–121.

95 Heitger, H.; Meyer, O. Improved method for the production of 4-(6-bromohexyloxy)butylbenzene. Patent *WO 03074458 A1*, **2003**.

96 Chen, F.-E.; Yuan, J.-L.; Dai, H.-F.; Kuang, Y. Y.; Chu, Y. Synthetic studies on *d*-biotin, part 6: an expeditious and enantiocontrolled approach to the total synthesis of *d*-biotin via a polymer-supported chiral oxazaborolidine-catalyzed reduction of *meso*-cyclic imide strategy. *Synthesis* **2003**, 2155–2160.

97 Federsel, H.-J.; Glasare, G.; Högström, C.; Wiestål, J.; Zinko, B.; Ödman, C. A convenient quaternization/rearrangement procedure for conversion of thiazoles to medium- and large-sized *N,S*-heterocycles. *J. Org. Chem.* **1995**, *60*, 2597–2606.

98 Knochel, P.; Dohle, W.; Gommermann, N.; Kneisel, F. F.; Kopp, F.; Korn, T.; Sapountzis, I.; Vu, V. A. Highly functionalized organomagnesium reagents prepared through halogen–metal exchange. *Angew. Chem. Int. Ed.* **2003**, *42*, 4302–4320.

99 Schiemenz, G. P. Diphenyl-*p*-bromophenylphosphine. *Org. Synth.* **1973**, *Coll. Vol. V*, 496–499.

100 Hargreaves, S. L.; Pilkington, B. L.; Russell, S. E.; Worthington, P. A. The synthesis of substituted pyridylpyrimidine fungicides using Pd-catalyzed cross-coupling reactions. *Tetrahedron Lett.* **2000**, *41*, 1653–1656.

Index

Side Reactions in Organic Synthesis. Florencio Zaragoza Dörwald

ISBN: 3-527-31021-5

d

l

m

t